Biofuels for Fuel Cells

Integrated Environmental Technology Series

The *Integrated Environmental Technology Series* addresses key themes and issues in the field of environmental technology from a multidisciplinary and integrated perspective.

An integrated approach is potentially the most viable solution to the major pollution issues that face the globe in the 21st century.

World experts are brought together to contribute to each volume, presenting a comprehensive blend of fundamental principles and applied technologies for each topic. Current practices and the state-of-the-art are reviewed, new developments in analytics, science and biotechnology are presented and, crucially, the theme of each volume is presented in relation to adjacent scientific, social and economic fields to provide solutions from a truly integrated perspective.

The *Integrated Environmental Technology Series* will form an invaluable and definitive resource in this rapidly evolving discipline.

Series Editor
Dr Ir Piet Lens, Sub-department of Environmental Technology, The University of Wageningen, P.O. Box 8129, 6700 EV Wageningen, The Netherlands (piet.lens@wur.nl).

Published titles
Biofilms in Medicine, Industry and Environmental Biotechnology:
    Characteristics, analysis and control
Decentralised Sanitation and Reuse: *Concepts, systems and implementation*
Environmental Technologies to Treat Sulfur Pollution: *Principles and engineering*
Phosphorus in Environmental Technology: *Principles and applications*
Soil and Sediment Remediation: *Mechanisms, techniques and applications*
Water Recycling and Resource Recovery in Industries: *Analysis, technologies and implementation*

Forthcoming title
Pond Treatment Technology

www.iwapublishing.com

Biofuels for Fuel Cells

Renewable energy from biomass fermentation

Edited by
**Piet Lens, Peter Westermann, Marianne Haberbauer
and Angelo Moreno**

Publishing
LONDON • SEATTLE

Published by IWA Publishing, Alliance House, 12 Caxton Street, London SW1H 0QS, UK

Telephone: +44 (0) 20 7654 5500; Fax: +44 (0) 20 7654 5555; Email: publications@iwap.co.uk
Web: www.iwapublishing.com

First published 2005, Reprinted 2007
© 2005 IWA Publishing

Printed in Great Britain by Lightning Source
Typeset by Gray Publishing, Tunbridge Wells, UK

ISBN 1 84339 092 2 / 978 1 84339 092 3

Front cover photographs
Polymer Electrolyte Membrane Fuel Cell (PEM-FC) pilot plant, which is connected to an agricultural
biogas plant, at the Institute of Technology and Biosystems Engineering of the Federal Agricultural
Research Centre (FAL) in Braunschweig (Germany). *Top*: Biological desulfurisation unit, inside and
outside, *Left*: PEM – FC test stacks, *Bottom*: Hydrogen Reforming System. Photographs courtesy of
Dr. Thornsten Ahrens (FAL, Germany).

Contents

Contents

Foreword

This book has been edited within the Network "Biomass Fermentation towards Usage in Fuel Cells" (BFCNet) that has been funded by the European Science Foundation. It has been running from January 2002 to December 2004 and involved nine partners from eight European countries in the field of biomass fermentation, biofuel upgrading and usage of biofuels in fuel cells.

The main activities of the network have been:

- Three Workshops have been organized: the first one was held in Genoa, Italy in February 2003 with the title "State of the Art and Perspectives in the Development of Less Contaminant Sensitive Fuel Cells Using Biomass Fermentation Fuels". It focused on the research on fuel cells in Europe and gave an overview of current research efforts on fuel cells. The second Workshop was held in Braunschweig, Germany and focused on "Biomass Fermentation as Basis for High Quality Fuel for Fuel Cell Applications – Fundamentals and Special Aspects". The third and final Workshop was held in Steyr, Austria with the title "Biomass Fermentation and Fuel Cells as Key to a Sustainable Decentralized Power Generation in Europe" with the main focus on biomass and fuel cell integration into the European energy supply.
- The BFCNet funded seven exchange visits of young scientists between European research institutes. The short-term visits enabled the benefiting researchers to learn more about the combination of biofuels and fuel cells which would not have been possible at their home institutions.
- The BFCNet has been operating a website where the most important data on the Network, about the participating institutions, events and calls for research projects in the respective field have been published.
- Finally, the Network prepared the present book where all member institutions of the Network contributed a chapter. The book described comprehensively the state of the art of biomass fermentation for the usage in different types of fuel cells and gives insight into the research activities in this field.

As chair of the BFCNet, finally I would like to thank the partners of the network for their cooperation: Thorsten Ahrens, Loreto Daza Bertrand, Marianne Haberbauer, Peter Holubar, Kevin Kendall, Piet Lens, Angelo Moreno, Åke Nordberg, Ewald Wahlmüller, Peter Weiland and Peter Westermann. Furthermore I would like to thank Svenje Mehlert and Stéphanie Pery representing the European Science Foundation for the financial and administrative support.

Werner Ahrer
Chair of the BFCNet

Preface

The current way to produce, convert and consume energy throughout the world is not sustainable. Our economic growth and social development can, however, only be implemented by means of appropriate availability of energy services. Due to their high efficiency, fuel cells are considered as a strategic technology for future energy supply systems. Biomass fermentation can be regarded as an energy-efficient technology for the mineralization of organic compounds in waste streams, as it results in the production of energy-rich compounds (biogas, hydrogen and ethanol). The unique and advantageous point in the combination of biomass fermentation and fuel cells results from the fact that biomass is a renewable source of energy which can be utilized most efficiently using fuel cells technology. This book discusses the optimal combination of biomass fermentation with energy production technologies using fuel cells for decentralised heat and power generation.

Due to limited amounts of fossil fuels and concern about the global warming, there is an increasing urge to develop more renewable energy sources. Biogas production via anaerobic digestion is an already established technology to produce biogas from renewable sources as biological wastes and energy crops. However, the potential of biogas is thus far not realized, as biogas is mainly produced locally, far away from the major energy consuming locations like urban and industrial areas. Also, there is a tendency towards centralization of plants in very large scale facilities to reduce costs and increase the earnings. Moreover, the electricity market was until a few years ago monopolised by a few companies and local energy production has been strongly discouraged. With the recent opening of the energy market, there is an incentive to develop alternative, low cost and on site electricity production technologies in an environmental friendly way. Thus, anaerobic conversion of organic wastes (biowaste, manure, biomass) gets a facelift from "dead end waste processing" into "energy production". This allows upgraded anaerobic digestion plants to be integrated in the energy cycle and thus contribute to sustainable development of our society, both in rural and industrialised areas.

The major bottleneck to upgrade anaerobic digestion to an "energy plant" is to link production and consumption of energy. In case the biogas quality is sufficiently high, it can be injected directly into the natural gas grid, as is already done in e.g. Sweden and Austria. Alternatively, methane (upgraded biogas) can be used as a car fuel or in combustion engines for local electricity production. A number of newly developed technologies, i.e. fuel cells and microturbines, can, however, be used to convert the biogas into electricity at a much higher efficiency than currently used technologies. Fuel cells have a high efficiency as they directly transform chemical energy into electricity. Electricity can be very easily transported from the production location to the consumption location. Fuel cells are easily scalable, i.e. they offer a high energy conversion efficiency independent of the size. Fuel cells thus make local high-efficiency electricity production possible. In this way, the use of fuel cells could facilitate the application of anaerobic digestion for the production of renewable energy at small scale. This scenario for renewable energy production is discussed in detail in the book volume.

The common end products of anaerobic digestion are biogas and a kind of compost. Biogas mainly consists of methane and carbon dioxide in an average ratio of 50–60% and 40–50% respectively. Such a ratio is theoretically well suited for fuel cells. As a matter of fact, pure methane cannot be used in fuel cells because of their inactivation by carbon production and pure methane needs to be spiked with carbon dioxide before it can be used in a fuel cell. Fuel cells enable an increased electrical output out of biofuels while reducing drastically the usual particle emissions in the exhaust gas. If using a biofuel, these advantages are joined by the fact that the system reduces the CO_2 emissions.

Several national and super national strategies have been outlined for a future in which hydrogen serves as a central energy carrier. As suggested by, e.g. DOE, a near term approach to the hydrogen economy is steam reforming of natural gas, followed by a mid and long term plan to use wind turbines, photovoltaics and biological production of hydrogen from biomass. Most studies of biological hydrogen production are performed at the laboratory scale with axenic cultures grown in laboratory media. Even in combined processes, the yield is still low and considerable developments are needed before the processes are economically exploitable. Several chapters of this book give an overview and discussion of the state of the art of this field.

Instead of concentrating on the biological production of only one energy carrier, a simultaneous production of hydrogen, methane and ethanol creates the possibility to optimise the exploitation of specific energy carriers to suit specific needs corres-ponding to the current fossil fuel use for specific purposes. Hydrogen can for instance be used in fuel cells for urban transportation. Ethanol can be used in fuel cells in rural areas, and methane can be used in fuel cells or microturbines for local electricity and heat production. Despite the obvious advantages of combining the production of different energy carriers, only a few concepts have so far been presented. In these concepts, the different processes are exploited in a sequential fermentation, transforming most of the energy available in the substrate to usable energy carriers. These so called biorefineries can be considered as more environmentally friendly processes since process water and nutrients from the different processes can be recirculated and waste production can be kept minimal. The biotechnological development and optimisation of the process conditions to produce these compounds from biological wastes and energy crops is an area of large interest, which is addressed in this book volume.

We wish to thank all the contributors for their enthusiastic support and timely submission of their manuscripts. This book has evolved out of the activities of the network "Biomass fermentation towards usage in fuel cells – BFCnet", see www.Bfcnet.info for more information. This network (2002–2004) was financially supported by the European Science Foundation. We are grateful to Alan Click and Alan Peterson of IWA Publishing for their help and editorial support in realizing this book.

Piet Lens
Wageningen, The Netherlands

Peter Westermann
Lyngby, Denmark

Marianne Haberbauer
Steyr, Austria

Angelo Moreno
Rome, Italy

Contributors

Francesca Accettola
Profactor Produktionsforschungs
GmbH
Im Stadtgut A2
4407 Steyr-Gleink, Austria

Thorsten Ahrens
Bundesforschungsanstalt für
Landwirtschaft (FAL)
Institut für Technologie und
Bioystemtechnik
Abteilung Technologie
Bundesallee 50
38116 Braunschweig, Germany

Birgitte Kiaer Ahring
Environmental Microbiology and
Biotechnology Group
BioCentrum-DTU, Bld. 227
Technical University of Denmark
DK-2800 Lyngby, Denmark

Galip Akay
School of Chemical Engineering
and Advanced Materials
Newcastle University
Newcastle upon Tyne NE1 7RU,
UK

Wolfgang E.Baaske
STUDIA – Studienzentrum für
internationale Analysen
Schlierbach 19
4553 Schlierbach, Austria

Vera L. Barbosa
Dept of Biochemistry,
Microbiology and Biotechnology
Rhodes University
PO Box 94
Grahamstown 6140, South Africa

Carmen Boeriu
Agrotechnology & Food
Innovations
Bornsesteeg 59
6708 PD Wageningen, The
Netherlands

Roberto Bove
Industrial Engineering Department
University of Perugia
Via Duranti 96
06125 Perugia, Italy

Miriam A.W. Budde
Wageningen UR
Agrotechnology & Food
Innovations
Biobased Products
Department Bioconversion
P.O.Box 17
6700 AA Wageningen, The
Netherlands

Burak Calkan
School of Chemical Engineering
and Advanced Materials
Newcastle University
Newcastle upon Tyne NE1 7RU,
UK

Omer Faruk Calkan
School of Chemical Engineering
and Advanced Materials
Newcastle University
Newcastle upon Tyne NE1 7RU,
UK

Pieternel A.M. Claassen
Wageningen UR
Agrotechnology & Food
Innovations
Biobased Products
Department Bioconversion
P.O. Box 17
6700 AA Wageningen, The
Netherlands

Brian P. Connor
Brian P. Connor and Associates
Ltd,
1 Richmond, Priests Road
Tramore
Co. Waterford, Ireland

Loreto Daza
Instituto de Catálisis y
Petroleoquímica (CSIC)
Campus Cantoblanco
28049 Madrid, Spain

Truus De Vrije
Wageningen UR
Agrotechnology and Food
Innovations
Department Bioconversion
P.O. Box 17
6700 AA Wageningen, The
Netherlands

Murat Dogru
School of Chemical Engineering
and Advanced Materials
Newcastle University
Newcastle upon Tyne NE1 7RU,
UK

Yanhai Du
Connecticut Global Fuel Cell
Center
44 Weaver Road, Unit 5233
Storrs, CT 06269-5233, USA

Maria José Escudero
CIEMAT – Centro de
Investigaciones
Energéticas Medioambientales y
Tecnologicas
Avda. Complutense 22
28040 Madrid, Spain

Hariklia N. Gavala
Laboratory of Biochemical
Engineering and Environmental
Technology
Department of Chemical
Engineering
University of Patras
26500 Patras, Greece

Frank Haagensen
Environmental Microbiology and
Biotechnology Group
BioCentrum-DTU, Bld. 227
Technical University of Denmark
2800 Lyngby, Denmark

Marianne Haberbauer
Profactor Produktionsforschungs
GmbH
Im Stadtgut A2
A-4407 Steyr/Gleink, Austria

Bert Hamelers
Wageningen University
Sectie Milieutechnologie
Bomenweg 2
6703 HD Wageningen, The
Netherlands

Joachim Hoffmann
MTU Motoren und Turbinen Union
Friedrichshafen GmbH
Abteilung Neue Technologien
Standort Ottobrunn
81663 Munchen, Germany

Peter Holubar
Institut für Angewandte
Mikrobiologie
Universität für Bodenkultur
Muthgasse 18
1190 Wien, Austria

Pawel Huczkowski
Forschungszentrum Jülich, IWV-2
52425 Jülich, Germany

Renato Iannelli
Università di Pisa
Dipartimento di Ingegneria civile
via Gabba 22
56126 Pisa PI, Italy

Leo J.L.A Jansen
Kerkeland 16
6883 HA Velp, The Netherlands

Anna Karlsson
Department of Water and
Environmental Studies
Linköping University
581 83 Linköping, Sweden

Piet N.L. Lens
Wageningen University
Sectie Milieutechnologie
Bomenweg 2
6703 HD Wageningen, The
Netherlands

Geert Lissens
Lab. Microbial Ecology and
Technology Faculty of Bioscience
Engineering
Ghent University
Coupure L 653
9000 Gent, Belgium

Piero Lunghi
Industrial Engineering Department
University of Perugia
Via Duranti 96
06125 Perugia, Italy

Gerasimos Lyberatos
Laboratory of Biochemical
Engineering and Environmental
Technology
Department of Chemical
Engineering
University of Patras
26500 Patras, Greece

Joan Mata-Alvarez
Dpt. Enginyeria Química
Universitat de Barcelona
Martí i Franquès 1, plata. 6
E-08028 Barcelona, Spain

Roel Meulepas
Wageningen University
sectie Milieutechnologie
Bomenweg 2
6703 HD Wageningen, The
Netherlands

Angelo Moreno
ENEA CR Casaccia
TEA – CCAT
Via Anguillarese 301
00060 S. Maria di Galeria (Roma),
Italy

Åke Nordberg
Swedish Institute of Agricultural
and Environmental Engineering
JTI – Institutet för jordbruks- och
miljöteknik
Box 7033
750 07 Uppsala, Sweden

Javier Pirón Abellán
Forschungszentrum Jülich, IWV-2
52425 Jülich, Germany

Willem J. Quadakkers
Forschungszentrum Jülich, IWV-2
52425 Jülich, Germany

Korneel Rabaey
Laboratory of Microbial Ecology
and Technology
Faculty of Bioscience Engineering
Ghent University
Coupure L 653
9000 Gent, Belgium

Wim H. Rulkens
Wageningen University
Sectie Milieutechnologie
Bomenweg 2
6703 HD Wageningen, The
Netherlands

Nigel Sammes
Connecticut Global Fuel Cell
Center
School of Engineering
University of Connecticut
44 Weaver Road
Storrs, CT 06269-5233, USA

Johan P.M. Sanders
Wageningen UR
Agrotechnology & Food
Innovations
Bornsesteeg 59
6708 PD Wageningen, The
Netherlands

Alfons Schulte-Schulze Berndt
RÜTGERS CarboTech Engineering
GmbH
Am Technologiepark 1
45307 Essen, Germany

Keith Scott
Department of Chemical and
Process Engineering
University of Newcastle upon Tyne
Newcastle upon Tyne NE1 7RU,
UK

Vladimir Shemet
Forschungszentrum Jülich, IWV-2
52425 Jülich, Germany

Ashok K. Shukla
Department of Chemical and
Process Engineering
University of Newcastle upon Tyne
Newcastle upon Tyne NE1 7RU,
UK

Lorenz Singheiser
Forschungszentrum Jülich, IWV-2
52425 Jülich, Germany

Ioannis V. Skiadas
Laboratory of Biochemical
Engineering and Environmental
Technology
Department of Chemical
Engineering
University of Patras
26500 Patras, Greece

Wim Soetaert
Laboratory of Industrial
Microbiology and Biocatalysis
Faculty of Bioscience Engineering
Ghent University
Coupure links 653
9000 Gent, Belgium

Jaime Soler
CIEMAT – Centro de
Investigaciones
Energéticas Medioambientales y
Tecnologicas
Avda. Complutense 22
28040 Madrid, Spain

Knut Stahl
RWE Fuel Cells GmbH
Gutenbergstr. 3
45128 Essen, Germany

Richard M. Stuetz
Centre for Water & Waste
Technology
School of Civil & Environmental
Engineering
The University of New South
Wales
Sydney, NSW, 2052, Australia

Bo H. Svensson
Department of Water and
Environmental Studies
Linköping University
581 83 Linköping, Sweden

Mads Torry-Smith
Novozymes North America, Inc.
77 Perry Chapel Church Road
PO Box 576
Franklinton, NC 27525-05, USA

Steven Trogisch
Profactor Produktionsforschungs
GmbH
Im Stadtgut A2
4407 Steyr/Gleink, Austria

Jan E.G. Van Dam
Wageningen UR
Agrotechnology & Food
Innovations
Bornsesteeg 59
6708 PD Wageningen, The
Netherlands

Ed W.J. van Niel
Department of Applied
Microbiology
Lund University
P.O. Box 124
SE-221 00 Lund, Sweden

Eric Vandamme
Laboratory of Industrial
Microbiology and Biocatalysis
Faculty of Bioscience Engineering
Ghent University
Coupure links 653
9000 Gent, Belgium

Adrie H.M. Veeken
Wageningen University
Sectie Milieutechnologie
Bomenweg 2
6703 HD Wageningen, The
Netherlands

Willy Verstraete
Lab. Microbial Ecology and
Technology Faculty of Bioscience
Engineering
Ghent University
Coupure L 653
9000 Gent, Belgium

Peter Weiland
Bundesforschungsanstalt für
Landwirtschaft (FAL)
Institut für Technologie und
Biosystemtechnik
Abteilung Technologie
Bundesallee 50
38116 Braunschweig, Germany

Peter Westermann
Biocentrum-DTU
Technical University of Denmark
Environmental Microbiology and
Biotechnology Group
Building 227
2800 Lyngby, Denmark

Loredana Zani
Institut für Angewandte
Mikrobiologie
Universität für Bodenkultur
Muthgasse 18
1190 Wien, Austria

PART ONE: Status and development of bioenergy

Section IA: Biofuels in our society

1

Transition management: from vision to action

L. Jansen

1.1 TOWARDS TRANSITION

Driving forces behind sustainable development mix up moralism and strategic pragmatism. The focus of moralism is on the benefit of community, future generations and bridging of the welfare gap. The focus of strategic pragmatism is on the benefit of the long-term continuity of economy, trade and industry. Business concerns for continuity of operations on the long term for private companies. Governments concern for long-term stable socio-economic development of the nations. Consumers concern for long-term availability of goods and services of acceptable quality at acceptable prices. Morality and strategic pragmatism increasingly mix up in the entanglement of people and organisations.

Sustainable development may be looked upon as an ongoing process, the origin of which was marked with warning signs like "Silent Spring" (Carson 1962) and "Limits to Growth" (Club of Rome 1972) and actions like the *First World Environment Conference* in Stockholm (1972) announcing the United Nations

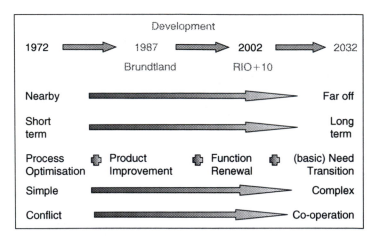

Figure 1.1   Increasing complexity in environment and development.

Environmental Programme (UNEP). In this process, the complexity and scope are continuously increasing (Figure 1.1).

Over time several modes can be recognised: cleaning up the environment and optimisation of existing consumption and production processes to begin with in the early 1970s. This was gradually paralleled in the early 1980s by the improvement of existing production and consumption within the existing structures with end-of-process and process-integrated measures. Later at the end of the 1980s by end-of-product measures like reuse and recycling, and in the early 1990s by product-integrated measures like redesign for the environment (Graedel *et al.* 1995) have been introduced.

The Brundtland Report *Our Common Future. World Commission on Environment and Development* (Brundtland 1987) confirmed the name of this transition process and broadened its view in time as well in its content: "Present and Future Generations" and integrating "Environment and Development". The report induced the next phase in the transition towards a society in sustainable development marked by attempts to make more fundamental approaches to the restructuring of the production and consumption system, and to make it more operational, like by UNEP, the World Council for Sustainable Development (WBCSD), the "factor approach" (Factor 10 Club 1997, Jansen 2002), Industrial Ecology (Allenby 1999) and in policies like the First National Environmental Policy Plan "To Choose or to Loose" in the Netherlands (NEPPs 1989, 1993, 1997, 2001), opting for a sustainable Netherlands within one generation. Holdren and Ehrlich (1974) gave the origin of the factor approach as early as in the early 1970s in the equation relating the pressure on the environment with the size of the world population, the welfare per head and the conversion metabolism from extractions out of the environment to products and services.

The challenge (Figure 1.2) given for the first half of the 21st century is a renewal of the production consumption system to improve the conversion mechanism dramatically by on the average a factor 10 or more expressed in the so-called eco-efficiency or the Materials Intensity Per Service Unit (MIPS) (Schmidt-Bleek 1994). This challenge induced research on the ways and means by which to achieve this goal in the interest of a sustainable society for the future generations.

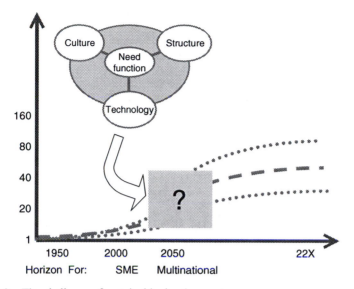

Figure 1.2   The challenge of sustainable development.

1.2   SOCIAL INVOLVEMENT

In parallel and in interaction with emerging environmental concern since the late 1960s, dramatic changes occurred in the involvement of civil society in the processes of societal and political decision-making. The 1968 movement induced democratisation measures and legislation to ensure the influence of workers. This development was supported by the creation of successive instruments of participation supporting the influence of citizens and other stakeholders (Figure 1.3). Illustrative of this are:

- Legislation on employees participation.
- Legislation on participation and right of appeal for citizens in physical planning and in (environmental and spatial) licence policies.
- Technology assessment policies, resulting in the Office of Technology Assessment in the USA (abolished) and the Dutch Office of Research on Technological Aspects (NOTA), later the Rathenau Institute.
- Constructive technology assessment.
- Public debates on energy policies, for example in UK, Austria and the Netherlands (Jansen 1985).
- Involvement of stakeholders in the preparation of the Dutch NEPPs (1989, 1993, 1997, 2001).

Towards the end of the 20th century this development created an attitude oriented at the involvement of stakeholders in major questions influencing their welfare and well-being. The concept of the Dutch Sustainable Technology Development (STD) Programme took advantage of these experiences (Jansen 1993).

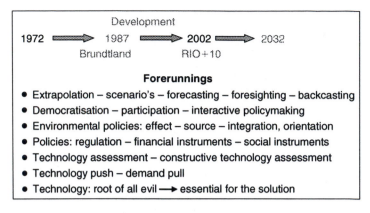

Figure 1.3   Increasing social involvement in the development of our society.

1.3   SYSTEMS RENEWAL: WAYS AND MEANS

Research delivered several approaches to develop new ways to meet the challenge of sustainable development (Robèrt *et al.* 2002). One of them, the STD approach, was investigated in a Dutch Inter-Ministerial Research Programme (Jansen *et al.* 2001; Weaver *et al.* 2000) from 1993 to 1998. The mission of this research programme was: *"To explore and illustrate, together with policy makers in industry and government, how technology development can be shaped and organised, from future orientation on sustainability and develop instruments to implement this".*

The architecture of the programme was based upon:

- Orientation on needs as the starting point for technology renewal.
- Backcasting from future images to the present and from needs to products on a systems basis, inspired by the work done in Sweden on Energy (Goldemberg *et al.* 1992), and originally suggested by Jantsch (1968) as normative forecasting.
- Future images on needs based on a factor approach with the "factor 20" (Weterings *et al.* 1992) as a target-metaphor for the improvement of the eco-efficiency to be reached within 50 years from a starting point in 1990. The factor 20 was based on the following assumptions made in 1992: a growth of the world population by a factor 2, an average growth of global welfare *per capita* by a factor 5 and a reduction of the environmental burden by a factor 2. As these figures differentiate depending on the environmental item, the region, the type of need, the number 20 has to be regarded as a metaphor to express the magnitude of the challenge of sustainable development.
- Iterative and interactive search.
- Participation of stakeholders.
- Interaction "culture–structure–technology".

In "learning by doing" research in a variety of projects covering domains of needs, a seven-step procedure was developed (Jansen *et al.* 1998) and tested as a

Table 1.1   Projects in the STD programme.

Domain	Illustrative process
Food supply	Novel protein foods Sustainable land use High-tech agroproduction Whole crop utilisation
Transport systems	Pipeline transportation of goods Computerised processing of transport demand Hydrogen for mobile applications
Shelter	Sustainable district renewal Sustainable office building
Water chain	Integrated sustainable urban/rural waterchain
Chemistry	Conversion of hydrocarbons (C1-chemistry) New cells for photovoltaic solar energy Whole crop utilisation Fine chemistry process technology Natural fibre-reinforced composite materials

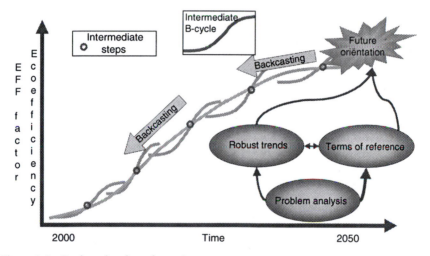

Figure 1.4   Backcasting from future images.

generalisable methodology for setting up and carrying through sustainability-oriented research and development (R&D) programmes (Table 1.1).

In an evaluation around some years after the completion of the STD programme it appeared that most of the initiatives were still followed up (Coenen 2000). The STD programme was the precursor of transition management approaches in the Netherlands at least. The programme delivered a distinct contribution in creation of approaches to transformation and to shaping a context for political formulation of the transformation policies.

The backcasting and systems approach (Figure 1.4) was also applied to the development of the national economy of New Zealand (Reeve et al. 2000) and served as a review on agricultural research (Jansen 2000).

Table 1.2   Organisations exploring future-oriented technology R&D.

STT	Foundation for future technology research (e.g. exploration future chances of nanotechnology)
Rathenau Institute Former NOTA	(Technology) assessments for parliament Dutch Office for Technological Aspects research (e.g. exploration of policy and public perceptions on genomics)
NRLO	Dutch Council for Agricultural Research Development of basics for renewal of the dutch agricultural system
DIMI	Dutch Institute for Management and Innovation Management of knowledge transfer and dissemination
STD	STD programme Generation of sustainable long-term future options for technology development

1.4 SYSTEMS RENEWAL AS A COMPLEX SOCIETAL PROBLEM

Around the turn of the century, several Dutch research organisations which developed future-oriented research programmes exchanged their experiences and observations. These organisations handle(d) complex explorative search processes (Table 1.2), which have the following characteristic in common: long-term future orientation for a variety of stakeholders in fields of conflicting interests in an environment with large uncertainties.

Apart from specificities related to the domain of research, there was one common element in the backbones of their approaches and procedures: an iterative and interactive process in which relevant stakeholders are involved. The iteration comprises creative generation of products and their assessment by stakeholders to improve the products as well as to gain stakeholder support in successive steps of a search process. The "products" comprise:

- a problem definition,
- generation of a vision, a strategy,
- derivation of options,
- formulation of selection criteria,
- making proposals (for R&D programmes).

As such these are "normal" steps in project management. The particularity lies in the nature of their management including the full involvement of stakeholders. Given the role of people in these processes in sharing their knowledge, the process was named: Knowledge fusion (Figure 1.5).

1.5 SYSTEMS RENEWAL FOR SUSTAINABLE DEVELOPMENT: EDUCATION AND PRACTICE

The ambition is to handle the tension between the urgency to renew numerous systems to fulfil people's needs and the inertia to processes of fundamental

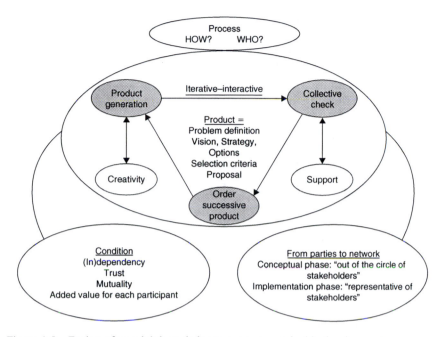

Figure 1.5   Fusion of people's knowledge to support sustainable development.

change. On the one hand, overacting and neglecting the inertia of change may be counter-productive. On the other hand, the pace of change has to be sufficient to achieve the timely renewal of major systems.

Given the urgency and the necessary scale of systems renewal for sustainable development, the proposal is to attain an ongoing process of systems renewal in the industrialised as well as in the developing nations within the next 15 years. In general, this approach consists of a process of interactive and iterative search in co-operative arrangements among private parties, science and technology and governmental parties that take the interaction of "culture–structure–technology" into account. The application of this approach must respect specific national and regional cultures and traditions.

The scale of a specific systems renewal process must be proportional to the scale of the system and its effects. Top-down and bottom-up approaches appear to be complementary.

The Copernicus Charter (a RIO 1992 initiative to integrate sustainable development in higher education worldwide) must be implemented within 10 years (by 2015) to guarantee sufficient capacity building. Experience with the embedding of sustainable development in educational systems gained up to now should be extended and practised. In the *First EESD Conference* (Engineering Education in Sustainable Development) at the Delft University of Technology succeeded by EESD 2 in Barcelona in October 2004, by EESD 3 in Lyon/St. Etienne in 2006 and by EESD 4 in Graz in 2008.

Backcasting as developed and practised in the STD programme (see § 1.3) suggests the step-wise approach based on exchange and parallel development

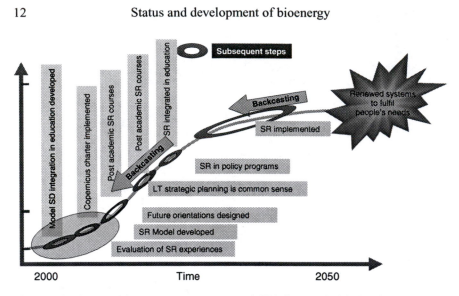

Figure 1.6.   Steps on the way to systems renewal (SR) for sustainable development in education and governance.

in the practice of systems renewal and educational renewal as shown in Figure 1.6.

1.6   RESEARCH AND TECHNOLOGY DEVELOPMENT FOR SUSTAINABLE DEVELOPMENT

In the late 1990s the question of "HOW" to design and support sustainable development through research and technology programmes was put on the agenda in the European Union's (EU) 5th Framework (e.g. in the EU, Joint Research Centre (JRC), STRATA programme). The core of a contribution to the European Conference on Cleaner Production in Lund, May 2001 (Jansen 2002), was presented earlier in a STRATA workshop in Maastricht in January 2000. Since then the Factor 10 Institute has published an analysis of complementary approaches to the "HOW" question (Robèrt et al. 2002). And the EU JRC has set up an inventory and analysis of sustainability-directed research programmes (Whitelegg et al. 2001). In the STRATA programme, the AIRP project was started in January 2002 with the goal of answering this question by mid-2003 (Hinterberger et al. 2001). This project aimed to deliver recommendations for setting up and evaluate new programmes on options for sustainable development and their contexts. The final report has been delivered to the EU.

The programmes (Table 1.3) are analysed (Table 1.4) and judged to gain experience in using the preliminary evaluation methodology on the following elements: outcomes (nature and quality), design and process characteristics (what, how and how well) and contextual conditions (dominant factors). The programmes selected had to meet some essential criteria like availability of information, orientation on sustainable development, and future and innovativity.

Table 1.3   Selected research, technological innovation and demonstration processes (RTD) in STRATA – AIRP.

Name	Abbreviation	Country
Eco-cycle Programme	Ecocycle	Sweden
Socio-Ecological Programme	SEP	Germany
Austrian Programme on Technologies for Sustainable Development	ATSD	Austria
Austrian Landscape Research Programme	ALR	Austria
STD Programme	STD	The Netherlands
Habiforum	Habiforum	The Netherlands
Human Resources and Employment in Fishery Sector	Mahre	Portugal
Consumer Unity and Trust Society	CUTS	India
Zero Emissions Research and Initiatives	ZERO	International

Table 1.4   Hypotheses in the analysis of RTD programmes.

Hypothesis I	Hypothesis II	Hypothesis III
Design and process characteristics of the programmes strongly influence the nature and quality of outcomes of a programme and are depending on contextual conditions.	The degree of correspondence to the sustainability orientation of design and process characteristics of the programmes strongly influence the nature and quality of outcomes of a programme.	Contextual conditions, design and process characteristics and outcomes of a programme are interconnected in a loop.

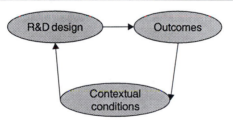

Confirmation of these hypotheses would identify tools to improve the ways and means of sustainability-directed research programmes. Contextual conditions, such as public and political awareness of the necessity of sustainable development, availability of definition of societal challenges, availability of scientific and social (human) capital, adequate organisation of science, could open opportunities for adequate process design and management.

1.7   TRANSITION FOR SUSTAINABILITY

Martens and Rotmans (2002) describe transition as "the result of developments in different domains, as a set of connected changes which reinforce each other but take place in several different areas, such as technology, the economy, institutions, behaviour, culture, ecology and belief systems". They regard transition as a "spiral that reinforces itself; there is a multiple causality and co-evolution caused by independent developments".

In successive Dutch NEPPs increasing attention was paid to the role of technology for sustainable development and broadening the scope to its interrelation with socio-economic, socio-cultural and institutional aspects:

- *NEPP 1 (1989)*: Recognition of the role of technology expressed in strategies expecting a contribution to sustainable development.
- *NEPP 2 (1993)*: Differentiation between systems optimisation, improvement and renewal and identifying different policy instruments.
- *NEPP 3 (1997)*: Recognition of the necessity of interdisciplinary co-operation, and Breakthroughs, and announcing the Dutch Sustainable Development Initiative (NIDO).
- *NEPP 4 (2001)*: Transition policies to cover persistent sustainability problems and to be initiated in the sectors: Energy, Agriculture and Transportation.

In its Fourth NEPP: "A world and a will, work on sustainability" (NEPP 4, 2001), the Dutch Government announces transition policies for traffic and transportation, agriculture and energy to cope with persistent problems and to step on the path to a sustainable future.

Freeman and Perez (1988) look upon successive technologies and sources for energy provision as the origin of the so-called Kondratieff waves. Another example is the changes as a result of successive transportation techniques in the 19th century (Grübler and Nakicenovic 1991). Shifts in energy systems and transportation systems from canal boats to trains, respectively, may very well be understood as successive transitions induced by changes in energy and transportation technology.

1.8   CONCLUSIONS

The main conclusions to be drawn on the process of transition of the current development to sustainable development are:

- The driving forces behind transition have to be based on moralism towards future generations and the environment as well as on strategic pragmatism for the concern of business, government and citizens/consumers.
- Three parallel trajectories cover the path towards sustainable development:
    (1) Optimisation in the short run.
    (2) Improvement in continuation of current environmental development in medium terms.
    (3) Renewal of production and consumption systems in the long run.
- Systems renewal appears to be a complex social process, which requires adequate process management.

REFERENCES

Allenby, B.R. (1999) *Industrial Ecology, Policy Frame Work And Implementation.* Prentice Hall, Englewood Cliffs, New Jersey, 308 pp.

Brundtland, G.H. (1987) *Our Common Future. World Commission on Environment and Development,* Oxford University Press, Oxford, New York.

Carson, R. (1962) *Silent Spring.* Houghton Mifflin, Boston.

Coenen, L. (2000) *Thesis University of Technology,* Eindhoven, July 7, 2000.

Curitiba and its visionary mayor, http://www.globalideasbank.org/BI/BI-262.HTML

EESD (2002) http://www.odo.tudelft.nl/conference.html

EESD (2004) http://congress.cimne.upc.es/eesd2004/

EU IPTS TECS (2001) Socio-economic evaluation of public RTD policies. *Thematic Network (EPUB/STRATA) Agenda First Workshop,* March 27–28, Seville (http://www.jrc.es).

Factor 10 Club (1997) *Statement to Government and Business Leaders.* Wuppertal Institute for Climate, Energy and Environment.

Freeman, C. and Perez, C. (1988) Structural crises of adjustment, business cycles and investment behaviour. In: *Technical Change and Economic Theory,* Pinter, London and New York.

Graedel, T.E. and Allenby, B.R. (1995) *Industrial Ecology,* Chapter IV, pp. 183–187. Prentice Hall, Englewood Cliffs, New Jersey.

Goldemberg, J., Johansson, T.B., Reddy, A.K.N. and Williams, R.H. (1985) An end-use oriented global energy strategy. *Annual Review of Energy* **10**, 613–688.

Grübler, A. and Nakicenovic, N. (1991) Long waves, technology diffusion and substitution. *Review* **XIV**(2): Spring 1991, pp. 313–342. IIASA, RR-91-17, October 1991, Novographic, Vienna, Austria.

Hinterberger, F. (2001) *Adaptive Integration of Research and Policy for Sustainable Development – Prospects for the European Research Area.* Project No. STPA – 2001-00007.

Holdren, J.P. and Ehrlich, P.R. (1974) Human population and the global environment. *Am. Scient.* **62**(3), 282–292.

Jansen, L. (1985) Handling a debate on a source of severe tension. In: *A Geography of Public Relations Trends.* Martinus Nijhoff Publishers (Kluwer Academic Publishers Group), Dordrecht/Boston/Lancaster, pp. 148–154.

Jansen, J.L.A. (1993) Towards a sustainable future: en route with technology! In: *The Environment: London, Towards a Sustainable Future,* pp. 497–523. Kluwer Academic Publishers, Dordrecht, The Netherlands.

Jansen, L., Bakker, C., Bouwmeester, H., Kievid, T., van Grootveld, G. and Vergragt, Ph. (1998) STD Vision 2040-1998, Technology, key to sustainable prosperity. ten Hagen & Stam, The Hague, December 1998, 80 pages (Dutch and English).

Jansen, J.L.A. (2000) Quality of life, sustainable and world wide: new challenges for agricultural research. In: *Towards an Agenda for Agricultural Research in Europe* (eds. Boekestein, A. *et al.*), pp. 227–237. Wageningen Pers, The Netherlands.

Jansen, J.L.A., Grootveld, G., van, Spiegel, E. and van, Vergragt, P.J. (2001) On search for ecojumps in technology: from future visions to technology programs. In: *Transdisciplinarity: Joint Problem Solving among Science, Technology and Society* (eds. Thompson, J. *et al.*), pp. 173–180. Birkhäuser Verlag Basel Boston, Berlin.

Jansen, J.L.A. (2002) The challenge of sustainable development. *J. Clean. Prod.* **11**(3), 231–245.

Jantsch, E. (1968) Integrative planning of technology, In: *Perspectives of Planning* (ed. Jantsch, E.). OECD, Paris 1969, pp. 179–185.

Martens, P. and Rotmans, J. (2002) *Transitions in a Globalising World.* Swets & Zeitlinger Publishers, Lisse, Abingdon, Exton (PA), Tokyo.

National Environmental Policy Plans I (1989), II (1993), III (1997), IV (2001). Ministries of agriculture, nature conservation and fishery, economic affairs, education and science, housing, spatial planning and environment, and transportation and water management. www.minvrom.nl/international.

Reeve, N. and Gandar, P. (2000) The New Zealand foresight project – an overview. In: *Towards an Agenda for Agricultural Research in Europe* (eds. Boekestein, A. *et al.*), pp. 101–110. Wageningen Pers, The Netherlands.

Robèrt, K.H., Schmidt-Bleek, F., Aloisi de Larderel, J., Basile, G., Jansen, J.L., Kuehr, R., Price Thomas, P., Suzuki, M., Hawken, P. and Wackernagel, M. (2002) Strategic sustainable development – selection, design and synergies of applied tools. *J. Cleaner Product* **10**(3), 197–214.

Schmidt-Bleek, F. (1994) *Wieviel Umwelt braucht der Mensch?* Birkhäuser Verlag, Berlin, Basel, Boston.

Weaver, P., Jansen, L., Grootveld, G. van, Spiegel, E. and van, Vergragt P.J. (2000) *Sustainable Technology Development.* Greenleaf Publishing, Sheffield, UK. 256 pp.

Weterings, R.A.P.M. and Opschoor, J.B. (1992) The ecocapacity as a challenge to technological development. Advisory Council for Research on Nature and Environment, Rijswijk.

Whitelegg, K. and Weber, M. (2001) EU JRC ESTO Project Report: National Research Activities and Sustainable Development, a survey and assessment of national research initiatives in support of sustainable development.

2

Biomass valorisation for sustainable development

C.G. Boeriu, J.E.G. van Dam and
J.P.M. Sanders

2.1 INTRODUCTION

2.1.1 The carbon economy

For many centuries, human civilisation depended almost exclusively on the availability of biomass to provide food, clothing, building materials and energy for heating, cooking and metal manufacturing. With the industrial revolution, the discovery of fossil resources and the continuous development of the technology for extraction and valorisation of fossil feedstock, the fate of biomass utilisation has changed significantly. In the industrialised countries, biomass has been replaced almost completely by coal, petroleum crude oil and natural gas as the source for energy and transport fuels, while in the developing countries biomass still contributes up to 35% of the primary energy supply. In the second half of the 20th century, fossil resources have become the world's main source of energy, while the contribution of biomass to energy decreased to about 7% of the total energy demand (Klass 1998).

A similar trend has been observed for the application of biomass for the production of chemicals. Up to the mid of the 19th century, biomass-derived resources were the primary source of chemicals and pharmaceuticals. Large amounts of ethanol, acetone and butanol were produced by fermentation of starches and sugars, while chars, methanol, acetic acid and acetone were produced by pyrolysis of hardwood. In the 20th century, due to the development of petrochemical industry, biomass has gradually been replaced by fossil fuel-based raw material as the preferred feedstock for most commodity organic chemicals. At present, biomass is used worldwide mainly as a source of many naturally occurring speciality chemicals and cellulosic and starch-based materials (Klass 1998; US-DOE 2003). Low-volume, high-value products such as vitamins and antibiotics, flavours and fragrances, antioxidants, dyes and pigments, oils, waxes, tannins, gums, rubbers, pesticides and speciality polymers are commercially extracted from or produced by conversion of biomass (US-DOE 2003). High-volume chemicals, including most organic compounds, polymers and plastics are mostly produced from fossil feedstock, and conversion of biomass is used only to a limited extent. Currently, 95.8% of all organic chemical compounds produced in Europe are derived from fossil resources (BACAS 2004).

In the 20th century, the industrial and societal development was driven by and dependent on the fossil resources and related technologies. The world economy turned from a bio-based economy into a petrochemical (i.e. hydrocarbon) economy.

2.1.2   Transition towards a sustainable economy

With a rapidly expanding world population and the increase of prosperity in less developed countries, it is expected that the world consumption of energy and resources will increase with a minimum factor of 3 by the years 2050, from 350 EJ in 2000 to approximately 1000 EJ in 2050 (Figure 2.1). This growth is associated with an estimated increase of the mean energy use per capita per year from 59 GJ in 2000 to 100 GJ in 2050.

Such an increase in resource consumption raises several issues of concern, including:

- *Fossil fuels reserves are a diminishing raw material.* The total reserve of fossil fuel was estimated at approximately 135×10^9 ton oil, $120 \times 10^{12} \, m^3$ gas and 850×10^9 ton coal, based on proven reserves (Klass 1998; Campbell and Laherrère 1998; Campbell 2003). An assumption based on the five times the global proven crude oil reserves of 6218 EJ as of 1 January 2000 and an annual growth in crude oil consumption of 2.3% indicates that oil production will approach a theoretical depletion time near 2070 (Klass 1998, 2003). At an annual growth of consumption of 3.2%, the theoretical depletion time for natural gas reserves is close to that of crude oil (Klass 2003). The depletion time of coal resources has been estimated to be more than 200 years.
- *Distribution of fossil resources.* Another factor that cannot be neglected is the uneven distribution of fossil resources. Half of the world's reserve of petroleum

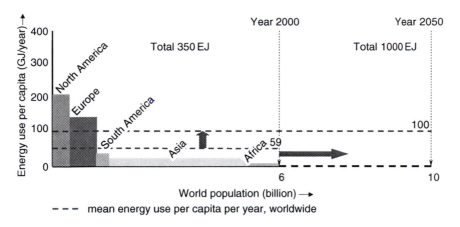

Figure 2.1   Trends in world energy consumption. Bars represent the energy use per capita in different regions of the world, in 2000.

crude oil and gas are sited in the Middle East and the former Soviet Union, while 70% of the coal reserves are located in North America, the Far East and the former Soviet Union (Klass 1998). This non-uniform distribution of fossil feedstock determines a strong dependency of the economy of many countries on resources and policy of a few other countries. Although the discovery of new reserves of oil, gas and coal, and the reduction of feedstock consumption due to technological developments might prolong the exhaustion time, the message that pertains is that fossil resources are not unlimited. Therefore, a sensible use of fossil fuel and the utilisation of complementary resources for the production of energy and materials is a must.

- *Climate change and the greenhouse effect.* CO_2 is the major "greenhouse" gas that accounts for global temperature increase, and all associated negative effects such as rise in sea level, the shift of agricultural zones, desertification and changes in polar ice caps. Although there is disagreement about the specific effects of greenhouse gases on global temperature, it is a fact that the concentrations of atmospheric CO_2 have increased about 30% in the past 150 years (Chamberlain *et al.* 1982). Fossil resources are the major source of the additional CO_2; the average global CO_2 emissions from fossil fuels combustion were as high as 22 Gton/year in the 1990s (Klass 1993).

Due to the large-scale use of fossil fuels, the concentration of CO_2 in the atmosphere continues to increase, despite the majority of the world countries have agreed in Kyoto that reduction of CO_2 emission is essential for our planet for maintaining the climatological and ecological balance. Therefore, it is essential to reduce the greenhouse effects and find sustainable solutions for the supply of food, energy and commodities to satisfy human needs in the future.

Up to 2012, the Netherlands have committed themselves to reduce the CO_2 emissions with 6% relative to the consumption in 1990. To prevent the increase of earth temperature with more than 2°C compared to the pre-industrial level, all the industrially developed countries should, in a common effort, lower the emissions with an average of 30% by 2020, as related to 1990 (VROM 2004). This implies the application of severe measures for the diminution of the use of fossil material. One possibility to achieve this target is the development and implementation of alternative energy sources. However, fossil resources are not only used for the production of energy. About 15% of fossil feedstocks are used for transport fuels, while the chemical industry (the petrochemical and inorganic industry together) consumes worldwide about 20% (Sanders *et al.* 2004). It is therefore reasonable to assume that the chemical industry can play an important role in the effort to reduce the consumption of fossil resources and simultaneously, in the utilisation of alternative "green" resources for the production of bulk chemicals.

In other words, transition to a sustainable bio-based economy as a consequence of fossil fuel depletion and the Kyoto protocol on global climate change includes a shift of feedstock for energy and chemical industries from petrochemical fossil reserves to renewable carbon and energy resources.

2.1.3   Biomass: a sustainable resource

During photosynthesis, the plants use solar energy, CO_2, minerals and water to produce primarily carbohydrates (Equation 2.1) and oxygen, and by further biosynthesis, a large number of less oxygenated compounds including lignin, triglycerides, terpenes, proteins, etc. The composition depends on the type of plant:

$$nCO_2 + nH_2O \xrightarrow[\text{Chlorophyll}]{h\nu} (CH_2O)_n + nO_2 \qquad (2.1)$$

On average, the capture efficiency of incident solar radiation in biomass is 1% or less, but it can be as high as 15%, depending on the type of plant (Klass 1998). Nevertheless, a total of about 170 Gton biomass is produced yearly through photosynthesis in green plants (Eggersdorfer *et al.* 1992), which accumulates a total energy equivalent to 2350 EJ/year (1 ton biomass = 15 EJ). The carbon (e.g. CO_2) and mineral (K, N, P) cycle is closed after decomposition of biomass or waste products, if disposed on land or after processing, consumption and degradation/combustion (Figure 2.2). Consequently, the life cycle of biomass as a renewable feedstock has a neutral effect on CO_2 emission.

Based on these facts, biomass is considered an intrinsically safe and clean material, with unlimited availability and high potential to be used as a renewable resource for the production of energy and alternative fuels, new materials in technical applications, and organic materials and chemicals. Medium- and long-term strategies have been developed and new programmes have been started aiming at enhancing the use of biomass for energy and chemicals. In 2000, the US Department of Energy stated its goal as: (a) the increase of biomass consumption for the industrial sector at an annual rate of 2% up to 2030, (b) the increase of transportation fuels from biomass from 0.5% in 2001 to 20% in 2030 and (c) the production of at least 10% of the basic chemical building blocks from plant-derived resources by 2020 and 50% by 2050 (US-DOE 1998; CLS-NRC 2000; BRDTAC 2002;

Duncan 2004). In Europe, there are current regulations only related to the substitution of petrochemical resources in the area of biofuels. The European Parliament approved in 2003 the first directive aiming at stimulation of biofuels (e.g. bio-ethanol and bio-diesel) in Europe, with the long-term objective that 20% of the total engine fuel consumption in Europe will be met by biofuels in 2020 (2003/30/EC).

The total contribution of biomass as carbon and energy source is currently marginal (Table 2.1). Assumptions are made that biomass could contribute for 15% of the total world energy demand in 2050, that is now estimated to come out at 1000 EJ/year (Figure 2.1) (Shell International 2001). An increase of the biomass contribution to 15% of the total by 2050 represents approximately 150 EJ equivalent fossil energy or 10 Gton/year of dry renewable resource. The contribution of biomass as carbon source for the chemical industry may then increase to about 20 EJ/year (1.3 Gton/year).

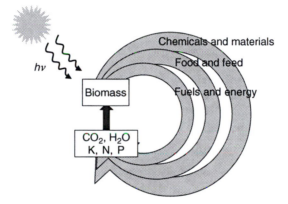

Figure 2.2   Carbon and mineral cycle during biomass growth and utilisation.

Table 2.1   Estimations (EJ/year) of the share of biomass used worldwide as carbon and energy source.

| | 2000 | | 2050 | | Biomass |
	Total	Biomass	Total	Biomass	share (%)
Fermentation/white biotechnology	1	1	8	8	100
Chemistry	70	<0.5	200[a]	20	10
Transport fuels	50	<0.5	150[a]	60	40
Electricity	70	3	200[a]	60	30
Rest (domestic)	160	_[b]	450[a]	_[b]	25
Total	350	4	1000	150	15

[a] Estimated threefold relatively to 2000.
[b] The biomass used for domestic consumption in the Third World countries are not included.

Emerging technologies for the production of energy and transport fuels from biomass are discussed in detail elsewhere in this book. Here, we present some considerations on the possibilities of the green biomass for the chemical industry.

2.2   BIOMASS FOR THE CHEMICAL INDUSTRY

2.2.1   Energy and economic considerations

Each year, about 6000 million tons of plant material are harvested worldwide, processed into food, feed and non-food materials, and chemicals (Eggersdorfer *et al.* 1992). Table 2.2 lists some of the most important biomass-derived chemicals. Besides polysaccharides (cellulose, starch, alginate, pectin, agar, inulin, etc.) a range of low-volume, high-value products are extracted from or produced by conversion of biomass. Biomass is, for example, still the only source for the production of most antibiotics.

High-volume chemicals, including most organic compounds, polymers and plastics are mostly produced from fossil feedstock, and conversion of biomass is used only to a limited extent. Nevertheless, biomass resources might become the source for commodity organic chemicals if they can provide energetic and economic advantages.

2.2.1.1   Energy considerations

Worldwide, the chemical industry uses about 20% of the fossil resources, of which 51% is used as carbon source. The rest is used to supply the process energy (e.g. heating, cooling), including also the manufacturing of process reagents such as sulfuric acid, chlorine, sodium hydroxide and others. Metzger and Eissen (2004)

Table 2.2   Primary building blocks from biomass and processing methods for the production of chemicals.

Building block	Processing*	Chemical product
C_5 sugars	Chemical	Furfural, xylitol
C_6 sugars	Fermentation	Ethanol, acetaldehyde, acetic acid, acetone, glycerol, 1,2-PDO, 1,3-PDO, n-butanol, n-butyric acid, amyl alcohols, oxalic acid, lactic acid, citric acid, amino acids, antibiotics, vitamins
	Chemical	Hydroxymethylfurfural, sorbitol
	Enzymatic	Fructose
Glycerides	Chemical + Enzymatic	Fatty acids, glycerol, surfactants, bio-diesel (lubricants, cosmetics)
Lignin	Chemical	Phenols, vanillin, lignosulphonates
Cellulose	Chemical Fermentation	Cellulose derivatives (esters, acids, nitrates, ethers, xanthogenates)
Ethanol (biofuel)	Chemical	Alginates, alkaloids, phenols, resins, rubber, saponins, sterols, tannins, terpenes, waxes, etc.

* Dominant processing methods.

have compared the gross energy requirement (GER) of some base chemicals. GER indicates the total primary energy consumption of the entire process chain for manufacturing of a chemical product, and includes the process energy and the non-energetic consumption representing the direct material use of fossil energy sources (Patel 1999). Significant differences have been observed in GER values for different bulk chemicals (Figure 2.3). Rapeseed oil, produced from natural feedstock, has the lowest GER (20 GJ/ton), with process energy as the main component, while manufacturing of functionalised chemicals such as adipic acid, propylene oxide and ethanol from fossil feedstock are characterised by high GER (80, 104 and 60 GJ/ton, respectively). It can be concluded that using bio-based feedstock for the production of upgraded chemicals, a significant reduction of fossil resource consumption can be achieved, provided that process energy is maintained at low levels.

Currently, theoretically, all organic chemicals can be obtained from bio-based resources using combinations of biorefining, chemical, fermentation and enzymatic technologies. Examples are given in Table 2.2. Moreover, many functionalised chemical compounds (e.g. compounds containing heteroatoms like oxygen, nitrogen, sulfur) can eventually be produced via less energy-demanding processes in plants, thus avoiding simultaneously the need for heat transfer as well as the use of process reagents like sodium hydroxide and sulfuric acid. Bulk chemicals such as urea or ε-caprolactam can in principle be produced from biomass. In such an approach, solar energy will account for most of the process energy, while carbon dioxide, water and minerals will be the atom source (Figure 2.4).

It appears from here that biomass is the preferred source for the production of functionalised chemical products, since it retains the functional groups already present in the raw material. Fossil resources, however, will remain the preferred choice for the underivatised hydrocarbons.

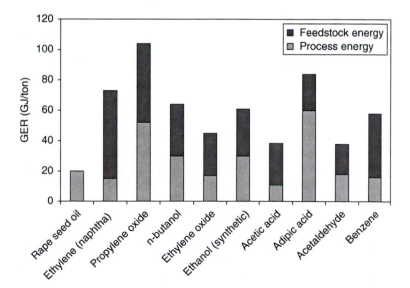

Figure 2.3   GERs for selected base chemicals (adapted from Metzger and Eissen 2004).

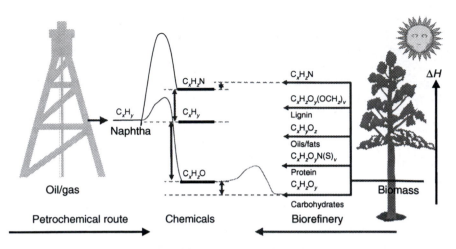

Figure 2.4   Biomass – a mixture of functionalised and non-functionalised compounds that can be produced with low enthalpy (ΔH) losses.

2.2.1.2   Economic considerations

Figure 2.5 shows raw materials and chemical products based on both renewable resources and petrochemicals of different upgrading levels as a function of the market price. Comparing the world market price for raw materials, the market price of renewable resources (e.g. corn, straw, tallow) and low-upgraded biomass derivatives (e.g. molasses, starch) appears to be in the same range or lower than that of crude oil. The current market price of biomass is determined by the caloric value of coal for generation of energy and amounts about 2 €/GJ. In other words biomass (dry weight) yields approximately 30–40 €/ton, representing about 18–22% of the price of crude oil (the average world market price of petroleum in the period 2000–2003 was 175 €/ton (BACAS 2004). Moreover, the market price of biomass-derived building blocks (cellulose, sugar, glucose, ethanol, vegetable oil) is comparable with the price of the major petrochemical-derived base chemicals (ethylene, propylene, benzene, methanol). In contrast, independent of the source, the market value of chemical products (e.g. organic acids, propylene oxide, capro-lactam, etc.) is high and depends mainly on the degree of upgrading. Consequently, the use of biomass as complementary resource for the production of bulk chem-icals offers potentially attractive economic perspectives.

The cost of the raw materials and the process chemicals together with invest-ment costs represent the most important economic factors for the production of chemicals from petrochemical resources. Significant costs are associated with heat losses and waste materials. Heat exchangers, for example, require high investment costs. High investments in waste water treatment and closed recycling systems have been major reasons for upscaling of chemical plants. The use of biomass can bring significant cost reductions, particularly related to the above-mentioned items, if functionalised compounds are the target products.

Since biomass is a complex mixture of hydrocarbon derivatives containing one or more heteroatoms in the molecule, such as oxygen, nitrogen, sulfur and others, it is

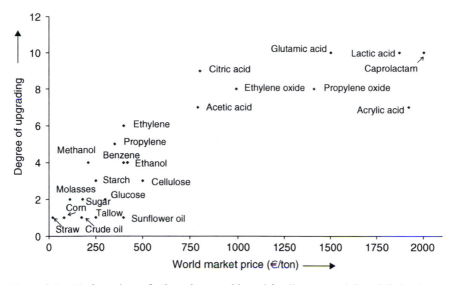

Figure 2.5   Market prices of selected renewable and fossil raw materials and derived chemicals. Market prices are at January 2004 (BACAS 2004).

possible to extract fractions enriched in or containing almost exclusively the desired chemical. A careful selection of the crop as well as of the associated biorefining technology will allow the isolation of both primary and secondary products, and their subsequent transformation into value-added products through enzymatic, fermentative or chemical routes. Industrial processes have already been developed for the production of many chemicals from selected crops or residues, such as ethanol from starch or sugar and bio-diesel from plant oils. However, the waste materials resulting from these processes contain high amounts of C_5 sugars and aromatic products such as lignin, tannins, flavonoids, etc., which can find applications as such or can be used for the production of other valuable chemicals, such as furan and vanillin.

Based on the production volume of bulk chemicals, it can be rationalised that the use of biomass as carbon source might be economically successful for the following categories of functionalised compounds:

- Chemical compounds with a small- (100,000 ton/year) to middle–high-production volume (1–5 Mton/year). These products can benefit the least, in terms of investment costs, from the *economy of scale*. For this class of products there are the biggest chances to identify the alternative competitive processes for their production from biomass in crops.
- Chemical compounds with a production volume of 100,000 ton/year or lower. The feasibility of such products through fermentation from carbohydrates has already been demonstrated. Examples are the production of 1,3-propandiol (PDO) from glucose (DuPont/Genencor) and polylactate (Cargill-Dow/Purac).

The approach described above will bring a significant shift in the use of plant crops as feedstock for the chemical industry from the production of low-priced food and bulk commodities to high-priced, specialised plant-derived products.

2.2.2   Strategies for implementation of renewable resources in the production of chemicals

2.2.2.1   Scenarios

Scenarios for the production of organic materials and chemicals from biomass have been designed (Morris and Ahmed 1992; US-DOE 1998; Okkerse and van Bekkum 1999; BRDTAC 2002). Two main strategies can be differentiated:

(a) use of biomass for the production of existing products;
(b) use of biomass for the development of new products with new properties.

Generally, bio-based building blocks, such as glucose, vegetable oils, lignins, etc., are used to produce high-value chemicals using advanced thermo-catalytic processes and biotechnologies. Two selected examples are discussed below (Section 2.2.2.2).

A different approach for the production of chemicals from biomass would involve the use of plants as bioreactors for the (a) direct improvement and modification of specialised constituents of plant origin and (b) manufacture in plants of atypical, non-plant constituents. Biotechniques, based on metabolic pathway engineering, are now available for modifying plant constituents that are used for food, energy and chemicals (Altman 1999; McKeon 2003). This approach has been used, for example, for the production of high-amylose or high-amylopectin starch (Båga et al. 1999), the modification of the ratio of saturated to unsaturated fatty acids in oils or to increase the content of specific valuable fatty acids like erucic acid, ricinoleic acid, etc. (Gunstone 1998). Recently, the plant production of polyhydroxyalkanoates, that is polyhydroxy butyrate for the production of thermoplastics has been investigated (Nawrath et al. 1994; Arai et al. 2002). A promising methodology will be the production in plants of new non-plant compounds that might be further used as new building blocks for the synthesis of chemicals. Such an approach requires, however, solutions to prevent the limiting factors for the use of plants as bioreactors and solutions to prevent this: reduction of the osmotic pressure of the desired chemical by decreasing its solubility, manipulation of osmotic stress tolerance via expression of osmoprotectants and compatible solutes, increase tolerance for feedback inhibition, etc.

2.2.2.2   Case studies

1,3-PDO is a chemical intermediate having potential utility in the production of polyester fibres and the manufacture of polyurethanes and cyclic compounds. 1,3-PDO is produced industrially by hydration of acrolein (Dow, BASF, DuPont) and by hydroformylation of ethylene oxide (Shell). DuPont in cooperation with Genencor has developed a biosynthetic route that produces PDO from glucose (corn-based) by fermentation (Figure 2.6). The organism used is *Escherichia coli*, modified by metabolic engineering to express the genes encoding the required enzymes (Nakamura and Whited 2003). The key-step of the process was the engineering of the 1,3-PDO pathway. In the process, glucose is converted to glycerol,

which is then transformed into PDO in a two-step reaction. The first step is dehydration of glycerol to 1,3-hydroxypropion aldehyde (by glycerol dehydratase), which
is then reduced to 1,3-PDO in a reaction catalysed by 1,3-PDO oxidoreductase
(DuPont 2004; Nakamura and Whited 2003). The bio-based 1,3-PDO will be manufactured by DuPont Tate & Lyle BioProducts, at the AE Staley Manufacturing
corn wet mill in Decatur, USA, and will be used for the production of DuPont's
polymer Sorona®, a polypropylene terephthalate (Fairley 2001; DuPont 2004).

Polylactic acid (PLA) is the homo-polymer of the biomass-derived lactic acid,
usually produced by a ring-opening polymerisation reaction from the lactide form
of the monomer. Lactic acid is an intermediate volume speciality chemical with
applications in food, pharmaceutical and cosmetic industries, and can be manufactured by chemical synthesis or carbohydrate fermentation (Figure 2.7). The
industrial chemical process is based on lactonitrile (Musashino, Japan and Sterling
Chemical Inc., USA) and produces a racemic mixture of lactic acid (Narayanan *et al.*
2004). Lactic acid in the form of L(+), D(−) or racemic mixture is produced by
carbohydrate fermentation with *Lactobacillus* strains (Narayanan *et al.* 2004). In
April 2002, Cargill-Dow opened a biorefinery plant in Blair Nebraska (USA) for the
production of PLA from corn-derived dextrose with an estimated capacity of
140,000 ton/year (Fairley 2001; US-DOE 2002). PLA is the first renewable polymer

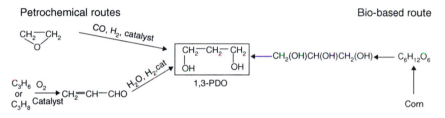

Figure 2.6   Petrochemical and bio-based route for the production of 1,3-PDO.

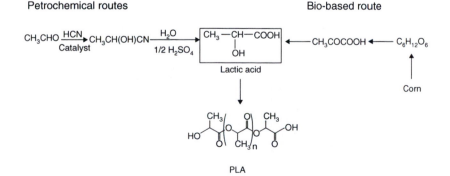

Figure 2.7   Petrochemical and bio-based route for the production of lactic acid.

produced on an industrial scale. The polymer has properties similar with the conventional polymers polyethylene and polypropylene, but it is completely bio-degradable (compostable).

2.3 BIOMASS AVAILABILITY

Yearly, a total of about 170 Gton biomass is produced through photosynthesis in green plants, of which only 3.5% (6 Gton) are being cultivated, harvested and used (Figure 2.8) (Eggersdorfer *et al.* 1992). Of this, 62% (approximately 3.7 Gton) are consumed for the production of food, and approximately 0.3 Gton are used by the non-food industry (Eggersdorfer *et al.* 1992; Röper 2000). Most of the biomass consumed for non-food applications goes to the paper and building industries. Significant amounts of biomass are also used as fuel for energy and housing.

If biomass should become available as source for the production of chemicals, as discussed here, and also for transport fuels and energy, high amounts of bio-mass must be supplied whole year round in sufficient amounts to sustain conversion plant operation. The following main categories of plant-based biomass raw materials can be identified:

- Agricultural residues from primary agricultural production, such as straw, foliage and hulls (about 5 Gton/year).
- Agroindustrial wastes generated in the food production at various links in the chain, as well as resources lost in the inefficiency of food uptake and animal production, vegetable, fruit and garden-waste and compost.
- Forestry residues and grass.
- Non-food and dedicated crops (e.g. energy and chemical crops), such as rape-seed, linseed, sugar beet, sugar cane, short-rotation forestry (willow, poplar), grass, cassava, algae, etc.

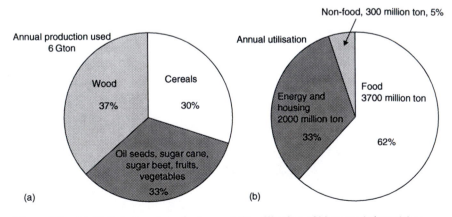

Figure 2.8 (a) Global annual production and (b) utilisation of biomass (adapted from Eggersdorfer *et al.* (1992) and Röper (2000)).

The question that arises is if there will be enough biomass for chemicals, while in the same time it will be needed to:

(1) ensure the food supply for 10 billion people and provide them with other required organic materials;
(2) contribute to the energy needs.

Calculations, based on an optimistically estimated maximal agricultural yield of 50 ton/ha, indicate that an amount of 50 Gton/year (80% non-food biomass, 10% forestry and 10% waste streams) could become available for non-food applications in 2040 (Okkerse and van Bekkum 1999). Even with less optimistic scenarios for the productivity of agriculture, it can be concluded that, in principle, there should be enough biomass to supply the yearly required amount of 10 Gton for industrial application (see Table 2.1).

The transition to a bio-based economy and consequently the enhanced demand for renewable resources should not increase the pressure on natural habitats or conflict with the preservation of biodiversity and disturb the ecological balance. So the biomass resources should preferentially be derived from the discarded residues from food production and the better exploitation of cultivated land by enhancing the biomass yields. This will also require reforestation and recultivation of suitable arable lands. Changing the agronomic management systems towards sustainable productivity and whole crop use demands education and support of non-governmental organisations (NGOs) and politics.

2.4   INTEGRATED TECHNOLOGY

In the past decades, intensive efforts have been invested by the chemical industry to develop more efficient technologies, with lower environmental impact. High-performance technologies are available for the production of chemicals from fossil resources. The conversion of biomass into high-value chemicals is confronted with the challenge to develop a platform of competitive processing technologies. This includes, amongst others, the development of efficient catalysts and enzymes for biomass pre-treatment, and modification and integration cascade reactions. These novel technologies should be implemented for the separation of the discrete compounds from a very complex mixture (biorefinery) and to process them, subsequently, to produce biofuels, chemicals and energy. Integration of plant biotechnology, processing, fermentation technology, enzyme catalysis, chemical synthesis and separation technology will lead to whole crop utilisation and "multi-product" valorisation (Figure 2.8).

Two important focus points (FP) have been identified (US-DOE 1998). The first one (FP-1, Figure 2.9) reflects the current and short-term developments in valorisation of existing crop parts and plant-based building blocks with existing or modified chemical and bio-processes. At present, many chemicals are industrially produced from biomass via fermentation, enzymatic, thermal or chemical conversion. Some examples were given (see Section 2.2.2.2), but many are described in the literature (US-DOE 2003). The second focus, which can be seen as a longer-term

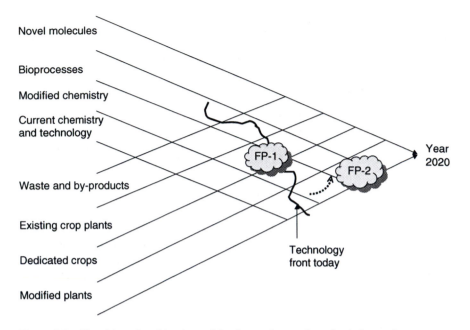

Figure 2.9   To achieve the objectives of the future, integration of existing and new technologies is required (adapted from DOE 1998).

perspective (FP-2 in Figure 2.8), requires a "quantum leap" for the production of novel molecules by the integration of modified plants with new technologies.

This innovation demands, besides knowledge development, conceptual developments in integrated processing technology. Generally, for the production of chemicals from biomass, only the part of the plant that is suitable for the envisaged transformation is used, the rest being mostly discarded. This useful raw material can be either a residue from food production or can be isolated from biomass, which is particularly enriched in certain compounds. Maximal utilisation of biomass in the chemical industry could be obtained only by the integration in a single facility, which supports various processing options to produce simultaneously chemicals, fuels and energy. Biomass supply (e.g. crop production and logistics) should also be considered. In such an integrated approach, all liberated components (e.g. C_5 and C_6 sugars, oils, lignins, proteins, new building blocks) will be looked as valuable platform intermediates and used in appropriate processes to yield higher-value products. Mineral recovery should also be considered with respect to return to the croplands as fertilisers. The biorefinery might, for example, produce one or more high-value chemical products and a low-value, but high-volume transportation fuel, while generating electricity and process heat for own needs and eventually for sale (Figure 2.10). Many different conceptual integrated system configurations are possible (Gravitis 1998; Klass 1998; Anastas and Lankey 2000; Kamm and Kamm 2004; NREL 2004; van Dam et al. 2005). The new biorefinery concept is analogous to the petrochemical refineries, but has the advantage of zero emissions.

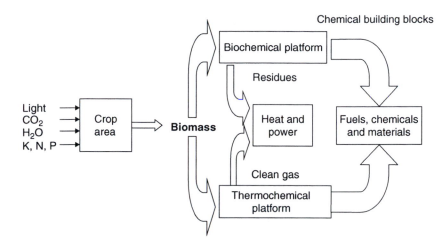

Figure 2.10   Example of an integrated biorefinery.

In conclusion, full exploitation of the potential of biomass as industrial chemical resource implies changing scenarios for bioconversion and accumulation of green chemicals. For a successful transition to a bio-based economy, a maximal implementation of the "green chemicals" concept in the petrochemical industry is needed, together with an integrated approach in which biorefinery, (bio)catalysis, fermentation (white biotechnology) and chemical technology supplement, complete and strengthen each other.

2.5   CONTRIBUTION TO SUSTAINABILITY

The use of plant/crop-based renewables as complementary resources to conventional feedstocks for the production of bulk and speciality chemicals, biofuels and energy is considered a sustainable and environmentally responsible approach. However, objective and complete quantification methods are needed to measure sustainability and allow sound comparisons between petrochemical and bio-based processes. In the past decades, intensive research has been performed to develop tools to assess the feasibility of products and processes for sustainable exploitation of natural resources. Petrochemical and renewable products have been compared by means of life cycle inventory (LCI) (Wightman *et al.* 1999) and complete life cycle assessment (LCA) (Hirsinger and Schick 1995; Bastiannoni and Marchetini 1996; Bendorichio and Jorgensen 1997; Jungmeier *et al.* 1998; Hovelius and Hanson 1999; Anastas and Lankey 2000). For example, LCA made for several surfactants and lubricants pointed to environmental advantages of oleochemical vs. petrochemical resources (Patel 2004). The production of polymers from renewable resources may result in significantly lower greenhouse gas emission and fossil energy use as compared to classical petrochemical polymers (Akiyama *et al.* 2003; Vink *et al.* 2003).

The analysis of the impact of products and processes on the environment has been extended to evaluate besides the quantity of energy used also the quality of the energy (i.e. the exergy). Exergy analysis applied for evaluation of biomass conversion processes to fuels (e.g. ethanol) (Jungmeier *et al.* 1998; de Vries 1999; Dewulf *et al.* 2000) and bio-diesel (Wall 1977) or energy (Szargut *et al.* 1998) provided quantitative information about the sustainability of the processes. It has been shown that the production of ethanol from biomass is more sustainable than the fossil production route, due to the higher renewable resource coefficient and higher environmental efficiency, although it has a lower-productivity factor (Dewulf *et al.* 2000). Such analysis, if applied to various complete production chains, might be able to allow the selection of the best technological option from the various approaches:

Existing crops vs. dedicated crops ⇔ existing technology vs. new technology
  ⇔ existing products vs. new products

In addition, exergy analysis can be used for the dynamic simulation of crop-based processes to target and improve the key-process components, as a predictive model, for comparison of different production routes or as a decision tool for the selection of the optimal combination of crop-process products.

2.6 CONCLUSIONS

The options for biomass valorisation should be urgently explored in more detail. The sustainability of the enhanced use of renewable resources for the chemical industry should be demonstrated by more economical viable examples. Integration of disciplines and "quantum leap" transitions towards a bio-based economy will be required to reduce the "ecological footprint" of mankind.

REFERENCES

Akiyama, M., Tsuge, T. and Doi, Y. (2003) Environmental life cycle comparison of poly-hydroxyalkanoates produced from renewable carbon resources by bacterial fermentation. *Polym. Degrad. Stabil.* **80**, 183–194.

Altman, A. (1999) Plant biotechnology in the 21st century: the challenges ahead. *Electr. J. Biotechnol.* **2**(2), 51–55.

Anastas, P.T. and Lankey, R.L. (2000) Life cycle assessment and green chemistry: the yin and yang of industrial energy. *Green Chem.* **2**, 289–295.

Arai, Y., Nahashita, H., Suzuki, Y., Kobayashi, Y., Shimizu, T., Yasuda, M., Doi, Y. and Yamagichi, I. (2002) Synthesis of a novel class of polyhydroxyalkanoates in *Arabidopsis* peroxizomes, and their use in monitoring short-chain-length intermediates of beta-oxidation. *Plant Cell Physiol.* **43**(5), 555–562.

BACAS (Belgian Academy Council of Applied Science) Report (2004) *Industrial Biotechnology and Sustainable Chemistry.* www.europabio.org

Båga, M., Repellin, A., Demeke, T., Caswell, K., Leung, N., Abdel-Aal, E.S., Hucl, P. and Chibbar, R.N. (1999) Wheat starch modification through biotechnology. *Starch–Starke* **51**(Suppl 4), 111–116.

Bastiannoni, S. and Marchetini, N. (1996) Ethanol production from biomass: analysis of process efficiency and sustainability. *Biomass Bioenerg.* **11**(5), 411–418.

Bendoricchio, G. and Jorgensen, S.E. (1997) Exergy as goal function of ecosystems dynamic. *Ecol. Model.* **102**, 5–15.

BRDTAC (Biomass Research and Development Technical Advisory Committee) (2002) Roadmap for Biomass Technologies in the United States. www.bioproduct-bioenergy.gov

Campbell, C.J. (2003) Industry urged to watch for regular oil production peaks, depletion signals. *Oil Gas J.* **101**(27), 38–45.

Campbell, C.J. and Laherrère, J.J. (1998) The end of cheap oil. *Scient. Am.* **278**(3), 78–83.

Chamberlain, J.C., Foley, H.M., MacDonald, G.J. and Ruderman, M.A. (1982) In: *Carbon Dioxide Review 1982* (ed. W.C. Clark), pp. 255, Oxford University Press, New York, USA.

CLS-NRC (Commission on the Life Sciences, National Research Council) (2000) *Biobased Industrial Products: Priorities for Research and Commercialization.* National Academy Press, Washington, DC.

de Vries, S.S. (1999) *Thermodynamic and Economic Principles and the Assessment of Bioenergy.* Ph.D. Thesis. OBL, The Hague.

Dewulf, J., Van Langenhove, H., Mulder, J., Van der Berg, M.M.D., van der Kooi, H.J. and de Swaan Arons, J. (2000) Illustrations towards quantifying the sustainability of technology. *Green Chem.* **2**, 108–114.

Duncan, M. (2004) U.S. Federal initiatives to support biomass research and development. *J. Ind. Ecol.* **7**(3–4), 193–201.

DuPont (2004) The miracles of science, Press release, May 26, 2004. http://www.dupont.com/NASApp/dupontglobal

Eggersdorfer, M., Meijer, J. and Eckes, P. (1992) Use of renewable resources for non-food materials. *FEMS Microbiol. Rev.* **103**, 355.

European Parliament and Council, Directive 2003/30/EC on the promotion of the use of biofuels or other renewable fuels for transport. *Official Journal of the European Union L123/42, 17.05.2003*, Brussels.

Fairley, P. (2001) Bioprocessing comes alive: no longer a field of dreams. *Chem. Week*, March 14. www.chemweek.com

Gravitis, J. (1998) A biochemical approach to attributing value to biodiversity – the concept of zero emissions biorefinery. Article presented at the *4th Annual World Congress on Zero Emissions in Windhoek*, Namibia.

Gunstone, F.D. (1998) Movement towards tailor-made fats. *Prog. Lipid Res.* **37**(5), 207–305.

Hirsinger, F. and Schick, K.P. (1995) A life cycle inventory for the production of alcohol sulphates in Europe. *Tenside* **2**, 128–139 LCI.

Hovelius, K. and Hansson, R.A. (1999) Energy- and exergy analysis of rape seed oil methyl ester (RME) production under Swedish conditions. *Biomass Bioenerg.* **17**, 279–290.

Jungmeier, G., Resch, G. and Spitzer, J. (1998) Environmental burdens over the entire life cycle of a biomass CHP plant. *Biomass Bioenerg.* **15**(4/5), 311–323.

Kamm, B. and Kamm, M. (2004) Biorefinery-systems. *Chem. Biochem. Eng.* **18**(1), 1–6.

Klass, D.L. (1993) Fossil fuel consumption and atmospheric CO_2. *Energy Policy* **21**(11), 1076–1078.

Klass, D.L. (1998) *Biomass for Renewable Energy, Fuels and Chemicals.* Academic Press, San Diego (California), USA.

Klass, D.L. (2003) A critical assessment of renewable energy usage in the USA. *Energ. Policy* **31**, 353–367.

McKeon, T.A. (2003) Genetically modified crops for industrial products and processes and their effects on human health. *Trend. Food Sci. Technol.* **14**, 229–241.

Metzger, J.O. and Eissen, M. (2004) Concepts on the contribution of chemistry to a sustainable development. *Renewable Raw Materials*, C.R. Chemie 7.

Morris, D. and Ahmed, I. (1992) *The Carbohydrate Economy: Making Chemicals and Industrial Materials from Plant Matter.* Institute for Local Self Reliance, Washington, DC [Abstract].

Nakamura, C.H. and Whited, G.M. (2003) Metabolic engineering for the microbial production of 1,3-propanediol. *Curr. Opin. Biotechnol.* **14**, 454–459.

Narayanan, N., Roychoudhury, P. and Srivastava, A. (2004) L(+) lactic acid fermentation and its product polymerization. *Electr. J. Biotechnol.* **7**(2), 167–179.

Nawrath, C., Poirier, Y. and Sommerville, C. (1994) Targeting the polyhydroxybutyrate biosynthesis pathway to the plastids of *Arabidopsis thaliana* results in high levels of polymer accumulation. *Proc. Natl. Acad. Sci. USA* **91**, 12760–12764.

NREL web page (2004) *Biomass Research.* www.nrel.gov

Okkerse, C. and van Bekkum, H. (1999) From fossil to green. *Green Chem.* **1**(2), 107–114.

Patel, M. and Wageningen, N.L. (1999) *Closing Carbon Cycles.* Ph.D. Thesis. University Utrecht.

Patel, M. (2004) Surfactants based on renewable materials: carbon dioxide reduction potential and policies and measures for the European Union. *J. Ind. Ecol.* **7**(3–4), 47–62.

Röper, H. (2000) Umwelt-und Ressourcen-schonende Synthesen und Processen: Perspektiven der industriellen Nutzung nachwachsender Rohstffe, insbesonders von Stärkae und Zucker. www.umweltchemie-gdch.de

Sanders, J., Boeriu, C. and Dam van, J. (2004) Milieu heeft chemie nodig. *Chemisch Weekblad* **100**(5), 16–19.

Shell International (2001) Energy needs, choices and possibilities – scenarios to 2050, global business development.

Szargut, J., Morris, D.R. and Steward, F.R. (1998) *Exergy Analysis of Thermal, Chemical and Metallurgical Processes.* Springer Verlag, Germany (3–4) 47–62.

US-DOE (US Department of Energy) (1998) Plant/crop based renewable resources 2020: a vision to enhance US economic security through renewable plant/crop-based resource use. www.science.doe.gov

US-DOE (US Department of Energy) (2002) The inauguration of Cargill Dow's Blair biorefinery marks new era in bio-plastics. *Newsletter Archive.* http://www.bioproducts-bioenergy.gov/news/NewsletterArchive/May2002.asp

US-DOE (US Department of Energy) (2003) Industrial bioproducts: today and tomorrow. www.science.doe.gov

van Dam, J.E.G., de Klerk-Engels, B., Struik, P.C. and Rabbinge, R. (2005) Renewable resources supply for changing market demands in a biobased economy. *Ind. Crop. Prod.* **21**, 129–144.

Vink, E.T.H., Rabago, K.R., Glassner, D.A. and Gruber, P.R. (2003) Applications of life cycle assessment to NatureWorks™ polylactide (PLA) production. *Polym. Degrad. Stabil.* **80**, 403–419.

VROM (NL Ministerie van Volkshuisvesting, Ruimtelijke Ordening en Milieubeheer) (2004) RVD Press Release, 16.04.2004.

Wall, G. (1977) *Exergy – A Useful Concept within Resource Accounting*, Report No. 77-42, Institute of Theoretical Physics, Calmers University of Technology and University of Goteborg, Sweden.

Wightman, P.S., Eavis, R.M., Batchelor, S.E., Walker, K.C. and Carruthers, S.P. (1999) Cost benefit assessment, including life cycle assessment, of oils produced from UK-grown crops compared with mineral oils. *Green Chem.* February, G6/G7 LCA.

Section IB: Biofuels from biomass

3

Biofuel production from agricultural crops

W. Soetaert and E. Vandamme

3.1 FOSSIL RESOURCES VERSUS RENEWABLE AGRICULTURAL RESOURCES

3.1.1 Environmental and strategic impact

Our society is using a large part of its raw materials to generate energy. According to the World Energy Council, about 82% of the world's energy needs are currently covered by fossil resources such as petroleum, natural gas and coal. Serious geopolitical implications arise from the fact that our society is so dependent on only a few resources such as petroleum, mainly produced in politically unstable oil-producing countries and regions. Also ecological disadvantages have come into prominence as the use of fossil energy sources suffers a number of ill consequences for the environment, including the greenhouse gas emissions, air pollution, acid rain, etc. (Wuebbles and Jain 2001).

Moreover, the supply of these fossil resources is inherently finite. It is assumed that we will be running out of petroleum within 50 years, natural gas within

65 years and coal in about 200 years at the present pace of consumption. With regard to the depletion of petroleum supplies, we are faced with the paradoxical situation that the world is using petroleum faster than ever before, and nevertheless the "proven petroleum reserves" have more or less remained at the same level since 30 years as the result of new oil findings (Campbell 1998). This fact is often used as an argument against the "prophets of doom", as there is seemingly plenty of petroleum around for the time being. However, those "proven petroleum reserves" are increasingly found in places that are poorly accessible, inevitably resulting in an increase of extraction costs and hence, oil prices. Campbell and Laherrère (1998), well-known petroleum experts, have predicted that the world production of petroleum will soon reach its maximum production level (expected before 2010). From that point on, the world production rate of petroleum will inevitably start decreasing. As the demand for petroleum is soaring, particularly to satisfy emerging countries such as China (by now already the second largest user of petroleum after the USA) and India, petroleum prices are expected to increase sharply. The effect can already be seen today, with petroleum prices soaring to over 50$/barrel at the time of writing (September 2004). Whereas petroleum will certainly not become exhausted from one day to another, it is clear that its price will continue to increase. This fundamental long-term upward trend may of course be temporarily broken by the effects of market disturbances, politically unstable situations or crises on a world scale.

All over the word, questions raise concerning our future energy supply. There is a continual search for renewable energy sources that will in principle never run out, such as hydraulic energy, solar energy, wind energy, tidal energy, geothermal energy and also energy from renewable raw materials such as biomass. Wind energy is expected to contribute significantly in the short term (Anonymous 1998). Giant wind parks are already being planned and built in the North Sea. In the long term, more input is expected from solar energy, for which there is still substantial technical progress to be made in the field of photovoltaic cell efficiency and production cost (Anonymous 2004). Bio-energy, the renewable energy released from biomass, is expected to contribute significantly in the mid- to long term. According to the International Energy Agency (IEA), bio-energy offers the possibility to meet 50% of our world energy needs in the 21st century.

3.1.2   Economic evaluation

In contrast to fossil resources, agricultural resources such as wheat or corn are becoming continuously cheaper because of the increasing agricultural yields, a tendency that will likely keep its pace for quite a while. New developments such as genetic engineering of crops will further promote this trend.

Agricultural crops such as corn, wheat and other cereals, sugar cane and beets, potatoes, tapioca, etc. can be processed in so-called biorefineries into relatively pure carbohydrate feedstocks, the primary raw material for most fermentation processes. These fermentation processes can convert those feedstocks into a wide variety of products, including biofuels such as bio-ethanol. Oilseeds such as soy, rapeseed (canola) and palm seeds can be equally processed into vegetable oils,

Table 3.1   Approximate average world market prices (€/ton) in 2003 of renewable and fossil feedstocks and intermediates.

Fossil		Renewable	
Petroleum	200	Corn	100
Coal	35	Straw	20
Ethylene	400	Sugar	180

that can be subsequently converted into biodiesel. Agricultural co-products or waste such as straw, bran, corn cobs, corn stover, etc. are lignocellulosic materials that are either poorly valorized or left to decay on the land and attract increasing attention as an abundantly available and cheap renewable feedstock. Estimations from the US Department of Energy have shown that up to 500 million tons of such raw materials can be made available in the USA each year, at prices ranging between 20 and 50$/ton (Clark 2004).

For a growing number of applications, the economic picture favours renewable resources over fossil resources as a raw material (Okkerse and Van Bekkum 1999). Whereas this is already true for a considerable number of chemicals, increasingly produced from agricultural commodities instead of petroleum, this is also becoming a reality for the generation of energy. The comparative price given in Table 3.1 is very informative in that respect. The prices given are the approximate average world market prices for 2003. Depending on local conditions such as distance to production site and local availability, these prices may vary rather widely from one place to another. Also, protectionism and local subsidies may seriously distort the price frame. As fossil and renewable resources are traded in vastly diverging measurement units and currencies, one needs to convert the barrels, bushels, dollars and euros into comparable units to turn some sense into it. All prices were converted into euros per metric ton (dry weight) for a number of fossil or renewable raw material as well as important feedstock intermediates such as ethylene and sugar, in order to allow an indicative cost comparison of fossil *versus* renewable resources.

From Table 3.1, one can easily deduce that on a dry weight basis, renewable agricultural resources cost about half as much as comparable fossil resources. Agricultural co-products such as straw are even a factor 10 cheaper than petroleum. At the present price of crude oil (>50$/barrel, corresponding to 300€/ton in September 2004), petroleum costs three times the price of corn. It is also interesting to note that the cost of sugar, a highly refined very pure feedstock (>99.5% purity), is about the same as petroleum, a very crude and unrefined mixture of chemical substances.

Also volume wise, agricultural feedstocks and intermediates have production figures in the same order of magnitude as their fossil counterparts, as indicated in Table 3.2.

It is clear that agricultural feedstocks are already cheaper than their fossil counterparts today and are readily available in large quantities. What blocks their further use is not economics but the lack of appropriate conversion technology. Whereas the (petro)chemical technology base for converting fossil feedstocks into a bewildering variety of useful products is by now very efficient and mature,

Table 3.2 Estimated world production and prices for renewable feedstocks and petrochemical base products and intermediates.

	Estimated world production (million tons per year)	Indicative world market price (euro per ton)
Renewable feedstocks		
Cellulose	320	500
Sugar	140	180
Starch	55	250
Glucose	30	300
Petrochemical base products and intermediates		
Ethylene	85	400
Propylene	45	350
Benzene	23	400
Terephthalic acid	12	700

the technology for converting agricultural raw materials is still in its infancy. It is widely recognized that new technologies will need to be developed and optimized in order to harvest the benefits of the bio-based economy. Particularly industrial biotechnology is considered a very important technology in that respect, as it is excellently capable to use agricultural commodities as a feedstock (Demain 2000; Dale 2003). The processing of agricultural feedstocks into useful products occurs in so-called biorefineries (Kamm and Kamm 2004; Realff and Abbas 2004). These are typically located in the countryside (close to the feedstock), contrary to port-based petrochemical refineries. Whereas the gradual transition from a fossil-based society to a bio-based society will take time and effort, it is clear that renewable raw materials are going to win over fossil resources in the long run. Particularly in view of the perspective of increasingly rarer, difficult to extract and more expensive fossil resources in contrast to ever cheaper agricultural crops.

3.2 COMPARISON OF BIO-ENERGY SOURCES

3.2.1 Burning of biofuels

Traditional renewable biofuels, such as firewood, used to be our most important energy source and they still fulfil an important role in global energy supplies today. The use of these traditional renewable fuels covered in 2002 is no less than 14.2% of the global energy use, far more than the 6.9% share of nuclear energy (IEA). In many developing countries, firewood is still the most important and locally available energy source, but equally so in industrialized countries. The importance is even increasing in several European countries, new power stations using firewood or straw have recently been put into operation and there are plans to create energy plantations with fast growing trees or elephant grass (*Miscanthus* sp.). A recent study has shown that, on the base of net energy generation per hectare, such energy plantations are the most efficient process to convert solar energy through biomass into useful energy (Garcia Cicad *et al.* 2003). An important factor in that respect is that such biofuels

Table 3.3   Energy yields of bio-energy crops in Flanders (Garcia Cicad *et al.* 2003).

	Yield (ton dry matter/ ha/year)	Biofuel type	Biofuel yield (ton/ha/year)	Gross energy yield	
				GJ/ton	GJ/ha/year
Wheat (cereal)	6.8	Bio-ethanol	2.29	26.8	61
Sugar beets (root)	14	Bio-ethanol	4.84	26.8	130
Rapeseed (seed)	3.1	Biodiesel	1.28	37	50
Willow/poplar (wood)	10.8	Firewood	10.8	18	194

have high yields per hectare (12 ton/ha and more) and can be burnt directly, giving rise to an energy generation of around 200 GJ/ha/year. The study is relevant to conditions in Western Europe and may be different for other regions of the globe.

3.2.2   Utilization aspects

The energy content of an energy carrier is however only one aspect in the total comparison. For the value of an energy carrier is not only determined through its energy contents and yield per hectare, but equally by its physical shape and convenience in use. This aspect of an energy source is particularly important for mobile applications, such as transportation. In Europe, the transport sector stands for 32% of all energy use, making it a very important energy user. There is consequently a strong case for the use of renewable fuels in the transport sector, particularly biofuels. Whereas in principle, we can drive a car on firewood, this approach is all but user friendly. In practice, liquid biofuels are much better suited for such an application. It is indeed no coincidence that nearly all cars and trucks are powered by liquid fuels such as gasoline and diesel. These liquid fuels are easily and reliably used in classic explosion engines and they are compact energy carriers, leading to a large action radius of the vehicle. They are easily stored, transported and transferred (it takes less than a minute to fill up your tank) and their use basically requires no storage technology at all (a simple plastic fuel tank is sufficient). Our current mobility concept is consequently mainly based on motor vehicles powered by liquid fuels that are supplied and distributed through tank stations.

The current strong interest for liquid motor fuels such as bio-ethanol and biodiesel based on renewable sources is strongly based on the fact that these biofuels show all the advantages of the classic (fossil-based) motor fuels. They are produced from agricultural raw materials and are compact, user-friendly motor fuels that can be mixed with normal petrol and diesel, with no engine adaptation required. The use of bio-ethanol or biodiesel therefore fits perfectly within the current concept of mobility. Current agricultural practices, such as the production of sugar cane or beets, rapeseed or cereals also remain fundamentally unaltered. The introduction of these energy carriers does not need any technology changes and the industrial processes for mass production of biofuels are also available.

Table 3.3 compares the energy yields of the different plant resources and technologies. For comparison, rapidly growing wood species such as willow or poplar as a classical renewable energy source, are also included.

It is clear that the gross energy yield per hectare is the highest for fast growing trees such as willow or poplar. However, a car does not run on firewood. Even if we restrict ourselves to the liquid fuels, there remain big differences between the different bio-energy options to be explained.

3.2.3   Energy need for biofuel production

At first sight, based on brut energy yield per hectare, bio-ethanol out of sugar beets would appear the big winner, combining a high yield per hectare and a high energy content of the produced bio-ethanol. Bio-ethanol out of wheat is lagging behind and biodiesel out of rapeseed comes last. Yet, biodiesel produced out of rapeseed is currently rapidly progressing in production volume in Europe. The comparison is clearly more complex than would appear at first sight, with several other facts to be considered.

3.2.3.1   The fossil energy replacement ratio

First of all, the energy input in the cultivation of the plants, transport as well as the production process itself needs to be taken into account. During the production of bio-ethanol, the distillation process is a big energy consumer. The amount of energy needed to produce the bio-ethanol is even close to the amount of energy obtained from the bio-ethanol itself. When all energy inputs are taken into account, the net energy yield can even be negative in poorly efficient production processes. It would then appear that more energy is being used than is produced. Ironically, this energy input often comes out of fossil energy sources. Obviously, this point is frequently used by opponents of bio-ethanol, they even consider it as an unproductive way to convert fossil energy in so called bio-energy, for the only sake of pleasing the agricultural sector. Whereas the data by Pimentel (2001), finding a negative energy balance, are outdated, Shapouri *et al.* (2003) have carefully studied the energy balance of corn ethanol and have concluded that the energy output : input ratio is 1.34. Sheehan *et al.* (2003) have determined the fossil energy replacement ratio (FER), the energy delivered to the customer over the fossil energy used. This parameter is important in relation to the emission of carbon dioxide (CO_2), the most important greenhouse gas. A high FER means that less greenhouse gases are produced (from the fossil fuel input) per unit of energy delivered to the customer. They have found an FER of 1.4 for bio-ethanol based on corn, and an FER as high as 5.3 for bio-ethanol based on lignocellulosic raw materials such as straw or corn stover. For comparison, the FER for gasoline is 0.8 and for electricity it is as low as 0.4.

3.2.3.2   Energy input for biofuel generation

Moreover, in order to properly evaluate this development, one must consider that bio-ethanol is a high-quality and energy-dense liquid fuel, perfectly usable for road transport. For its production, one needs mainly energy in the form of heat (for distillation), a fairly cheap, low-quality and non-portable energy source. The conversion of one energy form into another, especially if it becomes portable, is indeed a productive process. In the case of biofuel production, one converts cheap low-quality heat and biomass into high-quality portable liquid motor fuel, a relatively expensive but very convenient source of energy, particularly for transportation use. In the same way, cars do not run on petroleum either, but on the fuel that is being distilled out of it. The distillation, extraction and the long-distance transport of petroleum also requires a large energy input. The matter of the fossil energy input into bio-ethanol production becomes a non-issue altogether, when biomass is used as the source of heat, as is commonly practiced in Brazil where the sugar cane residue bagasse is burnt to generate the heat required for distillation. Similar production schedules may soon become a reality in the USA or Europe when, for example, ethanol is produced from wheat, with the wheat straw being burned to generate the heat for distillation.

Biodiesel has a lower energy input required for its production. However, the low yield of the crops from which it is produced per hectare dampens the perspectives for biodiesel. Concerning the difference in gross energy yield per hectare between sugar beet and wheat, it has also to be beared in mind that European farmers have traditionally obtained very high prices for their sugar beets. This high price is being maintained by the sugar market regulation (quota regulation) in Europe. Even if the yield per hectare is higher for sugar beets, it is more economical to produce ethanol out of wheat or other cereals with the current price structure. Unless the ethanol production can be coupled to the sugar production, a production scheme that offers technical advantages. However, the European sugar market regulation will be revised in 2006, probably with a sharp decrease of the sugar price as a result. The economic picture will then have to be reconsidered.

3.2.3.3   Bio-energy utilization

Concluding, liquid energy carriers are an (energetically) expensive but very useful energy carrier for mobile applications such as transportation. It is clear that energy sources for mobile applications should not only be compared on the basis of simple energy balances or costs, but also on the base of their practical usefulness, quality, environmental characteristics and convenience in use of the obtained energy carrier. It is interesting to note that Henry Ford, when designing his famous model T-car, presumed that ethanol would become the car fuel of the future. Although initially petrochemistry got the upper hand, it now seems as though Henry Ford was way ahead of his time and proven right in the long run.

3.3   BIO-ETHANOL

3.3.1   Bio-ethanol production process

Bio-ethanol (alcohol or ethylalcohol, the same molecule as potable alcohol) is mostly produced by microbial fermentation of carbohydrates (sugars), usually with the help of yeasts (e.g. *Saccharomyces cerevisiae*) as production micro-organisms. The generalized production scheme is depicted in Figure 3.1. The carbohydrate substrate can be obtained from several renewable sources, such as sugar beets or

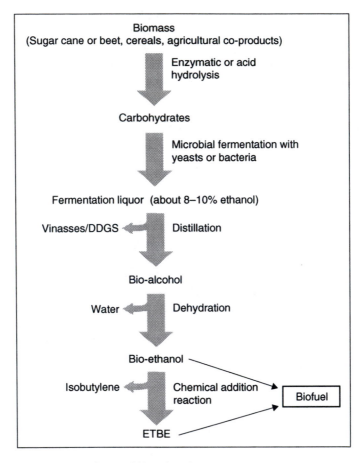

Figure 3.1   Production scheme of bio-ethanol.

sugar cane, cereals such as wheat and corn or agricultural co-products or waste streams. A sugar containing liquid substrate medium is inoculated with viable yeast and this converts the carbohydrates within a couple of days under the right circumstances into alcohol, typically in a concentration of around 8–10%. During the fermentation process, 1 kg of sugar is converted into approximately 0.5 kg ethanol and 0.5 kg CO_2. Although this mass balance suggests inefficiency, little energy gets lost during this transformation as ethanol (27 MJ/kg) has a higher energy content than sugar (17 MJ/kg). After the fermentation process, the alcohol is being extracted through simple distillation of the fermentation liquid.

Besides ethanol, a number of co-products are generated during production such as vinasses or cereal protein fractions such as distillers dried grain solubles (DDGS), which are used as animal feed or as a fertilizer. The concomitant production of CO_2 is very significant, as there is as much production of CO_2 as ethanol during fermentation. This kind of CO_2 is very pure and is often recovered for use in the chemical industry and for the production of soft drinks. Bio-ethanol is also used in the chemical, cosmetic and pharmaceutical industry and a small part is used for human consumption. Potable alcohol is in a certain way also a biofuel, it provides our bodies with 6.5 kcal/g.

3.3.2 Bio-ethanol use in the transportation sector

For use in mixtures with gasoline, the water in the alcohol (96% pure with about 4% water) must be removed, mostly with the help of regenerable absorption agents. The anhydrous alcohol, or bio-ethanol, can be used under different forms in motor fuels, typically in mixtures with normal gasoline. Anhydrous alcohol can be added directly to gasoline, a procedure that is practiced in Brasil. A practice more used in Europe and the USA is to react the bio-ethanol with the petrochemical intermediate isobutylene, and then add the obtained ethyl tertiary butyl ether (ETBE) to normal gasoline. Whereas European practice adds ETBE in low percentages, higher percentages are being used in the USA. A common blend in the USA is E15, containing 15% ethanol.

The application of bio-ethanol does not require an adaptation of the vehicle engine up to an addition percentage of 15%. Quite to the contrary, the addition of bio-ethanol or ETBE increases the oxygen value of the fuel, leads to a better combustion of the fuel mixture and acts as an octane enhancer. A fuel with 10–15 vol.% ETBE leads – in comparison with normal fuel – to a reduction of 20–25% of the carbon monoxide emission, 10–15% reduction of volatile hydrocarbons, 30% less soot, 20–30% less benzene and a strong reduction of the ozone content (in function of the weather conditions) (Anonymous 2000). It is also remarkable that ETBE acts as a lead substitute in fuel, so that strong ecological progress is obtained through the use of bio-ethanol in fuel blends. In view of its high purity, high energy content and easy storage, ethanol has also very good perspectives to drive fuel cells, instead of being used in a combustion engine (Chapter 17).

In the USA, ethanol production established a role as an octane enhancer as the Environmental Protection Agency began to phase out lead in gasoline. Later, ethanol production was strongly promoted with the passage of the Clean Air Act Amendments of 1990. Blending gasoline with ethanol has become a popular method for gasoline producers to meet the oxygen requirements mandated by the act. Initially the "oxygenate" methyl tertiary butyl ether (MTBE) was mainly used, produced out of methanol derived from fossil resources. Due to pollution of the ground and surface water by MTBE, strong preference is currently given to ETBE (Bekri and Pauss 2003).

3.3.3 Bio-ethanol world production

Table 3.4 indicates the world production figures for bio-ethanol in 2002. Globally 26 million tons of ethanol was produced worldwide of which 63% is used as biofuel. The production of ethanol is very unevenly distributed throughout the world. This is mainly due to the different policies and resulting regulatory and support measures that apply in different parts of the world. Whereas Brazil still ranks as the number one producer, the USA is quickly catching up with Europe seriously lagging behind, producing only 6% of the ethanol in the world.

Table 3.4   Annual production figures for bio-ethanol in 2002.

Country	Main feedstock used	Total production (million ton/year)	Biofuel (million ton/year)	Biofuel (%)
Brazil	Sugar cane	9.5	8.7	92
USA	Corn	6.4	5.7	90
Europe	Sugar beet and cereals	1.6	0.2	14
World	Diverse	26	16.4	64

3.3.4   Support policies for bio-ethanol utilization

In all cases, active support policies promote the use of biofuels. These support measures are generally motivated by different sets of arguments such as decreasing the dependency on imported petroleum, promoting sustainable development of our society and decreasing greenhouse gas emissions. An important aspect is that the use of biofuels produced from agricultural crops is a way of coping with the effects of agricultural overproduction and provides extra sources of income for farmers.

Brazil has supported the use of bio-ethanol from sugar cane as a biofuel since the mid-1970s (National Fuel Alcohol Program Proalcool), mainly motivated by its desire to reduce its dependency of foreign petroleum import. The programme was a huge success, with ethanol having over 50% market share already by 1988. This dominance has been lost in recent years to about 30% market share. Brazil currently requires 22–24% ethanol blends in all of its gasoline, and is the only country where 100% pure alcohol (aqueous, 96%) is used as a fuel in specially adapted cars. Brazil is still the leading bio-ethanol producer in the world and is presently a major bio-ethanol exporter.

The US bio-ethanol industry is of a later date and is nearly completely based on corn as the feedstock, with 7% of all corn in the USA being used for bio-ethanol production. Governmental support comes in the form of strong fiscal incentives for blending bio-ethanol into normal gasoline. As a result, an enormous boom in the production of bio-ethanol for biofuel use has been observed in the last 10 years in the USA. The growth percentages for bio-ethanol production in the USA amount to 10–12% per year during the last 5 years and this growth rate may even increase to 15% per year.

Although nearly all bio-ethanol in the USA is based on corn, a lot of money and effort is currently invested into the development of bio-ethanol production from agricultural co-products such as straw and corn stover (Kadam and McMillan 2003). A massive research effort is focusing on biotechnological research for cheaper production and improvement of the required cellulase enzymes to break down cellulose into the easily fermentable glucose. Whereas the cost of cellulases amounted to over 1.3$/l (5$/gal) in the past, specific research programmes have successfully reduced that cost by a factor of ten and lower (Goedegebuur 2003). At the same time, the development of recombinant organisms that can convert hemicellulosics into ethanol is also under development.

In Europe, the bio-ethanol industry is poorly developed. In 2002, 1.6 million tons of ethanol was produced in the European Union, of which only 225.000 ton was used as biofuel, mainly in Spain, France and Sweden. Most of the alcohol is produced out of sugar beets, wheat and other cereals, which provide an easily fermentable substrate. Only recently a number of stimulatory measures have been taken to promote the development of biofuels in Europe. Only in 2003, two guidelines have been approved that should stimulate the development of biofuels in Europe, particularly for the transport sector that accounts for 32% of the total energy use within the European Community. A first European guideline (2003/30/EC) lays down the objectives: by 2005, 2% of all fuel consumption for transport in Europe has to be covered by biofuels. This minimum percentage increases progressively in order to obtain a minimum percentage of 5.75% in 2010. By 2020, the European Commission wants to reach 20% substitution. Today this percentage is only 0.3% and is very unevenly divided amongst the different countries of the European Union. In order to obtain those objectives, Europe would need to produce no less than 9.3 million tons of bio-ethanol by 2010. This boils down to 3.7 million ha of wheat and sugar beets, a number to be compared with the 5.6 million ha of uncultivated farmland for which the European Commission pays its farmers in order to produce absolutely nothing!

A second European guideline (amendment 92/81/EC) sets the conditions for the use of biofuels, in particular the fiscal aspects. The fiscal aspects are extremely important since the taxation of petroleum-based motor fuels in Europe is traditionally very high, and amounts to more than half of the price paid at the pump. The guideline clears the road for a tax exemption of biofuels but leaves it up to the member states to choose for a full or partial tax exemption of biofuels, depending on the political will of the member state. Each member state has to find out the best way to reach the objectives set by the first guideline. The conversion of the European directives into national laws leads to strong differences in the various member states. Whereas Germany has already decided to exempt biofuels completely from taxation, other countries such as France have introduced partial detaxation. Other European member states have even taken no initiative at all so far.

3.4 BIODIESEL

3.4.1 Biodiesel production process

Biodiesel is produced from vegetable oils and fats through a simple chemical process. Except for lipids from recuperated oils (e.g. used frying fat), most of the biodiesel is produced from either soybeans (USA) or rapeseed (Europe). Biodiesel consists mainly of methyl esters of C_{16}–C_{18} fatty acids (FAME, fatty acid methyl ester). There are a number of production schemes, but the basic principles are always the same, as depicted in Figure 3.2. The oil fraction (triglycerides) is extracted from the oilseeds and then treated in the presence of an excess of methanol with alkalines. A trans-esterification occurs by which glycerol is released and FAMEs are formed. After the reaction, the reaction mixture can be physically separated in an oily biodiesel phase and a watery phase that contains the salts and glycerol. The excess of

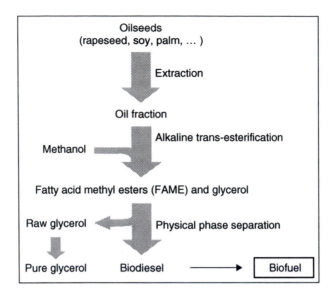

Figure 3.2   Production scheme of biodiesel.

methanol is recovered by distillation and reused. The obtained biodiesel fraction is a slightly yellow low-viscosity liquid, comparable with normal diesel oil. The mass balance is favourable since 1 kg of oil reacts with approximately 0.1 kg methanol, producing 1 kg of biodiesel and 0.1 kg glycerol. The co-product glycerol is a relatively expensive base chemical that is used in various applications in the cosmetic and chemical industry. Biotechnological alternatives for the current (chemical) production process are under development (Fukuda *et al.* 2001; Du *et al.* 2003). The use of enzymes, particularly lipases, can avoid the use of alkaline reaction agents and could make the production process considerably "greener".

3.4.2   Biodiesel use in the transportation sector

Biodiesel (37–41 MJ/kg) has an energy content comparable to ordinary diesel fuel (36–45 MJ/kg) and can be mixed with normal diesel fuel, usually up to 5%. In France, a mixture of 30% biodiesel is used in the so-called "Villes Diester", where a considerable share of vehicles runs on this "Diester". The application of biodiesel does not require any adaptation of the diesel engine. To the contrary, biodiesel addition is highly appreciated because of its good lubrication effect on the motor. In France, one out of every two diesel cars already runs with 2% biodiesel added to the normal diesel fuel, without even knowing it as it is not indicated at the pump. Pure biodiesel is also used, mainly in Germany and Austria. However, a small engine adaptation is then required. Problems can occur during cold winters when using pure biodiesel, due to crystallization at low temperatures. For the sake of completeness, it must also be mentioned that cars can run directly on pure vegetable oils instead of needing to be converted to biodiesel, provided the car is equipped with a special motor configuration such as the Elsbett motor.

In its environmental characteristics, biodiesel compares very favourable to normal diesel (Anonymous 2000). Biodiesel contains almost no sulfur, contrary to normal diesel that has a maximum sulfur content of 350 ppm in Europe. The presence of oxygen in biodiesel considerably improves the combustion, resulting in a strongly reduced emission of soot. Pure biodiesel can reduce the cancer risks by 93% (Anonymous 2000). In comparison to fossil-based diesel, biodiesel emits 56% less hydrocarbons, 55% less particulates and overall results in a reduction of 60–90% of air toxics. Only with regard to nitrous oxides, a slight 6% increase in emissions occurs.

3.4.3   Biodiesel production and support policies

The European production of biodiesel increases strongly and has exceeded 1 million ton in 2002, mainly produced in Germany, France and Italy. Especially Germany has shown impressive growth, due to a 100% tax exoneration for all biofuels. Presently biodiesel is exclusively produced from methanol, derived from natural gas and is cheaper than ethanol. It is possible to use bio-ethanol instead of methanol, obtaining fatty acid ethyl esters (FAEE), to produce a biodiesel completely derived from renewable resources. In the future, our cars may thus run on an interesting (bio-)chemical combination of sugar beets and rapeseed. The future can also bring other possibilities such as so called oxydiesel, a mixture of normal diesel and bio-ethanol, with considerably improved combustion characteristics, thanks to the presence of oxygen in bio-ethanol.

In Japan and the USA, the biodiesel production is much lower than in Europe. The USA interest in biodiesel was stimulated by the Clean Air Act of 1990 combined with regulations requiring reduced sulfur content in diesel fuel and reduced diesel exhaust emissions. The Energy Policy Act of 1992 established a goal of replacing 10% of motor fuels with non-petroleum alternatives by the 2000 and increasing to 30% by the year 2010. However, not much has happened, in part because in the USA nearly all cars run on gasoline and only trucks and buses run on diesel. In early 1995, Japan decided to explore the feasibility of biodiesel by initiating a 3-year study. A new biodiesel plant was to be constructed with its feedstock being recycled vegetable oils collected in the Tokyo area, estimated at 0.2 million ton annually.

3.5   CONCLUSION

The use of bio-ethanol and biodiesel derived from agricultural crops is a technically viable alternative for fossil-based gasoline or diesel. Moreover, their use fits perfectly in the present concept and technology of our mobility. Agricultural crops or organic waste streams can be efficiently converted into biogas and used for heat, power or electricity generation. Even as the discussion about the sense or nonsense of biofuels is ongoing, the transition process from a fossil-based to a bio-based society is clearly moving forward, with impressive growth in the USA, and Europe finally catching on. There is little doubt that in the medium term, we will all fill up our car with a considerable percentage of biofuels, probably unaware of it and without noticing the difference.

REFERENCES

Anonymous (1998) Renewable energy target for Europe: 20% by 2020. Report from the European Renewable Energy Council (EREC).

Anonymous (2000) Biodiesel: the clean, green fuel for diesel engines (fact sheet). NREL Report no. FS-580-28315; DOE/GO-102000-1048.

Anonymous (2004) Renewable energy scenario to 2040: half of the global energy supply from renewables in 2040. Report from the European Renewable Energy Council (EREC).

Bekri, M. and Pauss, A. (2003) Fate of fuel oxygenates in the environment. *Mededelingen Faculteit Landbouwkundige en Toegepaste Biologische Wetenschappen Universiteit Gent* **68**, 33–40.

Campbell, C.J. (1998) The future of oil. *Energy Explor. Exploit.* **16**, 125–152.

Campbell, C.J. and Laherrère, J.H. (1998) The end of cheap oil. *Sci. Am.* **278**, 78–83.

Clark, W. (2004) The case for biorefining. National Renewable Energy Laboratory. www.sae.org/events/sfl/pres-clark.pdf

Dale, B.E. (2003) "Greening" the chemical industry: research and development priorities for biobased industrial products. *J. Chem. Technol. Biotechnol.* **78**, 1093–1103.

Demain, A. (2000) Small bugs, big business: the economic power of the microbe. *Biotechnol. Adv.* **18**, 499–514.

Du, W., Xu, Y. and Liu, D. (2003) Lipase-catalysed transesterification of soy bean oil for biodiesel production during continuous batch operation. *Biotechnol. Appl. Biochem.* **38**, 103–106.

Fukuda, H., Kondo, A. and Noda, H. (2001) Biodiesel fuel production by transesterification of oils. *J. Biosci. Bioeng.* **92**, 405–416.

Garcia Cicad, V., Mathijs, E., Nevens, F. and Reheul, D. (2003) Energiegewassen in de Vlaamse landbouwsector. Publication 1, 94 p. STEDULA, steunpunt voor duurzame landbouw. Info@stedula.be

Goedegebuur, F. (2003) Improved cellulase products for biomass conversion. *Mededelingen Faculteit Landbouwkundige en Toegepaste Biologische Wetenschappen Universiteit Gent* **68**, 275.

IEA Bioenergy. www.ieabioenergy.com

Kadam, K.L. and McMillan, J.D. (2003) Availability of corn stover as a sustainable feedstock for bioethanol production. *Bioresour. Technol.* **88**, 17–25. NREL Report No. JA-510-34434.

Kamm, B. and Kamm, M. (2004) Principles of biorefineries. *Appl. Microbiol. Biotechnol.* **64**, 137–145.

Lynd, L.R. (1996) Overview and evaluation of fuel ethanol from cellulosic biomass: technology, economics, the environment, and policy. *Ann. Rev. Energ. Environ.* **21**, 406–465.

Okkerse, C. and Van Bekkum, H. (1999) From fossil to green. *Green chem.* **1**(2), 107–114.

Pimentel, D. (2001) The limits of biomass energy. *Encycl. Phys. Sci. Technol.* September 2001.

Realff, M.A. and Abbas, C. (2004) Industrial symbiosis: refining the biorefinery. *J. Ind. Ecol.* **7**, 5–9.

Shapouri, H., Duffield, J.A. and Wang, M. (2003) The energy balance of corn ethanol: an update. USDA Report no. 814.

Sheehan, J., Aden, A., Paustian, K., Killian, K., Brenner, J., Walsh, M. and Nelson, R. (2003) Energy and environmental aspects of using corn stover for fuel ethanol. *J. Ind. Ecol.* **7**, 117–146. NREL Report no. JA-510-36462.

Vandamme, E.J. and Soetaert, W. (2004) Industrial biotechnology and sustainable chemistry. BACAS-report, Royal Belgian Academy, 34p. http://www.europabio.org/documents/150104/bacas_report_en.pdf

Wuebbles, D.J. and Jain, A.K. (2001) Concerns about climate change and the role of fossil fuel use. *Fuel Process. Technol.* **71**, 99–119.

4

Biomass processing in biofuel applications

G. Akay, M. Dogru, O.F. Calkan and B. Calkan

4.1 INTRODUCTION

4.1.1 Biomass in sustainable energy technology and feedstock for chemicals

The most important and urgent global issue is the establishment of a sustainable energy technology. While serious concerns about global warming are becoming reality as a result of the current fossil fuel-based energy technology, others are concerned about the depletion of the crude oil and natural gas reserves in the next 50 years. Therefore, alternatives for fossil fuel-based energy and raw materials for chemicals industry must be established urgently while improving the efficiency of energy conversion and preservation processes. The major renewable sources of energy (wind and solar), provide raw material free technology compared with biomass. However, their availability and control are their main drawbacks. In addition, they cannot provide a sustainable feedstock source for chemicals and they are unsuitable for the needs of a car-based society. Biomass-based energy technology does not have these strategically important drawbacks of wind–solar energy and it makes the society a stakeholder in the production and use of energy for the reasons summarised below.

According to the US Department of Energy National Renewable Energy Laboratory (NREL), the energy equivalent of global biomass production is about eight times the world energy requirement every year. Currently, only a small fraction of this vast resource is utilised globally. Like all renewable energy sources, biomass displaces fossil fuel use and its associated impacts. However, unlike the other energy sources, biomass can be harmful to the environment if it is not utilised as renewable source of energy and raw materials. Unused biomass would otherwise be burnt, land-filled or accumulate as excess biomass in forests. Biomass landfill burial leads to leachates and greenhouse gas emission, soil and water contamination. Biomass accumulation in forests increases wildlife hazards, depresses watershed and productivity. On the other hand, control of combustion conditions, use of catalysts, gas and process water cleanup, reduces pollutant emission by two orders of magnitude and creates heat and power (Klass 1998).

However, biomass energy production has costs that are inherently high compared with the gas/oil/coal-fired electricity generation processes. This is due to:

(1) biomass fuels have low bulk density, they are expensive to gather, process, transport and handle;
(2) biomass power generators are smaller than conventional generators, which makes bioenergy generation economically disadvantaged;
(3) biomass energy technology is not as advanced and integrated with the needs of the society compared with oil/natural gas/coal.

Therefore, investment in biomass energy is strategic in order to achieve a sustainable global energy policy. This strategy has become more acceptable as a result of concerns over global warming, emission of toxins and carbon dioxide (CO_2), and oil/gas supply/distribution security and disruption to the very large, centralised power generators. Therefore, localised energy/power generation and consumption based on biomass using small- to medium-sized generators integrated with the biomass resources and environment should be the strategic goal of any integrated, combined heat and power generation.

The urgent need for the establishment of a carbon neutral sustainable energy and carbon reduction technologies are now well understood. One of the most important routes for such sustainable technologies is based on biomass, either in the form of energy crops or biomass waste. Despite reaching its saturation as regards the generation of novel platform technologies, the fossil fuel-based energy and chemical technologies are going to continue for some time, dictated by the emergence of sustainable alternatives. The negative impact of fossil fuels to the environment can be reduced by combining them with biomass which, in some cases, can also result in synergy.

In order to compete with fossil fuel-based energy technology without any tax support, biomass-based energy systems must achieve feedstock and processing advantages over fossil fuels. To a certain extent, this advantage is already present in feedstock logistics for small-scale power generation based on waste biomass. However, for biomass to replace fossil fuels, large-scale harvesting, and process integration of various strands of the technology are essential. In processing, step changes in the efficiency of conversion must be achieved through the use of state-of-the-art processing technology, biotechnology, nano-technology and post-genomics. These step changes can be achieved through the use of physical–chemical and bioprocess intensification and miniaturisation and genetic engineering of microorganisms and plants (Akay 2005; Akay et al. 2005a–c).

4.1.2 Significance of bioethanol in sustainable energy technology

Bioethanol is used today as an alternative for gasoline, or as fuel extender, an oxygenate and octane enhancer for gasoline. It has the potential to reduce vehicle CO_2 emission by 90% (Tyson et al. 1993) since CO_2 production during bioethanol fermentation and consumption is part of the global carbon cycle. Ethanol has also an important impact on automobile exhaust emissions since it reduces CO emission during oxidative combustion and therefore it can replace methyl ter-butyl ether as an oxygenate (Tyson et al. 1993; Putsche and Sandor 1996; Unnasch et al. 2001). Currently, up to 85% ethanol can be used to power vehicles without any extra cost (Rosillo-Calle and Cortez 1998; Sheehan 2001). Furthermore, bioethanol can be catalytically converted to hydrogen through steam reforming and therefore bioethanol can be used to store hydrogen (Marino et al. 2001).

Compared with lignocellulose-based bioethanol, biodiesel represents a partial short-term solution to a sustainable energy technology since the biodiesel feedstock is essentially a food grade material. Moreover, only a small fraction (ca. 5–10%) of the cultivated feedstock is converted into biodiesel and over 70% of biodiesel cost is associated with the raw material. However, the picture is more promising for bioethanol production from lingocellulose is considered. In the latter case, bioethanol product yield can be as high as ca. 20% and the cost of lignocellulose feedstock can be negative or negligible. Therefore, the cost of bioethanol is overwhelmingly associated with fuel processing which should be reduced by a factor of 10 in order to compete with the fossil fuels at current cost levels. Clearly, bioethanol production presents a great opportunity for novel processes to reduce cost.

There are several routes for bioethanol production which can involve biotechnology at several stages, i.e., lignocellulose conversion to sugars and fermentation to alcohol. However, it is also possible to produce bioethanol using catalyst and a mixture of carbon monoxide (CO)/carbon dioxide/hydrogen which constitute the main components of the gas produced in gasification (Inui and Yamamoto 1998). An integrated plant producing bioethanol, thermal energy and power from biomass represents an important route for a sustainable energy technology.

4.1.3 Strategic importance of gasification

Due to the logistics of biomass feedstock (often as biomass waste), biomass-based energy technology has to be locally supported and utilised. Therefore, such a technology will not have the scales of economy enjoyed by fossil fuel-based

energy technologies, but it will have the desired environmental and societal acceptance as well as economic competitiveness based on process integration, reduced
transport and delivery costs, reduced capital cost through process intensification
and miniaturisation (Akay 2005; Akay *et al.* 2005a–c) and being locally utilised.
Operated in relatively small scale (0.1–20 MWe) with locally available feedstock,
it will provide security of raw material and energy supply. In order to achieve this
economic competitiveness, types of biomass which can be easily integrated, have
potential for further development and reduce environmental impact of energy
production need to be examined. In this respect, the most important raw materials
are lignocellulosic feedstock (from agriculture, forests, energy crops), municipal
solid waste (MSW), sewage sludge and industry specific wastes such as animal
and oil/food waste.

   Irrespective of the route (biological or chemical) for transport fuel production,
chemical hydrogen storage, or in the production of biogas from landfilled biomass,
gasification is essential for the treatment of waste produced in these processes.
Alternatively, biomass can be directly converted primarily into carbon monoxide
and hydrogen through gasification. Therefore, in this work, the gasification characteristics of biomass/waste using a novel pilot plant scale gasifier (10–50 kWe
capacity) are given. This gasifier system has recently been scaled up to 1 MWe.
These gasifiers also integrate novel gas and process water cleaning systems and
the produced gas is suitable for electricity generation through internal combustion
engine, high-, intermediate- or low-temperature fuel cells or micro-turbines
provided that sufficient gas cleaning is achieved.

4.1.4   Gasification and fuel cells

High-temperature fuel cells; solid oxide fuel cells (operating at 650–1000°C)
(Kim and Virkar 1999; Fuel Cell Handbook 2002) and molten carbonate fuel cells
(operating at 600–650°C) (Lobachyov and Richter 1998; Fuel Cell Handbook 2002;
Kivisaari *et al.* 2002) or intermediate temperature (operating at 400–700°C) ceramic
fuel cells (Zhu *et al.* 2002) are particularly useful for integration with gasifiers since
suitable gas mixtures containing CO_2, CO, H_2 and methane (CH_4) can be used
and air can be utilised as an oxidant. Water gas shift reaction involving CO and
steam reforming of CH_4 produce H_2 for the anode reaction. The *in situ* generation
of hydrogen promotes rapid kinetics with non-precious catalyst materials and
high quality by-product heat for cogeneration. In molten carbonate fuel cells, CO_2
produced at the anode can be utilised at the cathode. Since the major components
of the product gas from the gasification are CO, CO_2, H_2, CH_4 only limited amount
of gas cleanup is necessary and unlike the low-temperature polymer electrolyte
membrane fuel cells, the gas composition is not critical. Furthermore, thermal
energy of the product gas also contributes to the higher efficiency of the high-
temperature fuel cells. However, high temperature of solid oxide fuel cells places
stringent requirements on the materials of construction.

   It is also important to note that, at present most commercially available high-
temperature fuel cells have capacity up to 60 MWe (Kivisaari *et al.* 2002).
Although this top fuel cell capacity is somewhat outside the range of downdraft
gasifiers which are suitable for localised power generation in the range 0.1–20 MWe,

capacity matching between the downdraft gasifiers and high-temperature fuel cells will not be a problem.

4.2   GASIFICATION OF BIOMASS

4.2.1   Gasifiers

Gasification is the thermal decomposition of solid fuel to a combustible gas, or product gas from the gasifier rich in carbon monoxide and hydrogen (Bridgwater 1995; Littlewood 1997). By using limited amount of oxidant (pure oxygen, air or steam) a partial oxidation (gasification) will take place. There are several types of gasifiers available; while fixed bed gasifiers have electrical energy output of up to a few MWe, fluidised bed gasifiers are suitable for multi-MWe operations. Locations of the biomass and air supply and product gas outlet are also important in the gasifier performance (Klass 1998; Lobachyov and Richter 1998; Dogru 2000; Wang *et al.* 2000; Midilli *et al.* 2001, 2002; Dogru *et al.* 2002a,b; Reed 2003a,b). In this chapter a fixed bed downdraft gasifier system performance using three different biomass fuels is presented.

4.2.2   Description of the gasifier used in present study

4.2.2.1   Gasifier development

An experimental pilot plant multi-mode gasifier system has been developed at Newcastle University over the last 5 years (Dogru 2000; Dogru and Akay 2004; Dogru *et al.* 2002a,b, 2003, 2004; Midilli *et al.* 2001, 2002, 2003, 2004; Akram *et al.* 2003) and this system has been scaled up from 10 to 50 kWe (for BLC Research, UK) to 250 kWe (for Anglian Water plc, UK) and finally in 2004, to 1 MWe (for Ecoprotech Ltd./ITI Ltd., UK). Although at each stage of development, the experience from the operation of the previous scale-up was utilised, the most radical changes were made in the ITI/Ecoprotech Gasifier System (Dogru and Akay 2004). Nevertheless, basic scale-up rules developed so far were validated in the subsequent gasifier designs and gas cleaning/water removal systems (Dogru and Akay 2004). The research reported here is based on the 10 kWe pilot plant gasifier which was also used as the template for the 50 and 250 kWe gasifiers described above. The 1 MWe Ecoprotech/ITI Plant is a down–updraft gasifier capable of handling different feedstock (Dogru and Akay 2004).

4.2.2.2   Process stages

The downdraft gasifier consists of a cylindrical reaction vessel, with a constriction near the base. Biomass fuel is admitted from the top and it proceeds by gravity down through the unit. Four process stages can be identified as: drying, pyrolysis, oxidation and reduction zones from top to bottom. Figure 4.1 is a diagrammatic illustration of the gasifier showing the reaction zones and the temperature profile. Total gasification system which consists of a gasifier, vortex scrubber and vacuum pump is shown in Figure 4.2 while the diagrammatical illustration is shown in Figure 4.3. The experimental setup and the procedure are discussed in

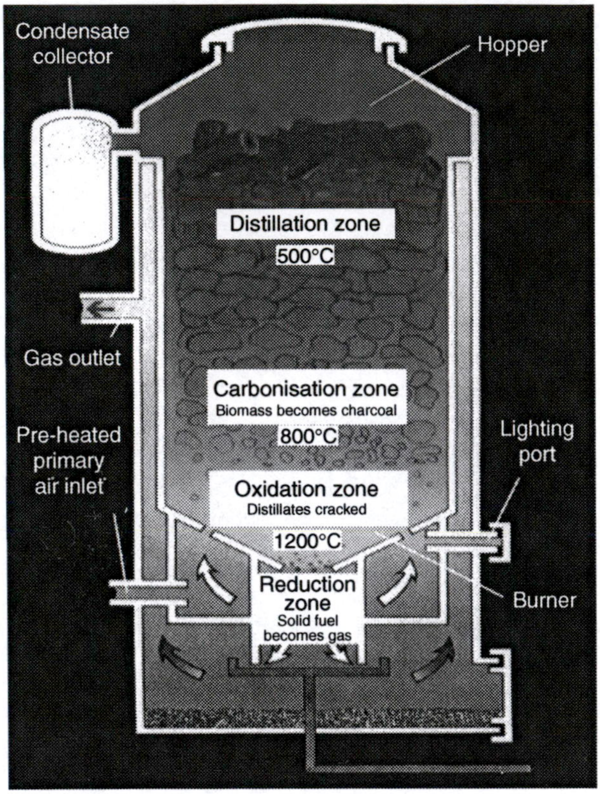

Figure 4.1   Simulated picture of downdraft fixed bed-throated gasifier. Height of the bed, $H = 810$ mm, diameter of the oxidation zone, $D = 450$ mm, diameter of the drying hopper zone, $L = 305$ mm, diameter of the throat, $d = 135$ mm.

detail by Midilli *et al.* (2001, 2002, 2003, 2004); Dogru *et al.* (2002a,b, 2003, 2004) and Akram *et al.* (2003).

In the drying zone, heat passes by convection and conduction within the gasifier upward to the fresh fuel charge. The temperature of the charge rises and the moisture content lowers. The temperature range and the height of the drying zone are *ca.* 70–200°C and 0.10 m, respectively. In the pyrolysis zone, the temperature of the fuel rises to the point where the volatile constituents are driven off. This takes place using the thermal energy released by the partial oxidation of the pyrolysis products at the zone below. The temperature range and the height of the pyrolysis zone are 350–500°C and 0.17 m.

In the oxidation zone, air is admitted via nozzles to the pyrolised biomass. The heat absorbed in all other stages is supplied by the exothermic reactions in this zone. Tar molecules are degraded in the gas phase to form CO_2 and H_2O at about 1200°C. The temperature and the height of the oxidation zone are approximately 1000–1200°C and 0.12 m. In the reduction zone, endothermic reactions occur between the char and the gases (including water) yielding mainly CO and H_2, with some CH_4.

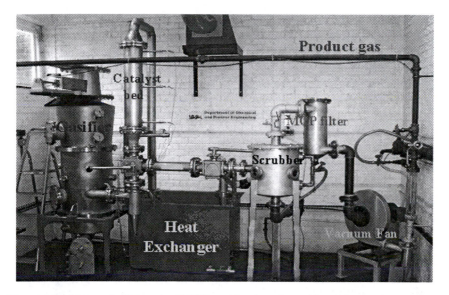

Figure 4.2   Newcastle University 50 kWe gasification plant with a 5:1 turn down ratio, illustrating the process flow diagram. This plant subsequently scaled up to 1 MWe for Ecoprotech/ITI.

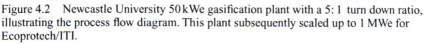

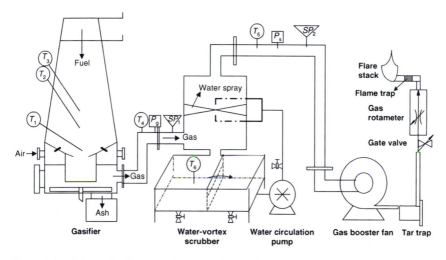

Figure 4.3   Schematic diagram of the experimental setup (T_1: throat (oxidation) zone temperature; T_2: gasifier outlet temperature; T_3: drying zone temperature; T_4: gasifier outlet temperature; T_5: scrubber outlet temperature; P_g: pressure drop across the gasifier; P_s: pressure drop across the scrubber; SP_1: gas sampling point at the gasifier outlet; SP_2: gas sampling point at the scrubber outlet).

4.2.2.3   Gas cleanup

The produced gas leaves the gasifier at a temperature range of 200–350°C. The produced gas needs to be cleaned as it is loaded with particulates and pyrolytic products, so called tar and steam. The cleanup unit should be designed according to the end use of the produced gas. Depending on the use of the gas, product gas may have to be cooled and water is removed. It is also important to remove heavy metals to achieve high final emission standards and prevent metal deposition in the proceeding process units.

The product gas was cleaned by a V-tex vortex scrubber (Dogru *et al.* 2002a,b). In the scrubber, the soluble inorganic compounds and some organic compounds in the gas are removed as the gas passes through a thin water film produced by the impingement of two water jets. Furthermore, the rapid cooling causes the deposition of high-temperature boiling tars from the gas. A gas booster fan is used to provide gas suction at the outlet to water scrubber before it is flared after ignition by a pilot burner.

4.2.3   Experimental procedure

Biomass fuel consistency is important in gasification in order to achieve continuous flow through the reactor and to provide reliable product gas composition and calorific value for the downstream energy conversion processes. Furthermore, densification of the biomass fuel reduces the reactor size, while shape and size of the densified fuel reduces fluctuations in gas and fuel flow rates as well as the subsequent product quality. For biomass fuels such as wood chips which are already dense, further densification is not crucial but desirable. However, for fuels with large variations in its content and low bulk density, such as MSW, densification is absolutely inevitable. Therefore, MSW, after the removal of metals and glass, is either heat treated to obtain cellulose rich powder or it is shredded and subsequently briquetted to densify the fuel.

The data acquired in the gasification experiments are: fuel and gas flow rates, gas composition, tar and particulate concentration in the gas, temperatures and pressures along the process line. Temperatures were recorded with an analogue to digital converter every 15 s for inlet air, drying zone, pyrolysis zone, throat and scrubber outlets. The pressure drops were measured at the gasifier and water scrubber outlet. The product gas flow rate was measured after the suction fan. The amounts of tar and condensate in the product gas were determined from gas samples taken at the gasifier and water scrubber outlets (Figure 4.3). Clean and dry wood chips and charcoal for use as filters are placed in the respective trays in the box filter (not shown in the diagram) between the scrubber outlet and gas booster fan. The gas flow meter is regulated to the required flow rate.

Two Pyrex U-tubes in series were used to collect tar and condensate, and to clean the gas samples for gas analysis. The first U-tube trap contained small spherical glass beads while the second trap contains silica. Produced gas passed through the U-tubes using a vacuum pump via a rotameter to measure gas flow rate from which tar content was evaluated. After each gasification experiment, gasifier was cleaned and amounts of ash, char and unused biomass were determined and the biomass flow rate was evaluated.

Gas chromatography (GC) was used to analyse the gas samples using helium as a carrier gas using a dual column system (chromosorp 101 and molecular sieve) with a thermal conductivity detector. Table 4.1 summarises the GC column operating

Table 4.1   GC specifications.

	Column 1	Column 2
Pressure (kPa)	130	95
Carrier gas velocity	40 ml/min	35 ml/min
Packing material	Molecular sieve	Chromosorb
Peaks for gases	H_2, O_2, CH_4, CO, N_2	CO_2, C_2H_6, C_2H_2

conditions and gas concentrations that are evaluated. Air and producer gas humidity were also measured in order to carry out accurate energy and mass balances.

4.2.4   Biomass characterisation

Based on the ultimate analysis, the high heating value (HHV) was calculated using the IGT method (Dogru 2000; Midilli *et al.* 2001; Dogru *et al.* 2002a) in which HHV is calculated from:

$$HHV = 0.341\ C + 1.323\ H \times 0.068\ S - 0.0153\ A - 0.120\ (O + N)$$

In designing thermochemical conversion systems, it is more realistic to use the low heating value (LHV) of the gas which excludes the latent heat of steam in the fuel as it does not contribute to the actual heating value of the fuel. LHV is calculated from (Dogru 2000; Midilli *et al.* 2001):

$$LHV\ (dry\ basis) = HHV\ (dry\ basis) - 0.218\ H$$

where, HHV and LHV are in kJ/kg; C, H, S, A, O, and N are the weight percentages of carbon, hydrogen, sulfur, ash, oxygen and nitrogen. Alternatively, the calorific value of the gas can be calculated from:

$$(CV)_{dry\ gas} = 1.055(12.1\ X_{H_2} + 11.9\ X_{CO} + 37.36\ X_{CH_4})$$

where CV is the calorific value, and X_{H_2}, X_{CO} and X_{CH_4} are the mole fractions of the main combustible gases, hydrogen, carbon monoxide and methane, respectively.

4.3   GASIFICATION OF SUGAR CANE BAGASSE

4.3.1   Background to sugar cane bagasse gasification

In the thermochemical decomposition of lignocellulosic compounds, emissions of NO_x and SO_x are low (Mathieu and Dubuisson 2002; McKendry 2002; Dogru *et al.* 2003). One source of lignocellulosic biomass is sugar cane, the world's largest agricultural crop (Garcia-Perez *et al.* 2001). Sugar cane is cultivated in 127 countries in both the tropics and subtropics with a global production in 2001/2 of 1300 million tons of cane. Sugar cane residues such as bagasse and cane trash (cane tops and leaves) can be used as an important source of biomass.

The bagasse used in this study was obtained from Barbados where 55,000 ton of sugar cane are harvested annually. Approximately 33 ton of bagasse are produced per 100 ton of cane in Barbados which results in an estimated 17,000 ton of bagasse. A portion of the bagasse is stored until the following year to provide fuel for start-up operations of the plant in the cane-harvesting season. Cane trash is left in the fields to provide soil cover and nutrients for soil.

4.3.2   Characteristics of sugar cane bagasse

The proximate and ultimate analyses of sugar cane bagasse are shown in Table 4.2. The HHV of bagasse was determined experimentally to be 18.14 MJ/kg (standard deviation 1.16) and theoretically calculated as 17.03 MJ/kg (standard deviation 0.46). This data is also in agreement with that of other workers (Suárez et al. 2000; Gabra et al. 2001). Typical biomass fuels for gasification have LHVs of approximately 15–17 MJ/kg. Wood waste, which has been the traditional fuel for biomass gasification, has a LHV in the range 17.8–20.8 MJ/kg. Based on the LHV value of 17.7 MJ/kg for bagasse, it can be concluded that this fuel is suitable for gasification.

The diameter and length of the briquetted bagasse were 63 and 130 mm, respectively. The absolute and bulk density were 679 and 642 kg/m^3. The fuel moisture content greatly effects both the operation of the gasifier and the quality of the product gas. The moisture content constraints for gasifier fuels depend on type of gasifier used. Higher moisture content fuels can be gasified in updraft gasifier systems but the upper limit acceptable for a downdraft reactor is generally considered to be around 40% dry basis (Dogru et al. 2003). Generally, moisture content of most biomass varies between 11 wt% and 18 wt%. Therefore most biomass are suitable and well within the range for an engine application (Reed 2003).

The absolute and bulk density of biomass is very important for process design in terms of handling and storage. Biomass with high bulk densities will require less reactor space for a given re-fuelling time. However, low bulk density fuels sometimes give rise to insufficient flow under gravity resulting in low gas heating value and possibly burning char in the reaction zone (Dogru et al. 2003). The bulk density of sugar cane bagasse is higher than that of wood chips (250 kg/m^3), and the experiments showed that there was no char burning in the reduction zone.

High volatile matter in biomass generally increases the tar content in the product gas and a substantial quantity of this tar should be removed before it is fed to an internal combustion engine (Reed and Das 1990). Sugar cane bagasse has lower volatile matter than wood chips (64.8%), resulting in high gas quality and lower tar content. The ash content of sugar cane bagasse is also low.

Table 4.2   Physical and chemical properties of sugar cane bagasse.

	Proximate analysis		Ultimate analysis (wt%)
Moisture (wt%)	6.4	Carbon	45.16 ± 0.14
Fixed carbon (wt%)	13.00 ± 2.85	Hydrogen	5.66 ± 0.01
Volatile matter (wt%)	39.89 ± 4.30	Oxygen	48.68 ± 0.06
Ash (dry basis) (wt%)	1.51 ± 0.32	Nitrogen	0.42 ± 0.06
HHV (dry fuel) (MJ/kg)	18.14 ± 1.16	Sulfur	0.02 ± 0.002
LHV (dry fuel) (MJ/kg)	17.70 ± 1.06	Chlorine	0.06 ± 0.02

4.3.3   Gasification characteristics

Wet product gas compositions, and mass and energy balance data for sugar cane bagasse gasification are listed in Table 4.3. The fuel flow rates varied between 7.44 and 2.87 kg/h in these runs. In Run 2 (fuel flow rate of 4.21 kg/h), bagasse was not briquetted which resulted in bridging and the gas composition refers to product gas during this period. As a result, the percentage of the combustible gases is reduced and the calorific value of the product gas is low. Otherwise, the main combustible gases are H_2, CO, CH_4, C_2H_2 and C_2H_6, and constituting approximately 25–27% of the total product gas. The LHV of the product gas from bagasse gasification ranged from 3.30 to 4.56 MJ/Nm^3 (wet gas). The volume of gas produced per unit weight of biofuel ranged from 2.19 to 4.57 Nm^3/kg of bagasse gasified.

The product gas burnt with a strong royal blue flame and was able to sustain the flame in the absence of the pilot burner for the remainder of the run (1½h). The gas samples collected every half-hour showed only small changes in gas composition and exemplified the steady nature of the flame. This also suggests good fuel flow characteristics within the reactor, which is expected from a high bulk density fuel. However, if bridging takes place, the variability of the gas composition increases rapidly. The variation of the product gas composition during gasification is shown in Figure 4.4 in the absence of bridging. During bridging the flame height decreased at the stack outlet and eventually production of combustible gas stopped as was seen by the disappearance of the flame. Within seconds of agitating the fuel bed, combustible gas was produced, as evidenced by the redevelopment of a strong royal blue flame.

Table 4.3   Gasification mass and energy balance data for sugar cane bagasse as the fuel (balance of the wet gas composition represents moisture which is approximately 3% by volume).

Parameter	Gasification runs				
	1	2	3	4	5
Fuel feed rate (kg/h)	7.44	4.21	4.43	2.87	3.98
Wet gas composition (vol.%)					
H_2	10.13	9.47	13.40	10.86	11.20
N_2	60.55	70.24	56.89	58.48	58.48
CO	8.46	5.66	15.56	16.45	12.16
CH_4	2.03	0.28	1.54	1.54	3.57
CO_2	16.11	12.09	9.98	10.19	11.36
C_2H_2	0.03	0.00	0.00	0.00	0.17
C_2H_4	0.46	0.03	0.03	0.15	0.52
C_2H_6	0.00	0.00	0.03	0.00	0.00
Mass flow rate of tar (kg/h)	0.057	0.024	0.055	0.058	0.086
LHV wet gas (MJ/Nm^3)	3.30	1.95	4.10	3.95	4.56
Mass balances (%)					
C	70.68	45.71	123.8	150.8	138.2
H	58.22	43.85	91.55	102.6	119.7
O	56.09	41.82	93.90	62.40	79.64
N	129.7	161.8	128.1	128.0	125.5

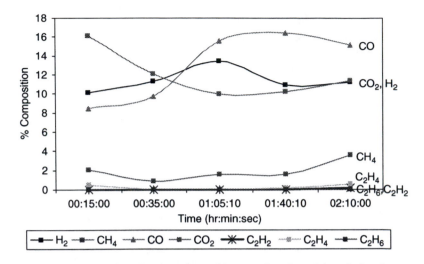

Figure 4.4   Variation of product gas composition as a function of time during the gasification of sugar cane bagasse. Biomass fuel flow rate is 4.43 kg/h (Run 3 in Table 4.3).

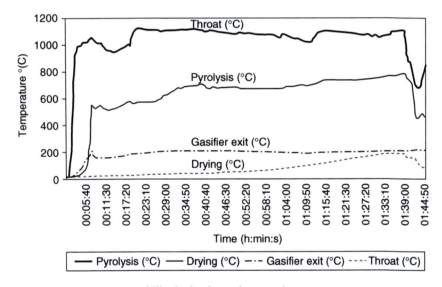

Figure 4.5   Temperature stability in the thermal conversion zones.

Bagasse is very similar in chemical composition to wood. Wood has been the traditional fuel in biomass gasification. A comparison of the average percentage composition of the component gases for gasification of both types of biomass using the downdraft gasifier was carried out. The comparison of the product gases from wood and bagasse indicate that the variations in gas composition are within the experimental variations for each biofuel.

The temperature profiles of the drying, pyrolysis, throat and gasifier exit are illustrated in Figure 4.5 using briquetted bagasse. Figure 4.5 clearly illustrates the

Table 4.4   Tar and particulate content of the product gas.

Gasification run	Gas flow rate (Nm^3/h)	Volume of product gas per unit weight of fuel (Nm^3/kg)	Gas flow rate per fuel flow rate (Nm^3/kg)	Tar + particulates (g/Nm^3)
1	17.14	2.16	2.30	3.35
2	9.24	2.05	2.19	2.63
3	17.14	3.40	3.87	3.20
4	13.11	4.00	4.57	3.83
5	17.14	3.80	4.31	5.00

rapid increase in temperature, which occurs after the throat has been lit, and the relatively stable temperatures in each of the thermal conversion zones even during the fuel-loading period.

The gravimetric analysis established that the tar and particulate matter content of the product gas ranged from 2.63–5.00 g/Nm^3 (Table 4.4). The lowest tar content of 2.63 g/Nm^3 corresponds to Run 2 during bridging. This tar range represents 0.13–0.50 wt% of the product gas. The procedure used in the analysis accounts for both tar and particulate matter, since both contaminants are entrained in the product gas. The concentration of tar alone was not determined.

The mass/energy balance on the gasifier provides a quantitative measure of the efficiency of conversion of fuel to product gas and ultimately to electricity using this particular type of gasifier. The mass balance analysis on the gasifier requires an evaluation of the inputs and outputs from the gasifier. It is not easy to obtain 100% closure, which illustrates the difficulties of obtaining such data. However, the mass balance closure (Table 4.3) for five runs was found to be 97.8%, which represents a relatively high efficiency.

4.4   GASIFICATION OF SEWAGE SLUDGE

4.4.1   Background to sewage sludge gasification

Sewage sludge is an important renewable biomass energy source for the developed and developing countries for fuel production. Although it is widely used as a fertiliser in many countries, there is excess capacity which can be utilised in the production of power and energy through gasification (Gruter et al. 1990; Midilli et al. 2001, 2002, 2003). Sewage sludge production and excess capacities are large: nearly 1 million m^3/year of sewage sludge dry solids are produced in the UK, 4.2 million m^3/year in Switzerland and 50 million m^3/year in Germany (Gruter et al. 1990; Midilli et al. 2001, 2002, 2003). This production will probably increase since treatment plants of municipal wastewater are continuously being built and these must comply with more stringent environmental standards (Lu et al. 1995). The production of sewage sludge in the European Union is expected to increase at least by 50% to the year 2005 (Littlewood 1997). The negative effects of sewage sludge on the environment and the living beings should be reduced using appropriate technologies for fuel production. This may be possible by converting sewage sludge to the fuel instead of landfill and/or incineration options which are likely to be limited in future. An alternative utilisation of sewage sludge is through gasification, where the produced gas can be used in an integrated gasification combined cycle (IGCC) power generation configuration (Bridgwater 1995).

4.4.2 Gasification characteristics of sewage sludge

Partially dried (moisture content 12 wt%) sewage sludge was supplied by Anglian Water plc (UK). Proximate and ultimate analysis of sewage sludge used in these study can be found in Midilli *et al.* (2001, 2002, 2003). The experiments were conducted in six runs from which, wet and dry gas mass flow rates were determined. Figure 4.6 presents the variation of mass flow rate of the product gas (wet or dry) with sewage sludge feed rate. As shown in Figure 4.6 mass flow rate of the product gas ranged from 4.98 to 10.9 kg/h which corresponded to dry product gas flow rates of 4.88 and 10.6 kg/h, giving 2–3 wt% water in the produced gas. It was observed that the amount of wet product gas increased by increasing sewage sludge feed rate. However, the flow rate of wet product gas was affected by char and ash from sewage sludge.

The variations of the sewage sludge fuel rate and calorific value of dry produced gas with the mass flow rate of dry produced gas are presented in Figure 4.7 and Table 4.5. When the mass flow rate of the product gas was less then *ca.* 7 kg/h, both the briquetted fuel flow rate and calorific value of the product gas remain constant. As the product gas flow rate increased, both variables increase. However, as the flow rate of the dry product gas increases to 10.6 kg/h, the flow rate of the solid fuel and calorific value of product gas both show a slight

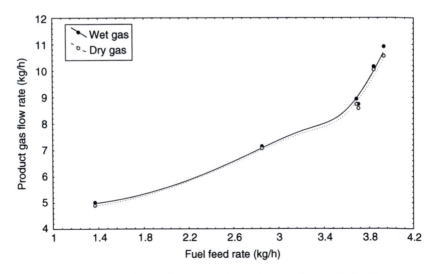

Figure 4.6　Variation of mass flow rate of the product gas (wet or dry) with sewage sludge fuel feed rate.

decrease; at this stage, nitrogen and carbon monoxide comprise almost 81 wt% of the dry product gas.

Figure 4.8 shows the variation of energy of the clean fuel with dry gas mass flow. Energy of the product gas increased with increasing fuel flow rate until it reaches a maximum value of *ca.* 29 MJ/h when clean fuel flow rate reaches *ca.* 10 kg/h. However, it appears that the energy value of the fuel decreases after this maximum as a result of the changes in the composition of combustible gases.

Dry product gas efficiency was investigated using three energy indicators: hot, cold and raw gas efficiency. If product gas is to be used directly in burners without any prior cleaning, it is possible to define the raw gas efficiency of the product gas to include the tar in the gas. It was observed that product gas efficiency increased with increasing wet product gas mass flow. However, at a wet product gas mass flow rate of 10.15 kg/h, product gas efficiency reached maximum and started to decrease afterwards. Maximum conversion efficiencies for the cold, raw and hot

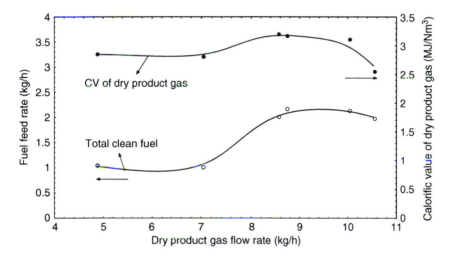

Figure 4.7   Variation of wet fuel feed rate and product gas calorific value with dry product gas flow rate.

Table 4.5   Characteristics of dry combustible gas produced from sewage sludge under different product gas flow rates. Flow rates (kg/h) are represented by F with subscript letters indicating the components of the combustible gases.

Dry product gas flow rate (kg/h)	10.6	10.1	8.74	8.56	7.05	4.88
CV of dry product gas (MJ/Nm³)	2.55	3.11	3.17	3.20	2.81	2.85
F_{H_2} (kg/h)	1.08	1.10	0.98	0.91	0.63	0.48
F_{CO} (kg/h)	0.66	0.72	0.71	0.93	0.66	0.46
F_{CH_4} (kg/h)	0.12	0.21	0.16	0.11	0.09	0.05
$F_{C_2H_2}$ (kg/h)	0.09	0.09	0.05	0.05	0.05	0.03
$F_{C_6H_6}$ (kg/h)	0.03	0.02	0.02	0.01	0.01	0.01
Briquetted solid fuel flow rate (kg/h)	1.98	2.14	2.12	2.11	0.94	1.03

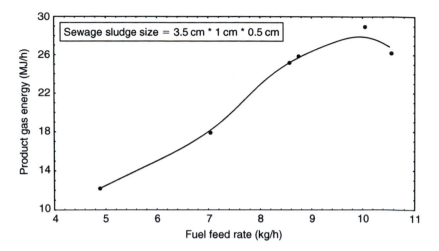

Figure 4.8   Variation of the product gas energy with fuel feed rate.

product gases are found to be 57%, 65% and 70%, respectively. Within the biomass flow rate range, the corresponding minimum conversion efficiencies were
48%, 55% and 60%, respectively.

4.5   GASIFICATION OF MUNICIPAL SOLID WASTE

4.5.1   Background to MSW gasification

Estimates in 2001 indicate that, in Europe, some 62.6% of the total waste is land-
filled, 21.9% is incinerated, 4.5% is composted and 11.0% is recycled (Akram *et al.*
2003). In 2001 it was reported that the total municipal solid waste (MSW)
production is about 32 million tons/year in the UK (Burnley 2001). After the
removal of metals and glass, some of the municipal waste is composted, but it is
mainly landfilled which is being severely curtailed due to the unavailability of
land and environmental concerns. The UK Landfill Directive on waste management strategies (Burnley 2001) forecasts that by 2015, 33% of household waste
should be recycled and therefore, very small (if any) growth in MSW generation
is predicted as shown in Table 4.6.

   MSW is generally defined as "household waste plus other waste of a similar
composition collected by (or on behalf of) the local authority". In practice, this
means that if the waste generated by a particular commercial business is collected
along the household waste the material is classed as MSW. The combustion of
MSW for energy production is an effective use of waste products that significantly
reduce problems of waste disposal (Demibras 2001). It has been claimed that lower
air emissions and more efficient energy recovery can be achieved with gasification
than with conventional mass burn technologies at similar cost (Niessen 1996).
The potential and the economical analysis of the clean fuels from MSW based on
metropolitan areas are available (Larson *et al.* 1996; Bjorklund *et al.* 2001).

Table 4.6   Strategies to meet the landfill directive and waste strategy targets (Burnley 2001).

Route	UK diversion required by 2020	
	Total mass (million tons per year)	Number of new plants required
Recycling	6.9–12.5	100–210
Organic waste composting	3.0–5.6	130–260
Incineration with energy recovery	10.9–30.1	35–110
Total diversion	20.8–48.2	–

Table 4.7   Physical and chemical properties (proximate, ultimate and calorific analyses gas calorific value) of MSW and wood.

	MSW	Wood chips
Absolute density (kg/m^3)	926	837
Bulk density (kg/m^3)	110	250
C (%)	45.60	42.70
H (%)	6.18	6.58
O (%)	24.70	47.77
N (%)	1.66	0.45
S (%)	0.46	0.37
Ash (%)	21.80	2.20
Moisture (%, wet)	8.65	21.10
Volatile matter (%)	53.70	70.20
Fixed carbon (%)	15.40	7.72
GCV (MJ/kg)	20.12	16.89

4.5.2   Characteristics of MSW

Black-bag bin waste as collected in refuse collection vehicle (RCV) was obtained from Slain Environmental Limited, UK. This material does not contain any metals or glass. It was briquetted prior to the gasification. Results of proximate and ultimate analysis of MSW are given in Table 4.7 where they are compared with those of wood (Dogru 2000). Bulk and absolute densities of both MSW and Wood Chips are given in Table 4.7. Although metals and glass were removed from MSW at the pre-gasification treatment process, samples still contained metal traces.

4.5.3   Gasification characteristics of MSW

Gasification characteristics of the briquetted MSW are summarised in Table 4.8 and compared with those of wood. Table 4.8 shows that percentage of carbon monoxide for MSW in the product gas is highest when gas flow is low and it decreases as the gas flow rate is increased. However for wood chips, CO content of the product gas first increases with the gas flow and then decreases as the flow rate of the gas increases which these changes are not significant. Tar content of the product gas was found to be between 4.55 and 5.75 g/m^3 for MSW. While for

Table 4.8   Product dry gas composition at different feed flow rates for MSW and wood.

	MSW	MSW	MSW	Wood	Wood	Wood
% ↓/Flow rate (nm^3/h) →	7.38	10.44	13.41	7.38	10.44	13.41
H$_2$	15.60	16.28	14.00	17.46	17.24	16.75
N$_2$	51.60	52.12	53.57	50.40	50.22	51.01
CH$_4$	1.98	1.93	1.80	2.04	2.21	2.13
CO	15.30	14.17	13.82	16.30	17.09	16.38
CO$_2$	14.90	15.03	16.17	13.20	12.54	13.09
C$_2$H$_4$	0.52	0.36	0.45	0.47	0.53	0.48
C$_3$H$_4$	0.10	0.11	0.19	0.13	0.17	0.16
Calorific value (MJ/m^3)	5.02	4.82	4.60	5.37	5.60	5.37

wood chips, tar content of the product gas was between 1.44 and 1.92 g/m^3. Net energy output varied from 0.62 to 0.76 kWe/kg of MSW and from 0.45 to 0.56 kWe/kg of wood chips. Both hot gas and cold gas efficiencies were in the range of 45% and 75%.

4.6   CATALYSTS FOR TAR CRACKING/GAS CLEANING AND GAS CONVERSION

Tar cracking, water shift reaction, conversion of producer gas to methanol and ethanol, or indeed reforming of these two important high energy density fuels/raw materials involve important catalytic reactions encountered in energy conversion processes (Simell et al. 1996; Delgado et al. 1997; Orio et al. 1997; Engstrom 1998; Inui and Yamamoto 1998; Hasler and Nussbaumer 1999; Wang et al. 2000; Nir et al. 2003). Spent catalysts can be subsequently used in high-temperature gas cleanup also (Furimsky and Biagini 1996). There are several techniques available for hot gas cleaning before the utilisation of the gasifier gas in internal combustion engine, fuel cells or gas turbine to generate electricity (Hasler and Nussbaumer 1999; Beer 2000; Wahlund et al. 2004).

Surface area and catalyst accessibility are important both in catalysis and gas cleanup (Kusakabe and Sotowa 2003; Akay et al. 2005a). Nano-structured micro-porous catalysts or catalyst supports offer intensified catalysis since they provide enhanced surface area which is accessible to the reactants and products through a network of channels feeding into the regions of catalytic activity (Akay et al. 2005a). In non-structured catalysts, although the surface area might be large, they are often not accessible as a result of surface fouling and the diffusion resistance can slow down the rate of reaction. Catalysts are either deposited as a thin film on a support or they are used as pellets. These two techniques have certain draw-backs. In coated systems, catalyst adhesion can be non-uniform and weak while the accessibility of the active sites within the interior of the catalyst pellets is hindered due to low porosity.

Recent developments in catalysts focus on nano-porous catalytic sites which are accessible through a network of arterial micro-pores. These catalysts are

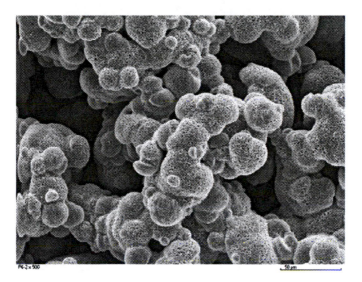

Figure 4.9   Micro-structure of nickel-based intensified catalysts showing the presence of arterial micro-pores.

obtained through a deposition of metals from solution on a micro-porous poly-meric template which is subsequently heat treated to obtain porous metallic struc-tures where the size of the pores range from tens of microns to tens of nano-meters, thus eliminating the problems of accessibility and rapid pore fouling and closure (Akay et al. 2005a). The technique differs fundamentally from the compression-based systems of micro-porous structure preparation where the porosity is reduced as a result of compaction. It also differs from the well-known wash-coating or chemical vapour deposition techniques where the penetration in to the micro-pores of the catalyst support is poor (see for example, Twigg 1989). Furthermore, the mechanisms of metal deposition within micro-pores and nano-structure forma-tion are novel. The importance and current fabrication techniques of porous mater-ials can be found in Akay (2005) and Akay et al. (2005a–c).

Figures 4.9–4.11 illustrate the micro-structure of these novel materials. The overall skeleton is formed by fused metallic grains of ca. 8 μm forming micro-pores of ca. 30 μm arterial passages (channels). As seen in Figure 4.9, surface of the grains are porous which can be clearly identified in Figure 4.10 from where the size of the surface pores can be evaluated to be ca. 200 nm. However, further detailed examination reveals that these materials have finer structures well below 50 nm. All of these parameters; grain, arterial channel and grain pore sizes can be controlled (Akay et al. 2005a). The examination of the grains shows the existence of porosity within the grains as shown in Figure 4.11. Therefore these structures provide a large available surface area. Details of material preparation, both the tem-plate micro-porous polymer and nano-structured metals are available (Akay et al. 2004, 2005b).

These materials are based on nickel which is useful as catalysts in many gas phase reactions, including those relevant to the tar cracking, ethanol production via

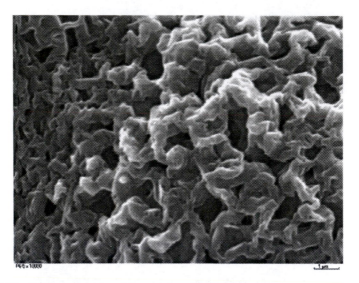

Figure 4.10   Surface structure of the metal grains illustrating the presence of the nano-pores.

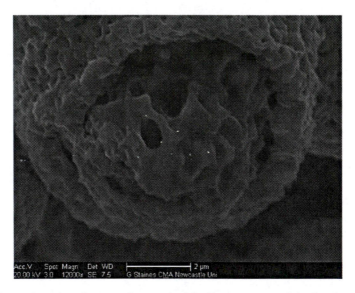

Figure 4.11   Scanning electron micrograph of the inner structure of the fused metal grains, which form the skeleton of the catalyst. Inner porosity can be controlled to provide large surface.

biomass gasification or hydrogen generation from ethanol (Inui and Yamamoto 1998; Wahlund *et al.* 2004). Initial experiments with these catalysts indicate that they can remove several impurities present in the gas stream. When the product gas from the gasification of wood (which is one of the cleanest biofuel) was passed through a bed of such catalyst (as described in Figures 4.9–4.11), tar and metal

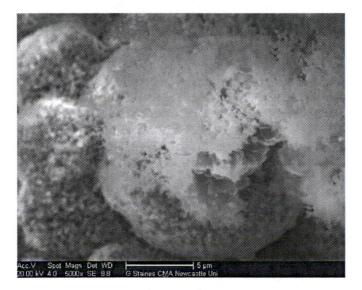

Figure 4.12   Scanning electron micrograph of the nickel-based catalyst surface showing tar deposition on the metal surface. The surface EDX analysis before and after tar deposition is shown in Table 4.9.

Table 4.9   EDX analysis of the nickel-based catalyst before and after testing when the product gas from the gasification of wood chips was passed over the catalyst. Total amount of gas passed was $2.4\,m^3$ at 350°C for 1 h.

Element (at.%)	Ni	O	P	C	K	Cl	S	Ca	Na	Mg	Si
Initial	78.01	21.39	0.60	0.00	0.00	0.00	0.00	0.00	0.00	0.00	0.00
Test	9.67	29.08	0.51	56.48	1.24	1.35	0.47	0.36	0.52	0.26	0.06

deposition was observed. These studies were carried out using scanning electron microscopy (SEM) equipped with energy dispersive X-ray micro-analysis (EDX) facility. Gas cleanup was conducted at 350°C through a line from the gasifier outlet at a rate of $2.4\,Nm^3/h$ for a period of 1 h. After the deposition experiment, the catalyst was examined under SEM with EDX facility to determine the composition of the deposit. The Scanning electron micrograph in Figure 4.12 shows the presence of deposition. The EDX surface analyses of the catalyst before and after the test are summarised in Table 4.9. It can be seen that several elements have been collected by the catalyst but the majority appear to be hydrocarbon.

4.7   CONCLUSIONS

In the replacement of fossil fuel-based energy technology and hydrocarbon feed stock by the corresponding sustainable technologies, biomass plays an important role since biomass storage is not an option for environmental reasons and it also provides a cheap raw material source. In a biomass-based sustainable energy technology, gasification has a pivotal role, irrespective of the route taken for transport fuel and power generation.

The most important biomass sources are wood, sugar cane bagasse, sewage sludge and MSW. Their gasification behaviour was examined using a 10 kWe pilot plant gasifier which has been successfully scaled up to 50, 250 and 1000 kWe capacity. The above cited biomass fuels can be gasified continuously and the product gas is rich in hydrogen and carbon monoxide, suitable for power generation through internal combustion engines and high-temperature fuel cells. In all cases, the calorific value of the gasifier gas was more than sufficient (usually two to three times the minimum required) to produce electricity using an internal combustion engine. Product gas calorific value can be further enhanced by using oxygen enriched air.

In order to be able to achieve continuous gasification, it is essential to have densified fuels which do not show wide variation in its composition. Therefore, pre-gasification densification of the above cited biomass is necessary. Fuel densification is part of the "process intensification" strategy adopted in the gasifier design as well as gas and process water cleanup and catalytic tar cracking processes, which make the gasification of biomass environmentally acceptable and economically sustainable. The application of the principles of process intensification (Akay 2005) to gasifier technology results in reduced plant size and capital/operating costs, enhanced gas quality and more responsive plant (Dogru and Akay 2004).

ACKNOWLEDGEMENTS

We are grateful to the UK Engineering and Physical Sciences Research Council (EPSRC), UK Department of Trade and Industry (DTI), Avecia/Cytec, BLC Research, BP Amoco, Exxon Mobil, Intensified Technologies Incorporated (ITI), Morecroft Engineers Ltd., Safety-Kleen Europe, Triton Chemical Systems and Willacy Oil Services Ltd. for their support. Help given by A. Midilli, M. Akram and C. Jordan in the experiments is gratefully acknowledged.

REFERENCES

Akay, G. (2005) Bioprocess and chemical process intensification. In: *Encyclopedia of Chemical Processing* (ed. S. Lee), Marcel Dekker, NY.

Akay, G., Birch, M.A. and Bokhari, M.A. (2004) Microcellular polyhipe polymer (PHP) supports osteoblastic growth and bone formation in vitro. *Biomaterials* **25**, 3991–4000.

Akay, G., Dogru, M., Calkan, B. and Calkan, O.F. (2005a) Flow induced phase inversion phenomenon in process intensification and micro-reactor technology: preparation and applications of nano-structured micro-porous polymers and metals. In: *Microreactor Technology and Process Intensification* (eds. Y. Wang and J. Halladay), Chapter 18, Oxford University Press, Oxford.

Akay, G., Bokhari, M.A., Byron, V.J. and Dogru, M. (2005b). Development of nano-structured materials and their application in bioprocess–chemical process intensification and tissue engineering. In: *Developments and New Trends in Chemical Engineering* (eds. M.A. Galan and E.M. Del Valle), Chapter 7, Wiley, London.

Akay, G., Erhan, E. and Keskinler, B. (2005c) Bioprocess intensification in flow through monolithic microreactors with immobilized bacteria, *Biotech. Bioeng.* **90**, 180–190.

Akram, M., Dogru, M. and Akay, G. (2003) Gasification of municipal solid waste (MSW). In: *Process Intensification and Miniaturization in Biological, Chemical, Environmental and Energy Conversion Processes* (eds. G. Akay and M. Dogru), DocQwise, York, UK.

Beer, J.M. (2000) Combustion technology developments in power generation in response to environmental challenges. *Progress in Energy and Combustion Sciences* **26**, 301–327.

Bjorklund, A., Melaina, M. and Keoloian, G. (2001) Hydrogen as a transportation fuel product from thermal gasification of municipal solid waste: an examination of two integrated technologies. *Int. J. Hydrogen Energ.* **26**, 1209–1221.

Bridgwater, A.V. (1995) The technical and economic feasibility of biomass gasification for power generation. *Fuel* **74**, 631–653.

Burnley, S. (2001) The impact of the European landfill directive on waste management in the United Kingdom. *Resour. Conserv. Recycl.* **32**, 349–358.

Delgado, J.M., Anzar, J.M.P. and Corella, J. (1997) Biomass gasification with steam in fluidized bed: effectiveness of CaO, MgO and CaO–MgO for hot raw gas cleaning. *Ind. Eng. Chem. Res.* **36**, 1535–1543.

Demibras, A. (2001) Biomass resource facilities and biomass conversion processing for fuels and chemicals. *Energ. Convers. Manage.* **42**, 1357–1378.

Dogru, M. (2000) *Fixed-bed Gasification of Biomass.* PhD Thesis, University of Newcastle, Newcastle upon Tyne, UK.

Dogru, M. and Akay, G. (2004) *Gasification.* International Patent Application, PCT/GB2004/004651.

Dogru, M., Midilli, A. and Howarth, C.R. (2002a) Gasification of sewage sludge using a throated downdraft gasifier and uncertainty analysis. *Fuel Process. Technol.* **75**, 55–82.

Dogru, M., Howarth, C.R., Akay, G., Keskinler, B. and Malik, A.A. (2002b) Gasification of hazelnut shell in downdraft gasifier. *Energy* **27**, 415–427.

Dogru, D., Akay, G., Calkan, O.F., Jordan, C.A. and Calkan, B. (2003) Gasification of sugar cane bagasse for power production using a throated downdraft gasifier. In: *Process Intensification and Miniaturization in Biological, Chemical, Environmental and Energy Conversion Processes* (eds. G. Akay and M. Dogru), DocQwise, York, UK.

Dogru, M., Midilli, A., Akay, G. and Howarth, C.R. (2004) Gasification of leather residues – Part 1: Experimental study via a pilot-scale air-blown downdraft gasifier. *Energ. Source.* **26**, 35–44.

Engstrom, F. (1998) Hot gas clean-up bioflow ceramic filter experience. *Biomass Bioenerg.* **15**, 259–262.

Fuel Cell Handbook (2002) EG&G Technical Services, Inc., DOE/NETL, Morgantown, WV, USA.

Furimsky, E. and Biagini, M. (1996) Potential for preparation of hot gas clean-up sorbents from spent hydroprocessing catalysts. *Fuel Process. Technol.* **46**, 17–24.

Gabra, M., Pettersson, E., Backman, R. and Kjellström, B. (2001) Evaluation of cyclone gasifier performance for gasification of sugar cane residue – Part 1: Gasification of bagasse. *Biomass Bioenerg.* **21**, 371–380.

Garcia-Perez, M., Chaala, A., Yang, J. and Roy, C. (2001) Co-pyrolysis of sugarcane bagasse with petroleum residue. Part 1: Thermogravimetric analysis. *Fuel* **80**, 1245–1258.

Gruter, H., Matter, M., Oehlmann, K.H. and Hicks, M.D. (1990) Drying of sewage sludge is important step in waste disposal. *Water Sci. Tech.* **22**, 57–63.

Hasler, P. and Nussbaumer, Th. (1999) Gas cleaning for IC engine applications from fixed bed biomass gasification. *Biomass Bioenerg.* **16**, 385–395.

Inui, T. and Yamamoto, T. (1998) Effective synthesis of ethanol from CO_2 on polyfunctional composite catalysts. *Catal. Today* **45**, 209–214.

Klass, D.L. (1998) *Biomass for Renewable Energy, Fuels and Chemicals*, Academic Press, New York, NY, USA.

Kim, J. and Virkar, A. (1999) In: *Solid Oxide Fuel Cells-VI* (eds. S. Singhal and M. Dokiya), Electrochemical Society, Pennington, NJ, USA.

Kivisaari, T., Bjornbom, P. and Sylwan, C. (2002) Studies of biomass fuelled MCFC systems. *J. Power Sources* **104**, 115–124.

Kusakabe, K. and Sotowa, K.-I. (2003) Microreactors for catalytic process intensification. *Catalysts Catalysis* **45**, 666–670.

Larson, E.D., Worrell, E. and Chen, J.S. (1996) Clean fuels from municipal solid waste for fuel cell buses in metropolitan areas. *Resour. Conserv. Recycl.* **17**, 273–298.

Littlewood, K. (1997) Gasification: theory and application. *Prog. Energy Combust. Si.* **3**, 35–71.

Lobachyov, K.V. and Richter, K.J. (1998) An advanced integrated biomass gasification and molten fuel cell power system. *Energy Convers. Mgmt.* **39**, 1931–1943.

Marino, F., Boveri, M., Baronetti, G. and Laborde, M. (2001) Hydrogen production from steam reforming of bioethanol using Cu/Ni/K/(γAl$_2$O$_3$ catalysts. Effect of Ni. *Int. J. Hydrogen Energ.* **26**, 665–668.

Mathieu, P. and Dubuisson, R. (2002) Performance analysis of a biomass gasifier. *Energ. Conver. Mgmt.* **43**, 1291–1299.

McKendry, P. (2002) Energy production from biomass-3: gasification technologies. *Bioresource Technol.* **83**, 55–63.

Midilli, A., Dogru, M., Howarth, C.R., Ling, M.J. and Ayhan, T. (2001) Combustible gas production from sewage sludge with a downdraft gasifier. *Energy Conv. Mgmt.* **42**, 157–172.

Midilli, A., Dogru, M., Akay G. and Howarth, C.R. (2002) Hydrogen production from sewage sludge via a fixed bed gasifier product gas. *Int. J. Hydrogen Energ.* **27**, 1035–1041.

Midilli, A., Dogru, M., Akay, G. and Howarth, C.R. (2003) Clean bio-fuel production from sewage sludge for future applications of fuel cells. In: *Process Intensification and Miniaturization in Biological, Chemical, Environmental and Energy Conversion Processes* (eds. G. Akay and M. Dogru), DocQwise, York, UK.

Midilli, A., Dogru, M., Akay, G. and Howarth, C.R. (2004) Gasification of leather residues – Part 2: Conversion into combustible gases and the effects of some operational parameters. *Energ. Sources* **26**, 45–53.

Niessen, W.R. (1996) *Evaluation of Gasification and Novel Processes for the Treatment of Municipal Solid Waste.* NREL/TP-430-21612, National Renewable Energy Laboratory Golden, Colorado, USA.

Nir, S.A., Pemen, A.J.M., Yan, K., van Gompel, F.M., van Leuken, H.E.M., van Heesch, E.J.M., Ptasinski, K.J. and Drinkenburg, A.A.H. (2003) Tar removal from biomass-derived fuel gas by pulsed corona discharges. *Fuel Process. Technol.* **84**, 161–173.

Orio, A., Corella, J. and Narvaez, I. (1997) Performance of different dolamites on hot raw gas cleaning from biomass gasification with air. *Ind. Eng. Chem. Res.* **36**, 3800–3808.

Putsche, V. and Sandor, D. (1996) Strategic, economic and environmental issues for transportation fuels. In: *Handbook on Bioethanol: Production and Utilisation* (ed. C.E. Wyman), pp. 21–35. Taylor and Francis, Washington, DC, USA.

Reed, T.B. (2003a) *Encyclopedia of Biomass Thermal Conversion: The Principals and Technology of Pyrolysis, Gasification and Combustion.* The Biomass Energy Foundation Press, Golden, Co., USA.

Reed, T.B. (2003b) Important advances in biomass thermal conversion, In: *Process Intensification and Miniaturization in Biological, Chemical, Environmental and Energy Conversion Processes* (eds. G. Akay and M. Dogru), DocQwise, York, UK.

Reed, T.B. and Das, A. (1990) Biomass downdraft gasifier engine systems handbook, SERI/SP-271-3022, USA.

Rosillo-Calle, F. and Cortez, L. (1998) Towards proalcohol II: a review of the Brazilian bioethanol programme. *Biomass Bioenerg.* **14**, 115–125.

Sheehan, J. (2001) The road to bioethanol: a strategic perspective of the US Department of Energy's National Ethanol Program. In: *Glycosyl Hydrolases for Biomass Conversion* (eds. M. Himmel, J.O. Baker and J.N. Sadler), pp. 2–25, ACS, Washington, DC, USA.

Simell, P., Kurkela, E., Stahlberg, P. and Hepola, J. (1996) Catalytic hot gas cleaning of gasification gas. *Catal. Today* **27**, 55–62.

Suárez, J., Luengo, C., Felfli, F., Bezzon, G. and Beatón, P. (2000) Thermochemical properties of Cuban biomass. *Energ. Source.* **22**, 851–857.

Twigg, M.V. (1989) *Catalyst Handbook*, Wolfe Publishing, London, UK.

Tyson, K.S., Riley, C.J. and Humphreys, K.K. (1993) Fuel cycle evaluation of biomass–ethanol and reformulated gasoline. Report no. NREL/TP-463-4950. National Renewable Energy Laboratory, Golden, CO.

Unnasch, S., Kaahaaina, N., Venkatesh, E., Rury, P. and Counts, R. (2001) Costs and benefits of a biomass-to-ethanol production industry in California. *California Energy Commission Docket P500-01-002.*

Wahlund, B., Yan, J. and Westermark, M. (2004) Increasing biomass utilisation in energy systems: a comparative study of CO_2 reduction and cost for different bioenergy processing options. *Biomass Bioenerg.* **26**, 531–544.

Wang, W., Padban, N., Ye, Z., Olofsson, G., Anderson, A. and Bjerle, I. (2000) Catalytic hot gas cleaning of fuel gas from an air-blown pressurized fluidized-bed gasifier. *Ind. Eng. Chem. Res.* **39**, 4075–4081.

Zhu, B., Bai, X.Y., Chen, G.X., Yi, W.M. and Bursell, M. (2002) Fundamental study on biomass-fuelled fuel cell. *Int. J. Energy Res.* **26**, 57–66.

PART TWO: Biomass fermentation

Section IIA: System design

5

Design methodology for sustainable organic waste treatment systems

H.V.M. Hamelers, A.H.M. Veeken and W.H. Rulkens

5.1 INTRODUCTION

Waste materials left to their own may cause serious environmental and health problems. To ensure proper treatment of wastes the existence of a so-called waste management infrastructure (WMI) is needed. A WMI may be defined as a set of coherent activities, institutions and technologies that together ensure an appropriate waste treatment (Dijkema *et al.* 2000). Privatisation, internationalisation and the societal demand for sustainability are strong societal trends that are drivers for changes in both organisational and technological WMI (Anonymous 2003).

These trends are an important incentive to develop innovative technology and (re)emphasising three important aspects for the design and engineering of waste management technology: cost effectiveness, sustainability and flexibility. These societal trends are broad and affect many sectors of the society, not only the waste sector. In the manufacturing sector these trends (amongst others) have lead to increased awareness of the importance of the design process (Tate and Nordlund

1995). Design, especially in mass produced products makes up only a small fraction of the production cost. However, from the perspective of total cost, design is important as it lays down the cost of the product during its whole life cycle. These total costs include the cost of raw materials, the cost of product parts, the manufacturing cost, cost made during consumption like water and energy, the external cost in terms of disposal, environmental damage, safety and health. It has been estimated that decisions made during design make up 70% of the cost of a product while it only makes up 6% of the product cost as such (Schlink *et al.* 2001). An omission or error made during design causes in general additional cost that surpass the design cost tremendously, a clear example are costly call-back operation of consumer goods, like cars.

Design is thus becoming more and more an important research item in engineering. Three different lines of research can be distinguished: design theory, design methodology and empirical design research. Design theory aims at developing concepts that help in understanding the nature of design. Design methodology aims at developing methods to improve the design process. These improvements include better products, reduced production cost and a more efficient design process itself. By using psychological research techniques, empirical design research tries to elucidate the actual design process. Of course all these lines are not developed autonomously but develop in interaction with each other.

Empirical design research indicates that design failure is clearly correlated to a neglect of key concepts in methodological design (Pahl *et al.* 1999). Although this research has mainly focused on mechanical engineering design, there seems to be no a priori reason why the advantages of rational design would not hold for engineering design in the field of waste management. This paper therefore explores the usefulness of the methodological design approach in the field of organic solid waste treatment. As the focus of this paper lies on the methodological design, elements of theory and empirical research are introduced only when needed.

5.2 WASTE HIERARCHY BASED DESIGN

5.2.1 Waste hierarchy paradigm

The uncontrolled spread of waste materials leads to health problems and environmental damage. Within policy the primary objectives of waste management is therefore to reduce the environmental burden of waste management. The principal guideline in the development of waste management schemes or systems is the so-called waste hierarchy paradigm (WHP) (Anonymous 1991, 2003; Barton *et al.* 1996; Sakai 1996).

This paradigm is a preference listing of the waste treatment options, with the top waste prevention and the bottom landfilling. Intermediate steps are reuse, recycling, energy generation and incineration. The waste hierarchy is an almost worldwide applied waste management paradigm (Sakai 1996). Waste strategies based on the waste hierarchy have in general the aim to reduce the reliance on landfilling by increasing the fraction of waste treated by options higher upon the ladder (Barton *et al.* 1996).

The waste hierarchy is typically used on the basis of the waste type. Thus, one considers all different types of waste one by one. For each of them it is separately determined in which way it must be treated, by applying the waste hierarchy. The hierarchy is used in a pragmatic way, this means that within the hierarchy the highest level with acceptable cost will be sought. The waste hierarchy guides the main development effort in waste technology design (Dijkema *et al.* 2000). The major influence of the WHP in design lies in directing the technology choice and development for specific types of waste by giving guidelines for evaluation. A clear example is the increased attention for anaerobic digestion, the waste hierarchy favours such a technology and as a result many different type of installations have been developed (Cecchi and Traverso 1988; Mata-Alvarez *et al.* 2000).

5.2.2   Current drivers in waste management

The broad societal trends of privatisation, sustainability and internationalisation are exerting their influence also in the field of waste management. The following effects of these trends on the design of waste management technology can be distinguished (Anonymous 2003).

5.2.2.1   *Privatisation*

Private enterprises play more and more an important role within a WMI, mainly in the fields of financing, operating and managing of the WMI. This is a break with the recent past in which the (often local) government was an almost exclusive player in policy making, planning, operation and control of the whole WMI. Involvement of private enterprises will lead to an increased cost awareness, the trade-off between cost and functionality of the technology will be more emphasised. This stronger limitation can either act as a brake on technology development but also as an incentive for the development of more cost-effective technology.

5.2.2.2   *Sustainability*

The concept of sustainability has become an important concept in shaping the thinking about production, consumption and the environment. Although the concepts misses a rigorous single definition, it does poses a challenge to the WMI, the waste is not something that should be safely stored but is a type of raw material that should be reused or, even better, should be prevented.

5.2.2.3   *Internationalisation*

While in the past national boundaries were almost impervious for waste, the international dimension is becoming important. Companies are operating across national boundaries and as result also waste will cross boundaries more and more. This international waste transport is a result of the search for the most cost-effective treatment. It also must be kept in mind that large part of the world, and especially the rapidly growing megacities, still lack access to appropriate waste management services. These are areas where current practices are not by definition the model to follow, again low-cost and flexible systems are needed. Flexibility is needed to cope with the rapidly changing social environment of the city.

The WHP has been developed in a response to prevent the environmental problems of waste and waste treatment. As such, it remains to be investigated whether the WHP remains a satisfying tool to cope also with the challenges of sustainability, flexibility and cost effectiveness.

5.3   KEY CONCEPTS IN METHODOLOGICAL DESIGN

5.3.1   Methodological design

Methodological design can be framed in the general idea of rational problem solving (Dorst and Dijkhuis 1995). Problem solving is defined as a process with at the start a problem (What to do with the organic waste?) and at the end a successful solution to the problem, being a plan for a technical installation like an incinerator or a composting facility. Rational expresses the assumption that by following some rules, this design process becomes more efficient, that is, better solutions are found and/or these are faster found. These algorithmic methods are referred to in this text as methodological design procedures. Some of the issues raised by methodological design are quite self-evident. However, the deeper implication of these self-evident issues are often overlooked. There is a large number of different design procedures described in literature (Tate and Nordlund 1995).

The procedure presented here is based on the work of Pahl and Beitz (Cross 2000). This method can be considered as a classic one and as such is widely applied, discussed and researched. To introduce and clarify this specific procedure an example is used. For the sake of brevity, this example is simplified both with respect to the method and the problem. The principal issues are however retained in the example and as such may serve as a tool for discussion later on. Figure 5.1 depicts a general overview of the design procedure. The actual procedure as described by Pahl and Beitz is more detailed; here only the main steps are presented introducing the principal aspects of methodological design in a coherent way.

Overall is the aim of the procedure to transform the problem into a successful technical solution. Four basic phases can be distinguished in this transformation: (1) task clarification, (2) conceptual design, (3) embodiment design and (4) detail design. Those phases that empirical research has shown to be more important are extensively treated. To make this discussion a less abstract the following example will be discussed:

As a result of national regulations a regional authority is obliged to separately collect the organic fraction of the household waste. The national regulations further stipulate that treatment of the waste should comply with the waste hierarchy. Based on a visit to a conference, the regional authority decides to build an anaerobic digestion plant for the organic waste treatment. The regional authority asks a consultant to come up with a design for the plant with methane utilisation as a biofuel.

5.3.2   Task clarification

The procedure starts with phase 1, task clarification analysing the problem as it is posed. A common situation is that the customer (in the example the regional authority) asks for the design of a specific solution. The customer is, however, not

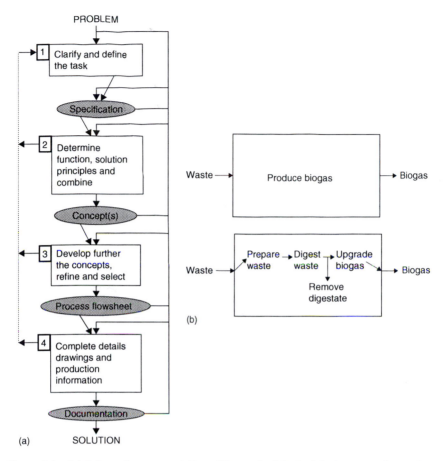

Figure 5.1   (a) Schematic representation of the methodological design procedure and (b) construction of the principal function and first sub-functions.

so much looking for this specific solution but is looking for a good solution to his problem.

In this example for instance, the designer should first get a good idea of the problem behind the question for a design of an anaerobic plant. Maybe the customer has a need of energy from the waste and the task should be how to get more energy from waste. It is important in this phase to distinguish between a problem and the solution. Going directly to a solution, without considering the problem first, it might and probably will leave better solutions out of the scope.

After a thorough problem formulation and analysis, objectives should be determined that describe what the design should achieve or bring about. If possible the objectives should be linked to performance specifications. For instance, the objective of the anaerobic digestion plant to produce biogas from organic waste, which objective could be specified as "producing at least $100 \, m^3$ of biogas per tonne of waste". Apart from the objectives also the relevant constraint must be identified at the start, like acceptable cost, product requirements, emission standards and safety standards. All these elements are recorded in the specification of the design process. Care should be taken again that this happens in a so-called solution neutral way, to keep all solutions' options open. A solution neutral description prevents exclusion of solutions without consideration. This is of course not always possible, for instance legislation may call for a specific solution, like a specific time–temperature duration for pathogen removal.

Empirical research has shown that neglecting a good task clarification is a good indicator of a bad design or cumbersome design process. Improper clarification forces the designer to go more often back to this initial step during the design or even to neglect important issues (Pahl *et al.* 1999).

5.3.3   Conceptual design

The aim of phase 2 is to identify a limited number of solution concepts. The first step in this phase is to identify the essential functions of the design object using the function analysis method (Cross 2000; Hubka and Eder 2001). This method aims at making a description of the functions that the design object should have, based on the input and output of the system. At the start, the design object is represented as a single black box with the appropriate inputs and outputs. Different types of inputs/outputs can be considered as mass, energy and information. Inputs can be determined from the problem formulation (the current situation) while the outputs can be determined from the objectives (the desired situation).

Functions can be found by describing a solution in a neutral way as the transformation from the input into the output. Functions thus relate the input to the output of the design object and typically consist of a noun and verb. The function thus focuses on what must be achieved and not on how this should be done. It should be emphasised that this process of determining functions is not unique. Different designers can come up with different descriptions.

In case of the example, the main input is the organic waste and the main output is the biogas. The choice of input and output depends of course on the specification made up in the previous phase. If for instance the problem has been rephrased as energy from waste, the main output would have been energy.

Starting from the highest level with one black box, one starts dividing the functions into sub-functions. This is necessary as most non-trivial design tasks have complex solutions. This division can be visualised by connecting blocks representing the sub-functions. The connections reflect the relationships between the sub-functions, in our case flows of energy, mass and information. By relating, these sub-functions the black box has been opened and made transparent. This process can be repeated until a certain level of refinement is reached.

In the example, the black box has a waste input and a biogas output. At the highest level one might state that, for example, the function of the design object is to "produce biogas", and that the associated input is the waste and the output is the biogas (Figure 5.1b). This function could be split up into three main sub-functions:

(1) "Prepare waste", preparation of the waste to make it useful for digestion.
(2) "Digest waste", digestion of the waste, the actual biogas formation.
(3) "Upgrade biogas", upgrading of the biogas to comply to the product standards.

There is a fourth function, "Remove Digestate", this function is necessary but does not directly support the main function and is thus called an auxiliary function. Of course whether a function is auxiliary or not depends on the main function and remains to a certain extent arbitrary.

Each of these sub-functions can be split up further. For instance, consider the sub-function "prepare waste", it might decided that the preparation consists of increasing the temperature, decreasing the particle size and increasing the moisture content of the waste (Figure 5.2a). This would mean that the sub-function "prepare waste" is again divided into three new sub-functions: "add water", "add heat" and "decrease particle size". These three new sub-functions can be structured in different ways and still retain the same overall functionality "prepare waste". Some examples of these different structures are given in Figure 5.2a. In a similar way one can proceed with the other sub-functions and refine them further.

This example makes clear that the number of potential structures that can be formed by combining the sub-functions is huge. If for each of the three main sub-functions ("prepare waste", "digest waste", "upgrade biogas") five alternatives

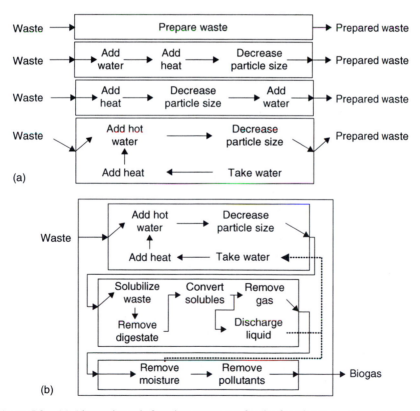

Figure 5.2   (a) Alternative sub-function structures for the function "prepare waste" and (b) selected overall structure.

can be generated, already in this phase 125 variants need to be considered. It is of course impractical to work out all of these variants in all detail and then select the best solution. Already, earlier in the design process decisions must be made.

Empirical research has shown that at this place experience is of great help. An experienced designer can rule out alternatives at the start more easily. Such a designer sees easier that first adding heat to the water and then adding the hot water to the waste is a better solution as heat transfer to a liquid is easier than to a slurry or semi-solid material.

After generating the alternatives some are selected as most promising. This evaluation is based on general insight in the processes and cannot be made on the strict specifications, as actual solutions are needed for specification evaluation, for instance for cost and environmental evaluation. It is good practice to store the discarded alternatives as they might be useful in a later phase.

As a result of the function analysis method, we end up with a number of function structures. Figure 5.2b shows an example of such a structure for our example of biogas production. For the sake of the example this figure is a still relatively crude representation. Further refinement would be necessary, notably internal transport and energy conversion should be further incorporated.

An important element of this function structure is that the water recirculation flows establish further linkages between the main sub-functions. These recirculation flows change the nature of the solution procedure. Without the recirculation the sub-functions can be further developed sequentially. However, with recirculation they now become linked and must be solved in parallel.

The next step is to transform function structure into a structure with working principles. Working principles are physical principles that can be used to achieve a certain function. A working principle is not yet an apparatus but a physical principle that is determining an apparatus family. For instance sieving is a principle that can be done with many different types of sieves.

The working principles can be selected using a morphological chart. Figure 5.3 shows as an example a chart for the main sub-function "Digest Waste". For each function different working principles are listed in a row left from the function. Most elements in the chart are well known, however also novel ideas (indicated by ??) could be brought in at this point. From this chart, it becomes clear that there are again many combinations of working principles possible, achieving the same functionality. By connecting working principles from each row, a set of working principles is selected that achieve the given sub-functionality "Digest Waste". Two possibilities are indicated by the two lines in Figure 5.3. Again a selection needs to be made, for further detailing. The choice for a certain combination of working principles might also lead to changes in the function structure itself. For instance, if a choice for chemical hydrolysis is made, attention needs to be paid to chemical choice, dosing and (re)use. The conceptual phase ends here when a (small) number of function structures together with working principles have been identified that are able to perform the objective of the design. The problem has been transformed into a structure of working principles. The structure itself is still solution neutral.

Function				
Waste solubilisation	Chemical acid hydrolysis	Chemical lye hydrolysis	Enzymatic hydrolysis	Microbial hydrolysis
Digestate removal	Sieving	Settling	Centrifuging	Flotation
Solubles conversion	Pure microbial conversion	Mixed microbial conversion	Catalysed chemical conversion??	Enzymatic conversion
Gas removal	Gravitational separation	Sub atmospheric separation	Membrane separation	

Figure 5.3   Morphological chart with two alternatives, indicated by the two lines.

5.3.4   Embodiment and detail design

In the next phase 3 of embodiment, a specific apparatus is chosen, embodying the function and working principle. The choice in this phase depends on the experience of the designer and the information at hand. Full accounting of the energy usage is, however, only possible if one fixes the type of composting, for example extensive or intensive composting. This phase is especially important, as evaluation for sustainability objectives is only possible at this level. The last phase 4 is the detailing phase and is concerned with preparing the documentation that will allow construction. This phase is important, as the final cost evaluation is possible at this level. At this point the design process has finished.

5.3.5   Design cycle

The design methodology as presented here characterises the procedure as a number of phases going from general to specific in an orderly way. This is however only one aspect of the complicated activity design. Within each phase a number of cyclic activities (design cycle) are performed, cycles repeated throughout the whole design process. This design cycle is another characteristic of the design process that should be well understood to achieve a suitable solution. This design cycle consists of the following sequential activities: analysis, synthesis, simulation and evaluation (Dorst and Dijkhuis 1995; Cross 2000). During analysis, the requirements for the solution are derived from the problem at hand. Synthesis means generation of a number of tentative solutions. Simulation is assessment of the expected performance of the tentative solution. Through evaluation of the expected performance, the best solution is then found. If this best solution meets the requirements, the design exercise is finished, otherwise a new effort has to take place.

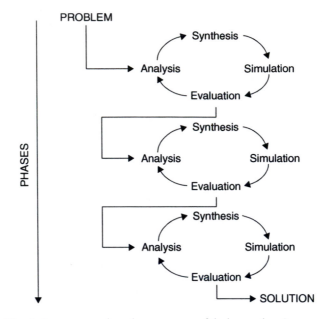

Figure 5.4   The design process viewed as sequence of design cycles. As an example three cycles are depicted, however more might be necessary.

This cycle can be recognised during several phases of the design methodology as outlined, for generating the specification, generating function structure, generating working principles, etc.

During each activity or phase, a decision has to be made to stop and proceed with the next activity, meaning that decision making is an important element of the design cycle. At these decision points, it also may be decided to go back to a previous activity, as the information from the previous activity was not complete. Iteration is thus another characteristic activity.

Figure 5.4 gives a pictorial representation of the design process viewed as a sequence of design cycles. The importance of this representation is that it emphasises the activities. It stresses more the strategy of designing while the stricter design methodology is more concerned about the content. Both aspects are important and should be taken care of in an actual design situation.

5.3.6   Design classification

It is useful to classify the type of design problem, at the start of each design process. The type of design problem influences strongly the way in which a certain methodology is used (Cross 2000).

Design can be firstly divided into three classes of design: novel, variant and evolutionary. Novel could be called problems that have not been solved, an important example could be a sustainable waste treatment system. Novel designs ask the most of intuition and creativity to generate ideas that are further developed in a rational way. Variant designs, relate to problems that are well understood and have well-defined solutions, for example a new aerobic wastewater treatment plant. Variant design relies heavily on the analysis of existing technical systems and literature. Evolutionary design relates to problems for which solutions are known but one is looking for a solution that contains novel solutions for some subproblem. For instance, using a fuel cell for electricity generation from biogas instead of a gas motor. Evolutionary design uses tools from variant design and novel design.

A second classification of the design is based on the scale of the problem. It makes a big difference whether one is asked to design a system like a WMI, a facility like an anaerobic plant or part of a composting plant like a biofilter.

In general, a design problem concerning a system or a more novel approach asks more efforts in phase 2, while a variant design problem asks for a more attention being paid to the embodiment design.

5.4 COMPARISON OF METHODOLOGICAL DESIGN AND WHP

This section explores the usefulness of the design methodology in the field of organic waste treatment by comparing it to the WHP. It is investigated in which way methodological design can help in coping with the challenges of sustainability, cost effectiveness and flexibility. This is done first, by a more general conceptual discussion and second, by discussing the topic of anaerobic technology for organic solid waste treatment.

5.4.1 Conceptual comparison

The design process is strongly determined by the scale and type of design. The WHP is a quite general listing of priorities that might be useful in conceptual design, but loses its function in phase 3. This makes the WHP less suitable for more specific design tasks as evolutionary design or design of unit operations like anaerobic digestion.

An important element of the task clarification is construction of a good problem description. Only in this way a solution free representation can be obtained. It is clear that the WHP gives no room for a renewed problem statement, in a sense it is already a listing of solutions. A solution free description is a crucial characteristic for any methodological design process. This step is introduced to prevent the design process narrowing at a too early stage. Jumping directly to a specific structural solution makes that other, maybe better, solutions are not considered. In this way, the WHP is expected to hinder innovative technology development.

From the problem definition objectives should be derived. As the example showed, the fact that the WHP only handles environmental aspects, tend to blur other objectives. This poses especially a problem if the WHP is applied outside the context in which it was developed. For instance in rapidly growing megacities in developing countries, objectives may be expected to be different, for instance job creation and economic development will be also of importance to waste management.

In general specification should be as concrete as possible, however the WHP hinders the development of a problem definition and objectives, a clear specification of the design task is not possible.

In phase 2, structure and workings principles are selected. At first sight the WHP seems to be very suitable in this phase, however the WHP at this level shows a peculiarity, on the one hand, the WHP contains functions (prevention, reuse) which are solution neutral, on the other hand it contains working principles. It is clear that such a inconsistent list cannot aid to making decisions. At least the WHP should be made consistent, preferably at the level of functions. As an example, this inconsistency leads to complex rules in the Dutch Waste Policy concerning the question whether incineration is considered as reuse or disposal (Anonymous 2003).

Evaluation is a crucial element in any design process. The designer has to find a good solution out of a vast wood of potential solutions. For environmental and sustainability issues often a life-cycle assessment (LCA) study will be performed (White et al. 1995; Barton et al. 1996). To make a sensible LCA, the embodiment and detail design has to be performed. A problem of course is that this type of evaluation is only possible at the moment the process is finished and already decisions have been made (Guinee et al. 2001). Of course, one could refrain from making decisions and take more options into account. This is however not practical, as the number of structures, working principles and devices that have to be taken into account are overwhelmingly big. At this point the WHP is very strong, as it is simple and straightforward. This might explain why it is so widely used. It is, however, an ambiguous tool, as it gives a number of alternatives with different ranking but does not give a tool for selection of a specific technology for a specific case.

5.4.2   Anaerobic waste treatment technology design

The development of anaerobic technology for organic waste is a clear example of an innovation in treatment technology resulting from the application of the WHP. The WHP clearly prefers the anaerobic technology over the aerobic technology for biowaste, which is corroborated by an LCA study (Kubler and Rumphorst 1999). However, in practice introduction of anaerobic technology lags behind with aerobic technology, a phenomena mainly attributed to the somewhat higher cost and limited practical experience (De Baere 1999). The reason for these higher cost lie partly in the occurrence of wastewater that should be treated and the additional cost for transforming the digestate into a marketable compost. This can be understood if one has a look at the functionality of the processes, as listed in Table 5.1.

Table 5.1 shows that the aerobic process has more functionalities especially in the field of composting and wastewater. This gives the aerobic process more flexibility, a property highly appreciated by facility management operating in a changing environment. The additional cost of the anaerobic system are also caused by the fact that additional measures have to be taken to produce an acceptable compost, cost which are not off-set by the additional value of the biogas.

Table 5.1 thus exemplifies the major flaw in the WHP, because of its limited scope. The WHP cannot handle objectives extraneously from environmental

Table 5.1 Comparison of the functionalities of the anaerobic and aerobic processes.

Functionality	Anaerobic	Aerobic
Produce energy	+	−
Clean water	−	+
Clean air	+	−
Stabilised compost	−	+
Dry compost	−	+
Odourless compost	−	+
Hygienic compost	−	+

+ : a Strong point; − : a relatively weak point.

concern, like product properties, flexibility and cost. Conversely, Table 5.1 can help in identifying strategies or circumstances that can be successful. For instance the success of the centralised anaerobic manure in Denmark might be explained that in such a case the functionalities that give composting an advantage are not needed. The farmer accepts a wet, not odour-free product (Hartmann *et al.* 1999).

5.5 CONCLUSIONS

This chapter investigated the usefulness of methodological design as compared to WHP in gearing technology development towards the objectives of cost effectiveness, sustainability and flexibility. This evaluation identified that the WHP has a number of inevitable shortcomings. A methodological design was proposed and adapted to the specific nature of waste management. To safeguard the strong points of the WHP, simple evaluation tools should be developed that can be used in different phases of the design process and retain the commitment to sustainability. Tools derived from the life cycle approach might give a good direction to look into.

REFERENCES

Anonymous (1991) Council Directive 91/156/EEC.
Anonymous (2003) Landelijk afvalbeheersplan 2002–2012, Vol. 1. Ministry of Spatial Planning and the Environment (VROM), The Netherlands.
Barton, J.R., Dalley, D. and Patel, V.S. (1996) *Waste Manage.* **16**, 35–50.
Cecchi, F. and Traverso, P.G. (1988) *Biomass* **16**, 257–284.
Cross, N. (2000) *Engineering Design Methods*, 3rd edn. John Wiley, New York.
De Baere, L. (1999) Anaerobic digestion of solid waste: state-of-the-art. *Proceedings of the II International Symposium on Anaerobic Digestion of Solid Waste* (eds. J. Mata-Alvarez, A. Tilche and F. Cecchi), Barcelona, Spain.
Dijkema, G.P.J., Reuter, M.A. and Verhoef, E.V. (2000) *Waste Manage.* **20**, 633–638.
Dorst, K. and Dijkhuis, J. (1995) *Design Stud.* **16**, 261–274.
Guinee, J.B. *et al.* (2001) Levenscyclusanalyse VROM.
Hartmann, H., Angelidaki, I. and Ahring, B.K. (1999) Increase of anaerobic degradation of particulate organic matter in full-scale biogas plants by mechanical maceration. *Proceedings of the II International Symposium on Anaerobic Digestion of Solid Waste* (eds. J. Mata-Alvarez, A. Tilche and F. Cecchi), Barcelona, Spain.

Hubka, V. and Eder, W.E. (2001) Functions revisited. *International Conference on Engineering Design ICED01*, Glasgow, UK.

Kubler, H. and Rumphorst, M. (1999) Evaluation of processes for treatment of biowaste under the aspects of energy balance and CO_2-emission. *Proceedings of the II International Symposium on Anaerobic Digestion of Solid Waste* (eds. J. Mata-Alvarez, A. Tilche and F. Cecchi), Barcelona, Spain.

Mata-Alvarez, J., Mace, S. and Llabres, P. (2000) *Biores. Technol.* **74**, 3–16.

Pahl, G., Badke-Schaub, P. and Frankenberger, E. (1999) *Design Stud.* **20**, 481–494.

Sakai, S. (1996) *Waste Manage.* **16**, 341–350.

Schlink, H., Schneider, H. and Hohne, G. (2001) The determination of function costs in engineering products. *International Conference on Engineering Design ICED01*, Glasgow, UK.

Tate, D. and Nordlund, M. (1995) Synergies between American and European approaches to design. *First World Conference on Integrated Design and Process Technology*, Austion, TX, USA.

White, P.R., Franke, M. and Hindle, P. (1995) *Integrated Solid Waste Management: A Lifecycle Inventory*. Blackie Academic & Professional, London.

6

Modeling of biomass fermentation: control of product formation

G. Lyberatos, H.N. Gavala and I.V. Skiadas

6.1 INTRODUCTION

Biomass is the term used to describe all biologically produced matter. Biomass consists on the average of 25% lignin and 75% sugars, after removing a total of approximately 5% of other compounds (mainly proteins and lipids). The world production of biomass is estimated at 146 billion metric tons a year, mostly wild plant growth. Some farm crops and trees can produce up to 49.5 tons of biomass per hectare annually. Some types of algae and grasses may produce 50 tons per year (U.S. ENERGY ATLAS 1980). The energy value of the annual production of bio-mass is estimated at 2.74×10^{19} Btus, which is 8 times as much as the annual worldwide energy consumption (3.4×10^{18} Btus). Today only 7% of the generated biomass is actually exploited (NREL 1998).

Biomass is a renewable energy source, which does not contribute at all to the greenhouse effect. Essentially solar energy is stored as energy of chemical bonds in the organic matter that forms biomass. Energy may be produced from biomass in

© 2005 IWA Publishing. *Biofuels for Fuel Cells: Renewable energy from biomass fermentation* edited by P. Lens, P. Westermann, M. Haberbauer and A. Moreno. ISBN: 1843390922. Published by IWA Publishing, London, UK.

two ways: chemical decomposition through thermal processes and through biologic conversion.

Chemical conversion includes incineration in excess of oxygen, pyrolysis (i.e. thermal conversion into gaseous, liquid and solid fuel products in the absence of oxygen) and gasification (similar to pyrolysis but with the use of sufficient oxygen to sustain exothermic reactions). Dried biomass has a heating value of 5000–8000 Btu/lb, with virtually no ash or sulfur oxides produced during combustion. Thermal processes such as incineration have the general disadvantage of causing atmospheric pollution, unless costly purification of the effluent gases is applied.

Biologic processes for the production of biofuels and chemicals are essentially anaerobic processes (fermentation and anaerobic digestion). The main processes that could be used to this end are the biologic production of ethanol (Chapter 10), hydrogen (Chapter 12) and methane (Chapter 7).

The objective of this chapter is to consider systematically the operating parameters that can be manipulated to steer a bioconversion process to the desirable product(s). Starting from a particular biomass source, one may envision a single bioconversion process, but also a sequence of multiple fermentations that lead to several biofuels and/or precursor chemicals.

A particular objective of this chapter is to consider the role that mathematical modeling of the fermentation processes and the subsequent use of optimization methods can play in the effort to maximize the process objective.

6.2   FERMENTATION BY PURE CULTURES

Microbial growth may be described as an overall reaction that converts various nutrients that include source(s) for carbon, nitrogen, oxygen, etc. into microbial biomass and products. Such overall microbial reactions are not simple reactions. This means that they do not represent molecular events, and as a result they cannot be described by simple mass action kinetics. They are overall reactions that are the results of summing up all biochemical reactions taking place inside the microbial cell.

6.2.1   Stoichiometry of microbial conversion

An immediate consequence of this fact is that the reaction stoichiometry; that is, the relative amounts of reactants and products is not fixed. The stoichiometry will vary depending on the particular conditions that prevail. To appreciate this, consider a simple system of two parallel first-order reactions:

$$A \rightarrow B$$
$$A \rightarrow C \tag{6.1}$$

taking place in a well-stirred reaction vessel. The overall reaction is:

$$A \rightarrow v_1 B + v_2 C \tag{6.2}$$

Assuming that initially there is only reactant A at a concentration C_A^0, and that the rates of the two reactions are $k_1 C_A$ and $k_2 C_A$, respectively, one can show that $v_1 = \frac{k_1 C_A^0}{k_1 + k_2}$ and $v_2 = \frac{k_2 C_A^0}{k_1 + k_2}$, respectively; that is, they depend on the relative reaction rate constants. Of course $v_1 + v_2 = 1$, but the exact distribution will depend on the rate constants, which in turn depend on the operating conditions (e.g. pH, temperature and pressure).

Similarly, microbial products will of course have to satisfy some overall stoichiometric restrictions, but the exact product distribution will depend on the relative rates of the microbial processes that take place. Important in this content are two general concepts, the yield and the selectivity. The yield has to do with the amount of the particular product of interest (say B in our example) obtained for a given holding time per mole of reactant being converted.

In our example case, the concentration of product B as a function of time t is:

$C_B(t) = \frac{k_1 C_A^0}{k_1 + k_2} [1 - e^{-(k_1 + k_2)t}]$ and the yield will therefore depend on the reaction

kinetics: high value of k_1 will give a high yield for a given holding time.

The other important concept, the selectivity, has to do with the relative amounts of B (desired product) and C (undesired products) generated. This in our case will be k_1/k_2 and consequently depends only on the relative rates of the wanted and unwanted reactions.

6.2.2 Microbial growth

Overall aerobic microbial growth (for the case of a single metabolic product in addition to microbial mass, growing on a single carbon and a single nitrogen source) takes the general form:

$$a CH_x O_y + b O_2 + c H_l O_m N_n \rightarrow CH_\alpha O_\beta N_\delta + d H_2 O + e CO_2 + f CH_v O_w \quad (6.3)$$

where $CH_x O_y$ is the carbon source, $H_l O_m N_n$ is the nitrogen source, $CH_\alpha O_\beta N_\delta$ is the cellular mass, and $CH_v O_w$ is the metabolic product. This description ignores all other elements that are pertinent to microbial growth besides C, H, O and N. These elements are of course conserved quantities and for a closed system, or an open system at steady state, an elemental balance shows that:

$$a = 1 + e + f$$

$$ay + 2b + mc = \beta + d + 2e + wf$$

$$nc = \delta \quad (6.4)$$

$$xa + lc = \alpha + 2d + vf$$

These four equations impose a restriction for the six unknown stoichiometric coefficients (a, b, c, d, e and f) if the cellular composition is considered as given and constant (which is not exactly the case). There are therefore two degrees of freedom; that is, the stoichiometric parameters may not be uniquely determined.

The exact stoichiometry will then depend on the rates of individual reactions, which will be a function of the growth conditions. The complication (and as a result the uncertainty) will increase further, if we allow for multiple carbon sources, and for multiple metabolic products. Furthermore, the chemical formula for the biomass has been assumed constant (coefficients α, β and δ in Equation 6.4), although in reality they will depend to some extent on the growth conditions.

6.2.3 Microbial yield

It is obvious from the above discussion that the biomass yield $Y_{X/S}$, defined by Equation 6.5, is a quantity that will strongly depend on the growth conditions. Similarly, the yield and the selectivity of specific metabolic products will strongly depend on the growth conditions (i.e. the nutrient concentrations, temperature, pH, redox potential, presence or absence of oxygen, etc.).

$$Y_{X/S} = \frac{\Delta X}{\Delta S} = \frac{r_x}{r_s} \tag{6.5}$$

where ΔX is the increase in biomass concentration and ΔS the corresponding decrease in substrate concentration.

Equation 6.3 is an overall microbial reaction, which is set up based on knowledge of the main nutrients and products for growth of a single microbial species. The Equation 6.4 was obtained based on elemental balances. Such an approach is of course the roughest possible approach in the effort of describing microbial growth as an overall reaction. A more sophisticated approach that allows direct accounting of the influence of some process parameters on the microbial reaction stoichiometry is based on energetics. This approach is very well described by Rittmann and McCarty (2001).

In this approach, microbial growth and metabolite production is considered as an overall redox reaction. Thus, in describing heterotrophic growth, an electron donor serves as the carbon and energy source, and provides the electrons that are used for growth and energy. A half reaction is written that shows the release of electrons. For example:

$$\frac{1}{8}CH_3COO^- + \frac{3}{8}H_2O \rightarrow H^+ + e^- + \frac{1}{8}CO_2 + \frac{1}{8}HCO_3^- \tag{6.6}$$

is the reaction that provides eight electrons per mole of acetate being utilized. This reaction gives $27.40\,kJ/e^-$eq at standard conditions ($\Delta G = -27.40\,kJ/e^-$eq). The electrons generated will then be distributed between two uses: biomass synthesis and energy generation. In particular, a fraction f_s will be used for biomass synthesis and a fraction f_e will be used for energy. Clearly $f_s + f_e = 1$. Depending on the electron acceptor, this reaction is then to be coupled with two more reactions, one describing cellular synthesis, and one describing energy generation. Thus, if oxygen is the acceptor, the half reaction is:

$$H^+ + e^- + \frac{1}{4}O_2 \rightarrow \frac{1}{2}H_2O \tag{6.7}$$

with $\Delta G = -78.72 \, \text{kJ/e}^- \text{eq}$. For cell synthesis, the half reaction is:

$$H^+ + e^- + \frac{1}{5}CO_2 + \frac{1}{20}HCO_3^- + \frac{1}{20}NH_4^+ \rightarrow \frac{1}{20}C_5H_7O_2N + \frac{9}{20}H_2O \quad (6.8)$$

where $C_5H_7O_2N$ is the empirical formula for the chemical composition of bacteria. This reaction requires energy. In order to obtain the overall microbial reaction, Equation (6.6) must be added with Equation (6.7) multiplied by f_e and with Equation (6.8) multiplied by f_s. This summation will lead to the elimination of the electrons and the protons giving the overall reaction:

$$\frac{1}{8}CH_3COO^- + \frac{f_e}{4}O_2 + \left[\frac{f_s}{20} - \frac{1}{8}\right]HCO_3^- + \frac{f_s}{20}NH_4^+$$

$$\rightarrow \frac{f_s}{20}C_5H_7O_2N + \frac{5f_e + 3f_s}{40}H_2O + \left[\frac{1}{8} - \frac{f_s}{5}\right]CO_2 \quad (6.9)$$

It is clear that the exact stoichiometry will be a function of the breakdown between f_s and f_e.

Rittmann and McCarty (2001) show how to calculate f_s (and hence f_e). This is done by considering the energy required to convert the carbon source (acetate in this case) to pyruvate and the energy required to synthesize biomass from pyruvate. Then, an energy balance is written that requires that the generated energy is used to drive cellular synthesis with certain efficiency. In the event of biomass decay, the fractions f_s and f_e are shown to depend on the microbial decay rate constant (b) and the retention time in the bioreactor. The influence of temperature is reflected through the dependence of the free energy change (ΔG) of the reaction on temperature.

6.2.4 Product formation

The above example considers no microbial product formation. In the event of microbial product generation, which is what happens under fermentation conditions, the generated electrons will be "dumped" to generate some products with CO_2 being the electron acceptor. The outlined procedure may be used to determine the stoichiometry in such cases as well. For mixed product fermentation, there are more degrees of freedom and the distribution between the various products will be a function of the operating conditions. *Escherichia coli* (*E. coli*), for example, generally ferments glucose to a mixture of acetate, ethanol, formate and hydrogen. In this case, there is a different half reaction for each product. Thus, the electrons provided by the breakdown of glucose will be distributed between biomass synthesis and end product formation. The exact breakdown will depend on the growth conditions in a complicated manner.

Here the fact that energetics present a more "educated" approach to obtaining the anticipated overall stoichiometry of microbial growth and product formation than simple elemental balances has to be stressed. However, bearing in mind that this overall process is really the makeup of a very large number of simple enzymatic reactions of cellular metabolism, its accuracy is still limited, as the relative rates of these reactions will depend strongly on the operating conditions.

Summarizing, elemental balances and energetics may be used to obtain anticipated stoichiometric relations for the apparent overall "microbial growth and product formation" reaction. The exact stoichiometry, however, will always be a function of the relative rates and consequently will depend on the microbial growth conditions.

6.2.5   Manipulation of product formation

Modification of the distribution among different microbial products may then be to some extent achieved through appropriate manipulation of the growth environment. In addition, realizing that optimal growth conditions may differ appreciably from optimal product formation, it becomes apparent that it is often possible to obtain superior results in a two-stage process; the first stage maximizing cellular mass formation, with the second stage maximizing the generation of the desired product.

In theory, given a particular organic substrate or substrate mixture and a particular microbial species (pure culture), it should be possible to determine the operating conditions that would maximize the production of a particular desirable fermentation product (expressed in moles per unit volume per unit time) or a fermentation product mixture of the highest possible value. The biochemical engineer can design the process with this aim, provided that an appropriate mathematical model has been developed.

The question then arises: could the natural barrier defined by the previous consideration be somehow overcome? The answer is positive if we allow modification of the microbial species through genetic engineering. In the recent years, the advent of metabolic engineering has precisely this aim: the directed improvement of product formation or cellular properties through modification of specific biochemical reactions or introduction of new ones with the use of recombinant DNA technology (Stephanopoulos 1999). The *in vivo* determinations of metabolic fluxes through material balances, measurement of labeled enrichment in selected secreted metabolites, analysis of the fine structure of nuclear magnetic resonance (NMR) spectra and measurement of isotopomer molecular weight distributions by gas chromatography–mass spectrometry represent invaluable aids in developing appropriate genetically modified organisms (GMOs) with desirable properties, through rational microbial modification and flux control.

As an example consider the work of Jeppsson *et al.* (2003) on developing appropriate recombinant yeasts for the production of ethanol from cellulose. *Saccharomyces cerevisiae* (*S. cerevisiae*) ferments hexose sugars to ethanol, but it cannot utilize the major pentose sugar xylose. It has been estimated that the fermentation of the xylose fraction in addition to the hexose sugars would reduce the ethanol production cost as much as 25% (Hinman *et al.* 1989). Recombinant strains generated by appropriate genetic manipulations (incorporating *Pichia stipitis* genes for xylose reductase expression) by Jeppsson *et al.* (2003) led to recombinant species that indeed utilize both glucose and xylose for ethanol production (see Table 6.1).

A major concern, however, when it comes to using metabolic engineering as a means for enhanced or optimized plant biomass utilization, stems from the fact that plant biomass utilization will generally be a non-sterile process owing to the

Table 6.1   Measured specific consumption rate (negative values), specific productivity (positive values: mmol $g_{biomass}^{-1}$ h^{-1}), biomass (g/l) and carbon recovery (mol-c out/(mol-c in)) from continuous cultivations with 20 g/l glucose + 20 g/l xylose, under anaerobic conditions at dilution rate 0.06/h for *TMB3001, TMB3260(PGK1p-XYL1), TMB3255(zwf1Δ)* and *TMB3261(PGK1p-XYL1, zwf1Δ)*. Results shown are the average values from two steady states with less than 5% deviation.

Strains	TMB3001	(PGK1p-XYL1) TMB3260	TMB3255 (zwf1Δ)	TMB3261 (PGK1p-XYL1, zwf1Δ)
Glucose	−3.39	−3.10	−4.00	−4.29
Xylose	−1.50	−2.29	−0.80	−0.84
Ethanol	5.44	5.78	7.10	7.22
Xylitol	0.64	0.96	0.10	0.11
Acetate	0.04	0.03	0.36	0.39
Glycerol	0.81	0.87	0.75	0.91
CO_2	6.59	7.51	7.56	8.41
Biomass (S.D.)	2.09(0.04)	2.28(0.05)	1.62(0.06)	1.61(0.01)
Carbon recovery	0.83	0.88	0.90	0.90

cost that a sterile process would imply. This then would certainly pose limitations because of the potential release of GMOs to the environment.

6.3   MIXED CULTURE FERMENTATION PATHWAYS

The discussion in this section hopefully allows an appreciation of the factors that may influence the particular distribution of fermentation products that can be observed for the biochemical utilization of plant biomass by a mixed microbial consortium.

6.3.1   Effect of microbial ecology on product formation

From the above discussion it should be clear that organic substrates could be fermented to a mixture of products that depends on the microbial species utilized and the operating conditions. In a mixed culture (non-sterile) environment, the conditions will influence not only the performance of an individual species, but also the exact populations that will prevail. The microbial ecology of the culture will depend on complex microbial interactions. Thus, a mixed consortium will be developed that contains various species growing in an interactive manner with commensalistic, ammensalistic, competitive and more complicated impacts of one species onto another. The varying number of the various species will result in a sum total or apparent "overall" fermentative pathway, which will not correspond to what could be obtainable from individual species. This possibility allows for a much larger number of degrees of freedom, as the activity of each metabolic step will be "attenuated" or "modulated" by the overall interspecies available enzymatic activity. This situation is exemplified in the sequel through the examination of the apparent kinetic behavior of a mixed anaerobic culture acclimated to two different organic substrates, a carbohydrate (lactose) and a protein (gelatine) (Gavala and Lyberatos 2001).

6.3.2 Effect of substrate on product formation

All apparent kinetics and calculated kinetic constants, coming from batch experiments with the lactose (LAC) and gelatin (GAC) pre-adapted inocula, are summarized in Figure 6.1 and Table 6.2. Acclimation has a significant effect on the maximum specific utilization rates of the various compounds and on their apparent consumption kinetics. A change in the activities of an anaerobic sludge was observed, depending on whether they are degrading mainly carbohydrates or proteins. A decreased acetogenic (from propionate) and methanogenic (from acetate)

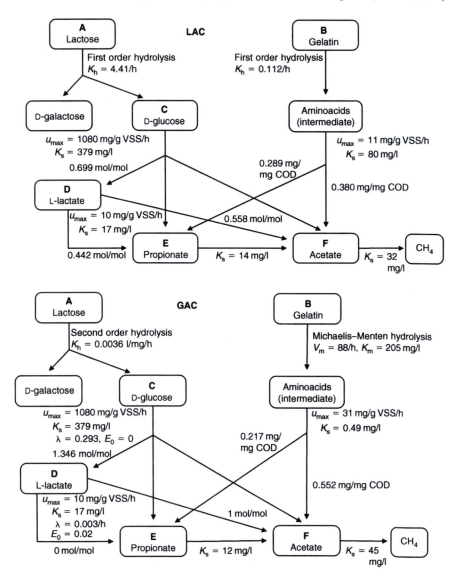

Figure 6.1 Kinetic characteristics of lactose (LAC) and gelatine (GAC) acclimated anaerobic cultures.

activity was noticed in the case of LAC. In every case (with propionate, acetate, L-lactate and gelatin as starting substrate), the maximum specific consumption rates of propionic and acetic acids are higher in GAC than in LAC.

On the other hand, second-order hydrolytic kinetics of lactose are observed for the GAC, whereas gelatin hydrolysis follows Michaelis–Menten kinetics for the GAC and first-order kinetics for the LAC. The adaptation process of the anaerobic cultures from one substrate to another was described by an appropriate mathematical expression. It is noticeable, however, that even if the anaerobic cultures were not exposed to a specific substrate over a long period of time (more than a year), they still kept the ability of hydrolyzing it (in case of lactose and gelatin) or degrading it (in case of D-glucose and L-lactate).

The acclimation has an unquestionable effect on the stoichiometry of the production of the volatile fatty acids (VFAs) from L-lactate and gelatin (aminoacids) metabolism. Specifically, the degradation of L-lactate and gelatin from the LAC lead to higher propionate and to lower acetate production per mole of substrate, as compared with the respective production from the GAC. Also the L-lactate production from D-glucose was different: the latter is much higher in the case of the LAC.

Independently of the acclimation, the maximum specific consumption rates of propionate and acetate showed a strong dependence on the starting substrate and lied between 1.110 and 4.100 mg propionate/g VSS/h and 5.970 and 9.611 mg acetate/g VSS/h for the LAC compared to 2.490 and 7.500 mg propionate/g VSS/h and 13 and 25 mg acetate/g VSS/h for the GAC (Gavala and Lyberatos 2001). Given that the various substrates serve as carbon and energy sources for the microorganisms and that the biomass is virtually constant during the described batch experiments, the change of the maximum specific consumption rates can be attributed to the change of the microbial energy demand; especially for the case of acetic acid, its maximum specific consumption rate may be influenced by the hydrogen partial pressure. Specifically, in the case of gelatin degradation, where an increase of hydrogen partial pressure is anticipated (Ramsay 1997), a decrease of u_{max} from 9.200 to 5.970 and from 25 to 13 for lactose and gelatin acclimated cultures, respectively, was observed.

The difference in the kinetics observed by the two differently acclimated cultures will certainly impact the observed stoichiometry in a bioreactor. This difference reflects the fact that different acclimation implies in principle selection of different microbial species and/or different enzymatic activities for each species.

Table 6.2 u_{max} values for propionic and acetic acid consumption in lactose (LAC) and gelatin (GAC) acclimated cultures depending on the initial substrate added.

Substrate	Propionate kinetics u_{max} (g/g VSS/h)		Substrate	Acetate kinetics u_{max} (g/g VSS/h)	
	LAC	GAC		LAC	GAC
Propionate	4100	7500	Acetate	9200	25
L-lactate	1110	2640	L-lactate	9610	15
Gelatin	1320	2500	Gelatin	5970	13

6.4   THE ROLE OF PRE-TREATMENT PROCESSES
FOR FERMENTATION OF BIOMASS

The first step of a biomass fermentation process is the hydrolysis during which both solubilization of insoluble particulate matter and biologic decomposition of organic polymers to monomers or dimers (i.e. simple sugars, amino acids, long-chain fatty acids and aromatic compounds) take place. These compounds can pass the cell membrane and are thus released in the reactor mixed liquor (Pavlostathis and Giraldo-Gomez 1991).

During hydrolysis, extracellular enzymes (hydrolases) carry out the biologic decomposition of organic polymers to monomers or dimers. The solubilization of insoluble particulate matter is not necessarily a biologic-enzymatic process, but could be a consequence of physico-chemical reactions (Gavala *et al.* 2003a). A number of pre-treatment processes have been developed in order to improve and enhance the disintegration and solubilization of solids. The choice of the pre-treatment method depends mainly on the solids structure, the process used for biomass fermentation and the desired microbial products. Representative examples where different pre-treatment processes are required are given below.

In the case of sludge generated in municipal wastewater treatment plants, pre-treatment processes prior to anaerobic digestion may include heat treatment in the temperature range of 40–275°C, freezing and thawing, chemical treatment using ozone, acids or alkali, mechanical treatment using ultrasound, mills or homoge-nizers, biologic hydrolysis with enzyme addition and combinations of two or more of the aforementioned methods (Müller, 2001; Gavala *et al.* 2003b). These processes aim mainly at enhancing the biodegradability of sludge solids and the efficiency of the overall methane production process.

Enzymatic pre-treatment of wastes/wastewater gained attention during the last decade especially in cases where recalcitrant organic pollutants were targeted in order to reach discharge standards or to facilitate subsequent treatment (i.e. polycyclic aromatic and phenolic compounds, chloro-organics, phthalic acid esters, aromatic dyes, surfactants, cyanide, etc.) (Aitken 1993; Karam and Nicell 1997; Durán and Esposito 2000; Gavala *et al.* 2004). In addition, enzymes have been used in the food processing industry in order to decrease food wastes with simultaneous formation of higher value byproducts and the production of renewable energy sources (i.e. pro-duction of ethanol from agricultural and forestry residues) (Galbe and Zacchi 2002).

In the latter case, wet oxidation (oxygen pressure, alkaline conditions and ele-vated temperature) of the lignocellulosic material has also been widely used in order to enhance the accessibility of the carbohydrates for fermentation to ethanol (McGinnis *et al.* 1983; Ahring *et al.* 1996; Bjerre *et al.* 1996; Schmidt and Thomsen 1998). Wet oxidation dissolves the hemicellulosic fraction and makes the solid cellulose fraction susceptible for enzymatic hydrolysis and fermentation.

One general word of caution is appropriate here. The use of energy-consuming pre-treatment processes such as wet oxidation may contrast the purpose of the production of alternative non-fossil fuels. Therefore, in order to justify using methods such as wet oxidation or ultrasound, it has to be evaluated if the energy required for biomass destruction is compensated by the increased yield in biofuels.

6.5 ALTERNATIVE PATHWAYS FOR BIOMASS FERMENTATION

Based on the discussion of Section 6.4, depolymerization (hydrolysis) is the first step towards the biologic conversion of biomass. This step may be accomplished biologically (either through the addition of enzymes or enzyme-producing microbial cells or through chemical and/or physical methods (non-biologic pre-treatment).

Once monomers have been generated, they may be microbially converted into a mixture of metabolic products such as fatty acids, hydrogen, alcohols, aldehydes, etc. Figure 6.2 presents an overall scheme that allows for the most usual conversions of a representative sugar, glucose, under fermentative (anaerobic) conditions. It should be clear from the previous discussion that what is produced and to what amount will depend on the operating conditions and the microbial communities that are available. For example, methane is usually produced from the combination of all pathways (1–12) shown in Figure 6.2. Depending on the microbial species present, one or more of the pathways could be missing. A representative example is discussed in the study of Gavala and Lyberatos (2001) where it was pointed out that an anaerobic culture acclimated to gelatin lacked the microorganisms responsible for bioconversion 5 (Figure 6.1, GAC case). Ethanol and lactate were produced almost exclusively from pathways 1, 2, 3 and 1, 4, respectively. Hydrogen was produced and consumed through a number of pathways. Therefore, successful selection of the microbial communities in relation to the environmental/operating conditions play a determinative role to a positive hydrogen balance and an effective hydrogen-producing process.

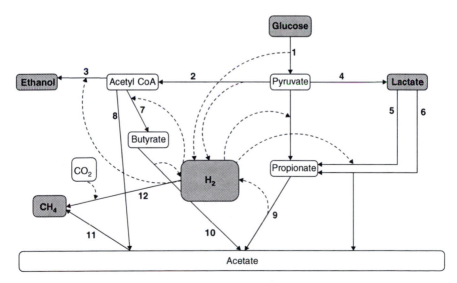

Figure 6.2   Overall scheme showing the most usual conversions of a representative sugar, the glucose, under fermentative (anaerobic) conditions. ▬: main substrate and products, ▭: intermediate products, →: flow of main products and/or substrates, ⇢: flow of side-products and/or co-substrates.

6.6   FERMENTATION OPERATING PARAMETERS

From the previous discussions, it should be appreciated that pre-treated or non-pretreated biomass may be fermented by an appropriate bacterial consortium to yield a variety of fermentation products, depending on the operating conditions. The different factors influencing the distribution of fermentation products are presented below.

6.6.1   Temperature, pH and alkalinity

In general, the temperature limits for biologic activity lie between slightly below 0°C to around 100°C. The maximum temperature at which biologic growth is possible depends mainly on the thermal stability of the components of the living cells, mainly the proteins and nucleic acids, which usually are deactivated rapidly from 50 to 90°C. The microorganisms have an optimal growth temperature at which they exhibit their highest metabolic and reproduction rates. Three temperature ranges are distinguished: the psycrophilic (<20°C), the mesophilic (20°C–45°C) and the thermophilic (>45°C) range. Extreme thermophiles or hyperthermophiles are organisms whose growth optimal temperature is above 65°C (Stanier *et al.* 1964). Archaea are the most often reported hyperthermophiles.

Temperature influences the metabolic activities of the microorganisms and generally higher temperatures result in higher metabolic activities. However, although bioreactions run faster at elevated temperatures, the growth rate of thermophilic microorganisms is considerably lower than that of mesophiles (Atlas and Bartha 1997). This is because a thermophile uses a considerable amount of energy to build up (produce) enzymes stable at high temperatures and replace thermally denatured enzymes and other cell components. For this reason, mesophiles are in general preferable to thermophiles in biotechnological applications such as the production of proteins or antibiotics. On the other hand, thermophiles are advantageous if the aim is degradation and bulk reduction, as is the case in composting, anaerobic digestion and other fermentation processes as bio-ethanol and bio-hydrogen production (Atlas and Bartha 1997).

Another critical factor that influences, not only the growth, but also the metabolic products of the microorganisms is the pH. Most microorganisms have a wide pH range (±3 pH units) at which they are able to grow and, in general, they cannot tolerate extreme pH values that irreversibly denature most proteins. The pH range at which most microorganisms grow and reproduce is 4.4–9.0 with the optimal pH being different from species to species.

The pH has a significant role in the performance and control of fermentation processes. For example, methanogenic archaea, which actually complete the anaerobic digestion process with the production of methane, can grow at a narrow pH range between 6.7 and 7.4 with the optimal being 7.0–7.2. If the pH value drops below 6.3, the anaerobic digestion process may fail. The pH value of a methanogenic digester depends mainly on the VFAs concentration and the alkalinity of the system, which is usually expressed in mg $CaCO_3$/l. VFAs generated during acidogenesis tend to lower the pH of the system. Under normal conditions, the pH is adjusted by the HCO_3^- ions and VFA consumption during acetogenesis and methanogenesis.

Under alkalinity deficient conditions and a sharp increase of VFA concentration or inhibition of methanogenesis, the buffering capacity of the system is exceeded, which has as consequence the drop of pH below 6.0 causing failure of the process. One method to keep the pH within the desirable range is to increase the alkalinity of the system by adding ammonia (NH_3), sodium hydroxide (NaOH) or $NaHCO_3$ (Bitton 1994). It has been proposed that the ratio of total VFA (expressed in mg acetic acid/l) to the total alkalinity (expressed in mg $CaCO_3$/l) should be less than 0.1 for a successful anaerobic digestion process (Sahm 1984).

As it has already been reported, pH affects also the metabolic products besides the microbial growth. A characteristic example is the fermentation of carbohydrates during which a mixture of acids and alcohols can be produced as shown in Figure 6.3. Both the overall conversion rate and the composition of the final product mixture are regulated by the relative availability of the reduced (NADH) and oxidized form (NAD^+) of the carrier molecule nicotinamide adenine dinucleotide, which is a function of the hydrogen partial pressure in the gas phase (P_{H_2}) and the pH (Equation 10) (Mosey 1983). Equation 10 is obtained under the assumptions that: (a) microorganisms maintain a constant internal pH value of 7.0 regardless of the variations in the pH value of their growth medium and (b) gaseous hydrogen diffuses both freely and rapidly into and out of the bacterial cells.

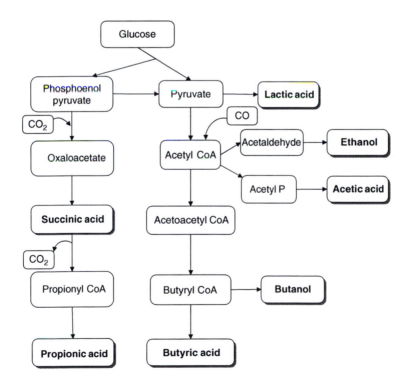

Figure 6.3   Anaerobic fermentation production of common acids and alcohols from carbohydrates.

$$\log\left(\frac{NADH}{NAD^+}\right) = P_{H_2} \times 10^{(pH-3.82)} \qquad (6.10)$$

This pH influence on the composition of the mixture of the microbial products can have a significant impact in the fermentative hydrogen production process as well (Fang and Liu 2002; van Ginkel *et al.* 2001). Controlling the pH can lead the microbial metabolism towards generation of acetic and butyric acids which are more favorable products for biologic hydrogen production than propionate or lactate.

6.6.2   Suspended versus attached growth (support media and reactor types)

6.6.2.1   Continuous stirred tank reactor

The reactor type to be used for fermentation comprises perhaps the most important decision to be made. The particular flow conditions that are established are decisive on the microbial populations that will prevail.

The most common configuration used for fermentation processes is the continuously stirred tank reactor (CSTR). The main advantage of this reactor is that it maintains constant and well-characterized conditions (homogeneity in time and space). The main problem of this reactor type is the fact that the active biomass grows in suspension and is continuously removed from the system thus leading to relatively long retention times and big volumes of reactors. In addition, the environment of a CSTR is a highly competitive environment which leads to washout of slower-growing organisms and hence a reduction of the reactor "biodiversity".

The disadvantages of a CSTR have been overcome through the development of high-rate systems based on the retention and accumulation of the active microbial biomass despite the applied high liquid flow rates.

High rate bioreactors have the advantages of:

(a) short hydraulic retention time (HRT) and high loading rate resulting in the reduction of reactor size, space requirements and capital cost,
(b) high efficiency,
(c) process stability,
(d) low or no requirement for mechanical mixing.

6.6.2.2   Biomass retention systems

The high retention time and concentration of the active microbial biomass in high-rate bioreactors is achieved by allowing for attached growth of the microbial cells, which form biofilms on the surface of solid materials present in the bioreactors, and/or the use of recirculation from specially designed settling devices located at the effluent of the bioreactors. According to Characklis and Marshall (1990), a biofilm consists of cells immobilized at a substratum and is a complex coherent structure of cells and cellular products like extra-cellular polymers. The substratum could be either a static solid surface (static biofilms, e.g. biofilters Figure 6.4) or suspended carriers (particle supported biofilms, e.g. expanded or fluidized beds Figure 6.5a). In the absence of a solid surface and under certain

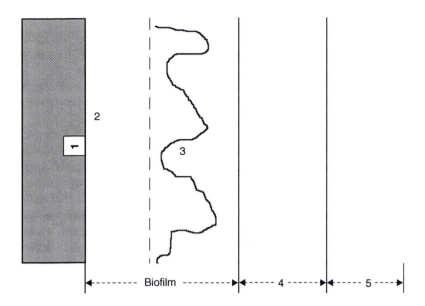

Figure 6.4   Schematic representation of a static biofilm including the following five compartments: 1) substratum, 2) base film, 3) surface film, 4) bulk liquid and 5) gas (Characklis and Marshall 1990). The base film and surface film constitute the biofilm.

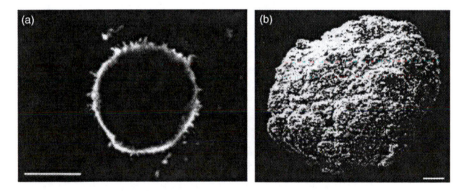

Figure 6.5   (a) Particle supported biofilm (Nicolella *et al.* 2000). Average diameter is 1.7 mm; bar = 1 mm and (b) scanning electron micrograph showing the surface topography of an entire granule (MacLeod *et al.* 1990), bar = 100 μm.

favorable flow conditions, the microbial cells can adhere to each other and form large, dense, self-supported biofilm particles usually called "granules" (Figure 6.5b).

Perhaps the most widely known high-rate bioreactor in anaerobic technology is the up-flow anaerobic sludge blanket reactor (UASB). The wastewater is fed from the bottom of the reactor and flows upward through a sludge bed. Despite the high up-flow liquid velocity and biogas production, the biomass is retained in the digester due to a gas–solids-separator located on the top section of the reactor.

In addition, the upward flow of the liquid creates a selection pressure for bacterial cells, which adhere to each other and form granules. The success of this system is the possibility of the high organic matter removal rate at short HRTs due to the biomass accumulation in the digesters. The UASB reactor can easily be operated at organic loading higher than $25 \, kg \, COD/m^3.d$ while conventional anaerobic digesters (CSTR) are usually operated at organic loading rates lower than $10 \, kg \, COD/m^3.d$ (Hall 1992; Lettinga 1995).

In comparison with suspended growth systems, attached growth systems have the additional advantage of the long retention time of the suspended solids. An example taken from the wastewater treatment technology is the anaerobic ammonium oxidation process (ANAMMOX). This novel process had been observed for the first time in a fluidized bed reactor treating the effluent from a methanogenic reactor (Mulder et al. 1995). Later studies showed that the microorganisms responsible of carrying out the ANAMMOX process grow extremely slowly (doubling time 11 days) (Strous et al. 1998; Jetten et al. 1999). Hence, it was the reactor configuration (high biomass concentration combined with long retention times of microbial cells) as well among other environmental conditions (pH, temperature, nitrate concentration, redox potential) that allowed for the growth and establishment of these types of microorganisms.

6.6.2.3   Hydraulic retention time

The HRT and/or the solids retention time (SRT) strongly affect the stability and the performance of a fermentation process and the yield and selectivity of specific metabolic products. The HRT is defined as Vr/Q, where Vr is the active volume of the fermentor and Q is the influent flow rate. The SRT is defined as the mass of solids in the fermentor divided by the solids effluent rate. It is obvious that for CSTR type fermentors, the HRT equals the SRT.

The dilution rate (D) is defined as the reciprocal of the HRT and it is the variable, which is mostly manipulated in order to control the performance of a continuous fermentation process. In CSTR systems, the growth rate of the microbial biomass must always be higher than D in order to prevent wash out. However, in cases of slowly grown biomass, the operation of CSTR fermentors at D higher than the biomass growth rate is necessary for practical reasons such as the reduction of the reactor volume. This is possible with (a) the increase of the SRT by the recirculation of concentrated biomass from the effluent of the reactor, (b) the retention of the active biomass in the reactor by the attached growth of the microbes on packing material or (c) the self-immobilization of the microbial cells (flocculation and/or granulation).

In practice, the manipulation of the HRT is only used during experiments with pilot or laboratory scale reactors aiming at the optimum selection of the operational parameters for specific reactor design or metabolic products. In full-scale units the influent composition is more or less fixed and therefore, the HRT is usually varied in case of an emergency to prevent total reactor failure. Typical example of an emergency is the overloading of an anaerobic digester during wastewater treatment processes. The overloading of an anaerobic digester leads to the accumulation of

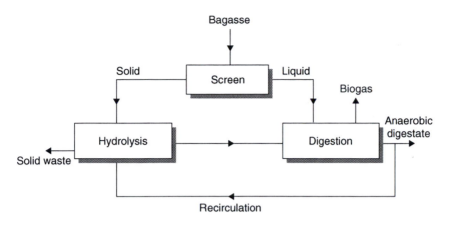

Figure 6.6   A two-stage system used for the anaerobic digestion of sorghum residues (Stamatelatou *et al.* 2003).

the VFAs produced by the fast-growing acidogenic bacteria, the subsequent inhibition of the slow-growing methanogens and the failure of the digester due to the washing out of the methanogens. This failure can only be avoided by the reduction of the HRT, which gives the methanogens the necessary time to recover and to consume the accumulated acids.

A problem of high-rate systems is that the feed has to be devoid of high solids contents. Therefore, such a process may not be a viable solution for biomass fermentation. This limitation may be overcome by two-stage processes that involve liquid extraction of soluble organics from the solid substrate, combined with separate hydrolysis of the solid fraction and subsequent high-rate bioconversion of the combined hydrolysate and leachate fractions (e.g. Figure 6.6).

6.6.3   Mixing and gas transfer

One characteristic of fermentation systems is that the overall conversion rate can be affected by external mass transfer resistance (i.e. mass transfer from the bulk liquid through the stagnant liquid film to the surface of microbial cells and/or cell aggregates/biofilms) and internal mass transfer resistance (i.e. mass transfer from the surface to the interior of an aggregate/biofilm). The mass transfer resistance is especially increased in systems where the transfer occurs between the gas and liquid phases. The gas molecules have additionally to be transferred from the bulk gas phase (e.g. rising gas bubble) through the stagnant liquid film around the gas–liquid interface to the bulk liquid. In continuously stirred systems, effective mechanical mixing can easily eliminate the influence of mass transfer resistance.

On the other hand, the performance of attached growth systems is strongly affected by transport phenomena that take place on and into the biofilm. The mass transfer resistance depends on the bulk substrate concentration, the degree of turbulence in the fluid, the morphology, the size and the cell density of the biofilm, the substrate saturation constant (K_S) and the maximum specific activity of the active microbial biomass (Dolfing 1985; Hulshoff Pol 1989; Schmidt and Ahring 1991). In general, diffusional resistance increases with increasing biofilm thickness, but whether the mass transfer is the limiting step of a substrate biodegradation depends on both the substrate bulk concentration and the morphology of the biofilm. According to Wu *et al.* (1995) who investigated the liquid film resistance of brewery anaerobic granular biofilms during acetate, propionate and ethanol utilization, a Reynolds number of 159,000 is necessary to ensure good mixing and minimize liquid film resistance. Beyond that point, the effect of liquid film on mass transfer can be neglected. In any case, modeling of a fermentation system has to take into consideration the mass transport phenomena and where the mass transport is a rate-limiting step, the model must include dynamic equations describing these phenomena.

Hydrogen is a low-solubility gas, which very often is involved in fermentation processes. The presence of hydrogen can play a very important role in the fermentation of biomass since it thermodynamically regulates the final products of the process. For example, the short-chained fatty acids other than acetate, such as propionate and butyrate, are oxidized to acetate, and H_2/CO_2 by H_2-producing acetogenic bacteria during the anaerobic digestion of biomass. Due to the unfavorable thermodynamics, oxidation of propionate and butyrate is only possible if the H_2 partial pressure is kept very low by the H_2 consuming methanogens. Propionate degradation is possible only below a partial pressure of 10^{-4} atm hydrogen (Bryant *et al.* 1967; Gujer and Zehnder 1982; Dolfing and Mulder 1985). A slight increase in the partial pressure of hydrogen results immediately in a decrease in the degradation rate of the two VFAs. Low hydrogen partial pressure can only be achieved by interspecies transfer of molecular hydrogen from hydrogen-producing bacteria to hydrogen-oxidizing methanogens in microcolonies (Schmidt and Ahring 1993; 1995). Thermodynamic and flux considerations have shown that the most effective degradation of propionate and butyrate will take place in microcolonies where the distance between the syntrophic bacteria is small (Pauss *et al.* 1990). The formation of microbial biofilm enhances the presence of such microcolonies and consequently, the disintegration of the biofilm (by e.g. increase of mechanical mixing in order to improve the gas–liquid transfer of hydrogen) leads to a decrease in the propionate and butyrate degradation rates (Schmidt and Ahring 1993, 1995) and subsequently to process instability and low performance.

6.6.4   Inoculation and acclimation of microbial populations

The ability of a heterogeneous, anaerobic microbial culture to ferment various organic compounds depends strongly on the particular mixture of the bacterial species present. The acclimation of a microbial culture is a very important process bringing significant changes to the microbial population and thus allowing the microbial population to be adapted to a specific substrate. In general, a well-acclimated culture is characterized by a better performance concerning the efficiency and the selectivity of the fermentation process compared to the initial inoculum.

It has been reported that a time period equal to 12 retention times in a continuous reactor is required for a microbial culture in order to be completely acclimated or adapted to a specific substrate (Toerien *et al.* 1967). The changes of the biologic characteristics of a microbial population during the acclimation process have been confirmed and quantified in a number of studies by conducting chemical and biologic analyses; for example, volatile suspended solids and their nitrogen content, proteins, DNA and enzymatic activities (Hattingh *et al.* 1967; Toerien *et al.* 1967; Breure *et al.* 1986a, b; Chynoweth and Mah 1971, 1977; Britz *et al.* 1994). Also kinetic studies represent a useful tool for the investigation of the changes brought to a microbial population after being acclimated to a specific substrate (Sorensen *et al.* 1991; Gavala and Lyberatos 2001). In other words, the acclimation of a heterogeneous anaerobic culture can be considered as a natural selection process resulting in a dominant microbial population that can be quite different from that in the initial inoculum.

The choice of the inoculum in a fermentation process should be based on both the biomass used as substrate and the desirable products. In case of complex substrates and processes, that is, anaerobic digestion of waste/wastewater, the presence of numerous microorganisms capable of biodegrading different organic compounds should be ensured in the initial inoculum. On the other hand, there are processes where the presence of certain groups of microorganisms is undesirable depending on the final product. A representative example is the biologic fermentative production of hydrogen where the inoculum should be free of hydrogen-utilizing methanogenic microorganisms. This can be accomplished by an acid/base (1 N hydrochloric acid/sodium hydroxide resulting in a pH of 3 and 10, respectively) (Chen *et al.* 2002) or thermal (15 min at boil temperature) enrichment of the inoculum (Lay *et al.* 1999).

6.7 MODELING, CONTROL AND OPTIMIZATION OF PRODUCT FORMATION

Hopefully the above discussions allow for an appreciation of what influences the performance of a biomass conversion process. It should be apparent that given a certain biomass source, the relevant question is the strategy that needs to be followed in maximizing product formation. The first question is of course an understanding of what are the possible products and what are the possible overall yields.

A product is possible if there exists at least one species that can generate it. If such an organism exists then it must be ensured that the microbial process has been inoculated either by a pure culture of this organism or by a mixed culture that contains it.

The second requirement is that practical operating conditions discussed in Section 6.6 can be established that allow for (a) the survival of the particular species and (b) the stability of the particular product under the particular operating conditions (i.e. need to make sure that the product of interest is not converted any further). This means that the pH and temperature are such that the particular organisms that generate the product can grow under such conditions. It also means that the SRT is low enough that the organism(s) of interest can grow. It also means that the particular conditions are such that the product can be generated by the organism (e.g. although *S. cerevisiae* can grow under aerobic conditions, no ethanol is generated under such conditions). Finally, it means that the conditions are such that undesirable organisms are excluded. This can be affected either by having a pure culture or by securing operating conditions that do not allow the persistence of the particular organisms.

Given the existence of appropriate microbial ecology in the bioreactor environment, the next concerns have to do with the possible product yield, selectivity and production rate. Depending on the composition of the feed, a maximum theoretical yield may be calculated. Thus, knowing the carbohydrate content in glucose equivalents, it can safely be assumed that for each mole of glucose, a maximum of 3 moles of methane, or 12 moles of hydrogen may be produced. This places an upper bound on what may be achieved based on stoichiometry. Energetics may be also used as discussed in Section 6.2 to obtain a better estimate of the energetically feasible yields that can be obtained.

Knowing the theoretically possible stoichiometry does not provide enough information for the design of a bioreactor. Knowledge of the bioreactor kinetics and their dependence on the process operating variables needs to be established. This requires an understanding of the individual microbial conversion processes that describe adequately the rate of production and consumption of all key intra- and interspecies metabolic intermediates. The basic rate processes may then be most conveniently portrayed in a tabular form as developed by the IAWPRC Task Force for activated sludge model development (IAWPRC 1987). Each row corresponds to a different key process and each column to a different key variable. Variables include original substrates, key intermediates, key microbial concentrations and end-products. A single rate expression corresponds to each row. Such an overall kinetic model may then be used to develop the proper material balances that describe the variation of all key variables, including the product(s) of interest.

Having the material balances that describe the state of a bioreactor, optimization methods may be used for determining the optimal bioreactor operating conditions. Optimization requires that a performance objective is defined. For a continuous bioreactor it may be the rate of a particular product generation per unit reactor volume and per unit time.

Consider for example the case of simple Monod kinetics describing the growth of a single species x on a soluble organic substrate S. Suppose that the objective is the production of biomass in a continuous well-stirred bioreactor. If the dilution rate is D (in h^{-1}), then the performance objective is simply $I = Dx$. In order to determine the dilution rate (1/retention time) that maximizes biomass production, the material balances for the two variables of interest, biomass and substrate are written as:

$$\frac{dx}{dt} = \frac{\mu_{max}[S]}{K_S + [S]} x - Dx \qquad (6.11)$$

$$\frac{d[S]}{dt} = \frac{1}{-Y_{X/S}} \frac{\mu_{max}[S]}{K_S + [S]} + D(S_o - [S]) \qquad (6.12)$$

At steady state the time derivatives become zero and the system may be solved to give the steady-state values for biomass and substrate as a function of the feed substrate concentration S_0 and the dilution rate. The dilution rate that maximizes the performance measure I is then easily found to be:

$$D_{opt} = \mu_{max}\left(1 - \sqrt{\frac{K_S}{K_S + S_0}}\right) \tag{6.13}$$

In principle, a complex pre-treated or non-pretreated biomass feed to a bioreactor will need to be described using multiple variables such as undissolved carbohydrates, undissolved proteins, etc., as well as soluble organic material. Then, the key hydrolysis processes will be responsible for converting these substrates to their soluble counterparts (e.g. simple sugars and aminoacids). These then may be converted to various fermentation products depending on the stoichiometry. Such a model was recently developed by the IWA Task Group for anaerobic digestion (IWA 2002).

Once such an overall model is expressed in matrix form, material balances may be written for a CSTR. Having such material balances, depending on our performance objective, determination of the operating conditions (e.g. retention time, temperature, pH, etc.) that will maximize process performance is a rather simple procedure.

The ability of such a modeling procedure to optimize the process performance, of course ultimately depends on the quality of the kinetic models obtained. Thus, in order to determine the optimal temperature, it is necessary that we have determined adequate expressions that describe the dependence of the kinetics of all key microbial processes on temperature. This is of course a rather impossible task. Consequently, in practice, all possible choices for temperature, pH, retention time, agitation system, etc. have to be screened based on literature findings, which are further narrowed down (focused) to something tractable and feasible from a practical point of view. This is precisely where the key challenge lies, when it comes to modeling and optimization for biomass conversion.

6.8 CONCLUSIONS

Biomass fermentation may generate a multitude of metabolic products such as ethanol, hydrogen and biogas that may be used as fuel. Given a particular biomass type, the question of optimizing product formation is raised. This chapter showed that specific product formation is determined the prevailing microbial ecology of the bioreactor sludge, which depends strongly on the bioreactor operating conditions.

Given a particular system, mathematical modeling followed by process optimization is in principle a useful systematic procedure for maximizing process performance. Such modeling must reflect a good understanding of the key rate processes (biochemical inter- and intra-species bioconversions and physicochemical processes) that govern the interrelations between substrates, microbial species and metabolic products. The difficulty lies in setting up appropriate models that contain the relevant information.

REFERENCES

Ahring, B.K., Jensen, P., Bjerre, A.B. and Schmidt, A.S. (1996) Pretreatment of wheat straw and conversion of xylose and xylan to ethanol by thermophilic anaerobic bacteria. *Bioresource Technol.* **58**, 107–113.

Aitken, M.D. (1993) Waste treatment applications of enzymes: opportunities and obstacles. *Chemical Engineer. J.* **52**, B49–B58.

Atlas, R.M. and Bartha, R. (1997) Physiological ecology of microorganisms: adaptations to environmental conditions. In: *Microbial Ecology. Fundamentals and Applications*, 4th edn. Benjamin/Cummings Science Publishing, California.

Bitton, G. (1994) Anaerobic digestion of wastewater and sludge. In: *Wastewater microbiology*. Wiley-Liss, Inc., New York, pp. 229–245.

Bjerre, A.B., Olesen, A.B., Fernqvist, T., Ploeger, A. and Schmidt, A.S. (1996) Pretreatment of wheat straw using alkaline wet oxidation and alkaline hydrolysis resulting in convertible cellulose and hemicellulose. *Biotechnol. Bioeng.* **49**, 568–577.

Breure, A.M., Beeftink, H.H., Verkuijlen, J. and van Andel, J.G. (1986a) Acidogenic fermentation of protein/carbohydrates mixtures by bacterial populations adapted to one of the substrates in anaerobic chemostat cultures. *Appl. Microbiol. Biotechnol.* **23**, 245–249.

Breure, A.M., Mooijman, K.A. and van Andel, J.G. (1986b) Protein degradation in anaerobic digestion: influence of volatile fatty acids and carbohydrates on hydrolysis and acidogenic fermentation of gelatin. *Appl. Microbiol. Biotechnol.* **24**, 426–431.

Britz, T.J., Spangenberg, G. and Venter, C.A. (1994) Acidogenic microbial species diversity in anaerobic digesters treating different substrates. *Water Sci. Tech.* **30**(12), 55–61.

Bryant, M.P., Wolin, E.A., Wolin, M.J. and Wolfe, R.S. (1967) *Methanobacillus omelianskii*, a symbiotic association of two species of bacteria. *Arch. Microbiol.* **59**, 20–31.

Characklis, W.G. and Marshall, K.C. (1990) Biofilms: a basis for an interdisciplinary approach. *Biofilms*. John Wiley and Sons, Inc., New York.

Chen, C.-C., Lin, C.-Y. and Lin, M.-C. (2002) Acid–base enrichment enhances anaerobic hydrogen production process. *Appl. Microbiol. Biotechnol.* **58**, 224–228.

Chynoweth, D.P. and Mah, R.A. (1971) Volatile acid formation in sludge digestion. In: *Anaerobic Biological Treatment Processes* (ed. F.G. Pohland). *Adv. Chem. Ser.* **105**, 41–54.

Chynoweth, D.P. and Mah, R.A. (1977) Bacterial populations and end products during anaerobic sludge fermentation of glucose. *J. WPCF* March, 405–412.

Dolfing, J. (1985) Kinetics of methane formation by granular sludge at low substrate concentrations. *Appl. Microbiol. Biotechnol.* **22**, 77–81.

Dolfing, J. and Mulder, J. (1985) Comparison of methane production rate and coenzyme F_{420} content of methanogenic consortia in anaerobic granular sludge. *Appl. Environ. Microbiol.* **49**, 1142–1145.

Durán, N. and Esposito, E. (2000) Potential applications of oxidative enzymes and phenoloxidase-like compounds in wastewater and soil treatment: a review. *Appl. Catal. B: Environ.* **28**, 83–99.

Fang, H.H.P. and Liu, H. (2002) Effect of pH on hydrogen production from glucose by a mixed culture. *Bioresource Technol.* **82**(1), 87–93.

Galbe, M. and Zacchi, G. (2002) A review of the production of ethanol from softwood. *Appl. Microbiol. Biotechnol.* **59**(6), 618–628.

Gavala, H.N. and Lyberatos, G. (2001) Influence of anaerobic culture acclimation on the degradation kinetics of various substrates. *Biotechnol. Bioeng.* **74**(3), 181–195.

Gavala, H.N., Angelidaki, I. and Ahring, B.K. (2003a) Kinetics and modeling of anaerobic digestion process. In: *Biomethanation, Advances in Biochemical Engineering/Biotechnology*, 81, 57–93, Springer-Verlag GmbH & Co.KG, Heidelberg.

Gavala, H.N., Yenal, U., Skiadas, I.V., Westermann, P. and Ahring, B.K. (2003b) Mesophilic and thermophilic anaerobic digestion of primary and secondary sludge. Effect of the pre-treatment at 70°C. *Water Res.* **37**(19), 4561–4572.

Gavala, H.N., Yenal, U. and Ahring, B.K. (2004) Thermal and enzymatic pretreatment of sludge containing phthalate esters prior to mesophilic anaerobic digestion. *Biotechnol. Bioeng.* **85**(5), 561–567.

Gujer, W. and Zehnder, A.J.B. (1982) Conversion processes in anaerobic digestion. *Water Sci. Tech.* **15**, 127–167.

Hall, E.R. (1992) Anaerobic treatment of wastewaters in suspended growth and fixed film processes. *Water Quality Management Library,* Vol. 7: *Design of Anaerobic Processes for the Treatment of Industrial and Municipal Wastes.* Technomic Publishing Company, Inc., Lancaster, Pennsylvania, U.S.A.

Hattingh, W.H.J., Kotzé, J.P., Thiel, P.G., Toerien, D.F. and Siebert, M.L. (1967) Biological changes during the adaptation of an anaerobic digester to a synthetic substrate. *Water Res.* **1**, 255–277.

Hinman, N.D., Wright, J.D., Hoagland, W. and Wyman, C.E. (1989) Xylose fermentation. *Appl. Biochem. Biotech.* **20/21**, 391–401.

Hulshoff Pol, L.W. (1989) *The Phenomenon of Granulation of Anaerobic Sludge.* Ph.D thesis, University of Wageningen, Wageningen, The Netherlands.

IAWPRC (1987) *Activated Sludge Model No. 1.* Scientific and Technical Report No. 1, London, England.

IWA (2002) *Anaerobic Digestion Model No. 1.* Scientific and Technical Report No. 13, by Task Group for Mathematical Modelling of Anaerobic Digestion Processes, IWA publishing, London, UK.

Jeppsson, M., Träff, K., Johansson, B., Hahn-Hägerdal, B. and Gorwa-Grauslund, M.F. (2003) Effect of enhanced xylose reductase activity on xylose consumption and product distribution in xylose-fermenting recombinant *Saccharomyces cerevisiae. FEMS Yeast Res.* **3**, 167–175.

Jetten, M.S.M., Strous, M., van de Pas-Schoonen, K.T., Schalk, J., van Dongen, U.G.J.M., van de Graaf, A.A., Logemann, S., Muyzer, G., van Loosdrecht, M.C.M. and Kuenen, J.G. (1999) The anaerobic oxidation of ammonium. *FEMS Microbiol. Rev.* **22**, 421–437.

Karam, J. and Nicell, J.A. (1997) Potential applications of enzymes in waste treatment. *J. Chemical Technol. Biotechnol.* **69**, 141–153.

Lay, J.-J., Lee, Y.-J. and Noike, T. (1999) Feasibility of biological hydrogen production from organic fraction of municipal solid waste. *Water Res.* **33**(11), 2579–2586.

Lettinga, G. (1995) Anaerobic-digestion and waste-water treatment systems. *Anton. Leeuwenhoek Int. J. Gen. M.* **67**(1), 3–28.

MacLeod, F.A., Guiot, S.R. and Costerton, J.W. (1990) Layered structure of bacterial aggregates produced in an upflow anaerobic sludge bed and filter reactor. *Appl. Environ. Microbiol.* **56**, 1598–1607.

McGinnis, G.D., Wilson, W.W. and Mullen, C.E. (1983) Biomass pretreatment with water and high pressure oxygen. The wet oxidation process. *Ind. Eng. Chem. Prod. Res. Dev.* **22**, 352–357.

Mosey, F.E. (1983) Mathematical modelling of the anaerobic digestion process – regulatory mechanisms for the formation of short-chain volatile acids from glucose. *Water Sci. Technol.* **15**(8–9), 209–232.

Mulder, A., van de Graaf, A.A., Robertson, L.A. and Kuenen, J.G. (1995) Anaerobic ammonium oxidation discovered in a denitrifying fluidised bed reactor. *FEMS Microbiol. Ecol.* **16**, 177–184.

Müller, J.A. (2001) Prospects and problems of sludge pre-treatment processes. *Water Sci. Technol.* **44**(10), 121–128.

Nicolella, C., van Loosdrecht, M.M. and Heijnen, S.J. (2000) Particle-based biofilm reactor technology. *Trend. Biotechnol.* **18**(7), 312–320.

NREL (1998) http://www.nrel.gov/research/industrial_tech/biomass.html

Pauss, A., Samson, R. and Guiot, S. (1990) Thermodynamic evidence of trophic microniches in methanogenic granular sludge-bed reactors. *Appl. Microbiol. Biotechnol.* **33**(1), 88–92.

Pavlostathis, S.G. and Giraldo-Gomez, E. (1991) Kinetics of anaerobic treatment. *Water Sci. Technol.* **24**(8), 35–59.

Ramsay, I.R. (1997) *Modelling and Control of High-Rate Anaerobic Wastewater Treatment Systems.* Ph.D. thesis, The University of Queensland, Australia.

Rittman, B.E. and McCarty, P.L. (2001) *Environmental Biotechnology: Principles and Applications*. McGraw-Hill International Editions, New York.

Sahm, H. (1984) Anaerobic wastewater treatment. *Adv. Biochem. Eng. Biotechnol.* **29**, 84–115.

Schmidt, A.S. and Thomsen, A.B. (1998) Optimization of wet oxidation pretreatment of wheat straw. *Bioresource Technol.* **64**, 139–151.

Schmidt, J.E. and Ahring, B.K. (1991) Acetate and hydrogen metabolism in intact and disintegrated granules from an acetate-fed, 55-degrees-c, uasb reactor. *Appl. Microbiol. Biotechnol.* **35**(5), 681–685.

Schmidt, J.E. and Ahring, B.K. (1993) Effects of hydrogen and formate on the degradation of propionate and butyrate in thermophilic granules from an upflow anaerobic sludge blanket reactor. *Appl. Environ. Microbiol.* **59**, 2546–2551.

Schmidt, J.E. and Ahring, B.K. (1995) Granulation in thermophilic upflow anaerobic sludge blanket (UASB) reactors. *Anton. Van Leeuwen.* **68**, 339–344.

Sorensen, A.H., Winther-Nielsen M. and Ahring, B.K. (1991) Kinetics of lactate, acetate and propionate in unadapted and lactate-adapted thermophilic, anaerobic sewage sludge: the influence of sludge adaptation for start-up of thermophilic UASB-reactors. *Appl. Microbiol. Biotechnol.* **34**, 823–827.

Stamatelatou, K., Dravillas, K. and Lyberatos, G. (2003) Methane production from sweet sorghum residues via a two-stage process. *Water Sci. Technol.* **48**(4), 235–238.

Stanier, R.Y., Doudoroff, M. and Adelberg, E.A. (1964) *General Microbiology*, 2nd edn. Macmillan, London.

Stephanopoulos, G. (1999) Metabolic fluxes and metabolic engineering. *Metab. Eng.* **1**, 1–11.

Strous, M., Heijnen, J.J., Kuenen, J.G. and Jetten, M.S.M. (1998) The sequencing batch reactor as a powerful tool for the study of slowly growing anaerobic ammonium-oxidizing microorganisms. *Appl. Microbiol. Biotechnol.* **50**(5), 589–596.

Toerien, D.F., Siebert, M.L. and Hattingh, W.H.J. (1967) The bacterial nature of the acid-forming phase of anaerobic digestion. *Water Res.* **1**, 497–507.

U.S. ENERGY ATLAS (1980) David J. Cuff & William J. Young. Free Press/McMillan Publishing Co. NY.

van Ginkel S., Sung S. and Lay J.-J. (2001) Biohydrogen production as a function of pH and substrate concentration. *Environment. Sci. Technol.* **35**, 4726–4730.

Wu, M.M., Criddle, C.S. and Hickey, R.F. (1995) Mass-transfer and temperature effects on substrate utilization in brewery granules. *Biotechnol. Bioeng.* **46**(5), 465–475.

Section IIB: Methane production

7

Methane production from wastewater, solid waste and biomass

R.J.W. Meulepas, Å. Nordberg,
J. Mata-Alvarez and P.N.L. Lens

7.1 INTRODUCTION

Since ancient times, it has been observed that "combustible gas" seeps out from different geological formations. The first anecdotal evidence of using this gas originates back to the 10th century BC in Assyria and the 16th century BC in Persia, where it has been suggested that this gas was used for heating bath water. The first recognition that this "combustible gas" resulted from organic matter degradation during anaerobic conditions is attributed to Volta, who showed in 1776 that this gas was formed from sediments in lakes, ponds and streams. A century later (1868), Bechamp showed that methane (CH_4) is formed by a microbiological process and Tappeiner in 1956 further affirmed the microbial origin of CH_4 in the so-called biogas (Barker 1956).

Biogas, consisting of CH_4 and carbon dioxide (CO_2), is the gaseous end product of the anaerobic microbiological degradation of organic matter. This conversion, as it occurs in natural environments, is engineered in the anaerobic digestion (AD) process. AD has become an established technology for environmental protection through treatment of solid wastes and wastewater. Currently, AD is receiving renewed attention as it can potentially reduce global warming via the utilization of, the CO_2 neutral, biogas as an energy source. This chapter aims to review the principles of AD, the AD technologies available, the substrates that can be utilized for biogas production and the quality of the produced biogas and digestate.

7.2 METHANOGENIC ENVIRONMENT

The outcome of organic matter degradation depends on the availability of the electron acceptors. Microorganisms are able to utilize oxygen, nitrate, iron (III), sulfate and CO_2 as electron acceptors for organic matter oxidation (Table 7.1). During these respiratory processes, microorganisms conserve energy for their metabolism. Microbial conversions occur in sequence from more to less energetically favourable reactions. If several electron acceptors are available, the most energetically favourable electron acceptor will be utilized.

Table 7.1 shows that oxygen (i.e. aerobic respiration) is the most favourable electron acceptor from an energetic point of view (i.e. more negative Gibbs free-energy value), and CO_2 reduction to CH_4 (i.e. methanogenesis) is the least favourable. Whereas sulfate reduction is energetically, only slightly, more favourable than methanogenesis, denitrification provides approximately six times more energy than methanogenesis during the degradation of an organic substrate under standard conditions (Gonzalez-Gil *et al.* 2002). In the absence of oxygen, nitrate, sulfate or other strong electron acceptors (anaerobic conditions), the end products of organic matter degradation are CH_4 and CO_2.

Natural methanogenic environments are encountered in freshwater sediments that are rich in organic matter, in swamps and in the intestinal tract of insects (e.g. termites) and animals (e.g. cows). In AD systems for the treatment of organic waste streams, an engineered methanogenic environment is applied. The principle advantage of methanogenic environmental biotechnology is that it does not require the

Table 7.1   Ecological process sequence according to the reduction potentials at conditions of natural waters, $E°_H$, and the Standard Gibbs free energy of the processes per electron transferred, $\Delta G°_H . n^{-1}$ (Schwarzenbach *et al.* 1993).

Process	Electron couple	Minimum $E°_H (v)$	Maximum $E°_H (v)$	$\Delta G°_H . n^{-1}$ (kJ/mol)
Aerobic respiration	O_2/H_2O	0.747	–	−78.3
Denitrification	NO_3^-/N_2	0.526	0.747	−71.4
Mn^{3+} reduction	Mn^{3+}/Mn^{2+}	−0.047	0.526	−50.2
Fe^{3+} reduction	Fe^{3+}/Fe^{2+}	−0.221	−0.047	4.6
Sulfate reduction	SO_4^{2-}/HS^-	−0.224	−0.221	21.3
Methanogenesis	CO_2/CH_4	–	−0.224	23.5

addition of an external electron acceptor. In addition, valuable products (biogas and a compost-like digestate) are generated (Gonzalez-Gil *et al.* 2002).

7.3 PATHWAY OF ANAEROBIC DIGESTION

AD has currently been recognized as a complex microbiological process involving a well-organized community of several microbial populations, which completely mineralize organic matter into CH_4 and bicarbonate or CO_2 (Kaspar and Wuhrmann 1978; Gujer and Zehnder 1983; Schink 2000). The digestion process can be divided into four steps: hydrolysis, acidogenesis (i.e. fermentation of organic monomers to form organic acids), acetogenesis (i.e. production of the methanogenic substrates acetate and hydrogen (H_2)/CO_2) and methanogenesis (Figure 7.1).

7.3.1 Hydrolysis

Complex particulate matter, such as cellulose, hemicellulose, lipids and proteins, are generally too large to be taken up by the bacteria and have to be degraded to soluble oligomers or monomers; that is, sugars, amino acids and long-chain fatty acids. The solubilization (hydrolysis) is performed by hydrolytic enzymes, which are produced and excreted by hydrolytic fermentative bacteria. Hydrolysis is often rate limiting for the AD process in cases where the substrate is in particulate form (Vavilin *et al.* 1996; Sanders *et al.* 2000).

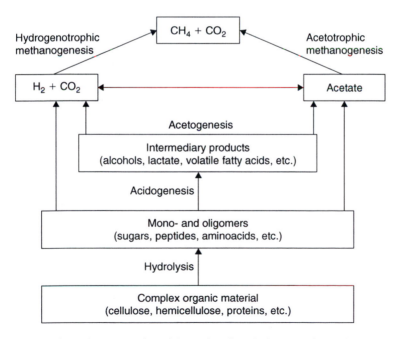

Figure 7.1 Schematic presentation of the carbon flow during AD of complex organic material to methane and CO_2.

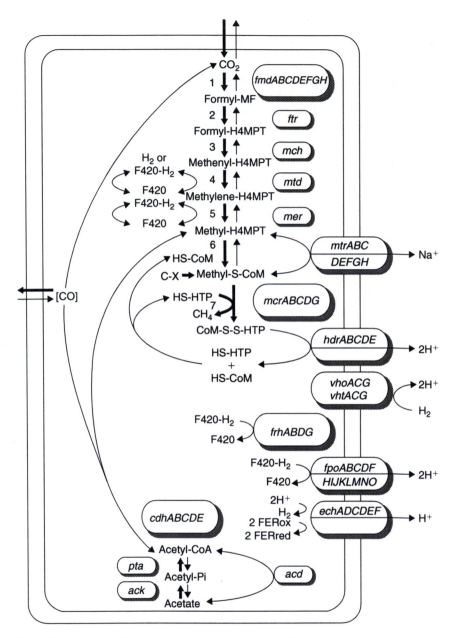

Figure 7.2   Combined pathway for hydrogenotrophic and acetotrophic methanogenesis (Hallam *et al.* 2004).

7.3.2   Acidogenesis

The solubilized monomers are taken up by different groups of fermentative bacteria and are subsequently metabolized to volatile fatty acids, alcohols, H_2 and CO_2 (acidogenesis). Due to the wide variety of starting compounds, this step is carried out by a diverse group of microorganisms (Britz *et al.* 1994; Hwu *et al.* 1998).

7.3.3   Acetogenesis

Lactate, alcohols and fatty acids longer than acetate are converted to acetate and H_2 by the obligate H_2 producing acetogenic bacteria (acetogenesis). These bacteria require a low H_2 partial pressure in order to conserve energy for growth (Dubourgier *et al.* 1988). Therefore, they have to grow in association with a H_2-consuming microorganism, such as a methanogen. This close relationship is referred to as syntrophy (Dong *et al.* 1994; Stams 1994). These facts emphasize the important role that the partial pressure of H_2 plays in the overall degradation of organic matter to CH_4 and stress the need for an intrinsic connection between fermentative bacteria and methanogens in order to achieve an efficient degradation of the different intermediates.

7.3.4   Methanogenesis

In the final step, acetate and H_2 are converted to CH_4 (methanogenesis) by acetotrophic and hydrogenotrophic methanogens. The proposed reactions and the Gibbs free-energy changes at standard conditions, calculated from Thauer *et al.* (1977), are as follows:

	$\Delta G^{\circ\prime}$ (kJ/mol)
$CH_3COO^- + H_2O \rightarrow CH_4 + HCO_3^-$	-31
$4H_2 + HCO_3^- + H^+ \rightarrow CH_4 + 3H_2O$	-136

Methanogens are strict anaerobes and belong to the Archaea (archaebacteria). Methanogenesis proceeds via a number of unique co-enzymes (Figure 7.2), which are exclusively found in methanogens (Blaut 1994). About 2/3 of the methane produced in an anaerobic digester comes from acetate (Gujer and Zehnder 1983). The methanogens have slow growth rates and are usually considered rate limiting for digestion of easy degradable materials.

7.4   ENVIRONMENTAL FACTORS AFFECTING AD

The AD process is influenced by different environmental factors such as temperature, available nutrients, toxicants, pH, alkalinity and water content.

7.4.1   Temperature

Temperature is an important factor for microbial activity. There are mainly three temperature intervals, which are considered as optimal for AD. These intervals are called psychrophilic, mesophilic and thermophilic (Table 7.2).

Mesophilic methanogenesis occurs at its best around 35°C. Thermophilic digestion has been reported to have several advantages over those at mesophilic temperature, such as higher reaction rates and pathogen-killing effect (Buhr and Andrews 1977). Psychrophilic digestion is commonly not used for large-scale applications, since the microorganisms within this temperature interval (ca. 15°C) have a rather low activity (Rebac *et al.* 1999).

Table 7.2   Temperature ranges and optimum for various anaerobic populations (Prescott *et al.* 1999).

	Temperature range (°C)	Temperature optimum (°C)
Psychrophilic	0–20	15
Mesophilic	15–45	35
Thermophilic	45–75	55

7.4.2   Nutrients

For optimal AD a number of substances are necessary. Carbon, nitrogen and phosphorus are fundamental for growth and multiplication. Macro- and micronutrient requirements are lower for anaerobic processes as compared with aerobic ones, due to the much lower biomass yield.

The nitrogen content of a substrate is, in addition to being essential for growth, important since digestion of nitrogenous compounds will contribute to the neutral pH stability by releasing ammonium cations. Different advice for C : N ratios of the substrate can be found in literature. An optimum C : N ratio for AD is suggested to be 20 : 1 to 30 : 1 (Hawkes 1980). However, non- or poorly degradable carbon should not be taken into account.

Micronutrients are required in AD as in any other microbial process. The most important needed to stimulate growth are sulfur, vitamins and traces of minerals (Zandvoort *et al.* 2005). Many functions of anaerobic bacteria are strongly dependent on the availability of trace elements, since they form part of the active sites of several key enzymes (Oleszkiewicz and Sharma 1990). Among the anaerobes, methanogens have a unique requirement for different trace elements, such as cobalt, nickel and molybdenum (Zandvoort *et al.* 2002a, b).

7.4.3   Toxicants

Biological processes are sensitive to a number of toxic compounds. These inhibitory substances can occur in the feedstock, but they can also be produced as a result of microbial activity, converting non-inhibitory substances to inhibitory substances. Toxic substances can be inorganic, such as heavy metal cations, hydrogen sulphide, salts (at high concentrations) and ammonia. Acetotrophic methane formation has previously been shown to be inhibited by ammonium concentrations above $1.7\,g\,NH_4^+$-N/l (Koster and Lettinga 1984). In addition, the inhibiting effect increases with increased pH due to the release of free ammonia (Wiegant 1986). The minimum inhibitory level reported for free ammonia is between 80 mg/l (de Baere *et al.* 1984; Koster and Lettinga 1984) and 150 mg/l (McCarty and McKinney 1961; Braun *et al.* 1981).

Toxic compounds can also be of organic origin. Waste originating from different agricultural products contains diverse assortments of natural polyphenolic compounds, which can be inhibitory to methanogenesis. Short exposure of simple phenols (catechol and pyrogallol) to oxidative conditions prior to feeding them to an anaerobic reactor leads to an increased toxic effect on methanogens during their methanogenic treatment (Field *et al.* 1989).

Oxygen is inhibitory to methanogens, which are obligate anaerobes. In addition, any highly oxidized material, as nitrate or nitrite, can exhibit inhibition (Jenney *et al.* 1999). In a mixed culture where facultative anaerobic bacteria are present, oxygen will be consumed by hydrolyzing and acidogenic bacteria.

Toxicity of the anaerobic process is a complex phenomenon (Kugelman and Chin 1971). Salts (e.g. sodium and potassium) can be stimulating at low levels but can be inhibitory at higher levels (de Baere *et al.* 1984; Feijoo *et al.* 1995; Kempf and Bremer 1998). In addition, the inhibitory effect can also be dependent on the concentration of other substances, which can have synergistic or antagonistic effects (McCarty and McKinney 1961). In many cases the toxicity is reversible and the methanogens can be acclimatized to tolerate the toxic substances (Speece 1996).

7.4.4   pH and alkalinity

To maintain a stable methanogenic activity, a pH between seven and eight is desired. The methanogenic process is alkalizing in itself, that is, it consumes hydrogen ions. Acidic biomass substrates (e.g. silage) can therefore be used without neutralization. However, unstable conditions caused by overloading or the presence of toxic compounds lead to the accumulation of volatile fatty acids, which causes a drop in pH. To avoid a pH drop sufficient alkalinity is needed, normally 1000–5000 mg/l (Tchobanoglous *et al.* 1993). Occasional addition of alkali (e.g. lime) during unstable conditions is feasible, but with the low value of the produced gas it is hardly economical to maintain a neutral pH with neutralizing agents. Recycling of digester effluent for dilution of incoming substrate can be a way of enhancing the buffer capacity, thus contributing to pH-stability (Nordberg 1996).

7.4.5   Water content

The water content of the substrate is important from several aspects. It is essential for biological activity, since nutrients must be dissolved in water before they can be assimilated (Forster and Wase 1989). In addition, water enhances the mobility of microorganisms facilitating their contact with the substrate. Water is also known to influence mass transport limitations. The low moisture content with difficult mass transport as well as poor penetration, diffusion and distribution of microorganisms throughout the substrate can complicate the digestion of solid wastes when using dry AD (Ghosh and Lall 1988).

7.5   SUBSTRATES FOR AD

7.5.1   Waste streams

In general, all wastes that contain organic matter can be utilized for methane production, easily degradable homogeneous concentrated streams are the most favourable. Table 7.3 summarizes sources of waste that are presently used for methane production, the waste streams are divided into solid wastes, waste slurries and wastewater.

Table 7.3   Origin of organic waste streams that can be utilized for the production of biogas (Adapted from Weiland 2000).

Solid wastes	Domestic	Separately collected vegetable, fruit and yard waste
		The organic fraction of source-sorted household waste
		Organic residual fraction after mechanical separation of integral collected household waste
	Agricultural	Crop residues
		Undiluted Manure
Waste slurries	Domestic	Primary and secondary sewage sludge
	Agricultural	Liquid manure
	Industrial	Slaughterhouses and meat-processing
		Fish-processing
Wastewater	Domestic	Sewage, Black
		Water Sewage
	Industrial	Dairy, sugar, starch, coffee processing, breweries and beverages, distilleries and fermentation, chemical, pulp and paper, fruit and vegetable processing

To minimize transport costs, wastes are often digested on site. However, in Denmark centralized co-digestions are widely used. Co-digestion is digestion of manure mixed with other organic waste streams, for example vegetable, fruit and yard waste (Holm-Nielsen and Al Seadi 2004).

Industrial wastewaters, which contain high concentrations of organic matter are very attractive waste streams for AD. Effluents from the food and beverage industries contain the highest concentration of organic compounds; the chemical oxygen demand of these waste streams can be up to 70–80 g/l. Anaerobic wastewater treatment is widely applied in this branch of industry as well as in the pulp and paper industry.

7.5.2   Energy crops

Besides for the treatment of wastes, AD can be used as an attractive energy conversion system for various types of biomass, for example energy crops like sugar cane (Lettinga 2001) or mais (Figure 7.3).

In general, these crops are specially cultivated for energy production, such as corn and grass silage. However, most of conventional agricultural crops can be applied for co-digestion with manure or other organic waste, since the methane yield per ton of organic matter is in the range of 350–450 m^3/ton organic matter.

Sugar cane is a crop with high biogas potential. Circa 60% of the energy from sugar cane can be converted to biogas (Lettinga and van Haandel 1992). Sugar cane has been used for the production of bioethanol for decades in Brazil (van Haandel 2001). Only 38% of the cane energy is converted into alcohol. However, out of the liquid waste stream of this process, an additional 100 kg of CH_4 per m^3 produced alcohol can be obtained when applying AD (van Haandel 2001). This makes the combined production of bioethanol and biogas from sugar cane more feasible.

Several energy crops have been considered for co-digestion of other farm wastes as manure, for example maize, rye, Jerusalem artichoke, oat, sunflower, triticale,

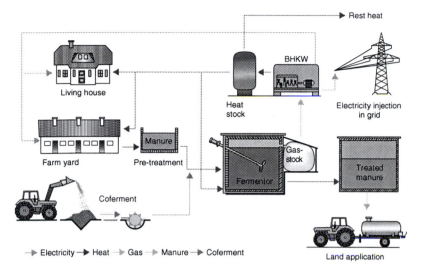

Figure 7.3   Co-digestion of manure and energy crops for improved biogas yields to optimize farm scale bioenergy production (www.kws.de).

rape and wheat (Ahrens and Weiland, 2004). The methane yield per input of organic matter rises in co-fermentation with manure, because of the positive influence of manure as a source of essential trace elements and buffering substances for the biogas process. The methane yield per liquid matter input decreases, because the water content of manure is up to 90% and higher (Ahrens and Weiland 2004). Co-digestion of manure can be done both decentralized (Figure 7.3) or centralized (Holm-Nielsen and Al Seadi 2004). In Germany, over 65% of the biogas plants utilize energy crops and crop residues for co-digestion with manure (Weiland *et al.* 2003).

7.6   AD TECHNOLOGIES

Table 7.4 presents examples of commercial available AD technologies used for wastewater treatment and biogas production. These AD systems can be divided into three categories: dry fermentation systems, wet fermentation systems and high-rate systems (Weiland 2000; Lissens *et al.* 2001). A more fully overview of different AD technologies and suppliers can be found at http://www.novaenergie.ch/iea-bioenergy-task37/.

7.6.1   Wet fermentation systems for the treatment of slurries

Wet fermentation techniques are used for the treatment of slurries and organic solid waste. In some cases, because these techniques require a low solid content, solid waste streams have to be diluted with recycled process water. One of the first wet type full-scale plants for the treatment of the organic fraction of municipal solid waste was built in Vaasa (Finland) in 1989, using the so-called Waasa technology.

The continuous stirred tank reactor (CSTR) is the most used reactor type to digest low solid waste streams. It is used for the anaerobic stabilization of sewage sludge

Table 7.4   Examples of AD technologies for biogas production.

Feedstock category	Commercial available systems	Reference
Waste streams diluted to form a slurry of less than 15% total solids (wet fermentation)	Accumulation System (semi-batch) Linde – KCA Ros-Roca VAGRON, Waasa (completely mixed) BTA (two-phase digestion process)	Zeeman (1991) Zeeman et al. (2000) Linde (2004) Korz (2003) Braber (1995) Vagron (2004) Kübler and Wild (1992) de Jong et al. (1993)
Waste streams with a total solid percentage of more than 20% and energy crops (dry fermentation)	Biocel (batch) Dranco (vertical plug-flow) Linde – BRV Kompogas (horizontal plug-flow) Valorga (vertical plug-flow, mixed by gas circulation)	de Jong et al. (1993) ten Brummeler (2000) de Jong et al. (1993) de Baere (2000) Linde (2004) Wellinger et al. (1992) de Jong et al. (1993) Fruteau de Laclos et al. (1997) Saint-Joly et al. (2000)
Waste water (high rate continuous systems)	Anaerobic filter Anaerobic fluidized bed Upflow anaerobic sludge bed (UASB) Expanded granular sludge bed (EGSB) reactor Internal circulation	Chua and Fung (1996) Chua and Fung (1996) Lettinga et al. (1980; 2000) Lettinga et al. (2001) Frankin (2001) Driesen and Yspeert (1999)

produced in wastewater treatment plants (Chynoweth and Pullammanappallil 1996) and for the treatment of piggery waste (Zeeman et al. 1988). These digesters consist of a void recipient stirred by biogas recirculation, liquid recirculation or mechanical means. In low solid waste digesters, the feedstock and the microorganisms have the same residence time. Therefore, the hydraulic residence time must be high enough to prevent a complete washout of microorganisms. In practice, this means that digesters should be designed with residence times above 10 days, which in turn means large reactor volumes.

In a batch system, the digester is filled with waste and some seed sludge. At the start of the process, the different processes like hydrolysis, acidification and methanogenesis will occur in sequence. A variation on this is the accumulation system, which is filled over a longer period. Such a system is suitable when storage of waste is required (Zeeman 2000).

Two-phase systems enable hydrolysis and methanogenesis to be spatial separated. The different processes have different optimal conditions, therefore higher conversion rates and biogas yields can be obtained when the processes are optimized separately in different reactors (Ghosh et al. 2000). The main advantage, however, is a greater biological reliability for wastes, which cause unstable performance in

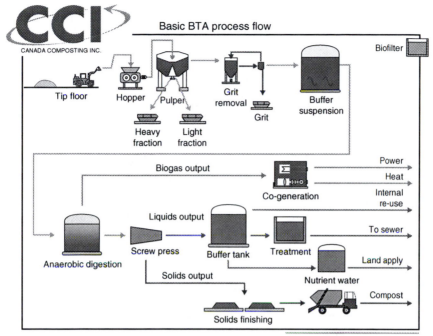

Figure 7.4   Two-stage wet type digester used in the BTA process (CCI, 2004).

one-stage systems (Vandevivere *et al.* 2002). Two-phase systems provide some pro-tection against the fluctuations of organic loading rates or against the presence of inhibitory substances. An industrial system using the two-phase approach is the BTA process (Figure 7.4), designed to treat the organic fraction of Municipal Solid Waste.

7.6.2   Dry fermentation systems for the treatment of solid waste

In dry systems, the fermenting mass within the reactor has a solid content of more than 20%. Therefore, dry AD digesters are smaller in size, require less process water and require less heating compared to wet fermentation systems (Mata-Alvarez and Macé 2004). However, due to the high viscosity of the waste streams, transport and handling requires special equipment. In addition, the dry fermentation systems are plug-flow reactors, contrary to wet systems where complete mix reactors are usu-ally used. The basic design of a plug-flow digester is a long tube, which enables hydrolysis and methanogenesis to be spatial separated (de Baere 2000).

Different reactor configurations were specifically developed for the digestion of the organic fraction of municipal solid waste. Figure 7.5 presents three successful approaches of plug-flow reactors: the dry anaerobic composting (Dranco) system, the Kompogas system and the Valorga system. In the Dranco and Kompogas a recycle stream of the digested paste is applied in order to inoculate the feed. In the

Figure 7.5   Dry type digesters: (a) the Dranco system, (b) the Kompogas system and (c) the Valorga system. Photos by courtesy of Dranco, Kompogas and Valorga, respectively.

Valorga process circulation of the gas phase is applied to achieve mixing. Mixing the incoming wastes with the fermenting mass is crucial to guarantee adequate inoculation and most of all to prevent local overloading and acidification.

It must be taken into account that for the treatment of municipal solid waste, some pre- and post-treatment steps are necessary. Pre-treatment steps include magnetic separation, comminution in a rotating drum or shredder, screening, pulping, gravity separation and pasteurization. As for post-treatment steps, the typical sequence involves mechanical dewatering, aerobic maturation and wastewater treatment (Mata-Alvarez and Macé 2004).

7.6.3   High-rate systems for the treatment of wastewater

In contrast to slurries, the organic matter in wastewater is highly diluted. Therefore, reactors are developed that allow high hydraulic loading rates without the washout of microorganisms. These high-rate reactors are divided into two categories:

(1) Systems with a suspended bacterial mass, where retention is achieved through external or internal settling.

(2) Systems with fixed bacterial films on solid surfaces.

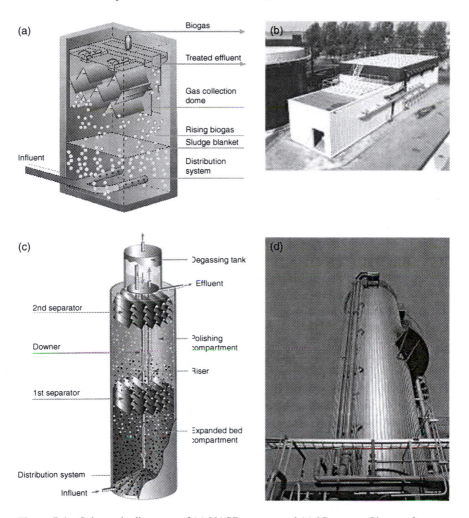

(a)

Biogas

Treated effluent

Gas collection
dome

Rising biogas
Sludge blanket

Influent

Distribution
system

(b)

(c)

Degassing tank

Effluent

2nd separator

Polishing
compartment

Downer

Riser

1st separator

Expanded bed
compartment

Distribution system

Influent

(d)

Figure 7.6   Schematic diagrams of (a) UASB reactor and (c) IC reactor. Photos of
full-scale plants of (b) UASB reactor and (d) IC reactor. Photos courtesy of Paques.

Among the first type are the contact digester and the UASB. The contact
digester comprises three units: a CSTR digester, a degasser and a settler. The
degasser is usually included to improve the settleability. The thickened sludge from
the settler is recirculated to the CSTR to maintain a high biomass concentration and
the clarified effluent is discharged. A biomass concentration over 25 g/l can be
obtained. Anaerobic contact systems have been installed in pulp and paper wastewater plants, and are the oldest system of this type. Today, they are not generally
used because there are many problems with the settler, as biomass still produces
gas and tends to float (Mata-Alvarez and Macé 2004).

The biomass retention in a UASB (Figure 7.6) is based on the development of
well-settable aggregates that are retained in the reactor called granules due to their
spherical form (Lettinga *et al.* 1983). The specially constructed internal settler

allows an effective degasification of the sludge. This makes solid retention times over 200 days at hydraulic retention times of only 6 h possible, in addition the biogas generation is enough to ensure adequate mixing. The internal circulation (IC) reactor (Figure 7.6) and the EGSB are variations on the UASB concept, which allow higher loading rates. Due to the higher hydraulic upflow velocities the sludge bed of the EGSB and IC reactors are more expanded and the mixing is better. These technologies are very successful for the treatment of industrial wastewater: hundreds of plants have been installed worldwide.

Successful examples of anaerobic digesters with fixed bacterial films are the anaerobic filter and the anaerobic fluidized bed. The anaerobic filter is similar to the trickling filters used in aerobic wastewater treatment. However, unlike the trickling filter the system is contained in an airtight vessel and the liquid movement is from the bottom upwards.

7.7 PRODUCTS OF AD

7.7.1 Biogas

Table 7.5 compares some biogas characteristics with those of natural gas and town gas. The quality of the biogas is defined by its composition. Biogas consists of CH_4 (50–70%), CO_2 (30–50%) and smaller amounts of hydrogen sulfide and ammonia. Trace amounts of hydrogen, nitrogen, carbon monoxide, non-methane volatile organic matter, halocarbons and oxygen are occasionally present in biogas. One cubic metre corresponds to approximately 10 kWh.

The exact composition of biogas depends on the feedstock composition, the process conditions and the type of digester used. An important process condition is the temperature. At a low temperature, more CO_2 is dissolved, resulting in a higher percentage of CH_4 in the gas. In addition, the fraction of carbohydrates is important, relative to that of proteins and fats. Digestion of carbohydrates results in 50% CH_4 and 50% CO_2. Pure protein and fat give, in theory, rise to biogas with a content of about 70% CH_4 and 30% CO_2 (Ahrens and Weiland 2004). The quality of biogas with respect to fuel cells is discussed in Chapter 21.

Table 7.5 Characteristics of natural gas, town gas and biogas (Wellinger and Lindberg 2000).

Parameter	Unit	Natural gas	Town gas	Biogas (60% CH_4, 38% CO_2, 2% others)
Calorific value (lower)	MJ/m^3	36.14	16.1	21.48
Density	Kg/m^3	0.82	0.51	1.21
Wobbe index (lower)	MJ/m^3	39.9	22.5	19.5
Maximum ignition velocity	M/s	0.39	0.7	0.25
Theory for air requirement	m^3 air/m^3 gas	9.53	3.83	5.71
Maximum CO_2 concentration in stack gas	vol%	11.9	13.1	17.8
Dew point	°C	59	60	60–160

7.7.2   Digestate

Anaerobic digestion draws carbon, hydrogen and oxygen from the substrate. Essential plant nutrients (N, P and K) remain largely in the residue (digestate). The digestate from a wet fermentation system can be dewatered into a solid and a liquid fraction. Most of the phosphorous will then end up in the solid fraction and most of the ammonia–nitrogen will end up in the liquid phase. In addition, the digestate contains bulky organic matter, more than compost derived from aerobic degradation (Brethouwer and Wierda 1993). Therefore, digestate is suitable as fertilizer and as soil conditioner. It can be used on farmland or gardens, to improve soil quality and fertilization (Brinkman *et al.* 1997; Hogg 2004).

The quality of the digestate depends on the quality of the feedstock. Impurities like metals and persistent organic contaminants will remain in the digestate. Physical impurities like plastic, glass and stones can be removed by mechanical pre-treatment (Monnet 2003).

7.8   CONCLUSIONS

Current energy prices and EU-targeted reduction of fossil fuel combustion in the coming years (as stated in the Directive EC/30/2003) will draw increasingly more attention towards AD. In the framework of the Kyoto agreements, many countries in Europe have agreed to subsidize the production of electricity from biogas in order to stimulate the production of methane from wastes and biomass. This can be achieved with a wide variety of biomass and organic solid and liquid wastes in a variety of reactor configurations.

REFERENCES

Ahrens, T. and Weiland, P. (2004) Electricity production from agricultural wastes through valorization of biogas. In: *Resource Recovery and Reuse in Organic Solid Waste Management* (eds. P. Lens, B. Hamelers, H. Hoitink and W. Bidlingmaier), pp. 395–410. IWA Publishing, London, UK.

Barker, H.A. (1956) *Bacterial Fermentations.* Wiley, New York.

Blaut, M. (1994) Metabolism of methanogens. *Anton. Van Leeuwen.* **66**, 187–208.

Braber, K. (1995) Anaerobic digestion of municipal solid waste: a modern waste disposal option on the verge of breakthrough. *Biomass Bioenerg.* 9(1–5), 365–375.

Braun, R., Huber, P. and Meyrath, J. (1981) Ammonia toxicity in liquid piggery manure digestion. *Biotechnol. Lett.* **3**, 159–164.

Brethouwer, T.D. and Wierda, C. (1993) GFT-compost als anti-stuifmiddel. NOVEM Report No. 9328, Utrecht.

Brinkman, J., Baltissen, T. and Hamelers, B. (1997) Development of a protocol for assessing and comparing the quality of aerobic composts and anaerobic digestates. RDA/SR-97001.

Britz, T.J., Spangenberg, G. and Venter, C.A. (1994) Acidogenic microbial species diversity in anaerobic digesters treating different substrates. *Proceedings of the 7th International Symposium on Anaerobic Digestion*, Cape Town, pp. 74–79.

Buhr, H.O. and Andrews, J.F. (1977) The thermophilic anaerobic digestion process. *Water Res.* **11**(2), 129–143.

CCI, Canada Composting Inc. (2004) www.canadacomposting.com

Chua, H. and Fung, J.P.C. (1996) Hydrodynamics in the packed bad of anaerobic fixed film reactor. *Water Sci. Technol.* **33**(8), 1–6.

Chynoweth, D.P. and Pullammanappallil, P. (1996) Anaerobic digestion of solid wastes. In: *Microbiology of Solid Waste* (eds. A.C. Palmisano and M.A. Barlaz). pp. 71–113. CRC Press Inc., Boca Raton.

de Baere, L.A. (2000) Anaerobic digestion of solid waste: state-of-the-art. *Water Sci. Technol.* **41**, 283–290.

de Baere, L.A., Devocht, M., van Assche, P. and Verstraete, W. (1984) Influence of high NaCl and NH₄Cl salt levels on methanogenic associations. *Water Res.* **18**, 543–548.

de Jong, H.B.A., Koopmans, W.F. and van de Kniff, A. (1993) Conversietechnieken voor GFT-afval; ontwikkelingen in 1992. Haskoning, Novem and RIVM, Nijmegen.

Dong, X., Plugge, C.M. and Stams, A.J.M. (1994) Anaerobic degradation of propionate by a mesophilic acetogenic bacterium in coculture and triculture with different methanogens. *Appl. Environ. Microbiol.* **60**(8), 2834–2838.

Driesen, W. and Yspeert, P. (1999) Anaerobic treatment of low, medium and high strength effluent in the agro-industry. *Water Sci. Technol.* **40**(8), 221–228.

Dubourgier, H.C., Archer, D.B., Algagnac, G. and Prensier, G. (1988) Structure and metabolism of methanogenic microbial conglomerates. In: *Anaerobic Digestion* (eds. E.R. Hal l and P.N. Hobson). Pergaman Press, Oxford.

Hwu, C.S., Lier, J.B. and Lettinga, G. (1998) Physicochemical and biological performance of expanded granular sludge bed reactors treating long-chain fatty. *Process Biochem.* **33**(1), 75–81.

Field, J.A., Kortekaas, S. and Lettinga, G. (1989) The tannin theory of methanogenic toxicity. *Biol. Waste.* **29**, 241–262.

Feijoo, G., Soto, M., Méndez, R. and Lema, J.M. (1995) Sodium inhibition in the anaerobic process: antagonism and adaptation phenomena. *Enzyme Microb. Technol.* **17**, 180–188.

Forster, C.F. and Wase, D.A.J. (1989) *Environmental Biotechnology*. John Wiley and Sons, New York.

Frankin, R.J. (2001) Full-scale experiences with anaerobic treatment of industrial wastewater. *Water Sci. Technol.* **44**(8), 1–6.

Fruteau de Laclos, H., Desbois, S. and Saint-Joly, C. (1997) Anaerobic digestion of municipal solid waste: Valorga full-scale plant in Tilburg, The Netherlands. *Water Sci. Technol.* **36**, 457–462.

Ghosh, S. and Lall, U. (1988) Kinetics of anaerobic digestion of solid substrates. In: *Proceeding of International Association on Water Pollution Research and Control, Anaerobic Digestion 1988* (eds. E.R. Hall and P.N. Hobson). Bologna.

Ghosh, S., Henry, M.P., Sajjad, A., Mensinger, M.C. and Arora, J.L. (2000) Pilot-scale gasification of MSM by high-rate and two-phase anaerobic digestion. *Water Sci. Technol.* **41**(3), 101–110.

Gonzalez-Gil, G., Kleerenbezem, R., Mattiasson, B. and Lens, P. (2002) Biodegradation of recalcitrant and xenobiotic compounds. In: *Water Recycling and Resource Recovery in Industry: Analysis, Technologies and Implementation* (eds. P. Lens, L.W. Hulshoff Pol, P. Wilderer and T. Asano), pp. 386–430. IWA Publishing, London, UK.

Gujer, W. and Zehnder, A.J.B. (1983) Conversion processes in anaerobic digestion. *Water Sci. Technol.* **15**, 127–167.

Hallam, S.J., Putnam, N., Preston, C.M., Detter, J.C., Rokhsar, D., Richardson, P.M. and DeLong, E.F. (2004) Reverse methanogenesis: testing the hypothesis with environmental genomics. *Science* **205**, 1457–1462.

Hawkes, D.L. (1980) Factors affecting net energy production from mesophilic anaerobic digestion. In: *Anaerobic Digestion* (eds. D.A. Stratford, B.I. Wheatley and D.E. Hughes), pp. 131–150.

Hogg, D. (2004) Cost and benefits of bioprocesses in waste management. In: *Resource Recovery and Reuse in Organic Solid Waste Management* (eds. P. Lens, B. Hamelers, H. Hoitink and W. Bidlingmaier), pp. 95–121. IWA Publishing, London, UK.

Holm-Nielsen, J.B. and Al Seadi, T. (2004) Manure-based biogas systems. In: *Resource Recovery and Reuse in Organic Solid Waste Management* (eds. P. Lens, B. Hamelers, H. Hoitink and W. Bidlingmaier), pp. 377–394. IWA Publishing, London, UK.

Jenney Jr., F.E., Verhagen, M.F.J.M., Cui, X. and Adams, M.W.W. (1999) Anaerobic microbes: oxygen detoxification without superoxide dismutase. *Science* **286**, 306–309.

Kaspar, H.F. and Wuhrmann, K. (1978) Kinetic parameters and relative turnovers of some important catabolic reactions in digesting sludge. *Appl. Environ. Microbiol.* **36**, 1–7.

Kempf, B. and Bremer, E. (1998) Uptake and synthesis of compatible solutes as microbial stress response to high-osmolality environments. *Arch. Microbiol.* **170**, 319–330.

Korz, D.J. (2003) Technology solutions of Ros Roca to meet the ABP requirements by digestion and composting. www.compostnetwork.info/downloads/abpworkshop/22_korz_abstracts.pdf

Koster, I.W. and Lettinga, G. (1984) The influence of ammonium–nitrogen on the specific activity of pelletized methanogenic sludge. *Agr. Wastes* **9**, 205–216.

Kübler, H. and Wild, M. (1992) The BTA-process high rate biomethanisation of biogenous solid wastes. In: *Proceedings of the International Symposium on Anaerobic Digestion of Solid Waste* (eds. F. Cechi, J. Mata-Alvarez and F.G. Pohland). Venice.

Kugelman, I.J. and Chin, K.K. (1971) Toxicity, synergism, and antagonism in anaerobic waste treatment processes. In: *Anaerobic Biological Treatment Processes* (ed. R.F. Gould). *Adv. Chem. Ser.* **105**, 55–90.

Lettinga, G. (2001) Digestion and degradation, air for live. *Water Sci. Technol.* **44**(8), 157–176.

Lettinga, G. and van Haandel, A.C. (1992) Anaerobic digestion for energy production and environmental protection. In: *Renewable Energy; Sources for Fuels and Electricity* (eds. T.B. Johansson *et al.*). Island press, California.

Lettinga, G., Hobma, S.W., Klapwijk, A., van Velsen, A.F.M. and de Zeeuw, W.J. (1980) Use of the upflow sludge blanket (USB) reactor concept for biological wastewater treatment. *Biotechnol. Bioeng.* **22**, 699–734.

Linde (2004) www.linde-process-engineering.com/en/p0001/p0052/p0054/p0054.jsp

Lissens, G., Vandevivere, P., de Baere, L., Biey, E.M. and Verstraete, W. (2001) Solid waste digesters: process performance and practice for municipal solid waste digestion. *Water Sci. Technol.* **44**(8), 91–102.

Mata-Alvarez, J. and Macé, S. (2004) Anaerobic bioprocess concepts. In: *Organic Solid Waste Management: from waste disposal to resource recovery* (eds. P. Lens, B. Hamelers, H. Hoitink and W. Bidingmeier), pp. 290–314. IWA Publishing, London, UK.

McCarty, P.L. and McKinney, R.E. (1961) Salt toxicity in anaerobic digestion. *J. Water Pollut. Cont. Fed.* **33**, 399–415.

Monnet, F. (2003) *An Introduction to Anaerobic Digestion of Organic Waste*. Remate, Scotland.

Nordberg, Å. (1996) *One- and two-phase anaerobic digestion of ley crop silage with and without liquid recirculation*. PhD. thesis. Swedish University of Agricultural Sciences, Uppsala.

Oleszkiewicz, J.A. and Sharma, V.K. (1990) Stimulation and inhibition of anaerobic processes by heavy metals – a review. *Biol. Wastes* **31**, 45–67.

Paques (2004) www.paques.nl/paques/webPages.do?pageID=200507

Prescott, L.M., Harley, J.P. and Klein, D.A. (1999) Microbiology, 4th edn.

Rebac, S., van Lier, J.B., Lens, P., Stams, A.J.M., Dekkers, F., Swinkels, K.Th.M. and Lettinga, G. (1999) Psychrophilic anaerobic treatment of low strength wastewaters. *Water Sci. Technol.* **39**(5), 203–210.

Saint-Joly, C., Debois, S. and Lotti, J.P. (2000) Determinant impact of waste collection and composition on anaerobic digestion performance: industrial results. *Water Sci. Technol.* **41**(3), 291–297.

Sanders, W.T.M., Geerink, M., Zeeman, G. and Lettinga, G. (2000) Anaerobic hydrolysis kinetics of particulate substrates. *Water Sci. Technol.* **41**(3), 17–24.

Schink, B. (2000) Principles of anaerobic degradation of organic compounds. In: *Biotechnology*, 2nd edn. (ed. J. Klein) 11b, pp. 169–192. Wiley-VCH, Weinheim.

Schwarzenbach, R.P., Gschwend, P.M. and Imbodem, D.M. (1993) *Environmental Organic Chemistry*. Wiley-Interscience, New York.

Speece, R.E. (1996) *Anaerobic Biotechnology for Industrial Wastewaters*. Archae Press, Nashville, Tennessee, TN, USA.

Stams, A.J.M. (1994) Metabolic interactions between anaerobic bacteria in methanogenic environments. *Anton. van Leeuwen. Intl. J. Gen. Molecul. Microb.* **66**(1–3), 271–294.

Tchobanoglous, G., Theisen, H. and Vigil, S. (1993) *Integrated Solid Waste Management: Engineering Principles and Management Issues.* McGraw Hill, New York.

ten Brummeler, E. (2000) Full scale experience with the BIOCEL-process. *Water Sci. Technol.* **41**, 299–304.

Thauer, R.K., Jungermann, K. and Decker, K. (1977) Energy conservation in chemotrophic anaerobic bacteria. *Bacteriol. Rev.* **41**, 100–180.

van Haandel, A.C. (2001) Resource recovery at alcohol distilleries for environmental protection and increased profitability. In: *Proceedings (Part 1) of the 9th World Congress: Anaerobic Digestion 2001* (eds. A.F.M. van Velsen and W.H. Verstraete), pp. 157–162. Technologisch Instituut, Antwerp, Belgium.

Vavilin, V.A., Rytov, S.V. and Lokshina, L.Y. (1996) A description of hydrolysis kinetics in anaerobic degradation of particulate organic matter. *Bioresource Technol.* **56**(2–3), 229–237.

Weiland, P. (2000) Anaerobic waste digestion in Germany – Status and recent developments. *Biodegradation* **11**, 415–421.

Weiland, P., Rieger, C. and Ehrmann, T. (2003) Evaluation of the newest biogas plants in Germany with respect to renewable energy production, greenhouse gas reduction and nutrient management. In: *Proceedings of the future of Biogas in Europe II, European Biogas Workshop*, 2–4 October 2003, University of Southern Denmark.

Wellinger, A. and Lindberg, A. (2000) In: *Proceedings of "Biogas Event 2000: Kick-Off for a Future Deployment of Biogas Technology".* 22–24 November 2000, Eskilstuna, Sweden.

Wellinger, A., Wyder, K. and Metzler, A. (1992) Kompogas – Ein neues Verfahren zur anaeroben Aufbereitung von organischen Abfällen. In: *Getrennte Wertstofferfassung und Biokompostierung Band 2.* (K.J. Thomé, Hrsg.), pp. 263–271. EF-Verlag, Berlin.

Wiegant, W.M. (1986) *Thermophilic anaerobic digestion for waste and wastewater treatment.* PhD. thesis. Agricultural University Wageningen, Wageningen.

Vagron (2004) www.vagron.nl/html/uk/vagron4.htm

Vandevivere, P., de Baere, L. and Verstraete, W. (2002) Types of anaerobic digesters for solid wastes. In: *Biomethanization of the Organic Fraction of Municipal Solid Wastes* (ed. J. Mata-Alvarez), pp. 111–140.

Zandvoort, M.H., Geerts, R., Lettinga, G. and Lens, P. (2002a) Effect of long-term cobalt deprivation on methanol degradation in a methanogenic granular sludge reactor. *Biotechnol. Progr.* **18**, 1233–1239.

Zandvoort, M.H., Geerts, R., Lettinga, G. and Lens, P. (2002b) Effect of nickel deprivation on methanol degradation in a methanogenic granular sludge reactor. *J. Ind. Microbiol. Biot.* **29**, 268–274.

Zandvoort, M.H., van Hullebusch, E.D., Gieteling, J., Lettinga, G. and Lens, P. (2005) Effect of sulfur source on the performance and metal retention of methanol-fed UASB reactors. *Biotechnol. Prog.* **21**(3), 839–850.

Zeeman, G. (1991) *Mesophilic and psychrophilic digestion of liquid manure.* PhD. thesis. Wageningen University, Wageningen.

Zeeman, G., Sutter, K., Vens, K. and Wellinger, A. (1988) Psychrophilic digestion of dairy cattle and pig manure: start-up procedures of batch, fed batch and CSTR-type digesters. *Biol. Waste.* **26**, 15–31.

Zeeman, G., Sanders, W.T.M. and Lettinga, G. (2000) Feasibility of the on-site treatment of sewage and swill in large buildings. *Water Sci. Technol.* **41**(1), 9–16.

8

Optimisation of anaerobic digestion using neural networks

L. Zani and P. Holubar

8.1 INTRODUCTION

8.1.1 Optimisation methods

The requirement for methods of optimisation arise from the mathematical complexity necessary to describe the theory of systems, processes, equipment and devices which occur in practice. Even quite simple systems must sometimes be represented by theory which may contain approximations or parameters which change with time. Although the theory is imperfect for many reasons, yet it must be used to predict the optimum operating conditions of a system such that a particular performance criterion is satisfied. At the best, such theory can predict only that the system is near the desired optimum. Optimisation methods are used to explore the local region of operation and predict the way that the system parameters should be adjusted to bring the system to optimum (Adby and Dempster 1974). In an industrial process, the criterion for optimum operation is often in the form of minimum cost where the product costs can depend on a large number of interrelated controlled parameters in the manufacturing process. In mathematics, the performance criterion could be to minimise the integral of the squared difference between a specified function and an approximation to it generated as a function of the controlled parameters. Both of these examples have in common the requirement that a single quantity is to be minimised by variation of a number of controlled parameters. The importance of optimisation lies not in trying to find out all about a system, but in finding out, with the least possible effort, the best way to adjust the system. If this is carried out well, systems can have a more economic and improved design, they can operate more accurately or at less costs.

The basic mathematical optimisation problem is to minimise a scalar quantity E, which is the value of n system parameters $x_1, x_2, \ldots, x_n$. These variables must be adjusted to obtain the minimum required:

$$\min (E) \quad \text{with } E = f(x_1, x_2, \ldots, x_n) \tag{8.1}$$

The value E of f embodies the designed criteria of the system into a single number which is often a measure of the difference between the required performance and the actual performance obtained. The function f is referred to as the *objective function*, whose value is the quantity that has to be minimised. In practice it is very difficult to determine if the minimum obtained by a numerical process is a *global minimum* or not. In most circumstances it can only be said that the minimum obtained is a minimum within a local area of search. A particular function may of course have several *local minima* (see Figure 8.1). One of these will be the global minimum but it is usually impossible to determine if a local minimum is also the global minimum unless all minima are found and evaluated.

Some optimisation methods require *gradient* information about the objective function given by $E = f(x)$. This is obtained in the form of the first derivative of f with respect to the n parameters. The Jacobian gradient vector $\bar{g}$ is defined as the transpose of the gradient vector ∇f which is a row matrix of first-order partial derivatives (Adby and Dempster 1974). The transpose of $\bar{g}$ is given by:

$$\bar{g}^{\mathrm{T}} = \nabla f = \left[\frac{\partial f}{\partial x_1} \frac{\partial f}{\partial x_2} \cdots \frac{\partial f}{\partial x_n} \right] \tag{8.2}$$

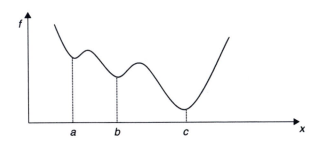

Figure 8.1   Minima of an objective function; for $x_i = a, b$ the function has local minima, for $x_i = c$ the function has a global minimum.

The $n \times n$ symmetric matrix of second-order partial derivatives of f is known as the *Hessian* matrix and is denoted by $\mathbf{H}$:

$$\mathbf{H} = \begin{vmatrix} \dfrac{\partial^2 f}{\partial x_1^2} & \dfrac{\partial^2 f}{\partial x_1 \partial x_2} & \cdots & \dfrac{\partial^2 f}{\partial x_1 \partial x_n} \\[2ex] \dfrac{\partial^2 f}{\partial x_2 \partial x_1} & \ddots & & \vdots \\[2ex] \vdots & & \ddots & \vdots \\[2ex] \dfrac{\partial^2 f}{\partial x_n \partial x_1} & \cdots & \cdots & \dfrac{\partial^2 f}{\partial x_n^2} \end{vmatrix} \tag{8.3}$$

The availability of the Jacobian vector and Hessian matrix simplify optimisation procedures. However, it is usually necessary to proceed without gradient information since derivatives may be uneconomic, impossible to compute, or do not even exist.

Optimisation methods are based on iterative processes thus problems of convergence arise. The first problem concerns the convergence to a minimum: Has the procedure reached a minimum and is this minimum a global or a local minimum? And, what was the speed of convergence? The value of E for successive iteration will be in most cases the only guide to the progress of the minimisation. When E does not reduce over a number of iterations some kind of minimum has been reached. It is almost impossible to predict if the minimum reached is the global minimum. Even extensive testing of all minima found is not a complete answer since a given technique may never converge to the global minimum of certain functions. Only when a function is derivable and the Hessian matrix is calculated one can have an idea about minima. When the Hessian is definite positive in each point of a region, then the function is convex in that region. In this case there is only one minimum in the region. But since most of the time the derivative cannot be calculated or does not exist, this technique cannot help (Monegato 1990).

A large number of optimisation problems are concerned with functions in which data are only available for a number of sample points. The basic optimisation problem in this case is posed in a slightly different form. The value E of the objective function is a vector whose elements are errors at the individual sample points. All elements of E must then be minimised simultaneously. If the individual errors are combined into one scalar quantity, standard methods of optimisation can be applied (Adby and Dempster 1974). The main problem with this error criterion is that the error E jumps discontinuously with the parameters. Derivatives of E with respect to the parameters are therefore undefined.

In many practical optimisation problems, there are constraints on the values of some of the parameters which restrict the region of search for the minimum. A common constraint on the variable x_i is in the form of inequality $x_{Li} \leq x_i \leq x_{Ui}$, where x_{Li} and x_{Ui} are fixed lower and upper limits to x_i (Adby and Dempster 1974). More generally inequality constraints (g) are formulated to specify relationships of the parameters involved in the constraint in the form:

$$g(x_1, x_2, \ldots, x_n) \leq 0 \tag{8.4}$$

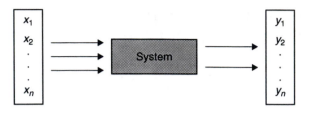

Figure 8.2   Multivariate input $\bar{x}$ producing multi-response output $\bar{y}$.

The region of search where the constraints are satisfied is termed *feasible region*, while the region in which constraints are not satisfied is termed the *infeasible region*.

8.1.2   Neural networks: basic concepts

Neural networks are mathematical models inspired by biological nervous systems (Zupan and Gasteiger 1993; Kasabov 1996). They are composed of simple interconnected elements operating in parallel and are basically a method for handling multivariate–multi-response data (Figure 8.2).

The relationship between a multivariate object $\bar{x}_s = (x_{s1}, x_{s2}, ..., x_{sm})$ and the response factor $\bar{y}_s = (y_{s1}, y_{s2}, ..., y_{sm})$ can be written in the following way (Zupan and Gasteiger 1993):

$$(y_{s1}, y_{s2}, ... , y_{sm}) = \mathbf{A}\,(x_{s1}, x_{s2}, ..., x_{sm}) \qquad (8.5)$$

where $\mathbf{A}$ can be a $(m \times n)$-variate matrix that linearly transforms vector $\bar{x}_s$ into vector $\bar{y}_s$ (s indicates different samples: s = 1, 2, ..., r), but actually is for many systems a very complex operator. In many applications $\mathbf{A}$ is not known and all that is available is a set of collected m-variate data $\mathbf{X}_s$ accompanied by a set of n-variate responses $\mathbf{Y}_s$. Information can be derived from these sets of data so that conclusions can be applied also to unknown samples.

Neural network are a mathematical tool that learns from examples and exhibit some capability for generalisation beyond the training data.

8.1.3   The neuron

Neural networks are composed of simple interconnected elements operating in parallel. Each single calculating element of a neural network is called neuron. Each neuron contains parameters, called weights, which multiply the input coming, in general, from measurement instruments and pass it through a transfer function to produce an output. Inputs and outputs can be scalar (Figure 8.3) or vectors (Figure 8.4). Both neurons have a scalar bias (b). One may view the bias as simply being added to the product ωp, or as shifting the function f to the left by an amount b. The bias is much like a weight, except that it has a constant input of 1. Both weight and bias are adjustable scalar parameters of the neuron. Thus, the network can be trained by adjusting weight or bias parameters (Demuth and Beale 2000).

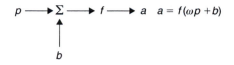

Figure 8.3   Scheme of single neuron having a scalar input and output (p = input, ω = weight, b = bias, n = net input to the transfer function = $\omega p + b$, f = transfer function, a = output).

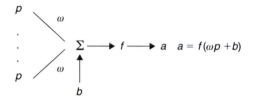

Figure 8.4   Scheme of single neuron having a vector input and a scalar output (p_i = input: r = elements column vector, ω_i = weight: r = elements row vector, b = bias, n = net input to the transfer function = $\omega p + b$, f = transfer function, a = output).

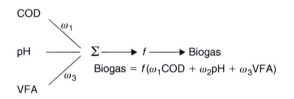

Figure 8.5   Example of a single neuron that takes a three-element vector as input and give a scalar (biogas) as output.

Let us make a practical example derived from biotechnology. If an anaerobic digestion has to be modelled measures of organic loading rate, pH, volatile fatty acids (VFAs) concentration and biogas production are taken. The data can be organised in input and output. The organic loading rate, pH and VFAs constitute the input of the model and the biogas production is the variable that has to be calculated by the model and compared with the real measure. The weights of the network are adjusted till when the calculated output is equal to the measured variable (Figure 8.5).

In many artificial neural networks the processing units are arranged in subsequential layers, called input-, hidden- and output-layer. The first layer receives a number of input signals and produces some output which is then fed to the next layer of processing elements and so on. The cascade of layers can be thought of as a black-box which maps input vectors to output vectors.

The knowledge of an artificial neural network is encoded in the values of its weights. The starting values of the weights are often chosen by the operator.

The process which adapts the weights of the network so that the network is able to reproduce the relationship between input and output is called training of the network and it takes place according to specific learning rules. Learning can be supervised or unsupervised. In supervised learning, the learning rule is provided with a set of examples (the training set) of proper network behaviour. As the inputs are applied to the network, the network outputs are compared to the targets. The learning rule is then used to adjust the weights and biases of the network in order to move the network outputs closer to the targets. In unsupervised learning, the weights and biases are modified in response to network inputs only. There are no target outputs available. Supervised learning is used for building models while the unsupervised method is used mainly for clustering (Demuth and Beale 2000).

8.1.4   Data collection

Before a network can be trained, data describing the process must be collected. The amount of data needed to train a network depends on the process characteristics and is often the limiting step in neural network applications.

When calibrating a model the number of data used must be bigger than that of parameters. The bigger is the training set the better is the generalisation capability of a network, where generalisation capability means ability to give a correct answer when an unknown input is presented to the network. For this aim the number of training cases must be a large and representative subset of all cases that have to be generalised. The importance of this condition is related to the fact that there are two possible generalisations: interpolation and extrapolation. Interpolation can be done quite reliably while extrapolation is notoriously unreliable (Sarle 1997). Hence it is important to have enough training data, which span the entire data-space, to avoid extrapolation. Moreover in training a network it is advisable to divide the number of data in three parts more or less of the same width. The first subset is the real training set which is used for computing the gradient and updating the network weights and biases. The second subset is the validation set whose error is monitored during the training process. The validation error will normally decrease during the initial phase of training. When the network begins to overfit the data the error on the validation set will typically begin to rise. When the validation error increases the training must be stopped and the weights and biases should then be reset at the minimum of the validation error. The third subset, the test set, is used to compare different models. This leads to the necessity of collecting a really huge amount of data.

8.1.5   Network architecture

As said before, when calibrating a model the number of data used must be bigger than the number of parameters. The network's parameters are the weights. The number of weights in a network depends on the network's structure (number of layers of neurons and number of neurons for each layer). Finding the best network for modelling a process is only matter of trial and error. One must keep in mind that a network that has too few neurons can fail to fully describe the input signal leading to underfitting and that a network with too many data may fit the noise, not just the signal, leading to overfitting (Sarle 1997).

8.2   USE OF NEURAL NETWORKS FOR ANAEROBIC DIGESTION MODELLING AND CONTROL

In this chapter, models are shown which were calibrated at the Institute of Applied Microbiology (IAM) (Universität für Bodenkultur – Vienna) with the purpose of predicting the behaviour of some lab-scale anaerobic digesters. The models were then embedded in a decision support system (DSS) built to optimise the reactor performance (Holubar *et al.* 2000, 2002).

8.2.1   Modelling

The decision of modelling the anaerobic fermentation with neural networks was taken because despite several attempts to control the process, the anaerobic digestion remains a kind of black-box (Dochain 1995; Steyer *et al.* 1995). Up to now classical mathematical modelling was only possible, when severe simplifications of the process representation were performed. The main reason for this situation is that the mechanisms ruling these processes are not adequately understood to formulate reliable nonlinear mathematical models (Schubert *et al.* 1994). As an alternative artificial neural networks are claimed to have a distinctive advantage over some other nonlinear estimation methods used for bio-processes, because they do not require any prior knowledge about the structure of the relationships that exist between important variables. Although the neural networks have the important advantage that there is no need to know the exact form of the analytical function on which the model should be built, all features known in standard model-generating techniques play an important role in the network training: choice of variables, experimental design, etc. (Zupan and Gasteiger 1993). Feed forward back propagation neural network have been used (Kasabov 1996). The number of layers and the number of neurons in each layer depend on the application for which the neural network is set up and is determined by trial and error. The most straight-forward method is to start from a small network and increase the number of neurons/layers until an acceptable model is attained. The acceptability can be assessed on the quality of the network fit to the validation data (Glassey *et al.* 1994).

The data to train the models and to test the DSS were achieved by operating four anaerobic continuous stirred tank reactors (CSTR) in steady-state conditions treating surplus- and primary sludge at organic loading rates (Br) of about $2 \, kg \, COD \, m^{-3} day^{-1}$, and disturbing them by pulse-like increase of Br up to $17 \, kg \, COD \, m^{-3} day^{-1}$. For the pulses, additional carbon sources like flour, sucrose, 1,2-diethylene glycol or vegetable oil. After a selection and optimisation process the best fitting models were found. (For details see Holubar *et al.* 2000, 2002, 2003.)

The model used in this experiments is a hierarchical structure of feed-forward neural networks (FFNN) (Zupan and Gasteiger 1993; Kasabov 1996). Nine input variables and six output variables were selected to describe the system. Two of the output variables (gas production and gas composition) were the output of the highest model level and were passed to a numerical optimisation algorithm. Three of the output variables (acetic acid concentration, propionic acid concentration and total VFAs concentration) were used to set constraint on the control signal. These

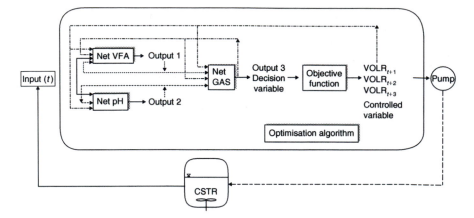

Figure 8.6   Control strategy scheme for an anaerobic digestion process.

five outputs plus the pH and some past values of the output variables themselves and of the control signal were then used as input to make the prediction two and three time steps forward.

The model was organised in three sub-models mainly to limit the number of parameters that had to be calibrated with the data set (Figure 8.6).

The three sub-models were: a model for biogas prediction, a model for the pH and a model for VFAs. The model for biogas production and composition prediction was a 9-3-2 FFNN. The number of parameters of such a model (with a bias in the hidden layer) is 44. The model for pH prediction was a 6-3-1 FFNN, the number of parameters is 28. The model for the VFAs prediction was a 9-3-3 FFNN, the number of parameters is 48. Thus, the model with the higher number of parameters is the last one, which has 48 weights. With this number of parameters several hundreds of measurements are required to calibrate it. The experience has shown that a number of data 10 times bigger than that of parameters should be used. Therefore, more or less 500 data were necessary (and actually taken) for the training set and, in theory, further 500 for the validation set. The number of data in the validation set was actually 352 so, in total, around 850 meaningful and well distributed data points were available. If only one model with 9 inputs and 6 outputs had been used, much more data should have been needed. Under the hypothesis of having 6 (5 + 1 bias) neurons in the hidden layer, the number of parameters would have been 90. Thus, the double amount of data would have been necessary. Since the anaerobic digestion has rather slow dynamics, no more than one set of measures per day was taken; 850 data were collected with four reactors in more than 1 year. This technique of using sub-models had been used by quite a number of authors who have preferred to use separate networks linked into a decision hierarchy, rather than one large network making all decisions simultaneously. Such designs are used either because inadequate computer resources or because not enough data are available to cover the entire variable space (Zupan and Gasteiger 1993).

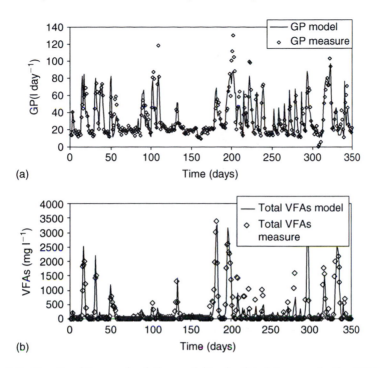

Figure 8.7   Results of the gas simulation model to simulate (a) gas production (GP) and (b) total VFA concentration.

Figure 8.7 shows the results obtained with the above-mentioned neural networks. The correlation coefficients of all simulated variables were between 0.8 and 0.9. It must be noticed that the correlations presented here are between calculated values and real measured data. In this situation a correlation around or higher than 0.8 is considered as very good. In literature correlations higher than 0.9 can be often found, but they are calculated as correlation between the results produced from the model under study and a reference model used to produce the target data (Alves *et al.* 1998), or training and validation data set are artificially expanded by interpolation methods. Models trained with data generated through a reference model or with artificially expanded series of data have little generalisation capability (Glassey *et al.* 1994).

8.2.2   Controlling

The controlled variable usually chosen in biotechnological process optimisation is the volumetric feed rate (Bv) (Alves *et al.* 1998) because, normally, different feed rates lead to different substrate concentration levels and, hence, to different metabolic activities (Schubert *et al.* 1994). However, other controlled variables could be used. The dilution rate (Bastin and Dochain 1990; Renard 1990) is often found in the literature. In case of aerobic processes, the control of the oxygen concentration in the mixed liquor is reported (Georgieva and Patarinska 1996).

In this study Bv was chosen as the control variable. The multi-components neural model was used to calculate the effects of different Bv on the biogas production and VFAs concentration. The gas composition and production predictions supplied by the model are passed to a numerical optimisation routine, which attempts to minimise the difference between the calculated values and some desired level over a time horizon of 3 days. The VFAs concentration is used to constrain the control variable. If the VFA concentration is too high, the Bv value is decreased and the system simulated again until the VFAs are below a certain threshold.

The main objective of an anaerobic treatment plant is the degradation of the organic materials present in the influent. The end products in the gaseous phase of this complex biological process are CO_2 and CH_4. The higher the amount of biogas ($CO_2 + CH_4$) produced, the lower is the amount of biodegradable substance left in the effluent. A high biogas production usually corresponds a low methane concentration in the biogas. Although in conventional surplus-sludge treatment the methane production is not the primary objective of the anaerobic digestion, it is very interesting to have a biogas rich in methane because this can be used as a source of energy. Since the methanogenic bacteria prefer a pH slightly basic it is normally very difficult to get both high biogas production rates and high methane concentrations in the biogas because high feed rate (which gives rise to high gas production) means also high fatty acids concentrations and thus a lower pH. A compromise between these two conflicting objectives must be found (Ryhiner et al. 1992).

The objective function of the optimisation problem is built as the weighted sum of these two objectives. The desired values of the two variables are compared with the predicted ones and the difference over a 3-day time horizon has to be minimised.

The optimisation algorithm contains several kinds of constrains. The maximum Bv allowed at each decision step (the upper boundary of the searching space) is limited, the maximum feed rate increase between two time steps is limited and there are thresholds on the higher level of total fatty acids and propionic acid concentration. The first constrain is set by a fuzzy logic-like algorithm which calculates the maximum allowable Bv on the basis of the last two Bv-measurements and the consequent gas productions. If the system is judged to be in a good condition, the last maximum Bv in memory is increased by one unit, otherwise it is decreased by one unit. The second limit is set as a percentage of the last Bv. This limit can be chosen by the programme user. During all the experiments, this percentage was fixed to 20%. Two kinds of constrains are imposed on the VFA, one on the predicted values and one on the measured values. If the VFA predicted as a result of a certain Bv application are above a certain threshold (T_1), the Bv is lowered and the VFA recalculated until they decrease under the fixed threshold. If the online measured VFA values were higher than a threshold T_2, the control algorithm even does not enter the optimisation iterations cycle but it sets directly the value of Bv to a low number. The threshold T_1 is actually split in two. A threshold T_{11} is fixed on the total VFA concentration and a threshold T_{12} is fixed on the propionic acid concentration.

The control tool was applied for 4 months. Figure 8.8a shows the loading rate over the course of the experiment. Three different periods can be distinguished in

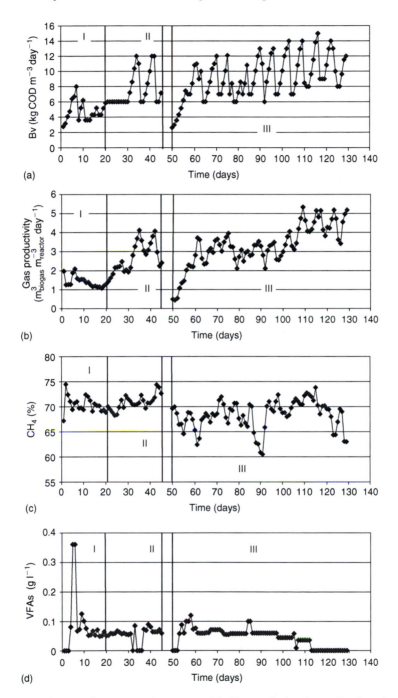

Figure 8.8   Course of parameters in an anaerobic digester during the application of the control tool: (I) start-up phase; (II) first phase and (III) second phase after reactor shutdown because of mechanical problems – (a) volumetric loading rate, (b) gas productivity, (c) gas composition and (d) VFA concentration.

the control experiment. Period I is the start-up phase. The regime phase is divided in two parts because after 2 weeks of rather high loading rates, a mechanical problem occurred. The system had to be shut down, and the pumps and tubes substituted with other ones are able to stand thick fluids. Period II is before the shutdown and period III goes from the new start-up till the end of the experiment 2.5 months later.

At the end of the start-up phase, the gas productivity was around $2 \, m^3_{biogas} \, m^{-3}_{reactor} day^{-1}$. During the period II the biogas productivity increased till $4 \, m^3_{biogas} \, m^{-3}_{reactor} day^{-1}$ with an average of $3 \, m^3_{biogas} \, m^{-3}_{reactor} day^{-1}$ (Figure 8.8b). During period I the percentage of methane in the biogas was oscillating around 70% and during the regime phase II it slightly increased till almost 75% (Figure 8.8c). The VFA concentration was kept very well under control by the optimisation tool. The total VFA concentration was always lower than $0.1 \, g \, l^{-1}$ with the only exception of 2 days of the first week when it reached the value $0.36 \, g \, l^{-1}$ (Figure 8.8d).

At the beginning of period III the maximum volumetric loading rate was set for the first day at $2 \, kg \, COD \, m^{-3} day^{-1}$ and then immediately left to the decision of the optimisation algorithm. As shown in Figure 8.8a, the volumetric loading rate increased for 12 days till a value of $11 \, kg \, COD \, m^{-3} day^{-1}$ and then the weekly oscillations started, triggered by the intentional reduction of Bv at the weekend. Nevertheless, a trend can be clearly seen in the Bv values: the minimum and the maximum values in each week increased with time. With the increase of the volumetric loading rate, which went once till $17 \, kg \, COD \, m^{-3} day^{-1}$ and never dropped below $6 \, kg \, COD \, m^{-3} day^{-1}$ (even $8 \, kg \, COD \, m^{-3} day^{-1}$ in the last 3 weeks of the experiment) the gas productivity was also increasing. After 1 week from the beginning of period III the gas productivity was again above $2 \, m^3_{biogas} \, m^{-3}_{reactor} day^{-1}$ and then it was always increasing and had an average values on the all period III of $3.3 \, m^3_{biogas} \, m^{-3}_{reactor} day^{-1}$ and several peaks above $5 \, m^3_{biogas} \, m^{-3}_{reactor} day^{-1}$. During all this time the VFA concentration was kept below $0.1 \, g \, l^{-1}$ and it can be seen from Figure 8.8d that it decreased with the time. The methane concentration in the biogas during period III was always between 65% and 72% with only two exceptions in 2 days probably due to a different feed composition.

8.3 CONCLUSIONS

This chapter showed that the anaerobic digestion of surplus sludge can be effectively modelled and controlled by means of a hierarchical system of neural networks. The control tool presented here could be improved and extended for instance including the effluent volatile suspended solid (VSS) as constrain variable and increasing the number of controlled variable with the aim of controlling not only the amount of influent but the feed composition as well.

Further modelling work in anaerobic digestion, using neural networks, will have to take account of the recently available mechanistic model of anaerobic digestion, the Anaerobic Digestion Model No. 1 (ADM1) (Batstone et al. 2002). By combining the advantages of both modelling approaches the black-box "anaerobic digestion" could be enlightened in the near future.

REFERENCES

Adby, P.R. and Dempster, M.A.H. (1974) *Introduction to Optimization Methods*. Chapman and Hall, London, UK.

Alves, T.L.M., Costa, A.C., Henriques, A.W.S. and Lima, E.L. (1998) Adaptive optimal control of fed-batch alcoholic fermentation. *Appl. Biochem. Biotechnol.* **70–72**, 463–478.

Bastin, G. and Dochain, D. (1990) On-line estimation and adaptive control of bioreactors. Elsevier, Amsterdam.

Batstone, D.J., Keller, J., Angelidaki, I., Kalyuzhnyi, S.V., Pavlostathis, S.G., Rozzi, A., Sanders, W.T.M., Siegrist, H. and Vavilin, V.A. (2002) *The IWA Anaerobic Digestion Model No 1 (ADM1)*. IWA Publishing, London.

Demuth, H. and Beale, M. (2000) *Neural Network Toolbox*. The Mathworks Inc. Natick, USA.

Dochain, D. (1995) Recent approaches for the modeling, monitoring and control of anaerobic digestion processes. *Proceedings of the International Workshop on Monitoring and Control of Anaerobic Digestion Processes*, France, December 6–7, 1995, Laboratoire de Biotechnologie de L'environnement INRA, pp. 23–29.

Georgieva, O. and Patarinska, T. (1996) Modeling and control of batch fermentation process under conditions of uncertainty. *Bioprocess Engg.* **14**, 299–306.

Glassey, J., Montague, G.A., Ward, A.C. and Kara, B.V. (1994) Enhanced supervision of recombinant *E. coli* fermentation via artificial neural networks. *Process Biochem.* **29**, 387–398.

Holubar, P., Zani, L., Hager, M., Fröschl, W., Radak, Z. and Braun, R. (2000) Modelling of anaerobic digestion using self organizing maps and artificial neural networks. *Wat. Sci. Tech.* **41**(12), 149–156.

Holubar, P., Zani, L., Hager, M., Fröschl, W., Radak, Z. and Braun, R. (2002) Advanced controlling of anaerobic digestion by means of hierarchical neural networks. *Water Res.* **36**, 2582–2588.

Holubar, P., Zani, L., Hager, M., Fröschl, W., Radak, Z. and Braun, R. (2003) Start-up and recovery of a biogas-reactor using a hierarchical neural network-based control tool. *J. Chem. Technol. Biotechnol.* **78**, 847–854.

Kasabov, N.K. (1998) Foundation of neural networks, fuzzy system and knowledge engineering. *A Bradford Book*. MIT Press, Massachusetts, USA.

Monegato, G. (1990) *Fondamenti di Calcolo Numerico*. Libreria Editrice Universitaria Levrotto & Bella – Torino (I).

Renard, P. (1990) Vérification expérimental de l'application de la théorie de l'estimation et du contôle adaptatif à des systèmes biologiques non linéaires. Thèse de doctorat en Science Naturelles Appliquées. Université Catholique de Louvain – Faculté des Sciences Agronomiques, Louvain-la-Neuve (Belgium).

Ryhiner, G., Dunn, I.J., Heinzle, E. and Rohani, S. (1992) Adaptive on-line optimal control of bioreactors: application to anaerobic degradation. *J. Biotechnol.* **22**, 89–106.

Sarle, W.S. (1997) *Neural Network FAQ, Part 1 of 7: Introduction*. Periodic posting to the Usenet newsgroup comp.ai.neural-nets.URL: ftp://ftp.sas.com/pub/neural/FAQ.html

Schubert, J., Simutis, R., Dors, M., Havlik, I. and Lübbert, A. (1994) Bioprocess optimization and control: application of hybrid modelling. *J. Biotechnol.* **35**, 51–68.

Steyer, J.P., Amouroux, M. and Moletta, R. (1995) Process modeling and control to improve stable operation and optimization of anaerobic digestion processes. *Proceedings of the International Workshop on Monitoring and Control of Anaerobic Digestion Processes*, France, December 6–7, 1995, Laboratoire de Biotechnologie de L'environnement INRA, pp. 30–35.

Zupan, J. and Gasteiger, J. (1993) *Neural Networks for Chemists*. VCH, Weinheim.

9

Quality function deployment as a decision support tool for the sustainable implementation of anaerobic digestion facilities

S. Trogisch, W. Baaske and B. Connor

9.1 INTRODUCTION

Using the quality function deployment (QFD) methodology as basis, a decision support tool has been developed in order to assess the sustainable integration of anaerobic digestion (AD) facilities in different European countries. The tool is based on implementation criteria that have been set up by various experts in the different areas concerning AD. QFD is a method widely used in industrial enterprises (especially the car industry) in Japan, USA and Europe for developing new products (Akao 1990; Akao and Mazur 2003; Mazur *et al.* 2003; Saatweber 1999) as well as regional services (Baaske and Lancaster 2004). This innovative approach detaches customer needs (represented by demand side requirements) and the customer satisfaction (represented by the specification of the technical solution to cover those demands). Therefore, it takes not only technical criteria into account, but also others as social, political and financial aspects. Also expected requirements and "kill-factors" (e.g. not enough organic waste, no area to spread out the digested material), which frequently are left aside are considered (Kano *et al.* 1984).

9.2   BACKGROUND

The potential for biogas plants seems to increase steadily all over Europe, especially with existing and forthcoming European Union (EU) directives that are relevant both in the industrial as well as in the agricultural sector. The EU Landfill Directive (1999/31/EC) regulates the disposal of organic wastes in landfills, so that AD can be an alternative, which has to compete with the long established conventional composting process. The AD process, has in contrast to composting, a positive energy balance. Additionally a valuable fertiliser is produced. This process is of industrial relevance (C.a.r.m.e.n., 1998), as it can be integrated within existing processes involving organic wastes and energy demand. A further relevant Directive is the EU Nitrates Directive on the protection of waters against the pollution, caused by nitrates from agricultural sources as in manure. AD from agricultural wastes contributes to a sustainable NO_3^- cycle if the digestate is brought out with the right technique (Graf, 1995; Rathbauer, 1992). The EU goes one step further with the forthcoming Groundwater Directive 91/676/EEC: sustainable agricultural practices, targeting the groundwater protection, are seen as the most effective approach. It further encourages to investigate the reduction of the impact of nitrates on the groundwater quality. Here again AD can play an important role as mentioned above. These two directives address mainly the agricultural biogas plants, which are generally smaller scaled as industrial AD facilities.

Therefore, AD is now attracting not only the interest of farmers, but also of large industries in the waste and agro-food sectors. They see (pressured by the mentioned Landfill Directive) in the "next AD generation" a realistic and economic solution of local waste treatment and disposal. Especially Eastern European countries can be regarded as a promising future market for AD due to their agro-food industries structures, and their need for an improvement of their environmental situation.

The expansion of AD is, however, not fulfilling the previsions due to different reasons as, for example, material corrosion, process failure in digester, total breakdowns, etc. The high costs for AD facilities and (in some cases) the insecure electricity tariffs for "green" electricity have in many cases hindered a financial breakthrough. Other factors have also influenced the failed implementation of AD: society was sceptical towards this technology; for example, they thought it was dangerous due to possible explosions. Also odour emissions from some facilities, due to wrong operation and doubtful construction proved discouraging. New legislation concerning the technical details of such facilities pressurise the operators into providing higher security standards (e.g. explosion protection).

The oil prices seem to pressure again the economies of most oil depending countries, giving AD the opportunity to prove itself, and play an important role in the waste, energy and environmental sectors. Therefore, it is essential to have a methodical approach for implementing AD facilities in the form of a reliable decision support tool, which takes all the relevant factors into account – to avoid failed projects, which create a negative image on the AD sector. A "holistic" sustainability can guarantee a sustainable rentability and operation. Therefore, not only technical, but furthermore social aspects have to be taken into consideration. The question is how to find which social factors in a systematic way: can a specific model or tool help to make an accurate assessment towards such factors, especially with a view to their influence on the sustainability?

9.3 METHODS

9.3.1 Basic concept

9.3.1.1 QFD for product design

The basic concept of QFD was presented by Yoji Akao in Japan in 1966 and was applied in the heavy industry from the late 1960s and early 1970s. QFD quickly found its way into the automotive development where it is used as a tool to develop products in a systematic way. There are many definitions on QFD, which reflect its broad application area. Deployment can be defined as "setting up" of all the parties involved in the whole process of designing a product or service. In combination with quality its means a "high-quality design" of a product (or service) from the beginning on of the whole process up to the using of the developed product or service by the customer. The main aim by using QFD is to define, develop, build, produce, deliver and install a product in such a way that the demands of the client are more than fulfilled. QFD is not a specific method, it is further a guide for a specific working style that has as aim the full satisfaction of the customers, thereby the know-how and experience of all employees contribute to this aim. QFD helps to transform the specific customers demands in the companies specific capability and to mobilise all the involved parties for finally producing a perfect product or service. This results in an effective design process where the customers wishes and demands are fully integrated in the final product (Figure 9.1). By detaching the customer needs from the customer satisfaction it is ensured that project engineers do not think in technical solutions first but in customer demands. Only in the second step, when the customer demands are clearly identified, a solution is developed to effectively cover those demands.

QFD can be defined in this context as converting the consumers' demands into "quality characteristics" and developing a design quality for the finished product by systematically deploying the relationships between the demands and the characteristics. The QFD approach expands the time it takes to define the product, shortens the time it takes to design the product by focusing priorities, promotes better documentation and communication during the process and virtually eliminates the need for redesign, especially on critical items (Figure 9.2). Therefore, the needs and characteristics of customers are collected and put into relation. Further the possible correlations and interdependencies of the different needs and characteristics are found. Starting from this structured information, the different direction of

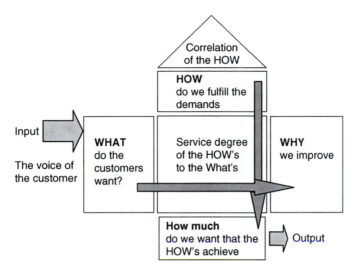

Figure 9.1   Structure of the house of quality of the QFD method.

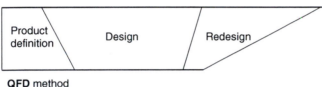

Figure 9.2   Time to market: QFD shortens the time to market considerably.

improvement can be identified and qualified. Figure 9.1 represents the house of quality of the QFD. It is apparent that QFD is a consequent asking and answering "game" that involves all players.

9.3.1.2   QFD for AD

For the adaptation of the QFD method to AD facilities, data was obtained from case studies, failed projects and existing biogas plants in order to obtain the necessary data for validating the basic QFD approach and integration in a support and implementation tool (SIT). Concerning the existing biogas plants, a questionnaire was developed to interview the operators of the biogas plants. It should serve as tool for collecting data from different biogas plants in the Upper Austrian region. It contained 32 blocks, from which each contained two questions. The first question refers to what the operator had expected of a specific aspect (as e.g. constant amount of manure). The second question refers to how the specific aspect turned out to be. The possible answers where: as expected, better or worse. These answers had to be codified and converted into numbers that were QFD compatible. Five different biogas plants where selected randomly for filling out the questionnaire. These where visited and their operators interviewed. The common denominator was the relative small size of the biogas plants: the electrical capacity of their respective combined heat and power (CHP) units was between 20 and 75 kW. Some of the biogas facilities were built during the early 1990s. Only one of them went into operation during the early summer of 2000. The extended age (2–10 years) of most of the biogas plants was a positive aspect for answering the questionnaire, as these operators had a lot experience with their biogas plants. The question topics were technology, economy and partnerships with other neighbouring enterprises.

The technical factors, which are usually narrowly linked to the economic factors, are well known. However, the social factors remained unknown. In order to find out which of these factors play a decisive role, five different locations where pinpointed. All of them where seriously interested in building an AD facility. All of these cases were thoroughly studied and special care was set on the social factors. A time dynamic opinion change tracking that is based on a review of expectations (What did you expect? – Has it been realised?) showed the impact of different factors on the possibilities of building an AD facility. For incorporating the collected data into the SIT, some changes were necessary. The customer demands kept the same, the indicators (relevant factors) were changed. This enabled an assessment of the data collected as well as of the biogas plants. New correlations had to be calculated, so that different values (percentages) turned up.

9.3.2   The support and implementation tool

Concerning the support and implementation tool (SIT), the results of the calculation provides interesting perspectives: different biogas plants could be ranked from the point of view of their sustainability. Further it was possible to see and analyse where the deviations (accentuated or attenuated by the QFD) of each biogas plant was. This was possible through a prioritisation of the consumer demands done by experts in the different areas of AD. Figure 9.3 shows the deviation of the "expectations versus facts" of the interviewed AD operators, calculated on basis of the QFD.

This is of relevance, as it enables to give these demands a weighing which classified them in comparison to each other. This adjustment, which was done to the QFD structure and which is part of the QFD philosophy (involvement of all relevant parties) leads to the mentioned attenuation or accentuation of the indicators. This finally shows the modified deviations, which represent a holistic view of the facility, which is its sustainable functioning. Slight deviations that are not that obvious in the questionnaires come immediately clear, so that it is possible to see in short time where the operator misjudged a situation. An interesting point is that all interviewed operators misjudged the same aspects.

The QFD model showed not only where the (factual) deviations were, but furthermore stressed the deviations of the most relevant indicators (=factors). A thorough analysis of each interviewed biogas plant operator showed that actually negatively marked indicators are the reason for the discontent of the operators, although this was never made clear during the actual interviews. Within these case studies, the following came clear: several factors were so strong that they could completely block the implementation of a biogas plant in the agricultural sector – existence of a natural gas grid, no possibility for selling or utilising the residual heat.

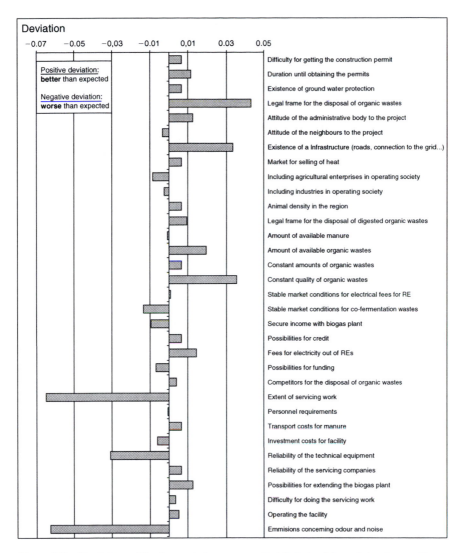

Figure 9.3   Deviation of the "expectations versus facts" derived from interviews with AD plant operators. This matrix was used to test out the SIT.

These are the so-called "kill-factors", that under normal circumstances can hinder the successful implementation no matter how good the other peripheral factors are.

Figure 9.4 shows a screenshot of the first dialogue window within SIT. The gained findings were integrated in the QFD method. That itself was integrated in a software tool that is a prototype of the decision support tool SIT. One of the reasons to integrate the adapted QFD methodology into a decision support tool was to make its usage more customer (user) friendly.

In order to adapt it to the needs of the end-users three main versions (or levels) have been created: the standard version, the short version and the experts (full)

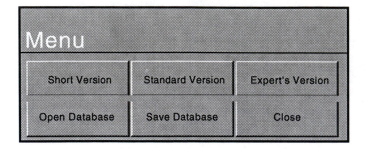

Figure 9.4   Screenshot of the first dialogue window within SIT.

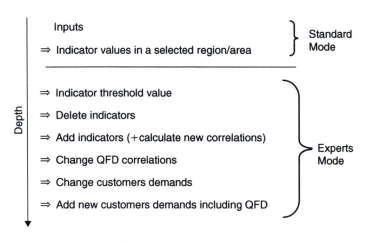

Figure 9.5   Options for using SIT.

version. The main difference between the versions is the degree of knowledge the user has as well as the amount of statistical data needed. The software provides the selection of the most promising regions/areas in which the possibilities of implementing successfully a biogas plant are the highest. It additionally provides a basis for an assessment of AD facilities and projects in a planning phase as well as a comparison of existing AD facilities keeping their sustainability in mind.

9.3.3   Options for SIT

Figure 9.5 shows the options for using SIT: the deeper the changes, the more individual will be the results of the tool. The selected approach shows that it is possible to adapt planning and design methods used in a completely different area (as the automotive is) for their use in the area of AD. This enables the utilisation of a "tool set" for planning and assessing, which has been in use for many years and is therefore fully developed and validated.

This first step to the decision support tool SIT has a realistic chance to have a positive impact on the sustainable integration of AD facilities in both the agricultural as well as in the industrial areas. SIT can further provide the welcome and needed link between these sectors. AD is of important relevance, especially for the rural area, as it can involve the industrial sector and give the agricultural AD sector an improved competitiveness by increased income. Therefore AD can additionally boost the local rural economy in AD development by opening a new employment market. This has been predicted by an ALTENER funded project, which expectations have been confirmed also by SAFIRE (new net full time employment): they both foresee a potential of 900,000 jobs (the job creation is expected to occur in all member states). This stresses therefore the importance and relevance of an objective decision support tool as SIT. It can contribute to achieving these ambitious aims.

By utilising tools as SIT, a positive effect on the sustainable implementation of AD facilities can be achieved. SIT should be supported by other measures that might be more AD operator orientated: seminars and similar activities should be standard in the AD area. This enables the operators to improve their knowledge on all topics related to AD, and would lead to an improved operating of the facility, less peripheral problems as odour emissions and at the same time improve the environmental impact.

9.4   CASE STUDIES

A QFD tool for AD integration was developed and tested in Ireland and Austria, as part of the European project "*Holistic integration of renewable energy*" (HOORAY). The clue for the application of QFD within the renewable energy (RE) area is the capability of combining the demands of the "customers" with the situation of the "suppliers". This means all factors (customer demands) can be taken in one single process into account, what was in the past not usual due to the one-sided planning process. With the QFD method social, technical and economical aspects are all being considered and taken into account. Like this and with certain adaptations an optimal region for the successful implementation of REs can be found having the certainty that all involved stakeholders will be satisfied. The work done on the QFD enables a smooth transition from the planning phase to the assessment phase. The adaptation and integration of QFD into a systematic decision support tool was based on an EXCEL surface, and virtual basic application (VBA) was used for the programming.

9.4.1   QFD-evaluation in Ireland

The QFD was tested in three phases:

Phase 1 – National survey using available statistical data
Phase 2 – Survey of Co. Kilkenny using available statistical data
Phase 3 – Survey of Freshford Village using data collected on site.

The first phase of this survey covered all the country. The objective was using available statistics to identify the most promising county in the country for the development of biogas plants. Ireland is divided politically into counties, each of which has its own local administration. Table 9.1 gives the range of data collected

Table 9.1   Agricultural relevant data range collected in Ireland.

Population	Population density per square kilometre for each county. The population density was then zoned per county.
Cattle	Cattle density per square kilometre for each county. The cattle density was then zoned for each county.
Pigs	Pig population per square kilometre per region. The pig density was then zoned per county.
Poultry	Poultry population per square kilometre per region. The poultry density was then zoned per county.
Farm size	Farms >50 ha per county. The counties were then zoned on this parameter.
Farms >50 ha	% Farms <50 ha per county were calculated.
GVE per hectare	This was calculated for each county. Counties were zoned in GVE per hectare.

in Ireland per county for statistical analyses using QFD. The data was interpreted using QFD tool and Co. Kilkenny was the county with the most suitable parameters for the development of a biogas plant.

The objective of Phase 2 was to select a part of the county that is best suited to the development of a biogas plant, for example, an area of say 20 km². While there are no administration areas within Co. Kilkenny, there are 113 District Electoral Divisions (DED). The statistical data was gathered on these 113 DED's with the intention of repeating the range of data that was collected in the National survey. All local bodies were requested to provide statistical data. The data collected gave statistical information on similar variables as above. QFD evaluation was applied to this statistical data and the following results were obtained. The report on Co. Kilkenny was forwarded to the Local Authority in Co. Kilkenny and also to the Local Leader Group B.N.S. Development Ltd.

Phase 3 was not planned, as the original intention was only to carry out Phase 1 and Phase 2 of QFD in Ireland. However, it came to light that the Village of Freshford, one of the top 10 areas for biogas identified in Co. Kilkenny using QFD has already a local development group (*Freshford 20/20*) actively considering biogas digester of farm and agro-industrial and sewage as an answer to some of their environmental and development problems. A number of meetings were held between the local group, *Freshford 20/20*, the local leader group, Barrow Nore Suir Rural Development Ltd and Kilkenny County Council staff. It was decided that under the direction of the HOORAY team, a pre-feasibility study of the construction of an AD Plant in Freshford would be carried out.

The questionnaire forms, as designed for use in Austria were modified and used in Ireland and a random sample of the population in Freshford were asked to complete the forms. A total of 43 forms were completed and these included from the following categories: administrative, neighbours, heat users and farmers who would supply organic material, industrial suppliers of organic material, farmers who would wish to construct on-farm biogas plants. Thus, the selection of Freshford as one of the top 10 locations in Ireland for AD development using the QFD tool has been shown to be very accurate as independently the local group *Freshford 20/20* had decided themselves that AD was a necessary facility for their area.

9.4.2   QFD evaluation in Austria

In Austria, unlike Ireland where the study was made on national level, the study area was confined to the Mühlviertel area and in particular to the areas supplied by a specific Power Company. In this area there are 83 municipalities. The study was carried out in two phases.

Phase 1 involved a survey of all 83 municipalities using available statistics. From the 83 original municipalities, the various indicators were obtained and included, among these, acceptance level among those responsible and potential project operators, or also the settlement structures and their potential as possible location for the production and for the consumption of heat. The statistical study using QFD produced a list of the 10 best municipalities and also the 10 following municipalities. The results of this first phase of the survey was that 13 municipalities were shortlisted and two further municipalities, which independent of the QFD survey, showed an interest in biogas were also taken into account. Figure 9.6 shows the construction of one of the core elements of a QFD-study: the so-called QFD-matrix. This matrix relates consumer requirements to product fulfilments. In our case, the "consumer" is the set of regional actors and stakeholder involved in an AD implementation project.

The "product" is the region, where such a facility should be or is planned to be implemented. Product fulfilments therefore are indicators related to the regions and reflecting the customer needs. For example, a constant supply of material could be reflected by the number of (expected) growth rate of farms with animal husbandry. The acceptance of an AD facility could be reflected by the inverse of population density. On a more work-intense level of research, the latter demand could also, or even better, be grasped with a specific survey designed for "neighbours". The HOORAY project has established such a list of demands and indicators. The QFD method has the potential to assign different weights to these demands, dependent on the type of customer, and thus identify the most appropriate region as well as the most critical factor for each region.

Phase 2 of the survey involved discussions with the mayors of the municipality by telephone. This was followed up by visits with the mayors who agreed to the meetings. The information sheet and the face-to-face discussion with mayors ensured that the principal of AD and an RE were fully understood and were passed on to the citizens. The conclusions for the integration of biogas was as follows:

- The construction of a biogas plant is to be carefully considered, for example, its effect on the landscape.
- Similar considerations must be made with the plant operator, agreement must be assured on the operational parameters.
- The argument of the improvement of air quality through the removal of digested slurry can be used with considerable public relations effect.
- While it is possible to achieve a positive effect for groundwater through the removal of digested slurry, this must on all accounts be made use of since there is a powerful public conscience on this issue.
- On the other hand in discussions with interested parties, no secret must be made of the fact that the nitrogen content of digested slurry is not at all lower than in the undigested.

Mühlviertel special indicator set comparison of potential regions
What customer's demands

Column headers (read vertically, left to right):

- Legal necessities for waste disposal
- Legal necessities for water protection
- Obligation to join distance heating
- Municipality with Climatic Alliance of Agenda 21
- Overnight stays in accommodation per km² per year
- Percentage of greens in local council
- Inhabitants per enterprise
- Median income per month
- Employees per enterprise
- Low number of farms necessary for 1000 GVE
- Share of animal farms with slurry storage tanks of at least 50 m³ storage capacity
- Large animal units (GVE) per community
- Large animal units (GVE) per farm
- Co-ferment availability
- Agricultural land per farm
- Large animal dung unit (DGVE) per ha agricultural land
- Low median income per month
- Unemployed per 1000 inhabitants
- No biogasplant in commune in neighbour commune
- Gas connections in commune
- No gas connections in commune
- Population growth rate in the last 10 years
- Population density
- Planed housing estates
- Existing and planed Industrial- and Technology estates
- ADDED CORRELATIONS

What customer's demands	Experts' priority	level 3 adapted to 2	ADDED CORRELATIONS
Permit is easy to get	2.1	2.1	0
Good legal conditions	2.4	2.4	0
Quick granting for the plant	1.7	0.6	18
Existing problems with the disposal of upcoming waste	2.6	1.0	21
Existing problems with the protection of ground water	2	0.8	15
Support by location	2.7	2.7	0
Positive attitude of the regional administration	2.4	0.3	30
Positive attitude of the neighbours	2.3	0.3	3
Positive attitude of the contractors	2.7	0.4	9
Positive attitude of the (end-)customers	2.4	0.3	9
Positive attitude of the visitors	1.6	0.2	18
Positive resonance of supraregional politics	2.4	0.3	24
Positive resonance of regional politicians, district mayor, local "inn"	2.6	0.4	30
(Informal) Integration of neighbours, contractors, customers	2.9	0.4	0
Positive structural circumstances	2.4	2.4	0
Appropriate site	2.6	2.6	0
Within easy reach	2	0.5	12
Good business location, heat consumption, laws	2.6	0.6	63
Appropriate current conductions, grid	2.7	0.7	27
Little competition with further own provided forms of energy	2.1	0.5	0
Connection to industrial gas pipeline possible	1.4	0.3	9
Good economic circumstances	3	3.0	0
Sufficient supply with raw material	3	3.0	0
Sufficient quantities of raw material	3	1.7	45
Constant quality	2.4	1.3	12
Good market	2.9	2.9	0
Constant income and consumption	2.7	0.7	33
High income	2.9	0.7	33
Secure income	2.9	0.7	30
Little expense (holistic)	2.9	0.7	9
Support on-site	2.7	2.0	0
Financial promotion	2	2.0	0
Small competition	2.1	2.0	27
Good potential for expansion and development	2.1	2.1	0
Rising demand for heat (thermoelectric energy)	2.3	2.3	0
Better client connection	1.6	0.6	3
Secure market share, raw materials, heat consumption, crowding out competition	2.4	0.9	21
Possibility of marketing the area	2	0.8	21
TOTAL	91		516

Column totals (bottom row, left to right): 15 | 15 | 48 | 12 | 9 | 12 | 21 | 24 | 18 | 3 | 6 | 9 | 12 | 3 | 63 | 21 | 21 | 3 | 12 | 9 | 21 | 18 | 45 | 36 | 51 | 516

Figure 9.6   QFD-matrix for the Mühlviertel area (Austria).

- It should be considered whether the contracted removal of slurry (slurry banks) should be pushed through with greater energy. In combination with biogas this could help to solve the groundwater problems of those questioned.

From the results of the questionnaire, some possible locations for biogas plants were suggested. It was remarkable that in some of the QFD-selected municipalities farmers and food industries already have taken initiative to implement AD facilities, although not succeeding yet.

9.4.3   QFD evaluations for biogas-driven fuel cells

QFD has been applied within the project "*Holistic integration of MCFC technology towards a most EFFECTIVE systems compound using biogas as renewable source of energy*" focusing on the integration of biogas-driven stationary fuel cells. Fuel cells are regarded as an efficient technology to produce electrical energy and (as a co-product) high-temperature heat. Other fuel cell features are low maintenance and related costs and low noise in operation. Combining fuel cells with biogas as an energy carrier, a decentralised RE supply system will be created, applicable in various situations, for example, for autarky. However, these features are somehow forthcoming. Although the technology is still improving, the costs of fuel cells are still high in comparison to a standard CHP. The present and the future application range therefore differ from an AD facility with a standard CHP: the heat could be more likely used for industrial processes rather than district heating, the control system must be more secure in order to capture the efficiency gains, and there should be other positive external effects as a co-product of the fuel cell integration. Examples for that are opportunities for research and development (R&D), for demonstration and training, and the related image transfer to sponsoring, promoting and operating enterprises and institutions. These external effects will be important promoting arguments for at least the next years, when fuel cells are still innovative.

The QFD system of the HOORAY project (1998–2001) was used as a starting point for improvements towards the fuel cell component. The range of customer demands had to be enlarged and other weights had to be applied. Furthermore, the situations in different countries (Austria, Spain, Slovakia) have been encountered. In Spain, different scenes have been analysed: food industry, communal services for wastewater treatment and solid waste disposal, large agricultural farms (pigs). Five target groups have been addressed:

(a) environmental experts,
(b) the food industry,
(c) city councils as landfills owners,
(d) sewage plant owners and operators,
(e) large agricultural facilities.

Six hundred questionnaires were sent out, 83 of them could be recollected. Out of these, 56 corresponded to specific sites (all except the target group of the environmental experts). QFD found out preferences within the observed 56 sites, but not between the groups. A more broad marketability study (Jaensch 2000) was made, evaluating the market potential, the external environment, the effects on related industries, costs and benefits as well as future scenarios. An important finding has been that there is a market potential of the fuel cell/biogas technology of 7.5 billion Euro in Spain. As a precondition, fuel cells have to be able to add more value to the conversion of biogas than internal combustion engines. The driving forces for dissemination will be R&D results to secure the expected features. Multinational corporations controlling the development of fuel cells will be the dominating market players, whereas the biogas technology is mostly nationally based. On the other hand, a dependency on political support during the transition period was discovered. This may be argued by international agreements (Kyoto protocol) as well as benefits for national small, medium and large sized companies.

9.5   CONCLUSIONS

The adapted QFD tool is able to provide an accurate statement concerning the sustainable integration of AD facilities. The approach enables the combination of all factors including the social aspects. This gives the base for a holistic view for the implementation of AD in the different sectors. The advantage of the followed approach is the possibility of finding out which factors and barriers can be a problem for existing AD facilities or for those that are in a planning phase. The SIT and the QFD method can be adapted to other sources of RE as well.

ACKNOWLEDGEMENTS

This work has been carried out within the EU Energy Marie Curie Fellowships scheme (TMR) and the EU projects HOORAY and EFFECTIVE.

REFERENCES

Akao, Y. (1990) *Quality Function Deployment – QFD – Integrating Customer Requirements into Product Design*. Productivity Press, USA.
Akao, Y. and Mazur, G. (2003) The leading edge in QFD: past, present and future. *Int. J. Qual. Reliabil. Manage.* **20**(1), 20–35.
ALTENER Project *The Impact of Renewables on Employment and Economic Growth*. Contract No. 4.1030/E/97/009.
Baaske, W.E. and Lancaster, B. (2004) *EmplocTool – Evaluating Local Commitment for Employment – Towards a Realization of the European Employment Strategy*. Linz 2004.
C.a.r.m.e.n. (1998) Biomass for energy and industry. *Proceedings of the 10th European Conference and Technological Exhibition*, Würzburg, 1–11 June 1998.
Council Directive 1999/31/EC on landfill waste. http://europa.eu.int/eur-lex/pri/en/oj/dat/1999/l_182/l_18219990716en00010019.pdf
Council Directive 91/676/EEC (Nitrates Directive) on the protection of waters. http://europa.eu.int/comm/environment/water/water-nitrates/91_676_eec_en.pdf

EFFECTIVE (2000–2004) *EU RTD Project: Holistic Integration of MCFC Technology towards a most EFFECTIVE Systems Compound Using Biogas as Renewable Source of Energy*. Contract No. ERK5-CT-1999-00007.

Graf, W. (1995) *Biogas für Österreich, 3*. Aufl., BMLFUW, Wien.

HOORAY (1998–2001) *EU RTD Project: Holistic Integration of Renewable Energies*. Contract No. JOR 3CT 980264.

Jaensch, V. (2000) *Marketability Analysis for an Innovative Fuel Cell/Biogas Technology in Spain*. MBA Thesis European Management. South Bank University, London.

Kano, Noriaki, Seraku, Takahashi and Tsuji (1984) Attractive quality and must-be quality. *Hinshitsu* **14**(2).

Mazur, G., Glenn, H. and Zultner, R.E (2003) *QFD Black Belt® Workshop*. Orlando, 15–19 December.

Rathbauer, J. (1992) Technische Verwertung und Aufbereitung von Gülle. *BLT Wieselburg Journal* 32/May.

Saatweber, J. (1999) *Kundenorientierung durch Quality Function Deployment*. München Wien, Hanser.

TMR Marie Curie Training Grant (1999–2000) *Sustainable Integration of Renewable Energies – Evaluation of the Quality Function Deployment Methodology towards Sustainable Biogas Integration*. Contract No. JOR3-CT98-5010.

Section IIC: Alcohol production

10

Fermentation of biomass to alcohols

F. Haagensen, M. Torry-Smith and B.K. Ahring

10.1 INTRODUCTION

This chapter summarizes the basic concepts in the research and development of bioethanol production based on lignocellulosic materials. The chapter seeks to present the different areas enclosed in the conversion of lignocellulosic materials into bioethanol and is as such not intended as a complete review of the abundant research done on the different aspects in bioethanol production.

10.1.1 Oil and energy reserves

… our fuels were produced millions of years ago and through geological accident preserved for us in the form of coal, oil and gas. These are essentially irreplaceable, yet we are using them up at a rapid rate. Although exhaustion of our fossil fuels is not imminent, it is inevitable.

Citation by Daniels, 1955 (Klass 1998a).

The consumption of fossil fuels has increased as a result of the industrialization throughout the 20th century, and although the above citation dates back to 1955

(Klass 1998a), the inevitable depletion of fossil fuels is still somewhat neglected. The exponential expansion of the human population and global industrialization creates worldwide problems such as pollution, increased energy requirements and food shortages. The worldwide consumption of oil has been increasing at a rate of 2.2% since the beginning of the 1990s. In 2000 the total consumption reached 75,600 million barrels per day, of which the USA alone uses one fourth (Birky *et al.* 2001). Leaning against the most optimistic assumptions about the remaining fossil fuel reserves combined with a prediction of low economic growth, the US Department of Energy (DOE) (Birky *et al.* 2001) estimates that half of the world's oil reserves will be depleted by 2040. However, based on the proven oil reserves a complete depletion could become a reality even sooner – as stated by Klass (1998a). With an estimated average growth rate in the consumption of petroleum at 1.2% (projected by US DOE until 2010), the proven oil reserves will be depleted by 2027. Evidently the future increase in energy consumption by the World's developing countries (Goldemberg 1998), as well as the rather unpredictable depletion of our oil reserves strongly emphasize the fact that alternative sources of energy need to be identified, developed and commercialized before depletion of the non-renewable energy pools occurs.

In 1999 Former US President, Bill Clinton, signed an executive order to increase the production of bio-based products and bioenergy by threefold before 2010. If these goals are met, the expected impact is an annual reduction in the use of fossil fuel of 4 billion barrels of oil (Taherzadeh *et al.* 1997b).

10.1.2 Renewable biofuels in transportation

The development of large amounts of domestically produced, clean burning and CO_2-neutral fuels is, according to Wyman (1996), important in the transportation sector if environmental and economic issues are to be addressed. The transportation sector contributes to a significant part (20–30%) of all CO_2 emission and projections are that the CO_2 emissions will increase even further: 17% by 2017 and 30% by 2050 (Fulton 2002). This has led to an increased focus from international organizations and national governments on ways to reduce the environmental impact from an ever-increasing transport sector.

In 2002, the European Union (EU) issued Directive 2003/30/EC *on the promotion of the use of biofuels or other renewable fuels for transport* as a mean to reduce the CO_2-pollution from the transportation sector. The EU Directive invites the Union's member countries to replace the fossil fuels by liquid biofuels with biodiesel and bioethanol as the main constituents. The Directive recommends an increase of biofuels to 2% in 2005 and 5.75% in 2010 of the total consumption of transportation fuels. Promotion of bioethanol has also been proposed in other parts of the world (e.g. USA and Brazil) – resulting in a significant increase in the requirement for bioethanol in the near future (Lichts 2001).

10.1.3 Benefits of fuel ethanol

Ethanol produced from biomass is an essentially unlimited renewable source of energy with a great potential of becoming the "green" liquid fuel of the future. In the USA, the production of ethanol has grown from a few million gallons in the 1970s to more than 1700 billion gallons in 2001 (Shapouri *et al.* 2002). In the 1980s, ethanol was introduced in the transportation sector by the Environmental Protection Agency (EPA) as an octane booster, as substitution for the undesired use of lead. In 1990, when the Clean Air Act Amendment was signed, the demand for fuel ethanol increased even further (Shapouri *et al.* 2002).

Using ethanol as a transportation fuel has a number of advances over gasoline. Ethanol produced from biomass is a CO_2-neutral energy source because the amount of carbon dioxide released at combustion corresponds to the amount originally fixed from the atmosphere by the biomass. A 10% substitution of ethanol in gasoline will decrease the net CO_2 emissions from the transportation sector by 10% – a major improvement that is easily implemented because normal combustion engines can use this blend without any adjustments. Except for the CO_2 savings obtained by implementing ethanol in the transportation sector, the combustion of ethanol gives lower emissions of nitrogen oxides, carbon monoxide and sulfur dioxides. As oxygenate in gasoline blends, ethanol ensures a more complete combustion, especially in old vehicles (Wyman 1996). The improved combustion decreases tail pipe emissions of carbon monoxide and the content of unburned hydrocarbons, which are constituents in ground-level ozone and particulate matter pollution (smog) (Wyman 1996).

Additional indirect benefits of introducing bioethanol are improvement and stabilization of the agricultural sector, creation of jobs in rural areas as a result of an agricultural boost, and reduced dependency of fossil fuel import (Wheals *et al.* 1999).

10.2   BIOETHANOL

10.2.1   History

Alcohol is reproduced in the cycle of the seasons; it is absolutely inexhaustible; it is made out of sunshine and air; and its composition does not lessen the value of the soil or the energy of the earth. Gasoline, on the contrary, represents a part of the stored energy of the earth; it exists only to the extent of about two percent in petroleum, and its supply, will in the future inevitably fail.

Citation by Duncan, 1907 (Klass 1998b).

The history of using ethanol as a fuel extender in gasoline dates back to the development of the internal combustion engine by Otto and Benz in the late 19th century. The first peak in production of fuel ethanol was seen all over the world, except in the USA, in the period following the First World War (1920–1930) where several countries passed legislations that supported the use of gasoline–ethanol blendings. However, as gasoline prices were significantly reduced during the 1930s most countries abandoned the legislations and fuel ethanol programs ended prior to the Second World War (Klass 1998b). The significant fluctuations in the oil prices during 1970s and early 1980s resulted in the launch of at least two major fuel ethanol programs: ProAlcool in Brazil and the US Fuel Ethanol Program – Gasohol (Hall *et al.* 1992). The programs were mainly launched to reduce the dependency on fossil fuel in the countries where import was the main option of oil supply. But also stabilization and further development of the agricultural sector were the reason for intensifying focus on biomass-based fuel alcohol.

As the history shows, the price of gasoline controls the production of fuel ethanol and will continue until technologies are developed in which fuel ethanol can be produced at a cost similar or lower to that of gasoline.

10.2.2   Bioethanol production

Production of bioethanol from lignocellulosic biomass has been attracting increased attention from a wide number of organizations, private companies and governments acknowledging the future potential of this renewable fuel. Currently in the USA, a subsidy has been given by the government to promote ethanol production, but by 2007 the financial support expires and the technology of producing bioethanol from biomass must have matured so that it can compete with other fuels on the fuel market (DiPardo 2000).

10.2.2.1   Conventional ethanol

Sugarcane, corn and grains comprise the main crops utilized in today's biomass-based ethanol industry for fuel alcohol. Sugarcane is primarily used in the Brazilian Fuel Program, whereas corn is the prime biomass material utilized in the USA and Canada (Figure 10.1).

Production of ethanol based on primary crops is a widespread technology at a high maturity level. However, a problem of using primary crops is that these biomass materials often are subject to competition for other production processes, which can lead to unstable prices. According to Wyman (1999), the world's sugar prices highly influence Brazilian ethanol production, resulting in supply problems and fluctuations in the production and prices. Furthermore, the fact that starvation and hunger are found around the world opposes the use of huge areas of farmland for production of crops solely for the purpose of producing energy. Growing crops

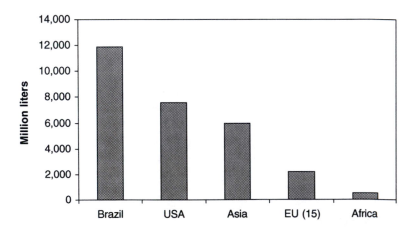

Figure 10.1   Ethanol production (in million liters) of selected countries and regions in 2001. Adapted from numbers reported by Lichts (2001).

exclusively for energy production can therefore only be considered as a non-sustainable temporary solution. Critics argue that the environmental problems caused by pesticides and fertilizers applied for production of crops for "green" energy must be included in the assessments of the sustainability of energy crops. The political and public acceptance of environmental hazards created from chemicals used in the production of food crops should not be taken for granted when looking at the production of biomass primarily for energy purposes.

Finally, it is important to realize that conventional ethanol has had a positive impact on the public awareness of this alternative fuel, and on developing an effective infrastructure for distribution. However, future projections predict a significant increase in the demand of fuel ethanol production induced by legislatives, directives, tax exemptions and international agreements. This necessitates an expansion of the existing production capabilities, the introduction of new biomass substrates and the development of commercially viable production technologies.

10.2.2.2   Lignocellulose bioethanol

A strong alternative to dedicated production of energy crops is the use of residual lignocellulose (waste) created from traditional primary production of food crops or forest industries. Using the newest technologies developed through a collaboration of scientists around the world, the lignocellulosic waste can be converted into bioethanol, biogas and other renewable energy carriers. Examples of lignocellulosic waste are listed in Table 10.1.

According to Himmel *et al.* (1997), the lignocellulosic waste and energy crops held in 1997 in the USA alone, there is a potential of 100,000 million gallons of ethanol. The global production of plant biomass, of which 4–10% is recoverable, is estimated to 200×10^9 ton/year (Szczodrak and Fiedurek 1996).

In order to make bioethanol from lignocellulosic biomass economically feasible, the production processes have to be more advanced than those found in the production of corn ethanol. To keep the production price at a minimum, it is crucial to apply technologies that enable an almost complete utilization of the biomass resource. For this reason, proposals of advanced concepts for ethanol production from lignocellulosic material seem to be based on the philosophy of maximizing

Table 10.1   Examples of lignocellulosic biomass material (primary crop, product or waste material) that can be used for bioethanol production.

Primary crop or product	Biomass residue
Corn	Corn stover, cob, etc.
Sugarcane	Bagasse, barbojo*
Grain (wheat, barley, etc.)	Straw
Rice	Rice straw, hulls, etc.
Green coffee	Husks (wet or dry)
Softwood (spruce, pine, etc.)	–
Hardwood (willow, aspen, etc.)	–
Energy crops (switch grass, etc.)	–

*Tops and leaves of sugarcane.

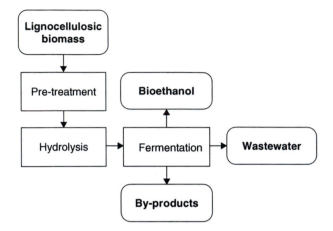

Figure 10.2   General outline of lignocellulose-based bioethanol production.

biomass conversions (Lynd *et al.* 1996b; Torry-Smith 2002). Although convincing facts for using bioethanol as a main transportation fuel exist, the political life and support for this alternative fuel is unpredictable (Himmel *et al.* 1997). To make the future of the lignocellulose bioethanol less sensitive towards political changes, focus has been on developing technologies that will enable the price of lignocellulose-based ethanol to compete against un-subsidized gasoline prices.

10.2.2.3   *Production concepts: lignocellulose*

The processes needed to convert lignocellulose to ethanol are comparable to those used in the commercial bioethanol production. However, the energy input, bio-mass complexity, enzymatic loadings and the diversity of sugars all affect the overall production cost of lignocellulosic bioethanol. The latter ultimately deter-mines the development of economically feasible technologies.

A general outline of the lignocellulose-to-ethanol production concepts is pre-sented in Figure 10.2. In order to convert the readily in-accessible sugars in the lignocellulosic biomass, disruption of the fibres are required in order to segregate the different fractions. Cellulose and hemicellulose holds the potential for ethanol production while lignin functions as the structural compound that also protects the fibres in contact with water from swelling, hereby getting vulnerable to microbial attack. Different concepts for bioethanol production from lignocellulose have been proposed (von Sivers and Zacchi 1995; So and Brown 1999; Wooley *et al.* 1999; Wyman 1999). All of these process set-ups face the problem of releasing the sugars from the cellulose and hemicellulose in an accessible form to microbial fermentation and to avoid lignin-related inhibition of the process, all at an acceptable price. Bioethanol is produced as the main product, but also various by-products have been suggested to add extra value to the overall process, potentially increasing the economic feasibility of the process.

In order to obtain a closed loop cycle of water, nutrients and un-utilized substrate, the reuse of the process wastewater has been assessed (Topiwala and Khosrovi 1978; Ahring *et al.* 1997; Larsson *et al.* 1997). As an example of the closed-loop approach,

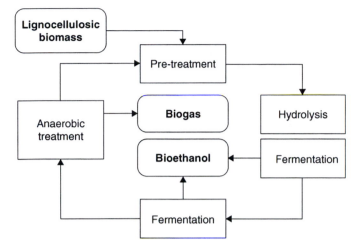

Figure 10.3   Simple outline of the DBC developed by the Technical University of Denmark (DTU) and RISØ National Research Laboratories (Haagensen 2004).

the Danish Bioethanol Concept (DBC) was developed in which purification of the wastewater was done using anaerobic treatment with the simultaneous production of methane as an additional energy carrier (Figure 10.3). The benefits of the anaerobic treatment and the DBC concept as such are given later on (see Section 10.6).

10.3   ETHANOL FROM LIGNOCELLULOSE

Lignocellulosic biomass waste from forest or agricultural biomass productions have a much higher potential of supply than dedicated food crops (Wyman 1999). Furthermore, these types of waste are rarely subject to competition and can therefore be considered a more price stable biomass substrate for ethanol production.

10.3.1   Lignocellulose

Lignocellulosic fibres are composed of three major fractions: cellulose, hemicellulose and lignin – typically distributed as 35–50% cellulose, 20–30% hemicellulose and 12–20% lignin (Wyman 1999). The lignocellulose also contains minor amounts of proteins, lipids and extractives (Sommer 1998). The cellulose and hemicellulose (together called holocellulose) comprise the sugars that can be converted into ethanol by fermentation. The main components of lignocellulose are depicted in Figure 10.4. The combination of the different fractions in lignocellulose results in a complex structure that protects the plant fibre from physical wear and microbial attack.

Cellulose consists of glucose units connected by β-1,4-glycoside bonds. Cellulose is closely linked to hemicellulose which is a heteropolysaccharide mainly consisting of pentose sugars. The characterization of hemicellulose is normally described by the dominating pentose sugar (typically xylose), oriented as a backbone with side chains containing different sugar substitutes (e.g. arabinose, glucose, galactose) (Sommer 1998).

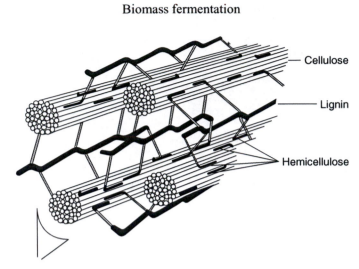

Figure 10.4   Illustration of the main components of the plant cell wall.
From Shleser (1994).

Lignin is the most complex compound in lignocellulose. It is a greatly branched high-molecular-weight heterogeneous compound based on a phenol–propene backbone (Wyman 1999). It is the tight association between lignin and the cellulose and hemicellulose that prevents fibres from reacting with water to swell, that conveys the protective effect of lignocellulose towards microbial and enzymatic attacks (Skaar 1984).

The use of lignocellulose as substrate for bioethanol production requires removal of lignin in order to liberate the cellulose and hemicellulose fractions to be accessible for fermentation.

10.3.2   Pre-treatment of lignocellulose

Pre-treatment is an important part of the lignocellulose-based ethanol production because the accessibility of the sugar fractions is significant for the ethanol yields and therefore for the overall process economy (Hsu 1996; von Sivers and Zacchi 1996). The pre-treatment basically hydrolyses the lignocellulosic material to depolymerize the components in order to achieve small compounds that are accessible for microbial conversion (Tong and McCarty 1991).

Pre-treatment can be carried out with a range of different processes based on a chemical, physical or biological approach, or a combination of these. According to Sun and Cheng (2002), it is important that the selected pre-treatment technology includes a number of requirements:

* The pre-treatment of the biomass should improve the enzymatic accessibility of the sugar compounds.
* The pre-treatment conditions should result in a minimum of degradation and loss of potential sugars.

- The technology should prevent the formation of compounds that are inhibitory towards enzymatic and microbial activity.
- The process must be economically sound in order to make the total lignocellulose-based ethanol production economically feasible.

Different concepts for pre-treatment have been proposed (McMillan 1994; Sun and Cheng 2002). These technologies must be low-cost processes that release the sugars in an accessible form for microbial fermentations, but at the same time minimize the release of inhibitory compounds, which are formed during the depolymerization of the lignocellulose. The physico-chemical pre-treatment technologies for bioethanol production can roughly be divided into two: acid-catalysed and alkaline processes. Advantages of acid-catalysed over alkaline pre-treatment are the higher solubilization and hydrolysis of the hemicellulose part of the lignocellulosics (McMillan 1994; Sun and Cheng 2002). From acid-catalysed steam explosion, the solubilization of hemicellulose was found to be more than 90% of the theoretical yield (Hsu 1996; Nguyen *et al.* 1999). Alkaline pre-treatment such as the acid-freeze explosion (AFEX) technology (Holtzapple *et al.* 1991) and the recently developed alkaline wet oxidation pre-treatment (Schmidt and Thomsen 1998) only partly solubilize the hemicellulose. The acid-catalysed processes, however, generate pre-treated suspensions that are high in potentially inhibitory compounds to the subsequent fermentation processes, hence washing or detoxification is necessary. High amounts of potentially inhibitory compounds are normally not produced during the alkaline pre-treatments (Klinke *et al.* 2002).

The recovery of cellulose and hemicellulose, the enzymatic convertibility, and the formation of potentially inhibitory compounds are traditionally taken into account when evaluating the efficiency of the specific pre-treatment applied. The efficiency of alkaline pre-treatment technologies has been reported to be affected by the lignin content of the biomass (McMillan 1994). Higher efficiencies have been shown with agricultural residues than with wood materials (Hsu 1996).

10.3.3   Enzymatic hydrolysis

The hydrolysate generated from the pre-treatment is composed of cellulose (with reduced crystallinity) contained in an insoluble fraction and hemicellulose (partly dissolved as oligomers or monomers and partly associated with the insoluble fraction). In order to make the sugars fermentable, hydrolysis of the sugar polymers are required. The hydrolysis can either be done by acid or enzymatic hydrolysis. The enzymatic hydrolysis has been identified as the most appealing alternative in the hydrolysis of cellulose and hemicellulose following pre-treatment, mainly due to high conversion yields and no major disposal problems (Himmel *et al.* 1996). The conversion of cellulose to glucose has attracted most attention in the research conducted on enzymatic hydrolysis, as the polymeric and crystalline structure of cellulose is more resistant to the pre-treatment methods applied compared to the hemicellulose fraction. However, in order to obtain the highest ethanol yield possible, the liberation of xylose through enzymatic or bacterial hydrolysis has also been targeted in research (Beg *et al.* 2001).

In general, it is necessary to treat the hydrolysates enzymatically to liberate the sugars in the polymeric cellulose and hemicellulose fractions that are not released during the pre-treatment. The price of cellulase and hemicellulase enzymes constitutes a considerable cost to the production of ethanol from lignocellulosic biomasses. Even though they are not more expensive to produce than enzymes used in the corn-ethanol industry, the activity is considerably lower (Nieves *et al.* 1998). To effectively hydrolyse the cellulose, an enzyme mixture with a combination of endoglucanases, exoglucanases and β-glucosidases is required (Pushalkar *et al.* 1995). The enzymatic hydrolysis of hemicellulose involves a larger array of enzymes compared to the hydrolysis of cellulose, due to a more complex structure of the sugar polymer. Depending on the main component of the hemicellulose, the required array of hemicellulases includes: xylanases, arabinases, mannases, glucanases, galactosidases, acetylesterases, etc. Occasionally hemicellulases are supplemented during the enzymatic conversion; however, some of the cellulase mixtures commercially available also contain hemicellulytic activities (Brigham *et al.* 1996).

10.3.4   Enzymatic inhibition

The efficiency of the enzymatic hydrolysis has been related to the composition and structure of the pre-treated biomass suspensions as follows (McMillan 1994; Hsu 1996):

- Porosity/accessible surface area.
- Crystallinity.
- Lignin content.
- Hemicellulose content and its degree of acetylation.

Some of these factors are interrelated, but the most prominent single factor affecting enzymatic hydrolysis seems to be the lignin content of the pre-treated suspensions. This is mainly due to the protective nature of lignin within the biomass (McMillan 1994). It has also been argued that the removal of hemicellulose (which has been identified as an enzymatic inhibitor) increases the saccharification of cellulose during enzymatic hydrolysis (Grohmann *et al.* 1985).

The lignin-related compounds found in hydrolysates might also interfere with the enzymatic hydrolysis by binding irreversibly to the enzymes or by blocking the enzymatic access to the polysaccharides (McMillan 1994; Larsson *et al.* 1999). Strong irreversible adsorption of lignin compounds to the enzymes reduces the enzymatic activity by preventing the enzyme from interacting with the target compounds (Zilliox and Debeire 1998). Zilliox and Debeire (1998) showed that strong bindings of xylanases on the lignocellulose structure of wheat straw occurred within 30 min. A subsequent washing with water did not release the adsorbed inactive enzymes.

The effect of the factors was shown to be governed by the biomass composition and the pre-treatment method applied, as the enzymatic hydrolysis was found to be both substrate- and pretreatment-specific (Hsu 1996).

10.3.5   Microbial inhibition

Independent of the pre-treatment method, formation of compounds inhibitory to enzymatic and microbial activity is inevitable (Ando *et al.* 1986; Taherzadeh *et al.* 1997b; Palmqvist and Hahn-Hägerdal 2000b).

The inhibitors found in lignocellulose-derived hydrolysates are a combination of weak acids (like acetate and formic acid) (Larsson *et al.* 1999), compounds originating from the breakdown of sugars (i.e. 5-hydroxymethyl furfural (HMF) and furfural) (Lee *et al.* 1997) and a wide array of lignin-related aromatic compounds (Delgenes *et al.* 1996). The inhibitory compounds can result in lower ethanol yields, lower productivity and even addition of extra process steps that eventually will reduce the economic feasibility of the overall process. Both the enzymatic hydrolysis and microbial fermentability (and the inhibition hereof) have been found to be specifically linked to the biomass type and pre-treatment technology (Hsu 1996).

10.3.5.1   Weak acids

Weak acids constitute a problem especially in fermentations at low pH where the negative effects can be a major burden on the process economy (Lawford and Rousseau 1998). Inhibition arises because a large portion of the weak acids is at an un-dissociated form at low pH. This configuration permits the acids to pass through the cell membranes. At neutral pH inside the cells, the acids dissociate releasing protons that un-couple the proton motive force, thereby resulting in a drop in ethanol productivity (Larsson *et al.* 1999; Palmqvist and Hahn-Hägerdal 2000b).

10.3.5.2   Sugar degradation compounds

The sugar degradation products, HMF and furfural originating from the breakdown of glucose and xylose, respectively, impose inhibition of fermenting microorganisms (Lee *et al.* 1997; Ranatunga *et al.* 1997). A clear relation between the furfural and HMF concentration, and the decrease in ethanol yield has been shown (Taherzadeh *et al.* 1997a; Larsson *et al.* 1999). Beck (1993) showed that fermentation of an acid hydrolysate could only be successful if the furfural concentration was less than 0.5 g/l.

10.3.5.3   Lignin-related compounds

Lignin-derived compounds, especially the small phenolic compounds, are inhibitors of microbial conversion of lignocellulose hydrolysates (Ranatunga *et al.* 1997; Palmqvist and Hahn-Hägerdal 2000a). Due to the highly amorphous structure of lignin, only limited information is available about this type of inhibition. However, the inhibitory effects from aromatic compounds have been suggested to greatly depend on the functional groups attached to the benzene ring (Ando *et al.* 1986). In general, the smallest lignin compounds have been found to be the most inhibiting. Lignin compounds with high molecular weight (1000–3000 g/mol) are too large to penetrate the cell walls of the microorganisms and cause inhibition (Sierra-Alvarez *et al.* 1994).

10.3.5.4   Reduction of microbial inhibition

In order to reduce the inhibitory effects of the hydrolysates different, means of detoxification have been evaluated (Palmqvist and Hahn-Hägerdal 2000a). The detoxification methods can be categorized as: physical (e.g. evaporation), chemical (e.g.

overliming) and biological (e.g. laccase enzymes). However, the inclusion of a detoxifying process step have been found to increase the ethanol production cost by 22% using a recombinant strain of *Escherichia coli* in conversion of willow into bioethanol (von Sivers *et al.* 1994).

In addition to detoxification, improvements of the fermentative microorganisms have been assessed in order to reduce the inhibitory impact of pre-treated biomass suspensions. Different approaches have been used to improve the ethanol yields from inhibitory hydrolysates in order to enhance the overall ethanol production as well as the economic feasibility of lignocellulose-to-ethanol technologies. Some of these approaches are described later (see Section 10.4.3.3).

10.4   ETHANOL FERMENTATION

10.4.1   The ethanol production process

Lignocellulosic hydrolysates contain two categories of sugars: the six-carbon sugars (hexoses) and the five-carbon sugars (pentoses). The two dominating monomeric sugars in lignocellulosic hydrolysates are glucose from the cellulose and xylose from the lignocellulose (Hayn *et al.* 1993; Wyman 1996). The glucose and xylose can be fermented to ethanol by separate fermentation or by co-culture fermentation. A wide array of microorganisms is able to ferment the sugars enclosed in lignocellulose hydrolysates into ethanol. The desired traits in a microorganism for commercial ethanol production are broad substrate utilization, high ethanol productivity and yield, high inhibitor tolerance, and a high cellulytic activity. Furthermore, it is also desirable that the organism is resistant to process imbalances, which are inevitable at full-scale operations (Table 10.2).

In order to maximize the ethanol yield from the lignocellulosic substrates, as many of the sugars as possible must be converted into ethanol because the biomass has been identified as the main contributor to the overall bioethanol production cost (Wyman 1999). The research concerning fermentations in lignocellulosic bioethanol production has been focused on glucose and xylose utilization, as these sugars are the most abundant in lignocellulose.

Table 10.2   Essential and desirable traits of the fermentative microorganisms in lignocellulose-to-ethanol technologies. From Zaldivar *et al.* (2001).

Essential traits	Desirable traits
Broad substrate utilization range	Simultaneous sugar utilization
High ethanol yields and productivity	Hemicellulose and cellulose hydrolysis
Minimal by-product formation	Comply with Food regulations (e.g. GRAS, generally regarded as safe)[*]
High ethanol tolerance	Recyclable
Increased tolerance to inhibitors	Minimal nutrient requirements
Tolerance to process hardiness	Tolerance to low pH and high temperature

[*]As defined by the US Food and Drug Administration (FDA) Agency.

10.4.2   Glucose fermentation

The most commonly used organism for ethanol production at industrial level is the glucose fermenting yeast *Saccharomyces cerevisiae* (Birol *et al.* 1998). The mesophilic yeast has many of the essential traits for ethanol production, being able to grow in high substrate and ethanol concentrations. Furthermore, the yeast has a growth rate that is relatively high compared to most anaerobic microorganisms (Banat *et al.* 1998). *S. cerevisiae* is a high-yielding organism and Lebeau *et al.* (1998) reports ethanol yields of 0.48 g-ethanol/g-sugar from a *S. cerevisiae* co-fermented with *Candida shehatae* in a medium containing both glucose and xylose.

Another candidate for hexose fermentation is the bacterium *Zymomonas mobilis* that reaches ethanol yields close to the stoichiometrical value of 0.51 g-ethanol/ g-sugar (Sommer 1998). None of the native strains of *S. cerevisiae* or *Z. mobilis* are able to convert xylose, but due to their high yields on glucose fermentation, they are preferred for fermenting this sugar in lignocellulosic hydro-lysates (Delgenes *et al.* 1996). Xylose fermenting yeasts like *C. shehatae* and *Pichia stipitis* are also able to ferment glucose, but have a lower conversion yield than dedicated glucose fermenting organisms like *S. cerevisiae* and *Z. mobilis* (see below).

10.4.3   Xylose fermentation

To maximize the utilization of raw material it is important to also ferment the pentose fraction in the hydrolysates (von Sivers and Zacchi 1996). One of the predominant pentose sugars derived from hemicellulose is xylose (Mohagheghi *et al.* 1998; Chandrakant and Bisaria 2000), which is the second most dominant sugar in wheat straw, corn stover, rice straw, birch wood and sugarcane bagasse (Hayn *et al.* 1993; Puls 1993; Ahring *et al.* 1996; Schmidt and Thomsen 1998; Nigam 2001). Thus, the conversion of the xylose in the hydrolysates of these biomasses significantly improves the overall ethanol yield and thereby the economic feasibility of the process.

10.4.3.1   Mesophilic xylose fermentation

The naturally occurring xylose fermenting yeasts *C. shehatae* and *P. stipitis* are not an optimal choice for commercial production of bioethanol, due to their requirement for exact oxygen levels during fermentation and their ability to re-assimilate the produced ethanol (Wyman 1996; Sommer 1998).

Other organisms for xylose fermentation are genetically modified strains of *S. cerevisiae* and *Z. mobilis* where genes coding for xylose conversion have been incorporated. Disadvantages of using these strains are high costs for safe discharge of wastewaters containing genetically modified organisms and the fact that the strains typically require detoxification of the hydrolysates in order to obtain satisfactory ethanol yields (Wyman 1996). Problems with rejection of plasmids carrying the xylose conversion genes have also been reported (Krishnan *et al.* 2000).

10.4.3.2   Thermophilic xylose fermentation

Thermophilic bacteria have been suggested as an alternative to the mesophilic strains. The use of thermophiles brings extra traits of advantage to the process feasibility. Ethanol fermentation at elevated temperatures has the advantages of high productivities and substrate conversions, low risk of contamination, facilitated product recovery, high-efficiency reactor systems and metabolization of a wide range of substrates (Lynd 1989; Sommer 1998; Torry-Smith 2002). Furthermore, the use of thermophilic strains in simultaneous saccharification and fermentation (SSF) reactor systems allows high-performance enzyme systems thus decreasing the reactor size. The prospects of using thermophilic processes have been emphasized by Mistry (1991), who estimated that the ethanol production cost of a thermophilic process with *Bacillus stearothermophilus* was 32% lower than a mesophilic process with *S. cerevisiae*.

The use of thermophilic bacteria in the conversion of lignocellulosic substrates, however, is still somewhat new and evaluation of its use in larger scale and in combination with other process steps needs to be further elucidated.

10.4.3.3   Improvement of fermentations

As mentioned previously, detoxification of the hydrolysates has been pursued as a way to improve fermentation yields and productivity. Additionally, different approaches have been assessed to improve the fermentative microorganism's ability to cope with inhibitory effects of the biomass hydrolysates. The different kinds of strain improvements include: adaptation, mutation and selection, and genetic modifications (Clausen 2000). In bioethanol production, most of the research conducted in improving fermentation has been focused on genetic engineering of *S. cerevisiae* and *Z. mobilis*, to include pentose utilization, or on the xylose fermenting bacteria, like *E. coli*, to reduce the formation of by-products (Clausen 2000; Zaldivar *et al.* 2001; Dien *et al.* 2003). Most of this research has been aimed at enhancing substrate utilization as well as the overall ethanol production. When strain improvement has been related to the content of inhibitory compounds in the hydrolysates, however, adaptation seems to be the most simple, less costly and less time consuming of these approaches.

10.4.4   Fermentation schemes

Many fermentation schemes exist for ethanol production, like batch and fed batch reactors and various types of continuous reactor systems (Sommer 1998). Of all reported fermentation technologies for ethanol production from lignocellulosic waste, the SSF is commonly regarded as the most promising technology from a process and economic point of view (So and Brown 1999). The simultaneous removal of the monomeric sugars by the fermenting organisms prohibits feedback inhibition of the enzymes and the gradual release of sugars prevents substrate shocks of the microorganism. However, the SSF technology also has some disadvantages. The most common organisms for fermentation like *Z. mobilis*, *S. cerevisiae* and *E. coli* all have an optimum growth temperature (30–37°C) below the optimal temperature for the enzymatic hydrolysis (45–55°C). This trade-off leads to an increased requirement for enzymes or longer hydraulic retention times.

10.4.5   Process water treatment

Production of bioethanol generates large amounts of wastewater that must be managed in an environmentally responsible and economically feasible way. But also the requirement of process water contributes to the total production costs and needs to be included in the overall evaluation.

Studies of bioethanol production plants having capacities from 40,000 to 700,000 ton of biomass input annually indicate that the ethanol price is inversely proportional to the plant size (von Sivers and Zacchi 1996). Plants of this size have a high requirement for process water. Hence, the reuse of the wastewater constitutes a potential saving on the production costs (Larsson *et al.* 1997).

10.4.5.1   Reuse of process water

One of the traditional approaches to handling process effluent water from bioethanol production is evaporation. This method is, however, highly energy consuming and the amount of live steam required is 3.6 times higher than the steam requirement for distillation, calculated per amount of ethanol produced (Larsson *et al.* 1997).

The continuous recycling of process water introduces the potential build-up of compounds inhibitory to the different fermentation processes. At a re-circulation rate of 80%, a compound that is not converted or removed will theoretically reach a steady-state concentration of 5 times the concentration in the inlet of the reactor. At 100% re-circulation, the theoretical concentration of the compound will not reach a steady state but will keep increasing. In order not to reduce the overall fermentation yields when process water recycling is used, it is therefore required that the compounds that accumulate are either not inhibiting or can be removed by microbial degradation, physical or chemical treatment.

10.4.5.2   Anaerobic treatment of bioethanol process water

The perspectives of incorporating a process water purification step in a bioethanol scheme for detoxification were studied by Torry-Smith *et al.* (2003). Treatment of bioethanol process water in a mesophilic upflow anaerobic sludge blanket (UASB) reactor showed that the methane consortium was able to remove 81% of the organic content (chemical oxygen demand, COD) in the process water at organic loading rates up to $29 \, \text{g COD.l}^{-1}.\text{day}^{-1}$. Furthermore, the purification step was assessed for the ability to remove a number of compounds identified as potential inhibitors in bioethanol production (Table 10.3). It was found that all of the tested compounds were removed to the analytical detection limit.

To investigate the long-term effect of a UASB purification step, it is required to perform a series of re-circulation experiments to get potential inhibiting compounds at low levels in the bioethanol effluent accumulated to a concentration that allows identification.

The methane produced in the purification step can be converted to electricity and heat for process purposes. The advantage from reusing process water and the profit from selling excess electricity can further improve the feasibility of the entire bioethanol production process (see below).

Table 10.3   Degradation of inhibitory compounds in UASB purification step (Torry-Smith *et al.* 2003).

Compound	Inlet* (ppm)	Outlet* (ppm)	Conversion (%)
2-furoic acid	5.6	0.0	100
4-hydroxybenzaldehyde	3.5	0.4	90
4-hydroxybenzoic acid	15.7	0.4	97
Vanillic acid	60.7	0.6	99
Homovanillic acid	25.1	0.7	97
Syringic acid	45.5	0.0	100
Syringol	7.4	1.0	87
Acetovanillone	5.4	1.1	79
Acetosyringone	28.1	1.0	97

*Measured by solid-phase extraction and gas chromatography.

10.5   ECONOMICS OF LIGNOCELLULOSIC BIOETHANOL

Corn- and sugarcane-based bioethanol production is a widespread technology at a maturity level that makes further research of optimization unlikely to noticeably reduce the production cost. Bioethanol produced from lignocellulosic biomass materials is a relatively new technology, which still is subject to massive research and development (R&D). Biomass waste holds a great potential if a suitable technology is developed, capable of producing bioethanol at a cost similar to the wholesale price of gasoline.

10.5.1   Production costs

A number of concepts have been economically evaluated focusing on the total production price of ethanol (Table 10.4). So and Brown (1999) evaluated three different production schemes, each designed to produce approximately 95,000 m^3 of ethanol per year. The three concepts were: fast pyrolysis and subsequent fermentation, acid pre-hydrolysis and SSF, and a dilute acid hydrolysis with a subsequent fermentation. The total capital investment of the plants was estimated to 64–69 million USD. Based on the economical evaluation, the authors found the SSF-based technology to be the most cost effective, rendering an ethanol production price of 0.34 USD/l (1.28 USD/gal). From an advanced model based on R&D data from the National Renewable Energy Laboratories (NREL), Wooley *et al.* (1999) estimated an ethanol price between 0.31 and 0.38 USD/l to be achievable with currently available technology. Based on the future research strategies and goals at NREL, the authors predict a further drop in ethanol production price of 0.11 USD/l within the next 10 years. Lynd *et al.* (1996b) created an advanced technology scenario in order to predict the potential advances in lignocellulosic conversion technologies and the expected savings in the ethanol production price. The scenario was constructed to simulate an ethanol production process at a maturity level comparable to an oil refinery. The technology was broken down in different components, which was examined and the expected features and abilities of a future

Table 10.4   Economic estimates on the ethanol production price and the total capital investments of various lignocellulose-based bioethanol technologies.

Annual production capacity (m³)	Estimated production price (USD/l)	Estimated capital investment (mio. USD)	Reference
90,000	0.32	73	Haagensen (2004)
~ 95,000	0.34	64–69	So and Brown (1999)
~ 197,000	0.31–0.38	234	Wooley et al. (1999)
Review	0.18–1.51	Review	von Sivers and Zacchi (1996)
~ 220,000	0.32	141	Wyman (1999)
~ 114,000	0.09–0.13	222	Lynd et al. (1996a)*

*Forecast of the likely costs of lignocellulosic technologies at a mature level.

mature technology were estimated. At these assumptions, it was found that, at a high maturity level, ethanol from lignocellulosic material could be produced at a price as low as 0.13 USD/l.

In order to achieve these goals, all parts included in the conversion of lignocellulose to ethanol have to be optimized. The economies of three different set-ups for lignocellulosic-based bioethanol production (enzymatic, dilute acid and concentrated acid process) were investigated in a study done by von Sivers and Zacchi (1995). In all three cases the sugar yields, energy integration and large production capability were pointed out as the most important factors affecting the ethanol production price.

In the majority of the bioethanol production schemes, the raw material was found to be the single most costly element (30–37%) of the total ethanol production price (von Sivers and Zacchi 1995; Wyman 1999). This fact strongly necessitates high recovery and conversion of the lignocellulosic sugars to achieve low-cost bioethanol production.

10.5.2   Sensitivity analysis

In order to determine the process parameters that make the ethanol production price most susceptible to variations, a sensitivity analysis is often made on the preferred process set-up. In these analyses, the cost of the biomass substrate was identified as one of the main constituents of the overall production cost of biomass-to-ethanol technologies (Wyman 1999).

A sensitivity analysis of the DBC (see Section 10.2.2.3) was also conducted. From this analysis, a 4.7% reduction in the ethanol production price occurs by a 10% decrease in the price of wheat straw (Haagensen 2004). This emphasizes the potential benefits by utilizing cheap sources of lignocellulosic biomass waste in bioethanol production technologies. The recovery of cellulose and hemicellulose was further identified to have a high impact on the ethanol production price. The cellulose recovery was shown to have a much higher impact on the ethanol production price compared to hemicellulose recovery. With a reduced cellulose recovery of 5%, the ethanol production price increased by 2.6%. When the hemicellulose recovery was reduced by 31%, the ethanol production price was only increased by 3.7%. The observed difference in the influence of cellulose and hemicellulose recoveries was related to the higher content of cellulose in the biomass substrate as well as the higher glucose conversion yields.

In production schemes where the biomass waste is only converted to one value product like ethanol, the variations in sugar yields will be reflected almost 100% in the ethanol production price, as found by von Sivers and Zacchi (1995). In the DBC, the loss in ethanol production due to reduced sugar yields was partly recovered as methane. Fibres that pass unchanged through the pre-treatment was separated and recycled to the wet oxidation step. Co-products, generating a profit, have been identified as an important factor to the overall plant economy in bioethanol production. Nguyen and Saddler (1991) constructed a simulation model of a bioethanol plant based on steam pre-treatment in order to evaluate the technology and the economic feasibility of the lignocellulose conversion process. Based on 500 metric ton of aspen wood per day, the selling price of lignin by-products was found to have a significant impact on the ethanol production price (Nguyen and Saddler 1991). Regarding co-products, Wyman (1999) further pointed out the importance of having a contracted market for these products in advance.

To evaluate the impact on the ethanol production price of different process parameters, a further sensitivity analysis was done using the DBC. Calculating the impact from these changes in various process factors (Figure 10.5) indicate that the relative change in the ethanol production price with changes in the different process parameters (Torry-Smith 2002). As can be seen, the cellulose recovery had the highest impact on the ethanol production price, resulting in a relative change in the ethanol production price of 48% relative to the change in cellulose recovery. Another important parameter was identified as the price of wheat straw. This coincides with the findings of Nguyen and Saddler (1991), who also found the raw material price to constitute an important factor of the ethanol production economy. The COD to methane efficiency of the biogas step, the number of plant production days and the ethanol yield from the xylose have less impact on the

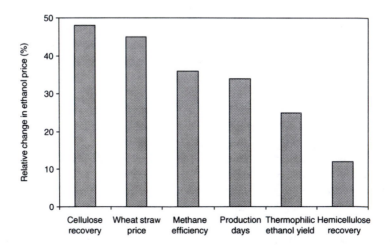

Figure 10.5   Relative change in ethanol price to changes in different process factors.

plant economy, causing a relative change in the ethanol production price between 25% and 28% of the percentage variation of the three factors. Since most of the ethanol production relates to the conversion of the cellulose fraction, the sensitivity of changes in the hemicellulose recovery was found to be the least important factor examined. As discussed earlier, the combination of different process steps, the production of multiple value products and an almost complete utilization of the raw materials renders the overall process economy less sensitive towards changes and variations in the different process parameters.

10.6   PERSPECTIVES FOR BIOETHANOL

Bioethanol production from lignocellulose holds a great potential as an alternative to organic waste disposal and incineration. If the technology could be further developed to effectively utilize organic household waste, various paper fractions and organic sludge, an enormous potential exists for sustainable energy production and disposal of organic waste, which could be implemented in both industrialized and developing countries.

10.6.1   Future bioethanol needs

Future bioethanol production plants should mainly be based on microbial and enzymatic processes to have a greater potential of increasing the process economy by feeding of new R&D achievements within the fields of strain development and lignocellulose-targeted enzyme systems. Improved bio-technological processes can substitute the present solutions without the requirement of re-designing the production facilities. This ability to take advantage of progress made within the applied technologies is an important trait from an industrial point of view. Further improvements could also be achieved by developing production concepts that enable high raw material utilization of which all products can be sold with profit. Additionally, it should be a low-cost technology with a minimum of energy requirement, non-inhibited fermentations and a sustainable solution for waste handling. Based on the future research strategies and goals at the NREL in Colorado (USA), the ethanol production price is expected to drop by 0.11 USD/l within the next 10 years (Wooley *et al.* 1999). From an advanced technology scenario attempting to predict the potential advances in lignocellulosic conversion technologies to a maturity level comparable to an oil refinery, it was estimated that ethanol from lignocellulosic material could be produced at a price of 0.13 USD/l (Lynd *et al.* 1996b).

  Future projections on the required fuel ethanol production evidently necessitates that utilization of biomass substrates other than corn, grain and sugarcane are introduced to meet the future needs (Lichts 2001). In the USA, the production of fuel ethanol is expected to increase to approximately 13 mio. m^3 in 2006 compared to the production of 7.6 mio. m^3 in 2000 as a result of national legislations such as the ban of the gasoline additive methyl tertiary butyl ether (MTBE) and a national strategy to further reduce the dependency of oil. The maximum production of corn-based ethanol has been estimated to 11–19 mio. m^3 (Lichts 2001).

This limit will already be reached within the next couple of years, emphasizing the need to develop commercially viable lignocellulose-to-ethanol technologies. From the new directive in promoting liquid biofuels in the EU, the requirement for bioethanol is also expected to increase remarkably from the current level of production in Europe. Based on projections made by the International Energy Agency (IEA) (Fulton 2002), the production of bioethanol will require an expansion of the existing level (2000) from 0.3 to 12 mio. m^3 by 2010 if bioethanol is to fulfil the EU Directive alone. Other major expansions in the market of fuel ethanol – especially in Asia (China, India and Thailand) – are expected to significantly expand the requirement and production of bioethanol.

10.6.2   Commercialization of lignocellulosic bioethanol technologies

Based on the expected increase in the production of fuel ethanol, development of economically feasible bioethanol technologies is needed in the near future. In order to make lignocellulose-based technologies commercially viable, construction of demonstration plants are required in order to evaluate the different technologies economically as well as technically. However, as heavy capital investments are needed in constructing demonstration plants in full scale, the attention of major investors has to be attracted. This has been particularly difficult with a stagnant world economy. A potential way to reduce the risk of investments is to include parts of the processes at existing bioethanol plants. Including cellulose-containing by-products generated during the starch-based technologies could promote the development and commercialization of lignocellulose-based bioethanol through different means:

(1) Adding value to the existing technologies.
(2) Obtaining new technological knowledge in lignocellulose to ethanol.
(3) Promoting capital investments.

If the different aspects listed above could eventually accelerate the commercialization progress of fuel ethanol based on lignocellulosic materials, commercial production of bioethanol requires large production facilities in order to assure a sound economic process. Therefore, in order to attract investors, to whom technology scale-up is a concern, comprehensive feasibility studies and risk assessments must be completed. The effect of the market price and availability of raw material is uncertain in scenarios of multiple full size bioethanol production facilities, which is also why market forecasts and life cycle analyses must be conducted. Furthermore, the construction of pilot and demonstration plants should be used as a basis for techno-economic evaluations of the novel technologies, greatly reducing the risks to the investors by preventing costly mistakes in full-scale production plants.

Introduction of ethanol to new geographic markets faces a complex barrier of absent demands and requires the co-operation of various entities. In countries without a tradition for fuel ethanol use in the transportation sector and in which the public is not familiar with the use of ethanol–gasoline blends, it is difficult to convince enterprises and the public to commit to bioethanol. Investors would not support a bioethanol production facility when there are no demands for ethanol because the public cannot buy vehicles able to run on high levels of fuel ethanol (e.g. flexible fuel vehicles, FFVs). And the car manufactures would not start producing FFVs if there is no demand and nowhere to fuel the vehicles. It is likewise difficult to convince the gas companies to give up gasoline tank storage and pump capacity to ethanol fuels when no cars on the street are capable of running on ethanol. Since all of the players on the market are somehow waiting for another player to be the pioneer, the entire process of implementing the ethanol technology is unable to move forward. However, organizations like the Swedish-based Bio-Alcohol Fuel Foundation (BAFF) founded in 1999 as a joint venture of public and private corporations, address these problems in order to promote the ethanol technology and co-ordinate the introduction of ethanol to the market. At present, it is possible to purchase an FFV in Sweden and drive it from north to south fuelling it only with bioethanol.

A lot of R&D effort is currently being focused on further development of well-known bioethanol process steps. This may not be the best and most economical long-term R&D strategy, since most of the existing technologies have already been developed to a level where future improvements will be of minor significance for the economic feasibility of bioethanol production. Instead, agreeing that production costs of bioethanol requires lowering to match the fossil fuel prices on the un-subsidized market, research should be directed at developing new technologies and approaches, allowing a substantial boost to the process economy. Scientists and researchers must be willing to give up years of research if a new technology emerges that has a greater potential. Another dilemma seen from a research perspective is the commercial interests. Attention and funding from the industry are very important in R&D, but often lead to patenting of new knowledge and technologies. Protection of property rights is necessary to attract investors, but at the same time it prevents the sharing and improving of new discoveries between researchers. This fact could mean that different technologies, each having the potential of being part of a promising combined bioethanol production concept, are kept concealed by property rights and the reluctance of investors to purchase licensed technology.

The issues discussed above need to be addressed in the near-term future if technologies for production of lignocellulosic bioethanol are to be developed to fulfil the prospected demand of fuel ethanol worldwide.

10.7 CONCLUSIONS

The inevitable limitation of fossil fuels deposits and concerns on the accumulation of CO_2 in the Earth atmosphere necessitates the development of renewable substitutes to the liquid fuels used in the expanding transportation sector. Bioethanol produced from primary crops are presently being used as a gasoline substitute. However, to reduce the CO_2 emissions from transportation even further, the development of economically viable technologies for production of bioethanol based on the vast generation of lignocellulosic biomass materials needs to be emphasized. These technologies should be able to convert the component of lignocellulose into bioethanol and valuable co-products in order to be competitive to gasoline. The development of bioethanol from lignocellulose will require not only technical improvements, political good will and societal acceptance will also have to be assist the development of sound technologies for bioethanol production based waste biomass and energy crops.

REFERENCES

Ahring, B.K., Jensen, K., Nielsen, P., Bjerre, A.B. and Schmidt, A.S. (1996) Pretreatment of wheat straw and conversion of xylose and xylan to ethanol by thermophilic anaerobic bacteria. *Bioresour. Technol.* **58**(2), 107–113.

Ahring, B.K., Thomsen, A.B., Nielsen, P. and Schmidt, A.S. (1997) Conversion of wheat straw to ethanol: a new concept. *Fifth Brazilian Symposium on the Chemistry of Lignins and Other Wood Components*, Curitiba, Brazil.

Ando, S., Arai, I., Kiyoto, K. and Hanai, S. (1986) Identification of aromatic monomers in steam-exploded poplar and their influences on ethanol fermentation by *Saccharomyces cerevisiae*. *J. Ferment. Technol.* **64**(6), 567–570.

Banat, I.M., Nigam, P., Singh, D., Marchant, R. and McHale, A.P. (1998) Ethanol production at elevated temperatures and alcohol concentrations. Part I: Yeasts in general. *World J. Microb. Biot.* **14**(6), 809–821.

Beck, M.J. (1993) Fermentation of pentoses from wood hydrolysates. In: *Bioconversion of Forest and Agricultural Residues* (ed. J.N. Saddler), pp. 211–229. CAB International, Wallingford.

Beg, Q.K., Kapoor, M., Mahajan, L. and Hoondal, G.S. (2001) Microbial xylanases and their industrial applications: a review. *Appl. Microbiol. Biot.* **56**, 326–338.

Birky, A., Greene, D., Gross, T., Hamilton, D., Heitner, K., Johnson, L., Maples, J., Moore, J., Patterson, P., Plotkin, S. and Stodolsky, F. (2001) *Future U.S. Highway Energy Use: A Fifty Year Perspective.* Office of Transportation Technologies, Energy Efficiency and Renewable Energy, US Department of Energy, Washington, DC.

Birol, G., Onsan, Z.I., Kirdar, B. and Oliver, S.G. (1998) Ethanol production and fermentation characteristics of recombinant *Saccharomyces cerevisiae* strains grown on starch. *Enzyme Microb. Technol.* **22**(8), 672–677.

Brigham, J.S., Adney, W.S. and Himmel, M.E. (1996) Hemicellulases: diversity and applications. In: *Handbook on Bioethanol – Production and Utilization* (ed. C.E. Wyman), pp. 119–141. Taylor & Francis, Washington, DC.

Chandrakant, P. and Bisaria, V.S. (2000) Simultaneous bioconversion of glucose and xylose to ethanol by *Saccharomyces cerevisiae* in the presence of xylose isomerase. *Appl. Microbiol. Biotechnol.* **53**(3), 301–309.

Clausen, A. (2000) *Improvement of the Ethanol Yield of Hemicellulose Degrading Bacteria.* Ph.D. Thesis. Department of Biotechnology, Technical University of Denmark.

Delgenes, J.P., Moletta, R. and Navarro, J.M. (1996) Effects of lignocellulose degradation products on ethanol fermentations of glucose and xylose by *Saccharomyces cerevisiae, Zymomonas mobilis, Pichia stipitis,* and *Candida shehatae. Enzyme Microb. Technol.* **19**(3), 220–225.

Dien, B.S., Cotta, M.A. and Jeffries, T.W. (2003) Bacteria engineered for fuel ethanol production: current status. *Appl. Microbiol. Biot.* **63**(3), 258–266.

DiPardo, J. (2000) *Outlook for Biomass Ethanol Production and Demand.* Energy Information Administration, US Department of Energy, Washington, DC.

Fulton, L. (2002) *Penetration of Alternative Fuels in EU and Potential Impact on Conventional Fuels Market.* IFQC European Automotive Fuels Briefing (November 21), International Energy Agency, Paris, France.

Goldemberg, J. (1998) Leapfrog energy technologies. *Energ. Policy* **26**(10), 729–741.

Grohmann, K., Torget, R. and Himmel, M.E. (1985) Optimization of dilute acid pretreatment of biomass. *Biotechnol. Bioeng. Symp.* **15**, 59–80.

Haagensen, F. (2004) *Production of Bioethanol from Various Lignocellulosic Residues.* Ph.D. Thesis. BioCentrum-DTU, Technical University of Denmark.

Hall, D.O., Rosillo-Calle, F. and Degroot, P. (1992) Biomass energy – lessons from case-studies in developing countries. *Energ. Policy* **20**(1), 62–73.

Hayn, M., Steiner, W., Klinger, R., Steinmüller, H., Sinner, M. and Esterbauer, H. (1993) Basic research and pilot studies on the enzymatic conversion of lignocellulosics. In: *Bioconversion of Forest and Agricultural Residues* (ed. J.N. Saddler), pp. 33–72. CAB International, Wallingford, UK.

Himmel, M.E., Adney, W.S., Baker, J.O., Nieves, R.A. and Thomas, S.R. (1996) Cellulases: structure, function, and applications. In: *Handbook on Bioethanol – Production and Utilization* (ed. C.E. Wyman), pp. 143–161. Taylor & Francis, Washington, DC.

Himmel, M.E., Adney, W.S., Baker, J.O., Elander, R., McMillan, J.D., Nieves, R.A., Sheehan, J.J., Thomas, S.R., Vinzant, T.B. and Zhang, M. (1997) Advanced bioethanol production technologies: a perspective. *Fuel. Chem. Biomass* **666**, 2–45.

Holtzapple, M.T., Jun, J.H., Ashok, G., Patibandla, S.L. and Dale, B.E. (1991) The ammonia freeze explosion (AFEX) process – a practical lignocellulose pretreatment. *Appl. Biochem. Biotech.* **28–29**, 59–74.

Hsu, T.-A. (1996) Pretreatment of biomass. In: *Handbook on Bioethanol – Production and Utilization* (ed. C.E. Wyman), pp. 179–212. Taylor & Francis, Washington, DC.

Klass, D.L. (1998a) Energy consumption, reserves, depletion, and environmental issues. In: *Biomass for Renewable Energy, Fuels, and Chemicals* 1st edn, pp. 1–28. Academic Press, San Diego, California, USA.

Klass, D.L. (1998b) Synthetic oxygenated liquid fuels. In: *Biomass for Renewable Energy, Fuels, and Chemicals* 1st edn, pp. 383–444. Academic Press, San Diego, California, USA.

Klinke, H.B., Ahring, B.K., Schmidt, A.S. and Thomsen, A.B. (2002) Characterization of degradation products from alkaline wet oxidation of wheat straw. *Bioresour. Technol.* **82**(1), 15–26.

Krishnan, M.S., Blanco, M., Shattuck, C.K., Nghiem, N.P. and Davison, B.H. (2000) Ethanol production from glucose and xylose by immobilized *Zymomonas mobilis* CP4(pZB5). *Appl. Biochem. Biotech.* **84–86**, 525–541.

Larsson, M., Galbe, M. and Zacchi, G. (1997) Recirculation of process water in the production of ethanol from softwood. *Bioresour. Technol.* **60**(2), 143–151.

Larsson, S., Palmqvist, E., Hahn-Hägerdal, B., Tengborg, C., Stenberg, K., Zacchi, G. and Nilvebrant, N.O. (1999) The generation of fermentation inhibitors during dilute acid hydrolysis of softwood. *Enzyme Microb. Technol.* **24**(3–4), 151–159.

Lawford, H.G. and Rousseau, J.D. (1998) Improving fermentation performance of recombinant *Zymomonas* in acetic acid-containing media. *Appl. Biochem. Biotech.* **70–72**, 161–172.

Lebeau, T., Jouenne, T. and Junter, G.A. (1998) Continuous alcoholic fermentation of glucose/xylose mixtures by co-immobilized *Saccharomyces cerevisiae* and *Candida shehatae. Appl. Microbiol. Biotechnol.* **50**(3), 309–313.

Lee, K.C.P., Bulls, M., Holmes, J. and Barrier, J.W. (1997) Hybrid process for the conversion of lignocellulosic materials. *Appl. Biochem. Biotechnol.* **66**(1), 1–23.

Lichts, F.O. (2001) World ethanol markets – analysis and outlook. Special Report no. 124, Agra Europe Ltd., London, UK.

Lynd, L.R. (1989) Production of ethanol from lignocellulosic materials using thermophilic bacteria: critical evaluation of potential and review. In: *Advances in Biochemical Engineering/Biotechnology* (ed. Fiechter A.), Vol. 38, pp. 1–52. Springer Verlag, Berlin.

Lynd, L.R., Elander, R.T. and Wyman, C.E. (1996a) Likely features and costs of mature biomass ethanol technology. *Appl. Microbiol. Biot.* **57–58**, 741–761.

Lynd, L.R., Elander, R.T. and Wyman, C.E. (1996b) Likely features and costs of mature biomass ethanol technology. *Appl. Biochem. Biotechnol.* **57–58**, 741–761.

McMillan, J.D. (1994) Pretreatment of lignocellulosic biomass. In: *Enzymatic Conversion of Biomass for Fuels Production*, pp. 292–324. American Chemical Society, Washington, DC.

Mistry, P.B. (1991) Comparative economic assessment of ethanol production under mesophilic and thermophilic conditions. *16th Conference on Energy from Biomass and Wastes*, Institute of Gas Technology, Chicago, Illinois, USA.

Mohagheghi, A., Evans, K., Finkelstein, M. and Zhang, M. (1998) Co-fermentation of glucose, xylose, and arabinose by mixed cultures of two genetically engineered *Zymomonas mobilis* strains. *Appl. Biochem. Biotech.* **70–72**, 285–299.

Nguyen, Q.A. and Saddler, J.N. (1991) An integrated model for the technical and economic-evaluation of an enzymatic biomass conversion process. *Bioresour. Technol.* **35**(3), 275–282.

Nguyen, Q.A., Tucker, M.P., Keller, F.A., Beaty, D.A., Connors, K.M. and Eddy, F.P. (1999) Dilute acid hydrolysis of softwoods. *Appl. Biochem. Biotech.* **77–79**, 133–142.

Nieves, R.A., Ehrman, W.S., Adney, W.S., Elander, R.T. and Himmel, M.E. (1998) Technical communication: survey and analysis of commercial cellulase preparations suitable for biomass conversion to ethanol. *World J. Microb. Biot.* **14**, 301–304.

Nigam, J.N. (2001) Ethanol production from wheat straw hemicellulose hydrolysate by *Pichia stipitis*. *J. Biotechnol.* **87**(1), 17–27.

Palmqvist, E. and Hahn-Hägerdal, B. (2000a) Fermentation of lignocellulosic hydrolysates. I: Inhibition and detoxification. *Bioresour. Technol.* **74**(1), 17–24.

Palmqvist, E. and Hahn-Hägerdal, B. (2000b) Fermentation of lignocellulosic hydrolysates. II: Inhibitors and mechanisms of inhibition. *Bioresour. Technol.* **74**(1), 25–33.

Puls, J. (1993) Substrate analysis of forest and agricultural wastes. In: *Bioconversion of Forest and Agricultural Plant Residues* (ed. J.N. Saddler), pp. 13–32. CAB International, Wallingford.

Pushalkar, S., Rao, K.K. and Menon, K. (1995) Production of beta-glucosidase by *Aspergillus terreus*. *Curr. Microbiol.* **30**(5), 255–258.

Ranatunga, T.D., Jervis, J., Helm, R.F., McMillan, J.D. and Hatzis, C. (1997) Identification of inhibitory components toxic toward *Zymomonas mobilis* CP4(pZB5) xylose fermentation. *Appl. Biochem. Biotechnol.* **67**(3), 185–198.

Schmidt, A.S. and Thomsen, A.B. (1998) Optimization of wet oxidation pretreatment of wheat straw. *Bioresour. Technol.* **64**(2), 139–151.

Shapouri, H., Duffield, J.A. and Wang, M. (2002) The energy balance of corn ethanol: an update. Agricultural Economic Report no. 814, U.S. Department of Agriculture, Washington, DC.

Shleser, R. (1994) *Ethanol Production in Hawaii*. Report prepared for the State of Hawaii Department of Business, Economic Development and Tourism, State of Hawaii.

Sierra-Alvarez, R., Field, J.A., Kortekaas, S. and Lettinga, G. (1994) Overview of the anaerobic toxicity caused by organic forest industry waste-water pollutants. *Water Sci. Technol.* **29**(5–6), 353–363.

Skaar, C. (1984) Wood–water relationship. In: *The Chemistry of Solid Wood* (ed. R. Rowell), pp. 127–172. American Chemical Society, Washington, DC.

So, K.S. and Brown, R.C. (1999) Economic analysis of selected lignocellulose-to-ethanol conversion technologies. *Appl. Biochem. Biotech.* **77–79**, 633–640.

Sommer, P. (1998) *Conversion of Hemicellulose and D-Xylose into Ethanol by the Use of Thermophilic Anaerobic Bacteria*. Ph.D. Thesis. Department of Environmental Science and Engineering, Technical University of Denmark, Denmark.

Sun, Y. and Cheng, J.Y. (2002) Hydrolysis of lignocellulosic materials for ethanol production: a review. *Bioresour. Technol.* **83**(1), 1–11.

Szczodrak, J. and Fiedurek, J. (1996) Technology for conversion of lignocellulosic biomass to ethanol. *Biomass Bioenerg.* **10**(5–6), 367–375.

Taherzadeh, M.J., Eklund, R., Gustafsson, L., Niklasson, C. and Liden, G. (1997a) Characterization and fermentation of dilute-acid hydrolyzates from wood. *Ind. Eng. Chem. Res.* **36**(11), 4659–4665.

Taherzadeh, M.J., Niklasson, C. and Liden, G. (1997b) Acetic acid – friend or foe in anaerobic batch conversion of glucose to ethanol by *Saccharomyces cerevisiae*? *Chem. Eng. Sci.* **52**(15), 2653–2659.

Tong, X. and McCarty, P.L. (1991) Microbial hydrolysis of lignocellulose materials. In: *Methane from Community Wastes* (ed. R. Isaacson), pp. 61–100. Elsevier Applied Science.

Topiwala, H.H. and Khosrovi, B. (1978) Water recycle in biomass production processes. *Biotechnol. Bioeng.* **20**, 73–85.

Torry-Smith, M. (2002) Optimization of Biological Processes Applied to Bioethanol Production. Ph.D. Thesis. BioCentrum-DTU, Technical University of Denmark, Denmark.

Torry-Smith, M., Sommer, P. and Ahring, B.K. (2003) Purification of bioethanol effluent in a UASB reactor system with simultaneous biogas formation. *Biotechnol. Bioeng.* **84**(1), 7–12.

von Sivers, M. and Zacchi, G. (1995) A technoeconomic comparison of three processes for the production of ethanol from pine. *Bioresour. Technol.* **51**(1), 43–52.

von Sivers, M. and Zacchi, G. (1996) Ethanol from lignocellulosics: a review of the economy. *Bioresour. Technol.* **56**(2–3), 131–140.

von Sivers, M., Zacchi, G., Olsson, L. and Hahn-Hägerdal, B. (1994) Cost analysis of ethanol production from willow using recombinant *Escherichia coli*. *Biotechnol. Progr.* **10**(5), 555–560.

Wheals, A.E., Basso, L.C., Alves, D.M.G. and Amorim, H.V. (1999) Fuel ethanol after 25 years. *Trends Biotechnol.* **17**(12), 482–487.

Wooley, R., Ruth, M., Glassner, D. and Sheehan, J. (1999) Process design and costing of bioethanol technology: a tool for determining the status and direction of research and development. *Biotechnol. Progr.* **15**(5), 794–803.

Wyman, C.E. (1996) *Handbook on Bioethanol: Production and Utilization*, Taylor & Francis, Washington, DC.

Wyman, C.E. (1999) Biomass ethanol: technical progress, opportunities, and commercial challenges. *Annu. Rev. Energ. Env.* **24**, 189–226.

Zaldivar, J., Nielsen, J. and Olsson, L. (2001) Fuel ethanol production from lignocellulose: a challenge for metabolic engineering and process integration. *Appl. Microbiol. Biot.* **56**(1–2), 17–34.

Zilliox, C. and Debeire, P. (1998) Hydrolysis of wheat straw by a thermostable endoxylanase: adsorption and kinetic studies. *Enzyme Microb. Technol.* **22**(1), 58–63.

11

The biorefinery for production of multiple biofuels

P. Westermann and B. Ahring

11.1 INTRODUCTION

Most approaches to combine biomass fermentation and fuel cell technology have focused on methane production in combination with fuel cells. Methane is, however, only one of many potential fuels derived from fermentation of biomass that might be used in present and upcoming fuel cell technology.

The political vision with respect to future energy supply is a hydrogen-based economy. This holds promising perspectives from an environmental point of view: reduced air pollution from fuel cell conversion of the hydrogen to electricity; low toxicity of the fuel; independency from politically problematic fossil fuel delivering states, etc. Although not considered important in current perspectives for future hydrogen production from renewable sources (United States Department of Energy 2002), fermentative hydrogen production from biomass could be a possible supplement to wind and photovoltaic produced hydrogen.

However, there are several problems associated with an energy economy based solely on hydrogen as energy carrier.

First, hydrogen in the near future has to be produced from fossil fuels (natural gas, coal) or from nuclear energy (United States Department of Energy 2002), which does not reduce the environmental problem although there might be local benefits of geographical separation of production and consumption of hydrogen. Carbon sequestration and capture might, however, alleviate the environmental burden, but an environmentally sound and low-cost technology for this is far from available.

Second, one can raise the question how suitable hydrogen is to match the future needs as a sole energy carrier. Currently, between 50% and 60% of the global oil consumption (69% in USA) are used for transportation (United States Department of Energy 2001). It is estimated that the transportation sector, especially passenger vehicles, will constitute the fastest growing consumer of fossil fuels in the future. However, numerous problems with respect to efficient transport and storing of hydrogen still remain to be solved.

Third, some of the reasons for implying a hydrogen-based economy address the reduction of greenhouse gas emission. Recent research has, however, pointed at hydrogen as a problematic gas with respect to both ozone destruction and greenhouse effect (Schultz et al. 2003; Tromp et al. 2003). This is an indirect consequence of the reaction between hydroxyl radicals and hydrogen leading to a prolonged residence time of the greenhouse gas methane in the troposphere as well as cloud formation and hence ozone destruction in the stratosphere (Schultz et al. 2003; Tromp et al. 2003).

The first problem has a natural time-based solution, since the amount of fossil fuels is limited, and all visions for the long-term hydrogen-based economy point at renewable energy as a source for hydrogen production. Hydrogen production from fermentative conversion of biomass could constitute a supplementary source, but as discussed later in this chapter, the current yield is low, and production of hydrogen as the only product of dark fermentation is thermodynamically not possible.

With respect to the second problem, much of the hydrogen-based economy concept takes its starting point in the utilization of fuel cell technology for clean and efficient energy conversion. Although the fuel cell technology was developed initially for molecular hydrogen as the energy source and is considered a cornerstone in the hydrogen economy, this technology is in rapid progression, and fuel cell systems dealing with more complex compounds such as ethanol are currently being developed (Lamy et al. 2001; Zhou et al. 2003). Also, highly efficient reformers capable of converting 1 mol ethanol to 4 mol hydrogen have recently been demonstrated (Deluga et al. 2004). This leaves the possibility to combine a less complex fuel-handling technology (ethanol instead of hydrogen) for transportation purposes with all the benefits of the hydrogen-based fuel cell technology. This further opens up for exploration of the metabolic diversity of biological fermentation processes of biomass for the production of different fuels, each having their advantage and purpose in the energy supply of the society: liquid fuels (ethanol) appropriate for transportation, gaseous fuel (methane) appropriate for heating, and hydrogen for local electricity production in a fuel cell stack.

This chapter will discuss the biological constraints of aiming fermentative transformation of biomass at a single product rather than exploiting the multitude of products offered by fermentative microorganisms. Instead of forcing through a hydrogen-based economy on purposes where it is not well suited, we argue for a renewable energy-based economy, where a wider diversity of energy carriers (including hydrogen) are utilized for each of their specific purposes.

11.2   ANAEROBIC BIOREFINING OF BIOMASS

The biorefinery concept is relatively new and parallels a traditional fossil carbon-based refinery in which the raw material has been exchanged with biomass, and where the majority of the processes exploited are biological and/or biochemical. A biorefinery, therefore, can be considered a factory including pre-treatments, separations and biological or biochemical transformation in multiple steps for the production of chemicals and fuels from biomass (Ohara 2003; Kamm and Kamm 2004). The biological part of the biorefinery will often include some of the fermentations outlined in Figure 11.1.

11.2.1   All roads lead to methane

The operation of a refinery based on biomass and biological processes, however, imposes several problems when specific products are the goal. In open anaerobic

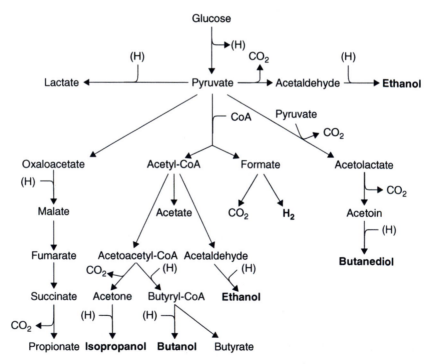

Figure 11.1   Fermentation pathways used by different fermentative bacteria. Products relevant to fuel cell utilization are in bold. One or several steps and intermediates have been deleted from most of the pathways to simplify the overview.

environments, in which the biomass is not pre-sterilized leaving access to ubiquitous microorganisms, the degradation of biomass normally follows rather well-defined pathways as described in Chapters 6 and 7. If no inorganic electron acceptors such as sulfate or nitrate are present, methane is the inevitable terminal biofuel product since all intermediates from the fermentative bacteria can be degraded to methane, carbon dioxide and water. During fermentation of biomass, energy bound in the reduced carbon compounds is made accessible to the microbes in a series of processes where electrons released from oxidation of carbon compounds are transferred to oxidized intermediate carbon compounds to maintain the electron balance (Figure 11.1).

This leads to the formation of carbon dioxide and reduced fermentation products (alcohols, volatile fatty acids, hydrogen) in which 90% of the energy of the converted biomass is conserved (Figure 11.2). Only 10% of the available energy is consumed by fermentative bacteria. In the terminal formation of methane, the process is finalized in that biomass carbon is sequestered completely to the most oxidized (CO_2) and most reduced (CH_4) state. The obligate biology leading to methane formation has an intrinsic stability governed by thermodynamics which ensures that methanogenesis proceeds within a wide spectrum of physical and chemical conditions although the energy yield of methanogenesis is relatively low (Figure 11.2).

11.2.2 The balance of fermentation

11.2.2.1 Fermentation products

Another problem in operating a biorefinery on biomass relates to the control of the variety of fermentation products that can be formed by fermentative bacteria. No single organism is capable of carrying out all the reactions shown in Figure 11.1.

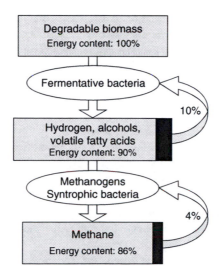

Figure 11.2   Available energy to the different microbial groups during anaerobic degradation of biomass.

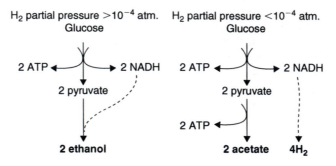

Figure 11.3   Effect of hydrogen partial pressure on product formation and ATP yield during fermentation of glucose. NADH is the electron carrier.

Most fermentative microorganisms are, however, characterized by the production of more than one fermentation product. The relative output of these fermentation products is regulated by the organism to optimize the energy yield and electron balance, and is therefore dependent on external concentrations of intermediate and end products.

Figure 11.3 shows one example of the effect of the hydrogen partial pressure on such a fermentation balance. At high hydrogen partial pressures, glucose is fermented to ethanol with a yield of 2 mol of ATP per mole glucose metabolized. At low hydrogen partial pressures, the reduction of free protons to hydrogen becomes exergonic leading to a change in fermentation from ethanol to acetate and hydrogen yielding 2 more ATP.

11.2.2.2   Reactor design

Although complex to some extent, most anaerobic digester configurations used are rather simple stirred fermentation tanks, which would not fulfil the requirements to specific fuel production other than methane. Numerous publications have demonstrated that it is possible to some degree to spatially uncouple fermentation and methanogenesis in two-step processes (Demeriel and Yenigun 2002; Yu *et al.* 2002; Raposo *et al.* 2004). The faster growth rate of fermentative bacteria compared to methanogens and syntrophic bacteria can be exploited by operating a fermenter at short retention time and low pH. The methanogenic step is then carried out in a subsequent fermenter at a longer retention time.

The possibility of controlling which compounds are formed in such a system is, however, limited since it still is open and continuously inoculated with bacteria from the feeding biomass. The only means of obtaining full control of product formation is by careful sterilization of the biomass followed by inoculation with specific microorganisms. This, however, increases the costs and also demands that the fermentation equipment can be sterilized and maintained free from contamination.

11.2.3   The low H_2 yield of biomass fermentation

Numerous fermentation processes are known leading to hydrogen formation, and might be coupled with other fermentation processes (Figure 11.1). Although the

Table 11.1 Energy yield from different hydrogen producing and consuming reactions.

Production	$\Delta G_0'$ (kJ/mol)
$C_6H_{12}O_6 + 12H_2O \rightarrow 12H_2 + 6HCO_3^- + 6H^+$	+3.2
$C_6H_{12}O_6 + 4H_2O \rightarrow 2CH_3COO^- + 4H_2 + 2HCO_3^- + 4H^+$	−206
Consumption	
$4H_2 + HCO_3^- + H^+ \rightarrow 2CH_4 + 3H_2O$	−135.6
$4H_2 + 2HCO_3^- + H^+ \rightarrow CH_3COO^- + 4H_2O$	−104.5

maximum theoretical hydrogen yield from complete conversion of 1 mol of glucose is 12 mol of hydrogen, this process is not thermodynamically feasible under standard conditions (Table 11.1). The maximum hydrogen yield from glucose fermentation is considered to be 4 mol per mole of glucose (for a further discussion of this, see Chapter 13). Paradoxically, a doubling of the hydrogen yield can be achieved by fermenting 1 mol of glucose to 2 mol ethanol and then reform the 2 ethanol to 8 hydrogen as described in Section 11.1 of this chapter.

Besides the low hydrogen yield, a major Achilles heel of fermentative hydrogen production is, similar to other fermentative reactions, hydrogen-consuming biological reactions such as methane and acetate formation. In these processes, hydrogen is inevitably converted into methane or acetate, respectively, unless methanogens and acetogens are excluded by sterilization of the biomass before fermentation and inoculation with specific hydrogen producing microbes, or the process is carried out under conditions adverse to the hydrogen utilizers.

11.2.4 Liquid biofuel production
11.2.4.1 Biofuel selection
Several of the fermentation products shown in Figure 11.1 are liquid and could serve as energy carriers in the transportation sector. The most relevant compounds are the alcohols, since they are more reduced and less corrosive than the corresponding acids. Although the heating value of the longer-chained alcohols is higher than for ethanol (butanol has a heating value of 33.1 MJ/kg compared to a heating value of 26.9 MJ/kg for ethanol) this is more than outbalanced by the stoichiometry of the fermentation reactions. Only 1 mol of the longer-chained alcohols is produced per mole of glucose compared to 2 mol of ethanol produced per mole of glucose. Also the boiling point of the longer-chained alcohols is higher than for ethanol (the 4-carbon alcohols even have a boiling temperature above water), which will complicate and increase the costs of their recovery from the fermented biomass. Ethanol is therefore the preferred fuel alcohol from biomass fermentation and already widely exploited globally. The annual production exceeds 31×10^9 l and numerous initiatives are taken these years to further increase the production of ethanol as a fuel additive as in E10 (10% ethanol, 90% gasoline) replacing the anti-knocking compound methyl tertiary butyl ether (MTBE) or as the major fuel part as in E85 (85% ethanol, 15% gasoline). The major producers are Brazil and USA, which account for 62% of the world production (Kim and

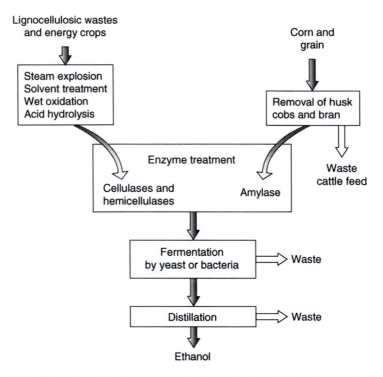

Figure 11.4 Flow sheet of main processes in the production of ethanol as single fuel from different types of biomass.

Dale 2004). As earlier mentioned efficient reformers and ethanol fuel cells are under development paving the way for ethanol use in fuel cells.

Most ethanol is currently produced by fermentation of grain, corn and sugar cane. The advantage of using these substrates is their high content of fermentable substrates, which only need simple physical pre-treatments followed by enzymatic treatment (wet milling or dry milling) (Figure 11.4). If a future production of biofuels for fuel cells shall constitute a significant part of the energy supply, it is, however, important to avoid conflicts between food use and industrial use of crops and also to maximize the biomass output from the limited arable land available. Other substrates such as household wastes, manure and sludges are far from sufficient and a considerable part of these wastes have low carbohydrate contents and are better suited for methane production in biogas plants. A further increase in biofuel contribution will therefore imply an exploitation of crop residues such as straw but also a specific production of energy crops as a biomass substrate for biological conversion to biofuels. Most plant material is, however, rich in lignin and lignocellulosic compounds, which are recalcitrant to microbial degradation (Lee 1997).

11.2.4.2 Biomass treatment

To make compounds entangled in the lignocellulose (cellulose, hemicellulose) accessible to biological transformation, it is necessary to perform a physical and/or a chemical pre-treatment of the biomass (Figure 11.4) as discussed in detail in Chapter 10. During this step, some of the hemicelluloses dissolve in water as monomeric pentose and hexose sugars and as oligomers and polymers. The temperature range of the pre-treatment is normally 150–200°C. The main processes investigated and used so far are steam explosion, treatment with ethanol/water mixtures (the Organosolv process), acid hydrolysis, and high-temperature/high-pressure treatment with acid alkalis, oxygen or both (Szczodrak and Fiedurek 1996; Sun and Cheng 2002). These treatments will on one side add to the cost of the fuel production, on the other side this is more than outbalanced by the lower price of lignocellulosic material compared to grain and corn, and the increased yield in the following fermentation process (Sun and Cheng 2002). A common feature and benefit of all the pre-treatment processes is that the biomass is sterile after treatment, which ensures that a controlled fermentation can take place.

To increase the yield, the physical–chemical pre-treatment is often followed by treatment with enzymes known as cellulases and hemicellulases, which hydrolyse cellulose and hemicellulose, respectively. The effectiveness of the enzymes depends on their origin (Thygesen *et al.* 2003), the nature of the previous treatment step, and the properties of the plant material, especially the degree of cellulose crystallinity and the amount and type of lignin. Pre-treatment using alkali and oxygen (wet oxidation) effectively removes lignin without producing toxic compounds and seems to give the best performance at the enzyme treatment stage when treating annual crops like wheat straw (Klinke *et al.* 2002, 2003).

11.3   CO-PRODUCTION OF LIQUID AND GASEOUS BIOFUELS

Conventional biogas plants only use a fraction of the added material. The remainder mainly consists of lignocellulosic materials, which make up a large proportion varying with the nature of the feed material. These materials are passing almost unconverted through the biogas plant. Current ethanol plants are on the other hand designed to work with starch or cellulosic materials. Lignins and other components, which cannot be turned into fermentable substrates, are considered as wastes requiring a further treatment process, or are at best burned as low-quality boiler fuels. Co-production of ethanol and gaseous biofuels would allow all the components of both plant biomass and animal manure to be used. The wastewater from the ethanol plant containing lignin and its oxidation products as well as by-products from the fermentation could serve as secondary substrates for the biogas reactor leading to a reduced cost price for ethanol due to the increased methane production.

The Danish Bioenergy Concept is an example of such a combinatory system approach (Figure 11.5): biomass is pre-treated to convert lignocellulosic compounds and to increase the availability of fermentable sugars, which are subsequently fermented to ethanol by yeast cells. Pentoses, which are not converted by yeasts, are then fermented to ethanol and hydrogen in a thermophilic fermentation process. After distillation, manure is mixed with the effluent from the pentose fermentation and methane is formed in an anaerobic digestion process. Process water from the biogas process is re-circulated to the pre-treatment process.

In the pre-treatment process, which takes a few minutes, lignocellulosic material is opened by wet oxidation using high pressure (12–20 bars) and addition of oxygen. The increased oxygen partial pressure leads to an exothermic auto-oxidation of mainly lignin resulting in a temperature increase to 170–190°C. The removal of protecting lignin and a partial decrystallization of cellulose and hemicellulose markedly increases the subsequent efficiency of polymer degrading enzymes (Klinke *et al.* 2003). The hydrolysate from the wet oxidation process contains 45–55 g sugars per 100 g straw added, and in contrast to most of the other pre-treatment processes, the wet oxidized material does not have to undergo detoxification (neutralization, solvent removal, etc.) before subsequent fermentation (Torry-Smith *et al.* 2003).

After cooling to 40°C, the hydrolysate is exposed to a simultaneous saccharification and fermentation (SSF) process. A commercial enzyme mixture consisting of pectinases, cellulases and β-glucosidases is added together with yeast (*Saccharomyces cerevisiae*). Monomeric sugars are released by enzymatic hydrolysis while the yeast simultaneously convert hexoses to ethanol.

In this second fermentation process, which is carried out under simultaneous distillation, ethanol and hydrogen are produced and recovered continuously from a thermophilic fermenter operating at 70°C. Pentoses (mainly xylose), which constitute a large part of the sugars released by the wet oxidation process, cannot be converted to ethanol by yeasts and act as primary substrates for immobilized cells of the thermophilic bacterium *Thermoanaerobacter mathranii*. As long as the

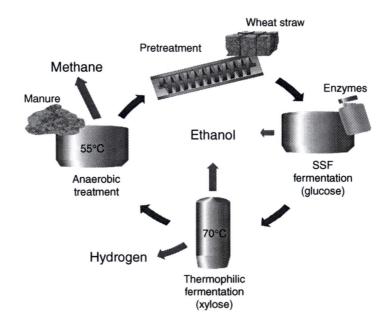

Figure 11.5   The Danish Bioenergy Concept.

available enzymes used in the SSF process are mesophilic, a two-step process using yeast followed by a thermophilic bacterium for converting all sugars into ethanol constitutes the best process scheme. However, with development of thermophilic enzymes, the use of a thermophilic bacterium alone will replace the two separate processes.

After ethanol fermentation a large part of the straw material will be left unused in the process water together with non-volatile fermentation products, which have not been removed during the distillation process for ethanol. Eighty per cent of the chemical oxygen demand in this stream can, however, be converted into methane in the third fermentation process of the biorefinery. The straw material is separated from the process water after the SSF fermentation and transferred to the anaerobic digester together with the effluent of the thermophilic fermenter. In this final process manure or other fermentable material is mixed with the effluent and straw fraction and co-digested in a conventional biogas process. During this anaerobic digestion process, all the different known lignin degradation products such as furanes and furanones are further converted allowing for re-circulation of 80% of the process water without accumulating inhibitory compounds (Torry-Smith et al. 2003). Solid residues are also re-circulated back to the wet oxidation process. Lowering the need for addition of fresh water to the system will have a major impact on the economy of the process. Production of methane will further decrease the cost of 1 unit of ethanol with more than 30% compared to a process without implementation of anaerobic digestion.

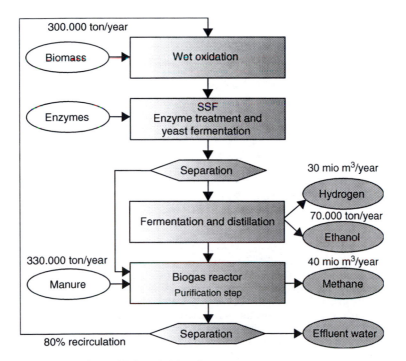

Figure 11.6   Flow sheet of full-scale biorefinery.

Figure 11.6 shows a mass budget for the production of ethanol, hydrogen and methane in a full-scale biorefinery based on an annual transformation of 300.000 ton wheat straw and 330.000 ton manure.

11.4 CONCLUSION

Based on the current state of art where bioethanol for transportation is produced from high-value crops such as grain and sugar cane, it is only possible to use ethanol as a supplement to gasoline replacing other additives such as MTBE. If a major substitution of fossil fuel for transportation is the goal, cellulose-rich biomass such as straw or specially grown energy crops is necessary.

The biorefinery concept outlined in this chapter is only one of many possible configurations in which known and new biotechnological processes are combined to optimize the transformation of low-cost substrates to valuable energy carriers. The production of hydrogen, methane and ethanol leaves the possibility to optimize the exploitation of the specific energy carriers to suit specific needs. Other possible configurations could be established where multiple lines of cell factories can perform different biochemical transformations leading to an array of products where the complete biomass is exploited.

REFERENCES

Deluga, G.A., Salge, J.R., Schmidt, L.D. and Verykios, X.E. (2004) Renewable hydrogen from ethanol by autothermal reforming. *Science* **303**, 993–997.

Demeriel, B. and Yenigun, O. (2002) Two-phase anaerobic digestion processes: a review. *J. Chem. Technol. Biotechnol.* **77**, 743–755.

Kamm, B. and Kamm, M. (2004) Principles of biorefineries. *Appl. Microbiol. Biotechnol.* **64**, 137–145.

Kim, S. and Dale, B.E. (2004) Global potential bioethanol production from wasted crops and crop residues. *Biomass Bioenerg.* **26**, 361–375.

Klinke, H.B., Ahring, B.K., Schmidt, A.S. and Thomsen, A.B. (2002) Characterization of degradation products from alkaline wet oxidation of wheat straw. *Biores. Technol.* **82**, 15–26.

Klinke, H.B., Olsson, L., Thomsen, A.B. and Ahring, B.K. (2003) Potential inhibitors from wet oxidation of wheat straw and their effect on ethanol production of *Saccharomyces cerevisiae*: wet oxidation and fermentation by yeast. *Biotechnol. Bioeng.* **81**, 738–747.

Lamy, C., Belgsir, E.M. and Léger, J-M. (2001) Electrocatalytic oxidation of aliphatic alcohols: application to the direct alcohol fuel cell (DAFC). *J. Appl. Electrochem.* **31**, 799–809.

Lee, J. (1997) Biological conversion of lignocellulosic biomass to ethanol. *J. Biotechnol.* **56**, 1–24.

Ohara, H. (2003) Biorefinery. *Appl. Microbiol. Biotechnol.* **62**, 474–477.

Raposo, F., Borja, R., Sanchez, E., Martin, M.A. and Martin, A. (2004) Performance and kinetic evaluation of the anaerobic digestion of two-phase olive mill effluents in reactors with suspended and immobilized biomass. *Water Res.* **38**, 2017–2026.

Schultz, M.G., Diehl, T., Brasseur, G.P. and Zittel, W. (2003) Air pollution and climate-forcing impacts of a global hydrogen economy. *Science* **302**, 624–627.

Sun, Y. and Cheng, J. (2002) Hydrolysis of lignocellulosic materials for ethanol production: a review. *Bioresource Technol.* **83**, 1–11.

Szczodrak, J. and Fiedurek, J. (1996) Technology for conversion of lignocellulosic biomass to ethanol. *Biomass Bioenerg.* **10**, 367–375.

Thygesen, A., Thomsen, A.B., Schmidt, A.S., Jørgensen, H., Ahring, B.K. and Olsson, L. (2003) Production of cellulose and hemicellulose-degrading enzymes by filamentous fungi cultivated on wet oxidized wheat straw. *Enz. Microbiol. Technol.* **32**, 606–615.

Torry-Smith, M., Sommer, P. and Ahring, B.K. (2003) Purification of bioethanol effluent in an UASB reactor system with simultaneous biogas formation. *Biotechnol. Bioeng.* **84**(1), 7–12.

Tromp, T.K., Shia, R.-L., Allen, M., Eiler, J.M. and Yung, Y.L. (2003) Potential environmental impact of a hydrogen economy on the stratosphere. *Science* **300**, 1740–1742.

United States Department of Energy (2001) *International Energy Outlook 2001.* http://tonto.eia.doe.gov/FTPROOT/forecasting/04842001.pdf

United States Department of Energy (2002) *A National Vision of America's Transition to a Hydrogen Economy – To 2030 and Beyond.* www.eere.energy.gov/hydrogenandfuel-cells/pdfs/vision_doc.pdf

Yu, H.W., Samani, Z., Hanson, A. and Smith, G. (2002) Energy recovery from grass using two-phase anaerobic digestion. *Waste Manage.* **22**, 1–5.

Zhou, W., Zhou, Z., Song, S., Li, W., Sun, G., Tsiakaras, P. and Xin, Q. (2003) Pt based anode catalysts for direct ethanol fuel cells. *Appl. Catalysis B: Environ.* **46**, 273–285.

Section IID: Hydrogen production

12

Dark fermentation for hydrogen production from organic wastes

B.H. Svensson and A. Karlsson

12.1 INTRODUCTION

Hydrogen gas is energy dense, renewable and also non-polluting, since its combustion produces only water vapour. These properties make hydrogen ideal to use as a fuel (Rao and Hall 1996). Moreover, the use of hydrogen as a fuel can make nations energy self-sufficient, which will improve their national strength. Renewable hydrogen can be produced biologically by dark fermentation of low-cost substrates or by photosynthetic splitting of water (Das and Veziroglu 2001). Conversion of organic wastes to hydrogen using dark fermentative processes is considered a promising near-term approach and the economics of hydrogen fermentation are regarded as favourable, even at yields lower than stoichiometrically possible (Archer and Thomson 1987). However, a major obstacle still remains, that is the low yield of hydrogen obtained so far (Benemann 1996, 2004; Hallenbeck and Benemann 2002). The basic problem of reaching the theoretical hydrogen yield during fermentation was addressed already by Thauer (1977).

© 2005 IWA Publishing. *Biofuels for Fuel Cells: Renewable energy from biomass fermentation* edited by P. Lens, P. Westermann, M. Haberbauer and A. Moreno. ISBN: 1843390922. Published by IWA Publishing, London, UK.

Another challenge is the need for separation of H_2 from other gaseous products that are produced during fermentation (Rao and Hall 1996). In this context the genetic modification of crucial bottlenecks of the central metabolic pathways leading to hydrogen is attractive (*cf.* Sode *et al.* 2001; Tanisho 2001; Penfold *et al.* 2003).

12.2   DARK FERMENTATIVE PRODUCTION OF HYDROGEN

12.2.1   Substrates for dark fermentation

Organic waste materials provide a wide spectrum of substrate components, which may form the basis for a renewable utilisation in hydrogen formation by microorganisms. Examples are slaughterhouse wastes, wet residues of sorted household wastes, waste from dairy and fish industry as well as residues from agriculture. The latter type of wastes are not dealt with in this article, while the reader is referred to Chapter 13 and the reviews by Wakayama and Mikyake (2001) and de Vrije and Claassen (2003). The wastes above have substantially larger fractions of both protein and fat compared to the, for methane and hydrogen production, more studied carbohydrate-rich residues from crops and their refinement. However, today, many of these fat- and protein-rich residues are used as co-substrates in methane-producing biogas processes. In principle, organic wastes used for the production of methane are potential substrates also for dark fermentative production of hydrogen. For each mole of methane formed, the theoretical yield is 4 mol of hydrogen under the assumption that all organic matter is oxidised to carbon dioxide, that is no acetate or other fermentation products are formed. Thus, as an example the potential hydrogen production from available organic wastes in Sweden corresponds to $1.5 \times 10^9 \, m^3$ as based on the calculated methane production possible to retrieve from the same waste pool according to Nordberg *et al.* (1998).

However, until now most waste materials used in investigations on their potential utilisation as substrates for hydrogen production are carbohydrates, that is starch, sucrose and to some extent cellulose (*cf.* Table 13.1 and references therein). One reason for this is that the fermentation of carbohydrates by many anaerobic bacteria includes the release of hydrogen gas, as a means to obtain the redox balance needed in their energy metabolism. Furthermore, the development of such bacterial populations have most likely occurred in experiments, where the inoculum has been preheated to temperatures promoting the development of spore-forming clostridia (*cf.* Lay 2000, 2001; Han and Shin 2004). The carbohydrate-dominated wastes used in these cases have likely led to the development of saccharolytic bacteria. Thus, these organisms are obvious targets to be explored for a future fermentative hydrogen production.

12.2.2   Pathways

Microbial hydrogen formation is possible as a part of mainly three fermentation pathways (Figure 12.1). The proportions of the fermentation products of these pathways are governed by the genetic diversity among the bacteria able to explore

them and are regulated by growth factors (e.g. nutrient availability, pH, temperature etc.) (Lay 2000; Lin and Lay 2004a,b). It should be noted that all three pathways generate products, which are still binding considerable amounts of hydrogen.
Thus, only a part of the potential hydrogen (2 out of 12 mol of H_2 mole of glucose)
that is formed during a complete biological oxidation of carbohydrate substrates
is released via two of the pathways in Figure 12.1.

In the first, exemplified by the mixed acid fermentation of glucose by
Escherichia coli to actyl co-enzyme A (Ac-CoA), hydrogen is formed via the

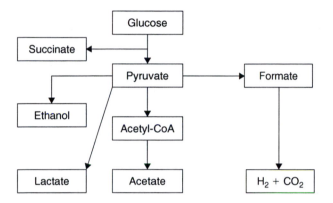

Figure 12.1(a)   Schematic picture of the mixed acid fermentation utilised by,
for example, *Escherichia coli*. The oxidation of glucose to pyruvate gives rise to NADH,
which is re-oxidised via the pathways leading to succinate, ethanol and lactate.
It should be noted that the formate split does not involve ferredoxin.

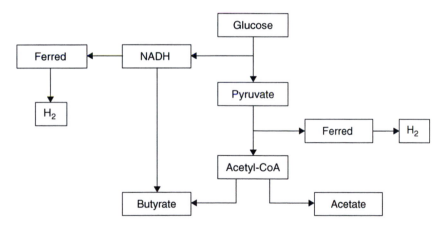

Figure 12.1(b)   Schematic picture of the butyrate–acetate fermentation utilised by
Clostridium butyricum. Observe that the ferredoxin-linked hydrogenase mediating the
hydrogen formation during the oxidative decarboxylation of pyruvate does not involve
a re-oxidation of NADH as is done by the NADH-ferredoxine oxidoreductase of the
pathway to the upper left. The strength of a hydrogen sink will govern the distribution
pattern among the fermentation products and also determine the extent of energy
efficiency allowed for.

splitting of formate to carbon dioxide and hydrogen by the activity of a formate–hydrogen lyase (Figure 12.1a). This membrane-bound enzyme complex is affected by the presence of redox-regulating compounds like nitrate and fumarate, but will form hydrogen in their absence. The second pathway for hydrogen formation is coupled to the oxidative decarboxylation of pyruvate involving a ferredoxin hydrogenase (below to the right in Figure12.1b). This reaction is common among the anaerobes (Thauer et al. 1974). It should be noted that the electron carrier $NAD^+/NADH$ is not involved in the hydrogen formation in any of these two reactions. The third option is exemplified by the butyrate–acetate fermentation utilised by, for example, Clostridium butyricum. This metabolic route makes use of the NADH–ferredoxine oxidoreductase, which oxidises the NADH formed in the Emden–Meyerhof–Parnas pathway oxidation via proton reduction and, thus, give rise to hydrogen (the upper left reaction in Figure 12.1b). However, in order to allow for a continuous hydrogen formation via this system the hydrogen partial pressure has to be low, that is the product has to be scavenged from the culture system. Under such cultivation conditions another 2 mol of those potentially available during anaerobic hexose oxidation may be extracted with a concomitant formation of 2 mol of acetate and carbon dioxide, respectively (cf. Thauer et al. 1974). Indeed, Hungate (1974) reported a complete transfer of the electrons to hydrogen during hexose fermentation. For the example given in Figure 12.1b a high hydrogen partial pressure will lead to a reoxidation of NADH at the expense of Ac-CoA giving butyrate as the product and accordingly less energy available for bacterial growth.

In many of the studies presented in Table 12.1, the efficiency of the hydrogen production as related to the substrate utilisation is compared under the assumption that 4 mol of hydrogen are available. However, in most cases the cultivation conditions used gave rise to hydrogen partial pressures, which would not allow for the NADH oxidation to occur. This is revealed by the formation of a wide range of fermentation products along with acetate and hydrogen. It may therefore be argued that any yield above two under most cultivation conditions used so far is due to some arrangement leading to a sink for hydrogen; for example, sparging with inert gases or co-cultivation with phototrophic organisms.

12.3 H_2 PRODUCTION YIELDS BY DARK FERMENTATION

Table 12.1 summarises some of the studies on hydrogen production during fermentation of carbohydrate-based wastes or model carbohydrates (e.g. starch, cellulose, sucrose, chitin and glucose). These studies reveal that the hydrogen yield from fermentation of organic waste, mainly composed of carbohydrates, ranges 0.07–3.0 calculated per mole of monosaccharide. In the study by Yokoi et al. (2002) a value of 3.7 may be argued for by including the yield obtained by letting the photosynthetic Rhodobacter sp. ferment the residual substrate from the co-cultured Clostridium and Enterobacter during exposure to light.

This means, however, that extra energy is supplied and that the basis for comparison is different. Still the yields above 2 mol of H_2 mole per monohexos unit

Table 12.1   Molar yields of hydrogen reported from experiments with different inocula and organic wastes as substrates.

Organisms or seed material	Waste substrate (main component)	Efficiency mol H_2/mol substrate units	References
Enterobacter cloacae	Various	0.95–6.0	Kumar and Das (2000)
Heat-treated soil	Potato starch	0.6	Logan *et al.* (2002)
E. aerogenes	Molasses (sucrose)	2.5[a]	Tanisho and Ishiawata (1994)
Mixed thermophiles	Cellulose	2.4[b]	Ueno *et al.* (1995)
Anaerobic waste water sludge	Sugar plant wastewater	2.6[b]	Ueno *et al.* (1996)
Clostridium paraputrificum	Chitin GlcNac	1.2–1.5 1.3–2.4	Evvyernie *et al.* (2001)
Heat-treated anaerobic sludge	Dining hall food waste	10–19% as part of COD reduction	Han and Shin (2004)
Thermotoga elfii	Paper sludge hydrolysate	46–48%	Kadar *et al.* (2003)
Heat-treated anaerobic sludge	Cellulose	0.07–0.3[b]	La (2001)
Acclimated municipal sludge	Sucrose	0.6–2.6[a]	Lin and Jo (2003)
Acclimated municipal sludge	Sucrose	4.8[a]	Lin and Lay (2004a)
Clostridium sp.	Cellulose hydrolysate	4.5[b]	Taguchi *et al.* (1996)
Mix *C. butyricum*, *E. aerogenes* and *Rhodobacter* sp.	Potato starch waste and corn steep liquor	1–3 + 4.5[c]	Yokoi *et al.* (2002)
C. beijerinckii	Glucose or starch	1.3–2.0 1.3–1.8	Taguchi *et al.* (1992)
Upflow reactor	Rice winery waste water	1.4–2.1	Yu *et al.* (2002)

[a] per mole sucrose.
[b] per mole glucose unit.
[c] 1–3 during repeated batch culturing of *C. butyricum* and *E. aerogenes*. Another 4.5 via *Rhodobacter* grown on the liquid used by the other two.

would suggest that an enough low partial pressure of hydrogen allowing for the oxidation of NADH by proton reduction occurred in the cultures. However, these values were mostly obtained in cultures with complex microbial substrates or substrates with organic waste, which will contain also other organic compounds than carbohydrates (*cf.* Tanisho and Ishiawata 1994; Taguchi *et al.* 1996; Kumar and Das 2000; Yokoi *et al.* 2002). Therefore, also other sources for hydrogen formation may have contributed to the yields observed and biased the comparisons.

12.3.1   Effect of environmental factors

Many factors may influence the potential hydrogen formation from dark fermentation of organic material. The most extensively studied relevant to the scope of this study, that is hydrogen production from waste material are pH and nutrient supply. The hydraulic retention time (HRT) has also been discussed in a few publications (see below). The temperature has in several studies been shown to effect the production of hydrogen, that is higher temperatures give higher hydrogen yields (*cf.* de Vrije and Claassen 2003). This is in agreement with the change in thermodynamics. However, most of these studies have been performed on pure cultures using carbohydrates (mainly glucose) as the substrate and will therefore not be further discussed here, but temperature is certainly a factor that should attract further investigations.

12.3.1.1   pH

The proton concentration affects both the rate and yield of hydrogen formation. Many studies show that a somewhat acid pH is optimal for hydrogen production (Taguchi *et al.* 1992; Kumar and Das 2000; Lay 2000; van Ginkel and Sung 2001), but not necessary for growth (Kumar and Das 2000). In most studies on transformation of organic wastes to hydrogen, no pH control except for the buffering capacity of the medium has been employed. However, Evvyernie *et al.* (2001) studied the hydrogen formation from chitinous wastes as well as from the hydrolyses product of such waste, that is *N*-acetyl-*p*-glucoseamine. They found that the optimal hydrogen production and yield from the monomer occurred at pH 5.8, which was slightly lower than that from the waste material (pH 6). The latter level was the one observed to be optimal for the chitinases of the bacterium used in these studies. This is an important observation pointing to the differential effects of the proton concentration on the processes, which are used to ferment organic waste materials.

Since many studies are done with "model" substrates, that is the monomers of polymeric materials in the wastes, an overestimation of the rates and the yields may take place. The study by Lay (2000) demonstrated that the pH-window for optimal hydrogen retrieval from carbohydrates may be narrow, that is a decrease of 50% from the optimum may occur within half a pH unit. Thus, as in many other biotechnology processes there is certainly a strong demand for a proper pH control to allow for optimal hydrogen-producing fermentations.

12.3.1.2   Nutrients

The balance of the energy and nitrogen sources was shown to be very important during the fermentation of sucrose to hydrogen by a mixed culture from sewage sludge (Lin and Lay 2004a). A five-fold increase of the formation rate occurred upon a rise of the C/N ratio from 40 to 47. The production rate stayed on this level at higher C/N ratios, as did also the specific hydrogen production rate. However, the hydrogen yield, which was doubled, was back on the initial levels at the higher C/N ratios. The same authors (Lin and Lay 2004b) also showed that phosphate and bicarbonate concentrations were critical for the same microorganisms. In their study of hydrogen fermentation on paper sludge hydrolysate by thermophilic anaerobes, Kádár *et al.* (2003) noticed that *Thermotoga elfii* produced more hydrogen at an increase of organic and inorganic nutrients, respectively, while *Caldicellulosiruptor saccharolyticus* responded to changes in the nutrient composition only.

12.3.1.3   HRT

The influence of the HRT has also been studied and most reports suggested that short HTRs give higher hydrogen yields (cf. Ueno et al. 1995, 1996, 2004). An HRT of half a day gave larger yields than HRTs of 1–3 days. Although these investigations revealed that it was possible to run continuous hydrogen production for more than 200 days, only a small fraction of the substrates was utilised (Ueno et al. 2004). This is likely explained by the high partial pressures of hydrogen (ca. 60%) occurring during these studies. Degradation coupled with hydrogen extraction might give a higher degree of degradation with increased retention times and, hence, higher hydrogen yields per gram chemical oxygen demand (COD) (see below).

12.4   DEVELOPMENT ROUTES TO EXPLORE

12.4.1   Substrate considerations

The brief literature survey on waste-related hydrogen production above show that the main emphasis has been (and still is) given to the fermentation of the carbohydrate fraction of waste materials. This is likely due to the fact that less quantities of protein- and fat-dominated wastes suitable as substrates for the production of biogas are available compared to carbohydrate wastes. Nevertheless, both protein and fat are potential feeds for the dark fermentation of organic wastes to hydrogen. Both are subjected to the same constraints as are the carbohydrates, that is their degradation is dependent on a low partial pressure of hydrogen and they give rise to fermentation products with the same restriction (cf. McInerney 1988; Örlygsson 1994; Örlygsson et al. 1995).

For many of the amino acids formed as a result of the protein hydrolysis, the initial deamination step requires a low partial pressure of hydrogen (Örlygsson et al. 1995). Such conditions also seem to enhance the degradation rate of amino acids, which are deaminated independently of the hydrogen concentration. This is likely due to the oxidative decarboxylation of the dicarboxylic acids formed as a result of the deamination (Örlygsson 1994). Amino acids found in proteins differ in size and structure; thus, they will consequently give rise to different amounts of hydrogen when completely degraded during anoxic conditions. Therefore, it is difficult to do a similar straightforward comparison of the hydrogen yields as can be made for carbohydrates based on the fermentation of glucose. Recently, however, Ramsay and Pullammanappallil (2001) summarised the amino acid composition of casein to average $CN_{0.23}H_{1.9}O_{0.51}$ and this formulation is used for comparison of hydrogen potentials below. Due to the similarity of their components, fats in the same way as carbohydrates lend themselves easier to a comparison of observed and theoretical yields.

In order to compare the hydrogen formation during anoxic oxidation of protein and fat compounds, and glucose, their carbon atom content was equalised. In the casein, case the unit formula was multiplied with six while caproic acid (C_6) was used as a rough estimator for fat. The results are shown in Table 12.2 and reveal that there is a minor difference between the hydrogen to be expected from proteins and carbohydrates, respectively. However, fat will yield about 50% more

Table 12.2   Hydrogen expected from carbohydrates, proteins and fats based on model monomer units containing six carbons.

Polymer	Monomer	Reaction	H_2/C
Carbohydrate	Glucose	$C_6H_{12}O_6 + 6H_2O \rightarrow 6CO_2 + 12H_2$	2
Protein	Average amino acid*	$6[CN_{0.23}H_{1.9}O_{0.51}] + 8.9H_2O$ $\rightarrow 6CO_2 + 1.4NH_3 + 12.5H_2$	2.1
Fat	Caproic acid	$C_6H_{12}O_2 + 10H_2O \rightarrow 6CO_2 + 16H_2$	2.7

* The average molecular formula given for casein by Ramsay and Pullammanappallil (2001).

hydrogen than the other polymer types, based on the same amount of carbon oxidised to carbon dioxide.

Therefore, in view of the theoretical yields, waste substrates with high fat content should be aimed at for dark fermentative production of hydrogen. However, due to the thermodynamics of the degradation of fat via the β-oxidation pathway, hydrogen production via this route would call for an efficient hydrogen sink. This is also valid for proteins, due to the low partial pressure of hydrogen needed for oxidative deamination and the degradation of the dicarboxy acids resulting from the deamination of amino acids. In this context, the results on protein degradation and low hydrogen retrieval observed by Noike and Mizuno (2000) and Okamoto et al. (2000) should be mentioned. The latter attributed the low H_2 retrieval to the interaction by the Stickland reactions for amino acid degradation.

12.4.2   Hydrogen extraction methods

To enhance the hydrogen yield from dark fermentation of organic matter, culturing methods enabling low or preferably very low hydrogen partial pressures have to be developed (Mizuno et al. 2000; Nielsen et al. 2001; van Groenesteijn et al. 2002). Alternatively, energy has to be supplied by, for example, photosynthesis (cf. de Vrije and Claassen 2003) or by other means. Benemann (2004) suggested that a partial oxygen-limited respiration mechanism could, at least theoretically, fulfil such needs. Thus, gas sparging with either external gas or internally produced gases recirculated in the system are possible ways. The recirculation demands removal of hydrogen from the gas phase (see below). This gives rise to low partial pressures of hydrogen resulting in which has shown to increase the hydrogen yield in several cases (Mizuno et al. 2000; van Niel et al. 2002). Also the removal of carbon dioxide has been discussed as a means to increase the hydrogen production by avoiding the formation of succinate in mixed acid fermentation (Tanisho et al. 1998). Their results showed an increase in acetate formation, which would also lead to a higher production of hydrogen. Due to the low economy of using external inert gases to flush out the hydrogen, van Groenesteijn et al. (2002) presented a concept utilising water vapour at low pressure from the growth medium at thermophilic conditions to decrease the hydrogen pressure. Their set up is planned to arrive at hydrogen levels allowing for the complete fermentation of carbohydrates to acetate and carbon dioxide. Other options are to combine dark fermentation of organic matter to ideally hydrogen, carbon dioxide and acetate with the utilisation of photosynthetic bacteria.

The latter will then further metabolise the acetate to hydrogen and carbon dioxide (*cf.* Akkerman *et al.* 2002; de Vrije and Claassen 2003). Also the combination of hydrogen production by dark fermentation as above with a methanogenic utilisation of the acetate and other fermentation products has been forwarded as an alternative (Hawkes *et al.* 2002) and is tried at pilot scale in Japan (Sawayama and Yokota 2004).

Nielsen *et al.* (2001) investigated the possibility to utilise gas flows in combination with hydrogen-specific membrane extraction techniques (see below) in order to lower the partial pressure of hydrogen in culture systems for hydrogen production with the purpose to enhance the hydrogen yield from dark fermentation of waste materials. This has included batch cultures with liquid substrates as well as municipal solid waste (MSW) and basically a flow of gas through the cultures either directly or through a thin Teflon tubing permeable to hydrogen submerged in the culture. A couple of different reactor designs with gaseous flushing have been evaluated including the solid bed MSW system described by Nielsen *et al.* (2001), MSW in continuously stirred reactors and recirculation of the biogas after hydrogen extraction by Karlsson *et al.* (manuscript). Recirculation of the fermentor gaseous phase including hydrogen extraction by the technique presented by Nielsen *et al.* (2001) showed that this is a possibility to keep the hydrogen partial pressure low enough to promote the hydrogen yield. Generally, enhanced hydrogen production rates and higher yield were achieved in all hydrogen extraction experiments (Karlsson *et al.* 2005). Thus, this adds to the experiences of other researchers (*cf.* Mizuno *et al.* 2000; Liang *et al.* 2002; van Niel *et al.* 2002).

12.4.3   Separation of biologically produced hydrogen

Palladium membranes possess catalytic activity and have a high selectivity to hydrogen (Kikuchi 1997; Shu *et al.* 1991). It is also possible to extract hydrogen from organic molecules present in the gas, for example methanol by dehydration over a palladium membrane (Amandusson *et al.* 1999). However, other compounds may poison the membrane surface. For example, carbon from dissociated organic molecules can block available adsorption sites and, thus, hamper the permeation over the membrane. Sulfides are well-known poisons for palladium surfaces (Antoniazzi *et al.* 1989; Gravil and Toulhoat 1999; Nielsen *et al.* 2001). Results in our laboratory show that this can be overcome by the use of a ZnO-trap. However, this technique is expensive by demanding heat and vacuum extraction, and more economically viable alternatives have to be developed. One promising way was forwarded by Teplyakov *et al.* (2002). They investigated different silicon-containing and polysulfone membranes for cleaning and concentrating the hydrogen formed from fermentation. In another approach, Liang *et al.* (2002) used a hollow fibre silicone membrane for hydrogen extraction from a fermentor, which increased the hydrogen formation rate and yield by 10% and 15%, respectively.

12.5   CONCLUSIONS

It is clear that there is a substantial potential for hydrogen production from organic waste materials. However, many obstacles have still to be overcome to reach a full-scale production of high yields. The development of such systems demands the close co-operation of microbiologist, physicist and engineers to solve the main problems of hydrogen extraction and separation as well as optimisation of operating conditions (e.g. temperature, pH, etc.).

ACKNOWLEDGEMENTS

The Swedish Energy Agency, The Swedish Natural Science Research Council and The Energy Research Fund of the Nordic Council of Ministers are greatly acknowledged for funding of the research group at Linköping University to which the authors belong.

REFERENCES

Akkerman, I., Jansen, M., Rocha, J. and Wijffels, R.H. (2002) Photobiological hydrogen production: photochemical efficiency and bioreactor design. *Int. J. Hydrogen Energ.* **27**, 1195–1208.

Amandusson, H., Ekedahl, L.-G. and Dannetun, H. (1999) Methanol induced hydrogen permeation through a Pd membrane. *Surf. Sci.* **442**, 199–205.

Antoniazzi, A.B., Haasz, A.A. and Stangeby, P.C. (1989) The effect of adsorbed carbon and sulphur on hydrogen permeation through palladium. *J. Nucl. Mat.* **162–164**, 1065–1070.

Archer, D.B. and Thomson, L.A. (1987) Energy production through the treatment of wastes by micro-organisms. *J. Appl. Bact. Symp. Suppl.* 59S–70S.

Benemann, J. (1996) Hydrogen biotechnology: progress and prospects. *Nature Biotechnol.* **14**, 1101–1106.

Benemann, J.R. (2004) A photobiological hydrogen production process. In: *Abstracts IEA Bioenergy 2004 WHEC 15*, June 29 to July 1, Yokohama, 29E-08, 9 pp.

de Vrije, T. and Claassen, P.A.M. Dark hydrogen fermentations. (2003) In: *Bio-methane & Bio-hydrogen* (eds. J.H. Reith, R.H. Wijffels and H. Barten). Dutch Biological Hydrogen Foundation. Smiet Offset, The Hague, pp. 103–123.

Das, D. and Veziroglu, T.N. (2001) Hydrogen production by biological processes: a survey of the literature. *Int. J. Hydrogen Energ.* **26**, 13–28.

Evvyernie, D., Morimoto, K., Karita, S., Kimura, T., Saaka, K. and Ohmiya, K. (2001) Conversion of chitinous wastes to hydrogen gas by *Clostridium paraputrificum* M21. *J. Biosci. Bioeng.* **91**, 339–343.

Gravil, P.A. and Toulhoat, H. (1999) Hydrogen, sulphur and chlorine coadsorption on Pd(111): a theoretical study of poisoning and promotion. *Surf. Sci.* **430**, 176–191.

Hallenbeck, P.C. and Benemann, J.R. (2002) Biological hydrogen production, fundamentals and limiting processes. *Int. J. Hydrogen Energ.* **27**, 1185–1193.

Han, S.-K. and Shin, H.-S. (2004) Biohydrogen production by anaerobic fermentation of food waste. *Int. J. Hydrogen Energ.* **29**, 569–577.

Hawkes, F.R., Dinsdale, R., Hawkes, D.L. and Hussy, I. (2002) Sustainable fermentative hydrogen production: challenges for process optimisation. *Int. J. Hydrogen Energ.* **27**, 1339–1347.

Hungate, R.E. (1974) Potentials and limitations of microbial methanogenesis. *ASM News* **40**, 833–838.

Kádár, Z., de Vrije, T., Budde, M.A., Szengyel, Z., Réczey K. and Claassen, P.A.M. (2003) Hydrogen production from paper sludge hydrolysate. *Appl. Biochem. Biotechnol.* **105–108**, 557–566.

Karlsson, A., Vallin, L. and Ejlertsson, J. (2005) Effects of temperature, HRT and H_2 extraction rate on hydrogen production from fermentation of food industry residues and manure. *Biores. Technol.* (submitted).

Kikuchi, E. (1997) Hydrogen-permselective membrane reactors. *CATTECH* **March**, 67–74.

Kumar, N. and Das, D. (2000) Enhancement of hydrogen production by *Enterobacter cloacae* IIT-BT 08. *Process Biochem.* **35**, 589–593.

Lay, J.-J. (2000) Modeling and optimization of anaerobic digested sludge converting starch to hydrogen. *Biotechnol. Bioeng.* **68**, 269–278.

Lay, J.-J. (2001) Biohydrogen generation by mesophilic anaerobic fermentation of microcrystalline cellulose. *Biotechnol. Bioeng.* **74**, 281–287.

Liang, T.-M., Cheng, S.S. and Wu, K.-L. (2002) Behavioral study hydrogen fermentation reactor installed with silicone rubber. *Int. J. Hydrogen Energ.* **27**, 1157–1165.

Lin, C.-Y. and Jo, C.-H. (2003) Hydrogen production from sucrose using an anaerobic sequencing batch reactor process. *J. Chem. Technol. Biotechnol.* **78**, 678–684.

Lin, C.Y. and Lay, C.H. (2004a) Carbon/nitrogen-ratio effect on fermentative hydrogen production by mixed microflora. *Int. J. Hydrogen Energ.* **29**, 41–45.

Lin, C.-Y. and Lay, C.H. (2004b) Effects of carbonate and phosphate concentrations on hydrogen production using anaerobic sewage sludge microflora. *Int. J. Hydrogen Energ.* **29**, 275–281.

Logan, B.E., Oh, S.-E., Kim, I.S. and van Ginkel, S. (2002) Biological hydrogen production measured in batch anaerobic respirometers. *Environ. Sci. Technol.* **36**, 2530–2535.

McInerney, M.J. (1988) Anaerobic hydrolysis and fermentation of fats and proteins. In: *Biology of Anaerobic Microorganisms* (ed. A.J.B. Zehnder). John Wiley and Sons, New York, pp. 373–415.

Mizuno, O., Dinsdale, R., Hawkes, F.R., Hawkes, D.L. and Noike, T. (2000) Enhancement of hydrogen production from glucose by nitrogen gas sparging. *Bioresourc. Technol.* **73**, 59–65.

Nielsen, A.T., Amandusson, H., Bjorklund, R., Dannetun, H., Ejlertsson, J., Ekedahl, L.-G., Lundström, I. and Svensson, B.H. (2001) Hydrogen production from organic waste. *Int. J. Hydrogen Energ.* **26**, 547–550.

Noike, T. and Mizuno, O. (2000) Hydrogen fermentation of organic municipal wastes. *Water Sci. Technol.* **42**, 155–162.

Nordberg, Å., Lindberg, A., Gruvberger, C., Lilja, T. and Edström, M. (1998). *Biogas Potential and Future Plants in Sweden*. JTI-Rapport, Kretslopp & Avfall nr 17. JTI, Uppsala (In Swedish).

Okamoto, M., Miyahara, T., Mizuno, O. and Noike, T. (2000) Biological hydrogen potential of material characteristic of the organic fraction of municipal solid wastes. *Water Sci. Technol.* **41**, 25–32.

Örlygsson, J. (1994) *The Role of Interspecies Hydrogen Transfer on Thermophilic Protein and Amino Acid Metabolism*. Dissertation. Department of Microbiology, Swedish University Agricultural Science Report No. 59. Uppsala.

Örlygsson, J., Houwen, F. and Svensson, B.H. (1995) Thermophilic anaerobic amino acid degradation: deamination rates and end product formation. *Appl. Microbiol. Biotechnol.* **43**, 235–241.

Penfold, D.W., Forster, C.F. and Macaskie, L.E. (2003) Increased hydrogen production by *Escherichia coli* strain HD701 in comparison with the wild-type parent strain MC4100. *Enz. Microb. Technol.* **33**, 185–189.

Ramsay, I.R. and Pullammanappallil, P.C. (2001) Protein degradation during anaerobic wastewater treatment: derivation of stoichiometry. *Biodegradation* **12**, 247–257.

Rao, K.K. and Hall, D.O. (1996) Hydrogen production by cyanobacteria: potential, problems and prospects. *J. Mar. Biotechnol.* **4**, 10–15.

Sawayama, S. and Yokota, N. (2004) Development of highly efficient hydrogen–methane fermentation process using organic wastes. In: *Abstracts IEA Bioenergy 2004 WHEC 15*, June 29 to July 1, Yokohama, 29E-02, 5 pp.

Shu, J., Grandjean, B.P.A., van. Neste, A. and Kaliaguine, S. (1991) Catalytic palladium-based membrane reactors: a review. *Can. J. Chem. Engineer.* **69**, 1036–1060.

Sode, K., Yamamoto, S. and Tomiyama, M. (2001) Metabolic engineering approaches for the improvement of bacterial hydrogen production based on *Escherichia coli* mixed acid fermentation. In: *Biohydrogen II. An approach to Environmentally Acceptable-Technology* (eds. J. Mikyake, T. Matsunaga and A. San Pietro), pp. 195–204. Pergamon, Amsterdam.

Taguchi, F., Chang, J.D., Takiguchi, S. and Morimoto, M. (1992) Efficient hydrogen production from starch by a bacterium isolated from termites. *J. Ferm. Bioeng.* **73**, 244–245.

Taguchi, F., Yamada, K., Hasegawa, K., Saito-Taki, T. and Hara, K. (1996) Continuous hydrogen production by *Clostridium* sp. strain no. 2 from cellulose hydrolysate and an aqueous two-phase system. *J. Ferm. Bioeng.* **82**, 80–83.

Tanisho, S. (2001) A scheme for developing the yield of hydrogen by fermentation. In: *Biohydrogen II. An approach to Environmentally AcceptableTechnology* (eds. J. Mikyake, T. Matsunaga and A. San Pietro), pp. 131–140. Pergamon, Amsterdam.

Tanisho, S. and Ishiawata, Y. (1994) Continuous hydrogen production from molasses by the bacterium *Enterobacter aerogenes*. *Int. J. Hydrogen Energ.* **19**, 807–812.

Tanisho, S., Kuromoto, M. and Kadokura, N. (1998) Effect of CO_2 removal on hydrogen production by fermentation. *Int. J. Hydrogen Energ.* **23**, 559–563.

Teplyakov, V.V., Gassanova, L.G., Sostina, E.G., Slepova, E.V., Modigell, M. and Netrusov A.I. (2002) Lab-scale bioreactor integrated with active membrane system for hydrogen production: experience and prospects. *Int. J. Hydrogen Energ.* **27**, 1149–1155.

Thauer, R. (1977) Limitation of microbial H_2-formation via fermentation. In: *Microbial Energy Conversion* (eds. H.G. Schlegel and J. Barnea), pp. 201–216. Pergamon Press, Oxford.

Thauer, R.K., Jungermann, K. and Decker, K. (1974) Energy conservation in chemotrophic anaerobic bacteria. *Bact. Rev.* **42**, 100–180.

Ueno, Y., Kawai, T., Sato, S., Otsuka, S. and Morimoto, M. (1995) The biological production of hydrogen from cellulose by natural anaerobic microflora. *J. Ferm. Bioeng.* **79**, 395–397.

Ueno, Y., Otsuka, S. and Morimoto, M. (1996) Hydrogen production from industrial wastewater by anaerobic microflora in chemostat culture. *J. Ferm. Bioeng.* **82**, 194–197.

Ueno, Y., Fukui, H. and Goto, M. (2004) Hydrogen production from organic waste. In: *Abstracts IEA Bioenergy 2004 WHEC 15*, June 29 to July 1, Yokohama, 29E-03, 9 pp.

Wakayama, T. and Mikyake, J. (2001) Hydrogen from biomass. In: *Biohydrogen II. An Approach to Environmentally AcceptableTechnology* (eds. J. Mikyake, T. Matsunaga and A. San Pietro), pp. 41–51. Pergamon, Amsterdam.

van Ginkel, S. and Sung, S. (2001) Biohydrogen production as a function of pH and substrate concentration. *Environ. Sci. Technol.* **35**, 4726–4730.

van Groenesteijn, J.W., Hazewinkel, J.H.O., Nienoord, M. and Bussmann, P.J.T. (2002) Energy aspects of biological hydrogen production in high rate bioreactors operated in the thermophilic temperature range. *Int. J. Hydrogen Energ.* **27**, 1141–1147.

van Niel, E.W.J., Budde, M.A.W., de Haas, G.G., van der Wal, F.J., Claassen, P.A.M. and Stams, A.J.M. (2002) Distinctive properties of high hydrogen producing thermophiles, *Caldicellulosiuptor saccharolyticus* and *Thermotoga elfii*. *Int. J. Hydrogen Energ.* **27**, 1391–1398.

Yokoi, H., Maki, R., Hirose, J. and Hayashi, S. (2002) Microbial hydrogen production starch-manufacturing wastes. *Biomass Bioenerg.* **22**, 389–395.

Yu, H., Zhu, Z., Hu, W. and Zhang, H. (2002) Hydrogen production from rice winery wastewater in an upflow anaerobic reactor by using mixed anaerobic cultures starch-manufacturing wastes. *Int. J. Hydrogen Prod.* **27**, 1359–1365.

13

Utilization of biomass for hydrogen fermentation

P.A.M. Claassen, M.A.W. Budde, E.W.J. van Niel and T. de Vrije

13.1 INTRODUCTION

In view of the detrimental effect of fossil fuel utilization on the environment, the need to employ renewable resources for the supply of energy is now globally acknowledged. The production of hydrogen from biomass is one of the options to meet this demand. Besides being derived from a renewable resource, the utilization of hydrogen offers several additional advantages and as a result, hydrogen is often given the epithet "fuel of the future". First of all, utilization of hydrogen as a fuel results in water as the end-product. Secondly, hydrogen is the fuel by choice to feed fuel cells which offer additional advantages such as, very high energy conversion efficiency, low maintenance and low noise. Thirdly, there are many different ways to produce hydrogen from renewable resources (biomass or hydropower, wind, solar or geothermal energy), all contributing to the large-scale introduction of this new fuel.

There are presently several initiatives, all in the Research and Development (R&D) stage, to establish the production of hydrogen from biomass. In general terms, the thermochemical conversion of biomass to "syn" gas is most suited for dry biomass. For wet biomass, research is focussing on supercritical water gasification and biological conversion. Between these two technologies, supercritical water gasification allows the utilization of all biomass components whereas biological conversion is restricted to the fermentable part of the biomass. On the other hand, the gas produced by supercritical water gasification may be mixed with traces of contaminants (CO or higher alkanes) whereas biological conversion is expected to deliver pure hydrogen and CO_2 only.

As stated above, the initiatives are in the R&D stage. Presently there are still many variables, for example, physical and chemical biomass composition, biomass availability, logistics etc., which will have their own specific contribution to the optimization of the hydrogen production from biomass and application chain. It is of prime importance to leave room for exploration of all alternatives and to evaluate and compare research results as soon as these become substantial.

In this chapter the focus is on biological conversion of biomass to hydrogen, with the emphasis on dark hydrogen fermentation, and a case-study is presented with results obtained using potato steam peels as feedstock.

13.2 HYDROGEN PRODUCTION FROM BIOMASS

13.2.1 Energy carriers form biomass by fermentation

There are four options for the biological conversion of biomass to energy carriers: anaerobic digestion to biogas with methane as energy carrier; acetone, butanol and ethanol (ABE) fermentation with butanol as the prime energy carrier; ethanol fermentation and hydrogen fermentation. Table 13.1 shows the amount of energy which is theoretically available in the end-products of an optimal fermentation with glucose as a model substrate (Claassen et al. 1999).

All these biological processes have their own advantages and disadvantages and, worldwide, extensive research is in progress to improve the respective drawbacks. Anaerobic digestion is now a well-established technology for conversion of a wide range of different feedstock to biogas (Chapter 7). However, specific conversion rates are rather low and the energy conversion efficiency is the lowest in comparison to other fermentation products (Table 13.1). The ABE fermentation also has the advantage of being suited for a great variety of feedstock. However,

Table 13.1 Comparison of energy yields in products from fermentative conversions with glucose as the substrate.

Fermentation product (mol)	$\Delta G^{\circ\prime}$ in product (kJ/mol glucose)
3 methane	−2281
ABE	−2397 (average)
2 ethanol	−2464
12 H_2	−2673

As comparison: $\Delta G^{\circ\prime}$ glucose $= -2699$ kJ/mol.

the inhibition of the fermentation by butanol is quite restrictive in industrial applications (Dürre 1998). At fairly low concentrations, butanol impairs fermentation thus necessitating efficient, and till now costly, product removal and recovery (Maddox *et al.* 1993). The conventional and already industrially applied ethanol fermentation enables high product concentrations (Lynd *et al.* 1996). The drawback here is the current inability of industrial yeasts to convert pentoses to ethanol. This way, the range of feedstock is limited to the fairly expensive feedstock derived from sugary or starchy biomass and therefore it is not surprising that the initiatives to augment the potential of yeasts by genetic modification are extensive and worldwide (Hahn-Hägerdahl *et al.* 1994). The variety of feedstock for fermentative hydrogen seems again quite extensive as reported by de Vrije and Claassen (2003). However, in the case of hydrogen fermentation, the complete conversion of glucose to 12 mol of hydrogen does not occur freely. The thermodynamic characteristics of the envisaged reaction are shown in Table 13.2.

The complete oxidation of glucose to 12 mol of hydrogen is hampered by unfavourable thermodynamics ($\Delta G^{\circ\prime} > 0$). This means that this reaction will not occur under standard conditions (Claassen *et al.* 1999). Furthermore, no metabolic energy becomes available by performing this reaction. In order to establish the production of 12 mol of hydrogen from 1 mol of glucose, two consecutive fermentations need to be joined to make one bioprocess (Figure 13.1). The reactions which form the basis of this bioprocess are shown in Table 13.2. The first fermentation enables the conversion of biomass to hydrogen and organic acids. Since many thermophilic bacteria, growing at 70–80°C, oxidize glucose to acetate as the lowest reductive state, the highest conversion efficiency with respect to hydrogen production is obtained using these bacteria (van Niel *et al.* 2002; de Vrije and Claassen 2003). In contrast, during anaerobic mesophilic fermentation, mixtures of acids and/or alcohols are produced and the hydrogen yield is lower (Table 13.3).

The conversion of the end-product of the thermophilic fermentation is hampered by unfavourable thermodynamics (Table 13.2). However, phototrophic purple, non-sulfur bacteria are able to overcome this barrier by employing energy from light during the utilization of acetate, which, like lactate, is a prime carbon source for these bacteria. Thus, complete conversion of hexoses, pentoses, oligosaccharides or starch, is established by coupling a thermophilic heterotrophic fermentation to a consecutive, photo-heterotrophic fermentation.

Table 13.2 Hydrogen production from glucose or acetic acid.

$C_6H_{12}O_6 + 6H_2O \longrightarrow 6CO_2 + 12H_2$		$\Delta G^{\circ\prime} = +3.2\,kJ$
$C_6H_{12}O_6 + 2H_2O \longrightarrow 2CO_2 + 2CH_3COOH + 4H_2$		$\Delta G^{\circ\prime} = -206\,kJ$

(hyper)thermophilic bacteria

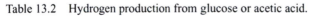

$h\nu$

$CH_3COOH + 2H_2O \longrightarrow 2CO_2 + 4H_2$ $\qquad \Delta G^{\circ\prime} = +104\,kJ$

purple, non-sulfur bacteria

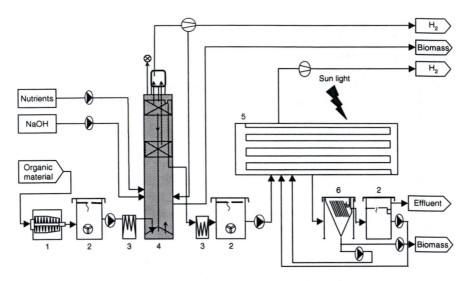

Figure 13.1   Simplified flow sheet of a bioprocess for hydrogen production. The first reactor is for thermophilic heterotrophic fermentation; the second reactor is for photo-heterotrophic fermentation; 1: extruder; 2: tank; 3: heat exchanger; 4: thermoreactor; 5: photoreactor; 6: tilted plate settler.

Table 13.3   Hydrogen production by heterotrophic bacteria.

Micro-organism(s)	T (°C)	Substrate	mol H_2/mol monosaccharide	Reference
Thermotoga elfii	65	Glucose	3.3	van Niel *et al.* 2002
Caldicellulosiruptor saccharolyticus	70	Glucose, sucrose	2.7–3.3	van Niel *et al.* 2003
Clostridium spp.	36	Glucose	1.4–2.4	Taguchi *et al.* 1995
Enterobacter aerogenes	38	Glucose	0.6–1.0	Rachman *et al.* 1998
Mixed population in sewage sludge	35	Glucose, sucrose	1.7	Lin and Chang 1999; Chen *et al.* 2001

13.2.2   Important parameters in hydrogen production

As stated above, the highest yield in H_2 from glucose is obtained when the substrate is oxidized to 2 mol of acetate and 2 mol of CO_2 (Table 13.2). However, this equation has to be modified with biomass production since the oxidation of glucose to hydrogen is growth related during heterotrophic growth. As a result, the observed hydrogen production will amount to 75–80% of the maximum theoretical efficiency, which is usually observed in thermophilic cultures (Table 13.4). Similarly in photo-heterotrophic cultures, part of the carbon source is used for biomass production and again the equation in Table 13.2 has to be modified to account for biomass synthesis. Presently, the observed ranges of conversion efficiency in photo-heterotrophic cultures are wide. This is partially due to the contribution of

Table 13.4   Typical fermentation parameters obtained during heterotrophic and photo-heterotrophic hydrogen fermentation.

	Enzyme for H_2 production	H_2 production efficiency (% of theoretical maximum)	Light energy conversion efficiency (%)	H_2 production rate (mmol/g DW h)	Critical H_2 in gas phase (kPa)
Caldicellulosiruptor saccharolyticus	Hydrogenase	74–80	n.a.	29	20–56
Rhodopseudomonas sp.; *Rhodobacter capsulatus**	Nitrogenase	26–87	1–2	0.8–1.6	≥90

*Light limitation; n.a.: not applicable.

the photochemical efficiency, which is an important parameter during photo-heterotrophic growth as discussed by Akkerman *et al.* (2003).

In thermophilic bacteria, during heterotrophic growth electroneutrality is maintained by reducing protons to hydrogen through the action of the hydrogenase enzyme. Unfortunately, hydrogenase is inhibited by its own product, hydrogen. This leads to a critical hydrogen concentration in the gas phase that is dependent on the growth phase, indicated by the range of critical hydrogen concentrations shown in Table 13.4. As a result, hydrogen must be removed as soon as it is formed. This confers a great challenge for bioreactor design because, in general terms, high volumetric productivity is important for favourable cost balances. After evaluating several alternatives, amongst others adsorption to palladium slurries, the application of a trickle bed reactor has been shown to offer a great potential with respect to cost effectiveness (van Groenestijn *et al.* 2002). In this case, gas is the continuous phase and bacteria are growing in a thin biofilm allowing a maximal gas–liquid interface.

Purple non-sulfur bacteria employ the nitrogenase enzyme for hydrogen production instead of the hydrogenase enzyme (Akkerman *et al.* 2003). As a result, the fermentation is not inhibited by hydrogen and high hydrogen concentrations are allowed. This way, high volumetric productivities in dense cultures can be envisaged to improve the specific hydrogen production rate this far established (Table 13.4). However, high density cultures will conflict with sufficient light penetration so again, there is a severe demand on elaborate bioreactor design. Although high conversion efficiencies with respect to the carbon source have been achieved, light energy conversion efficiency and hydrogen production rate are parameters that necessitate further research and development (Table 13.4).

13.2.3   Perspectives

Several feedstocks have been investigated with respect to their potential application for the production of hydrogen and organic acids by thermophilic bacteria. Table 13.5 shows the occurrence and composition of potential feedstock throughout Europe. Generally, the feedstock that confers a competitive edge has high sugar or starch and high moisture content because of high conversion efficiencies.

Table 13.5   Production and composition of biomass for hydrogen production (w/w%).

Raw material	Production, dry weight	Starch	Cellulose	Hemi-cellulose, Pectin	Sucrose	Extractives	Lignin	Ash
Miscanthus	15–25 ton/ha, NL	–	38	24		7	25	2
Sorghum bicolor	30–40 ton/ha, GR	–	18	10	61	–	7	2
Paper sludge	8.5 Mton/a, EU	–	36	4	–	–	18	35
Potato steam peels	0.6 Mton/a, NL	51	n.d.	n.d.	–	28	n.d.	8
Domestic organic waste	1.6 Mton/a, NL	–	24	20	–	13	12	16

Extractives are soluble sugars, protein, organic acids, lipids; n.d.: not determined; –: not applicable.

This is the case with potato steam peels and the juice of sweet sorghum, which is obtained after pressing the sucrose-rich stalks of the plants. However, in line with the search for cheap biomass for energy production, also lignocellulosic biomass, derived from energy crops or agro-industrial waste streams, as feedstock for hydrogen production has been applied (Claassen *et al.* 2002). When using *Miscanthus*, the residue of sweet sorghum stalks, paper sludge or domestic organic waste, pretreatment and hydrolysis is required to mobilize the sugars in the (hemi)cellulose. This far, the industrial application for converting lignocellulosic biomass to fermentable feedstock, is hampered by either high environmental burden or high cost for environmental friendly procedures such as enzymatic hydrolysis. This problem is shared with other initiatives for biofuel production, such as ethanol. Progress in this respect has been recently achieved but further decrease in pretreatment and hydrolysis costs is still required (www.novozym.com).

Thermophilic bacteria offer the advantage of the ability to metabolise hexoses and pentoses simultaneously, producing hydrogen from both substrates (de Vrije *et al.* 2002) but in anaerobic systems lignin remains untouched. Since several initiatives for hydrogen production from biomass are currently being researched, an obvious development would be to make an alliance with a thermochemical method to convert the non-fermentable biomass to hydrogen. This way, the moist fermentable part of the biomass would be substrate for fermentative conversion to hydrogen whereas the drier part can be transported to large-scale installations for thermochemical conversion to hydrogen.

13.3   HYDROGEN PRODUCTION FROM POTATO STEAM PEELS

Potato steam peels form a highly viscose slurry obtained as a by-product in the potato processing industry. The current use of this by-product is as component of

Table 13.6   Concentration of substrates and products after thermophilic fermentation by *Caldicellulosiruptor saccharolyticus* and subsequent photo-heterotrophic fermentation by *Rhodobacter capsulatus*, of potato steam peels hydrolysate.

mM	Glucose	H_2	Acetate	Lactate	CO_2
Start	63	0	7	16	0
End of thermophilic fermentation	18	131	75	22	67
End of photo-fermentation	0	280	0	0	n.d.

n.d.: not determined.

wet feed in the fodder industry. Because of the low N over C balance, mixing of potato steam peels with other wet by-products from, for example, the food industry is needed to achieve a nutritious feed. Due to several international developments in the feed industry as well as the energy sector, there is a current interest to convert this by-product to biofuel.

The main component in potato steam peels is starch (Table 13.5). Even though thermophilic bacteria are able to convert starch to hydrogen, liquefaction is desirable in view of adequate rheological properties. Besides, separation of the liquid hydrolysate and the solid residue results in a secondary by-product that is enriched in protein and possesses improved properties for processing to fodder.

13.3.1   Proof of principle

Potato steam peels hydrolysate, with glucose as its main carbohydrate component, is suited for hydrogen fermentation by *Caldicellulosiruptor saccharolyticus*. In Table 13.6 results of an experiment are shown of which the purpose was to demonstrate the complete bioprocess, that is the combination of a thermophilic heterotrophic and a photo-heterotrophic fermentation. As a result, ammonium ions were omitted from the substrate mixture for the thermophilic fermentation and this has led to the incomplete and relatively slow utilization of the substrate. Consumption and production of substrate and products, respectively, at the end of the batch fermentation in a submerged culture are shown. Hydrogen was continuously removed by stripping with nitrogen gas. The concentration of hydrogen is presented as cumulative hydrogen and was calculated from on-line measurements in the gas phase where the partial concentration was maximally 1.5%. The effluent of the thermophilic fermentation was transferred to a cylindrical photobioreactor and inoculated with *Rhodobacter capsulatus*. Hydrogen production was fairly slow but very efficient with respect to acetate conversion as 87% of the substrate was used for hydrogen production.

The achieved yield of hydrogen from glucose and organic acids in this two-stage bioprocess amounted to 47%, which is quite promising as compared to the 69% being the maximum achievable yield. This maximum achievable yield is derived from two separate fermentations that operate at 80% conversion efficiency. In the first fermentation one third of the hydrogen is produced, in the second the remaining two thirds. As a result, the total achievable conversion efficiency of the bioprocess becomes 69%.

Table 13.7   Techno-economic evaluation of a conceptual design
for biological hydrogen production from potato steam peels.

	€/h
Depreciation, maintenance, insurances and overhead	114.94
Personnel	10.00
Potato steam peels	39.37
Enzymes for hydrolysis	0.02
Caustic	6.27
Electricity	7.33
Total cost	177.93

13.3.2   Economic evaluation of a conceptual design

On the basis of results obtained and improvements that are deemed feasible on the
short to medium term, the production costs of hydrogen in an industrial plant have
been calculated. The assumed conditions in the conceptual design are a capacity
of 17 and 40 kg hydrogen/h in the thermophilic and photo-heterotrophic fermentation, respectively, amounting to 57 kg hydrogen/h in total. The required volume
of the trickle bed reactor used for the thermophilic fermentation is $450\,m^3$. This
reactor is run at 70°C and a reduced pressure of 0.5 bar. The main dimension of
the tubular photobioreactor is its surface area which amounts to 12 ha in total. The
photobioreactor operates at 35°C and 2.5 bar. The dry off gas from the thermophilic fermentation contains 50% hydrogen whereas the off gas from the photobioreactor contains >85% hydrogen.

Most apparatus required for the industrial plant (reactor vessels, compressors,
heat exchangers, etc.) are commercially available with the exception of the tubular photobioreactor. The cost of the available apparatus has been derived from the
handbook of the Dutch Association of Cost Engineers and using a Lang factor of 4.
The cost of the photobioreactor has been estimated on the basis of an experimental
installation ($400\,m^2$) employed for cultivation of other phototrophic micro-organisms.
For operation of the industrial plant continuous operation of 8000 h per year was
assumed with two operators working on an 8 h/day shift.

Potato steam peels were used as biomass, to be acquired at a cost which is
presently in competition with the amount paid by the fodder industry in the
Netherlands.

Table 13.7 shows a preliminary estimate of the operating cost of the plant in
€/h. The total production cost of hydrogen amounted to €3.10/kg which is approximately three times the amount currently paid for hydrogen produced from fossil
fuels in large-scale installations.

13.4   CHALLENGES FOR BIOLOGICAL
HYDROGEN PRODUCTION

Biological hydrogen is aimed at providing a clean biofuel for use in fuel cells
of small-scale installations. As such it meets all the societal demands for clean

environment, sustainable energy production, independence of foreign countries and development of rural communities (see www.biohydrogen.nl). Notwithstanding, even though it seems realistic that a cleaner environment will need to be paid for, decrease in hydrogen production cost is the main challenge. Since the presented bioprocess is still in the early stages of development, there appears to be sufficient room for optimization of all process units such as, reactor design, and increase of system efficiency.

The development of sustainable hydrogen production systems is associated with the development of fuel cells. Pure hydrogen is the feed by choice for proton exchange membrane (PEM) fuel cells with an operating temperature of around 90°C. On the other hand, molten carbonate fuel cells (MCFC) or solid oxide fuel cells (SOFC) that operate at much higher temperatures (600–900°C), enable the application of methane as feed. Presently, no fuel cells have reached the market yet with competitive prices. It is still obscure which fuel cell will fulfil best the future demands of cost-effective sustainability in the automotive sector or the stationary grid.

In spite of the uncertainties described above, there is one great, globally acknowledged, certainty with respect to the need for sustainability to decrease emissions as described in the Kyoto protocol. As such, it is of prime importance to further develop and meet the challenges inherent to the introduction of new energy carriers such as hydrogen, which enable the most efficient conversion of renewable resources.

ACKNOWLEDGEMENTS

The results of this chapter have been produced by participants in the Biological Hydrogen Production project, supported by the Dutch Programme Economy, Ecology, Technology, a joint initiative of the Ministries of Economic Affairs, Education, Culture and Sciences, and Housing, Spatial Planning and the Environment (EETK99116).

REFERENCES

Akkerman, I., Janssen, M., Rocha, J.M.S., Reith, J.H. and Wijffels, R.H. (2003) Photobiological hydrogen production: photochemical efficiency and bioreactor design. In: *Bio-methane and Bio-hydrogen* (eds. J.H. Reith, R.H. Wijffels and H. Barten), pp. 152–155. Dutch Biological Hydrogen Foundation, Smiet Offset, The Hague.

Chen C-C., Lin C-Y. and Chang J-S. (2001) Kinetics of hydrogen production with continuous anaerobic cultures utilizing sucrose as the limiting substrate. *Appl. Microbiol. Biotechnol.* **57,** 56–64.

Claassen, P.A.M., van Lier, J.B., Lopez Contreras, A.M., van Niel, E.W.J., Sijtsma, L., Stams, A.J.M., de Vries S.S. and Weusthuis R.A. (1999) Utilisation of biomass for the supply of energy carriers. *Appl. Microbiol. Biotechnol.* **52,** 741–755.

Claassen, P.A.M., Budde, M.A.W., van der Wal, F.J., Kádár, Zs., van Noorden, G.E. and de Vrije, T. (2002) Biological hydrogen production from biomass by thermophilic bacteria. *Proceedings of 12th European Conference and Technology Exhibition on Biomass for Energy, Industry and Climate Protection*, Amsterdam, 17–21, June 2002, pp. 529–532.

de Vrije, T. and Claassen, P.A.M. (2003) Dark hydrogen fermentations. In: *Bio-methane and Bio-hydrogen* (eds. J.H. Reith, R.H. Wijffels and H. Barten), pp. 150–152. Dutch Biological Hydrogen Foundation, Smiet offset, The Hague.

de Vrije, T., de Haas, G.G., Tan, G.B., Keijsers, E.R.P. and Claassen, P.A.M. (2002) Pretreatment of Miscanthus for hydrogen production by *Thermotoga elfii*. *Int. J. Hydrogen Energ.* **27**, 1381–1390.

Dürre, P. (1998) New insights and novel developments in clostridial acetone/butanol/ethanol/isopropanol fermentation. *Appl. Microbiol. Biotechnol.* **49**, 639–945.

Hahn-Hägerdahl, B., Jeppson, H., Skoog K. and Prior B.A. (1994) Biochemistry and physiology of xylose fermentation by yeasts. *Enzyme Microbiol. Technol.* **19**, 933–943.

Lin, C-Y. and Chang R-C. (1999) Hydrogen production during the anaerobic acidogenic conversion of glucose. *J. Chem. Technol. Biotechnol.* **74**, 498–500.

Lynd, L.R., Elander R.T. and Wyman C.E. (1996) Likely features and costs of mature biomass ethanol technology. *Appl. Biochem. Biotechnol.* **57/58**, 741–761.

Maddox, I.S., Qureshi, N. and Gutierrez N.A. (1993) Utilization of whey by clostridia and process technology. In: *The clostridia and biotechnology* (ed. D.R. Woods) pp. 343–370. Butterworth-Heinemann, Stoneham.

Rachman, M.A., Nakashimada, Y., Kakizono, T. and Nishio, N. (1998) Hydrogen production with high yield and high evolution rate by self-flocculated cells of *Enterobacter aerogenes* in a packed-bed reactor. *Appl. Microbiol. Biotechnol.* **49**, 450–454.

Taguchi, F., Mizukami, N., Saito-Taki, T. and Hasegawa K. (1995) Hydrogen production from continuous fermentation of xylose during growth of *Clostridium* sp. strain no.2. *Can. J. Microbiol.* **41**, 536–540.

van Groenestijn, J.W., Hazewinkel, J.H.O., Nienoord, M. and Bussmann, P.J.T. (2002) Energy aspects of biological hydrogen production in high rate bioreactors operated in the thermophilic temperature range. *Int. J. Hydrogen Energy* **27**, 1141–1147.

van Niel, E.W.J., Budde, M.A.W., de Haas, G.G., van der Wal, F.J., Claassen, P.A.M. and Stams, A.J.M. (2002) Distinctive properties of high hydrogen producing extreme thermophiles *Caldicellulosiruptor saccharolyticus* and *Thermotoga elfii*. *Int. J. Hydrogen Energ.* **27**, 1391–1398.

van Niel, E.W.J., Claassen, P.A.M. and Stams A.J.M. (2003) Substrate and product inhibition of hydrogen production by the extreme thermophile, *Caldicellulosiruptor saccharolyticus*. *Biotechnol. Bioeng.* **81**, 255–262.

PART THREE: Fuel cells

Section IIIA: System design

14

Fuel cell principles and prospective

N.M. Sammes, Y. Du and R. Bove

14.1 FUEL CELL BACKGROUND AND CLASSIFICATION

A fuel cell (FC) is an electrochemical device that converts the chemical energy of gaseous (e.g. hydrogen, natural gas, and biomass derived gas) or solid (mainly coal) fuels directly into electrical energy (and heat) via an electrochemical process. The basic physical structure of a fuel cell consists of an electrolyte layer in contact with a porous anode and cathode on either side. In contrast to conventional methods, fuel cells offer a fundamentally different way of generating electrical power from a variety of fuels. For example, natural gas, coal gas, and biomass derived gas may be used, as well as pure hydrogen. A fuel cell can be regarded as a battery with an external rather than internal energy source. Although a fuel cell has components and characteristics similar to a battery, it has several differences. Prominent among these differences is the fact that a battery is an energy storage device, while a fuel cell is an energy conversion device. All batteries will eventually run down or require recharging but a fuel cell generally can produce electricity as long as the fuel and oxidant are supplied to the electrodes. In reality, the life of a fuel cell is limited by component degradation, primarily from corrosion. As for batteries, in a fuel cell

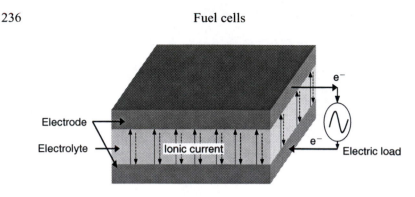

Figure 14.1   Schematic representation of a fuel cell (dashed arrows are referred to positive ions, while solid to negative ions).

an ionic current flows from one electrode to the other through an electrolyte, while, simultaneously, an electronic current flows through an external electric circuit, thus generating an electric current (Figure 14.1). According to the specific technology, ions flowing through the electrolyte can be negative or positive, thus flowing from the cathode to the anode or vice versa.

The reactions that occur inside a fuel cell depend on the type of fuel cell and the type of gas provided. However, the following reaction is common for all the fuel cells fed with a hydrogen-rich gas mixture:

$$H_2 + \frac{1}{2}O_2 \rightarrow H_2O \qquad (14.1)$$

Reaction (14.1) is exothermic, presenting an enthalpy of reaction of 285.8 kJ/mole under standard conditions (EG & G Service Parsons 2002). Depending on the efficiency of the fuel cell, part of this energy is converted into electricity, while the remaining is converted into heat.

Fuel cells are normally classified according to the material used as the electrolyte:

- alkaline fuel cells (AFC);
- phosphoric acid fuel cell (PAFC);
- proton exchange membrane fuel cell (PEMFC);
- molten carbonate fuel cell (MCFC);
- solid oxide fuel cell (SOFC).

AFC have the advantages of presenting very high electrical efficiency, but they can work only on very pure hydrogen and oxygen. This limitation makes AFCs very expensive, and so their applications are limited to some niches, where the efficiency has a primary role, compared to the cost. The first AFC system was developed in 1952 by the group of F.T. Bacon at the University of Cambridge, that realized a 5 kW system, operating on H_2 and O_2 (Carrette *et al.* 2001). The development of the AFC, however, drastically accelerated during the 1960s and 1970s, when NASA decided to equip the Apollo space vehicle with an AFC, to provide on board electricity. Currently, United Technologies Corporation (UTC) Fuel Cells produces AFCs for the on board power of the space shuttle orbiter (Figure 14.2).

Figure 14.2   The kilowatt AFC system used for the space shuttle, www.utcfuelcells.com

PAFCs have now reached commercial maturity. The PC 25, produced by UTC Fuel Cells provides 200 kW and is designed to run on natural gas. The system has a remarkable reliability (99.9999% using the probabilistic risk analysis) and an electrical efficiency of about 37%. However, the capital cost is too high compared to traditional systems. The reasons for the high capital cost are both the type of technology (i.e. the expensive materials) and the complex stack design (Cropper 2002). Since PAFC present a reliability that is well above traditional systems, for some applications they represent the answer to electricity quality and availability needs. Table 14.1 reports some examples of these applications and the relative costs of 1 h of electricity interruption (Pehnt and Ramesohl 2003).

Table 14.1   Example of hourly cost of electricity outage in USA (Pehnt and Ramesohl 2003).

Premium power user	Typical cost for 1-h interruption (US $)
Cellular communication	41,000
Telephone ticket sale	72,000
Air reservation system	90,000
Semiconductor manufacturer	2,000,000
Credit card operation	2,580,000
Brokerage firm	6,480,000

Other PAFC developers include Fuji Electric (Japan) that recently launched a second-generation commercial 100 kW system, labeled FP-100F, and the Indian Bharat Heavy Electrical Limited.

PEMFCs are able to generate high power density, thus they are attractive for both stationary and automotive applications. One of the main advantages of a PEM is that the electrolyte is solid, thus avoiding problems related to liquid management. The system range varies from some hundred watts to several kilowatts. PEM was the first fuel cell to be used in space, where the Gemini program employed a 1 kW system to provide auxiliary power. Nevertheless the membrane used at that time was not stable enough, and so NASA decided to switch to the AFC (Carrette *et al.* 2001). A major breakthrough in PEM research came with the introduction of Nafion® membrane material developed by DuPont. Nafion® is still the most widely used material for the PEM membrane, although in 1987, the Canadian company Ballard introduced the Dow® membrane, that allows the cell to operate at four times higher current densities than the standard Nafion®, maintaining the same voltage (Carrette *et al.* 2001). A detailed review of the PEM evolution is given in Appleby and Yeager (1986) and Gottesfeld and Zawodnisky (1997). At the present time, there are several companies involved in PEM development, for both stationary and automotive applications, among them UTC Fuel Cells (USA), Nuvera Fuel Cells (Italy/USA), Arcotronics (Italy), Plug Power (USA), and Ballard (Canada).

PEMFCs are described in more detail in Chapters 16 and 17.

MCFCs are composed of two nickel-based porous electrodes and an electrolyte in the liquid form, embedded in a matrix. The electrolyte presents an acceptable ionic conductivity at a temperature range of 600–700°C. The typical operating temperature of MCFC is about 650°C.

MCFCs are in the pre-commercial phase and power plant units, in the kilowatt and megawatt sizes have been realized all around the world (mainly USA, Europe, and Japan). Although MCFC has shown very high conversion efficiency, fuel flexibility and good response to electric load change, there are still technical issues that must be overcome, to successfully enter the market. High costs and short lifetime currently represent the two main drawbacks of MCFC. The operating experiences, related to the installed power plants, however, are expected to provide useful feedback for further technology development. The main developers are Fuel Cell Energy (FCE) and GENCELL in the USA, MTU Friedrichshafen and Ansaldo Fuel Cells in Europe, KEPCO, Marubeni and Ishikawajima-Harima Heavy Industries in Asia.

SOFCs employ a solid oxide material as the electrolyte, thus leakage related problems are avoided. On the other hand, due to the high operating temperature (typically 800–1000°C) the solid structure has cracking problems, because of the different thermal expansion of the electrodes, electrolyte, and current collectors. SOFC are currently manufactured using at least two different main designs: planar and tubular. Due to the mechanical stress distribution in a tubular geometry, large size fuel cell systems have been realized only employing tubular cells. In particular, the Siemens–Westinghouse has been employed for stand alone, as well as hybrid power plants realization. The main drawback of tubular SOFC, however, is that the performance is expected to be lower than that of planar, because of the current path. In order to solve the thermal-expansion-related problems, several companies, and research centers are trying to lower the operating temperature.

There are several companies that are currently developing SOFC, each of them usually have proprietary cell configurations, such as anode, cathode or electrolyte supported cells, all ceramic, metal interconnected, etc.

Among the others, the most important SOFC developers are FCE, Siemens–Westinghouse, Delphi, SOFCo-ESF, and General Electric in North America, Rolls-Royce and Sulzer-Hexis in Europe, Mitsubishi Heavy Industries, Tokyo Gas Co and Osaka Gas Co for Asia, and the Australian company, Ceramic Fuel Cells Limited (CFCL).

MCFCs and SOFCs operate at higher temperature than AFC, PAFC, and PEM and for this reason they are also called high temperature fuel cells.

Operating at high temperature leads to several advantages, such as:

- No expensive catalysts are needed for the electrochemical reactions.
- CO is not harmful for the cell, but it represents an additional type of fuel.
- Outlet gas temperature is high enough for combined heat and power (CHP) applications, or to recover the thermal energy through a gas turbine or steam turbine.

On the other hand, the main issues of high temperature are currently:

- long start-up due to the high temperature;
- sealing (SOFC);
- corrosion (MCFC).

MCFC and SOFC description is the object of Chapter 18. Table 14.2 summarizes the main characteristics of each technology.

Fuel cells are sometimes classified according to other criteria, such as:

- The operating temperature, distinguishing between high- and low-temperature fuel cells.
- The way they reform the fuel. Some types of fuel cells internally reform the fuel, while others need an external fuel processor. The first are usually called internal reforming fuel cells, while the second, external reforming.
- The type of fuel used. According to this classification, there are natural gas fuel cells, hydrogen fuel cells, direct methanol fuel cells, and direct carbon fuel cells. When the hydrogen is produced through water electrolysis, using a renewable source of energy (typically wind or sun), the fuel cell is usually called a regenerative fuel cell.

The fuel cell effects and principles were discovered by the Swiss professor Christian Friedrich Schoenbein (1777–1868) at the University of Bassel in 1838 (Bossel 2000). After initial experiments in 1839, Sir William Robert Grove (1811–1896), a London lawyer with a strong engineering background, invented the fuel cell concept and in 1845 demonstrated an apparatus for replacing batteries. This was regarded as the first fuel cell in the world (Grove 1839). The Grove cell, as the new fuel cell was called, operated at room temperature with dilute sulfuric acid as the electrolyte and platinum (Pt) electrodes. In 1889, Mond and Langer used porous electrodes to improve the fuel cell performance to $3.5 \, mA/cm^2$ at 0.73 V, while operating on hydrogen and oxygen (Acres 2001). In 1933, Bacon, the first recipient of the Grove Medal, started developing a fuel cell system capable of delivering a power density of $1000 \, mA/cm^2$ at 0.8 V (Acres 2001). Bacon's cell was operated on hydrogen and oxygen at elevated pressure.

Over the years, there have been many attempts to develop fuel cells as power sources. In the early 1960s, AFC were developed as electrical generators and a source of drinking water for Apollo spacecraft; which could be considered as a significant milestone for fuel cell technology development (Cacciola et al. 2001). The recent drive for more efficient and less polluting distributed power-generating

Table 14.2   Main characteristics of each fuel cell technology.

	PEMFC	AFC	PAFC	MCFC	SOFC
Operating temperature	60–100°C	60–120°C	160–220°C	600–650°C	600–1000°C
Electrolyte	PEM	Potassium hydroxide	Liquid phosphoric acid	Molten carbonate	Ceramic
Charge carrier	H^+	OH^-	H^+	CO_3^-	O^-
Catalyst	Pt	Pt	Pt	Ni	Perovskites/Ni
Prime cell components	Carbon-based	Carbon-based	Graphite-based	Stainless steel-based	Ceramic
CO impact	Poison	Poison	Poison	Fuel	Fuel
S impact	Few studies	Unknown	Poison	Poison	Poison
External reformer for CH_4	Yes	Yes	Yes	Yes/No	Yes/No
Co-generation	No	No	Yes	Yes	Yes
Efficiency (LHV)	30–45%	30–50%	30–45%	45–60%	45–75%
Major advantages/disadvantages	Quick start-up H_2 preferable heat and water management issues	Pure H_2 good performance expensive	Lower performance than AFC expensive	Wider fuel choices co-generation liquid electrolyte	All ceramic wider fuel choices co-generation durability
Example of installed units	250 kW (Ballard)	12 kW (for space shuttles) (ZeTek)	PC25 > 40.000 h (UTC), 1.2 MW 5000 h (Milan)	2 MW 4000 h (FCE)	300, 220 kW; 100 kW-17,500 h (SWPC)
Major applications	Transportation stationary	Space stationary transportation	Stationary	Stationary transportation (APU)	Stationary transportation (APU)

LHV: low heating value; APU: auxiliary power unit; SWPC: Siemens–Westinghouse Power Corporation.

technologies has resulted in substantial resources being directed into fuel cell development (Badwal and Foger 1996; Department of Energy 2003).

14.2 MERITS OF FUEL CELLS

Fuel cells offer a new alternative to conventional electrical power sources for many applications. Singh *et al.* (2001) have shown that 33% of carbon dioxide emissions come from industrial use, 32% from transportation, 20% from residential, and 15% from commercial use. Fuel cells, have lower emissions than an equivalent power plant using natural gas. Table 14.3 gives an emission comparison between a state of the art natural gas combined power plant with a UTC PC-25 PAFC.

The comparison of Table 14.3 is very conservative, in fact, while the UTC PC-25 system presents a nominal power of 200 kW, combined power plants, for achieving good performance, must be realized in the range of several hundred megawatts. If traditional energy systems are considered in the range of the UTC PC-25 nominal power, traditional systems' emissions are significantly higher. In a distributed scenario of energy generation, it is impossible to achieve the performance given in Table 14.3 for traditional systems.

Fuel flexibility is another benefit of fuel cell technology. Depending on the type of fuel cells and the fuel processing/reformer, any of the following fuels can be used: hydrogen (H_2), methane (CH_4), methanol (CH_3OH), ethanol (C_2H_5OH), propane, anaerobic digester gas, natural gas, coal gas, landfill gas, gas from gasification, gasoline, or even diesel. At present, gaseous H_2 is the fuel used for the most applications due to its high reactivity when suitable catalysts are used, its ability to be produced from hydrocarbons for terrestrial applications, and its high energy density when stored cryogenically for closed environment applications such as in space. However, alternatives such as CH_4 can be electrochemically oxidized in a fuel cell, while air is the commonly used oxidant. To use hydrocarbon fuels, such as natural gas or even gasoline, a fuel processor or reformer is often needed. This can be either an external reformer (where the fuel is processed outside the fuel cell) or an internal reformer (where the fuel is processed inside the fuel cell).

As implicitly stated above, fuel cells have the relevant advantage of presenting the same performance at full as well as at partial load. The low difference in performance is mainly due, not to the fuel cell, but to the components of the system. As a consequence, the values of Table 14.3 can be considered valid, with an acceptable tolerance, for all fuel cell systems' installed power.

Table 14.3   Emissions comparison between a combined power plant and a UTC PC-25 PAFC.

	Combined Plant (Brayton+Rankine)	FC UTC PC-25
NO_x (mg/kW-h)	249.6	8
CO (mg/kW-h)	57.4	15
SO_2 (mg/kW-h)	25.4	0
VOC (mg/kW-h)	5.6	2.5

Other benefits of fuel cells, include:

- High energy conversion efficiency, and consequently, reduction of fuel consumption.
- Very low noise emissions, related only to the balance of plant (BoP), and not to the fuel cell module (where no parts are in movement).
- Improved reliability, due to the absence of moving parts.

On the other hand, the drawbacks of fuel cells include:

- Expense.
- Public unfamiliarity with the technology.
- Lack of fuel infrastructure if fed with hydrogen.

14.3   FUEL CELL BASICS

As it is well known from thermodynamic principles, a reaction occurs in the direction that minimizes Gibbs free energy (G). A positive free energy associated with a generic reaction indicates that this cannot occur spontaneously. The contrary is true, when ΔG is negative. When the reaction is at equilibrium, instead, the change in G is zero.

The variation of G also quantifies the maximum work obtainable. When a reaction is conducted at constant pressure and temperature, ΔG is expressed by:

$$\Delta G = \Delta H - T\Delta S \tag{14.2}$$

where H represents the enthalpy, T the temperature, and S the entropy.

According to the first thermodynamic law for open systems at steady state conditions, $\Delta H = Q - W$, where Q is the thermal energy supplied to the system and W is the work produced by the system. If $W = 0$ (combustion) the heat released is quantified by the enthalpy variation. In analogy with internal combustion engines (ICE), where ΔH represents the thermal energy obtainable from total combustion of a fuel, the efficiency of a fuel cell is given by the ratio between the work produced and the change of enthalpy; that is, in an ideal case (Kordesh and Simader 1996):

$$\eta_{id} = \frac{\Delta G}{\Delta H} \tag{14.3}$$

This definition, however, can lead to a misunderstanding of the first law of thermodynamics. For some reactions, in fact, the entropy change is negative, thus ΔG is higher than ΔH. According to definition (14.3), the resulting ideal efficiency is larger than one. This is the case, for example, for the following reactions of carbon:

$$C + \frac{1}{2}O_2 \rightarrow CO \tag{14.4}$$

$$C + O_2 \rightarrow CO_2 \tag{14.5}$$

that, under standard conditions, present, respectively, ΔG equal to 137.3 and 394.6 kJ/mol and ΔH equal to 110.6 and 393.7 kJ/mol, thus giving an ideal efficiency of 1.24 and 1.002, respectively. In this case, however, thermal energy must be provided to the fuel cell, and this quantity is not taken into account in expression (14.3).

However, definition (14.3) is useful to compare the efficiency of ICE and fuel cells, as explained in Section 14.4.

Considering a general reaction, that can take place in a fuel cell:

$$\alpha A + \beta B \rightarrow \chi C + \delta D \tag{14.6}$$

the relative ΔG can be calculated through:

$$\Delta G = \Delta G^0(T) + RT \ln \frac{P_C^\chi P_D^\delta}{P_A^\alpha P_B^\beta} \tag{14.7}$$

where $\Delta G^0(T)$ is the change in free energy at standard pressure, R the universal gas constant and P_i the partial pressure of the ith gas. From expression (14.7), the reversible voltage of a fuel cell can also be calculated (Chen 2003):

$$E = \frac{\Delta G}{2F} = E^0 + \frac{RT}{2F} \ln \frac{P_C^\chi P_D^\delta}{P_A^\alpha P_B^\beta} \tag{14.8}$$

where F is the Faraday constant. Expression (14.8) is the Nernst equation.

If hydrogen is used as the reactant, and oxygen as the oxidant, Equation (14.6) can be re-written in the form:

$$H_2 + \frac{1}{2} O_2 \rightarrow H_2O \tag{14.9}$$

and relation 14.8 becomes:

$$E = E^0 + \frac{RT}{2F} \ln \frac{P_{H_2O}}{P_{H_2} P_{O_2}^{0.5}} \tag{14.10}$$

When a fuel cell operates, however, the actual voltage is lower than that computed using relation (14.8), due to activation, diffusion and ohmic losses, as well as internal or external cross-over.

The reduction of the voltage is proportional to the current circulating through the cell. Figure 14.3 shows a typical voltage–current density of a fuel cell, while a full description of the phenomena is given in Chapter 15, where fuel cell operation models are presented.

The effect of a pressure variation can be assessed through:

$$\left(\frac{\partial E}{\partial P} \right)_T = \frac{-\Delta v}{nF} \tag{14.11}$$

where Δv is the change in volume of the reaction. Increasing the total pressure causes the volume to be reduced. Moreover, an increase in the total pressure leads

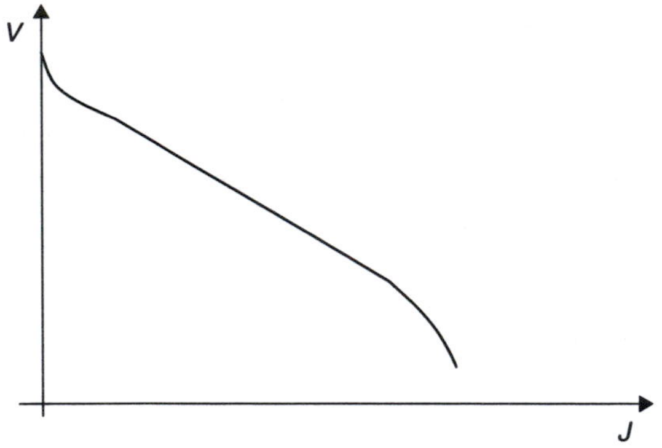

Figure 14.3   Typical voltage–current density curve of a fuel cell.

to an efficiency enhancement (EG & G Service Parsons 2002). However, pressurizing the fuel cell section, implies a compliance for the system.

The effect of the temperature variation is given by:

$$\left(\frac{\partial E}{\partial T}\right)_P = \frac{\Delta v}{nF} \tag{14.12}$$

When reaction (14.9) is considered, the entropy variation is negative, thus an increase of T leads to a reversible potential reduction. The effect of an increase in the temperature, however, is different for the real voltage. When the temperature is raised, in fact, the losses are reduced. The overall effect is an increase in the real voltage (EG & G Service Parsons 2002).

14.4   COMPARISON WITH INTERNAL COMBUSTION ENGINES

The main difference between fuel cells and thermal engines is that in fuel cells no combustion occurs, since the reactant is electrochemically oxidized, releasing electrons. As a consequence, fuel cells emit much less pollutants, in particular NO_x and SO_x, which are related only to the balance of plant (BoP). Moreover, the absence of moving parts, as mentioned in Section 14.2, reduces noise emissions and increases the reliability.

The absence of combustion in fuel cells, however, has widely led to the misconception that fuel cells are more efficient than ICE because, while thermal energy's efficiency is limited by the Carnot principles, fuel cells are not. One of the main reasons of the higher efficiency of fuel cells, compared to ICEs, is not the Carnot principle, but the fact that combustion is a highly irreversible process.

To clarify this concept, consider an ICE operating between a thermal source at T_A and a thermal sink at T_B. The maximum efficiency is obtainable through the irreversible Carnot cycle, that is:

$$\eta_{rev} = 1 - \frac{T_B}{T_A} \tag{14.13}$$

The reversible efficiency is increased if the maximum temperature (i.e. the operating temperature) is increased.

For a fuel cell, combining Equations (14.2) and (14.3), the reversible efficiency is given by:

$$\eta_{rev} = 1 - T\Delta S \qquad (14.14)$$

If ΔS is positive, as for the case of hydrogen oxidation, an increase in the fuel cell operating temperature decreases the ideal efficiency of the fuel cell, while it increases that of the ICE (i.e. independent from the gas type, but depends only on the thermal level). The opposite trends of fuel cell and ICE reversible efficiencies as a function of temperature, leads to a break-even point, where the two values are the same. Hammes (2004) quantified this temperature value to be around 1200 K for SOFC operating on hydrogen. The consequence is that an SOFC operating at 1000°C on pure hydrogen presents a lower reversible efficiency than a Carnot cycle operating at the same temperature.

In real conditions, however, fuel cell efficiencies (Table 14.2) are significantly higher than those of ICE.

14.5   FROM SINGLE CELLS TO SYSTEMS

A single fuel cell, when operating at nominal current density, typically produces a voltage that is lower than 1 V. If directly connected to an external electric load, even a very low resistance would let the voltage drop to zero. According to Ohm's law, in fact, voltage and current are related to the following:

$$\Delta V = IR \qquad (14.15)$$

where ΔV is the voltage drop, I is the current flowing through the external load, characterized by a resistance R. Considering, for example, a single cell with an active surface of 100 cm^2, and operating at 100 mA/cm^2, a resistance of 0.1 Ω would cause a voltage drop of 1 V.

For this reason, and in order to obtain a power unit capable of producing a convenient electrical power, single cells are usually connected in series and parallel. Multiple single cells are called a "stack", because in the case of planar single cells, the connection is achieved by stacking the cells; that is, forming a pile. An example of a fuel cell stack is shown in Figure 14.4.

Although a stack can provide electricity, a fuel cell system requires several other components that form the so-called BoP. As schematically shown in Figure 14.5, the three main parts composing the system are: the fuel processing, the power section, and the power-conditioning units. The fuel processor is needed to obtain a hydrogen rich gas mixture, with specific requirements that depend on the fuel cell technology. If the fuel is natural gas, for example, the fuel processor is usually composed of a desulfurizer, to reduce the amount of sulfur to a desired concentration, and a steam reformer reactor, which converts natural gas into a mix of H_2, CO, CO_2, and H_2O.

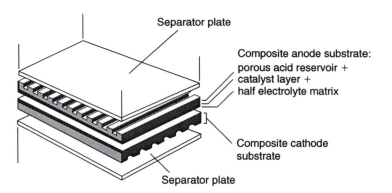

Separator plate

Composite anode substrate:
porous acid reservoir +
catalyst layer +
half electrolyte matrix

Composite cathode
substrate

Separator plate

Figure 14.4   Schematic of a fuel cell stack.

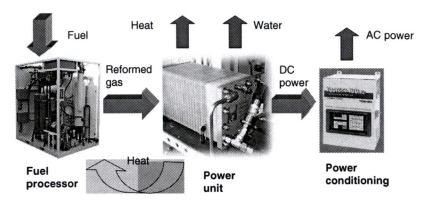

Figure 14.5   Schematic representation of a fuel cell system.

In the case of PEM, the fuel processor also contains one or more shift reactors that convert the CO coming from the reformer into CO_2 and H_2. The power section is composed of one or more fuel cell stacks and it is usually called the fuel cell module. Finally, the power-conditioning system is needed to convert the DC electric current into AC, with specific frequency, active and reactive power.

The BoP, currently accounts for about 2/3 of the total cost of the system, and so, in order to reduce system costs, a strong research and development activity is required for BoP optimization and cost reduction.

14.6   CONCLUSIONS

Fuel cells are electrochemical devices capable of converting the energy content of a gaseous, liquid, or solid fuel directly into electricity, thus avoiding the efficiency reduction, associated with combustion. Different types and solutions have been proposed, since the first fuel cell was created in the 19th century. Although each technology presents advantages and disadvantages, all fuel cells proposed so far, present high capital cost that make them non-competitive with traditional energy conversion systems.

REFERENCES

Acres, G.J.K. (2001) Recent advances in fuel cell technology and its applications. *J. Power Sources* **100**, 60–66.

Appleby, A.J. and Yeager, E.B. (1986) Solid Polymer Electrolyte Fuel Cells (SPEFCs). *Energy* **11**(1–2), 137–152.

Badwal, S.P.S. and Foger, K. (1996) Solid oxide electrolyte fuel cell review. *Ceram Int.* **22**, 257–265.

Bossel, U. (2000) *The birth of fuel cells*. European Fuel Cell Forum Publishers, Oberrohrdorf, Switzerland.

Cacciola, G., Antonucci, V. and Freni, S. (2001) Technology update and new strategies on fuel cells. *J. Power Sources* **100**, 67–79.

Carrette, L., Friederich, K.A. and Stimming, U. (2001) Fuel cells-fundamentals and applications. *Fuel Cells Fundament. Syst.* **1**(1), 5–39.

Chen, R. (2003) Thermodynamics and electrochemical kinetics. In: *Fuel Cell Technology Handbook*. CRC Press, USA.

Cropper, M. (2002) *Why is Interest in Phosphoric Acid Fuel Cells Falling?* Fuel cell today's knowledge bank article. www.fuelcelltoday.com

Department of Energy (2003) *Fuel Cell report to Congress*. ESECS EE-1973.

EG & G Service Parsons (2002) *Fuel Cell Handbook*, 6th edn. US Department of Energy.

Gottesfeld, S. and Zawodnisky, T. (1997) *PEFC chapter*. In: *Advances in Electrochemical Science and Engineering*, Vol. 5, pp. 195–301, Wiley-VCH, New York.

Grove, W.R. (1839) On voltaic series and the combination on gases by platinum. *The London and Edinburgh Philosophical Magazine and Journal of Science*, 127–130, Richard and Taylor, London, UK.

Hammes, K. (2004) Fuel cell dogmas revisited. *Proceedings of the Second International Conference on Fuel Cell Science, Engineering and Technology*, Rochester, NY, June 14–16.

Kordesh, K. and Simader, G. (1996) *Fuel Cells and Their Applications*. VCH, Germany.

Pehnt, M. and Ramesohl, S. (2003) *Fuel Cells for Distributed Power: Benefits, Barriers and Prospective*. IFEU, Wuppertal Institut, Final Report.

Singh, P., Pederson, L.R., Simner, S.P., Stevenson, J.W. and Viswanathan, V.V. (2001) Fuel cell power generation systems. *Proceedings of 36th Annual Intersociety Energy Conversion Engineering Conference Advance Program*, 953–958.

15

Modeling of fuel cell operation

P. Lunghi and R. Bove

15.1 INTRODUCTION

In order to become commercially attractive, fuel cells need to be at least as cost effective as the traditional energy conversion systems. The cost is not only due to the investment, but also to operating the system itself. The high fuel cell efficiency allows to significantly reduce fuel consumption.

However, on the other hand, the costs related to the operation and maintenance (O&M), and to the stack purchasing are still relevant, due to technological obstacles that must be overcome, such as lifetime and, in some case, reliability and durability. Any of the above-mentioned requisites (i.e. efficiency, reliability and durability) are very important for fuel cells to be economically feasible.

As an example, the phosphoric acid fuel cells (PAFCs) have reached a remarkable proven reliability (99.9999% using the probabilistic-risk analysis) and an electric efficiency of about 37%, but the high capital cost is too relevant for it to be a competitor to the traditional systems. The reasons for the high capital cost are due to both the types of technology (i.e. the expensive materials), and to the complex stack design (Cropper 2002).

PAFC applications are limited to particular applications where the system reliability plays a critical role, as for example, power supply for hospitals, casinos and electronic business transactions (banking, credit card, ATM, etc.).

Proton exchange membrane (PEM) fuel cells, solid oxide fuel cells (SOFCs) and molten carbonate fuel cells (MCFCs) are nowadays considered to be the most promising fuel cell technologies, in terms of performance and costs. To make this promise come true, an intense research activity coupled with demonstration projects are needed. The lesson learned with the PAFCs suggests to carefully guide the design phase, making use of models, before creating the finished product. The final goal is to obtain a product that is a compromise between high performances (in terms of reliability and efficiency) and low capital cost. Even if the models are a guide for the design, the experimental activity generally provides the most significant feedback in terms of operation performance and limitations. However, experimental activities are time and (human and economic) resource consuming. Consequently, beyond the experimental activity, numerical simulations are generally performed to reduce the costs related to the experiments.

Nevertheless, the two activities must be integrated; that is, numerical results must be the guide for prototype realization and design of experiments, while numerical models and simulations need experimental verification.

15.2   ELECTROCHEMICAL AND THERMODYNAMIC MODELING

Numerical models for the electrochemical and thermodynamic analysis are designed and created, according to specific investigation goals; that is, the mathematical structure of the model must be driven by the specific necessity in the modeling process. In other words, if the investigation goal is, for example, to find a good cell geometry for electrochemical loss reduction, all the phenomena involved with a geometry variation must be accurately taken into account, as for example, the in-plane and cross-plane losses or the electrodes polarization phenomena. When the main purpose of the model is to predict the fuel cell performance, in order to analyze the whole system, the physical–chemical variables variation (such as gas concentration, temperature, pressure and current density) is not relevant. However the performances, in terms of power, heat and input requirements are very important.

According to this point of view, fuel cell models can be classified into:

- detailed or micromodels,
- analytical or macromodels.

The first type generally uses differential equations, integrated in a two- or three-dimensional (2D or 3D) domain. The resulting information is usually very detailed and related to the inside conditions of the part under analysis. Detailed models can be developed at a stack, single-cell or component (electrode, membrane, etc.) level. The results generally provide useful indication for an optimal design. Since the models are usually based on differential equations, the solution is found by making use of numerical integration methods, like finite element (FEM), finite volume and finite difference. The solution convergence is not always guaranteed and the related calculation time can be relevant.

Even if they produce less information than the detailed models, analytical models can be solved in a closed form or by using simple numerical methods. The main advantages are related to the very short computation time. Even if the analytical models are based on much easier assumptions, compared to the detailed models, the accuracy of the results can be still acceptable, although such models generally provide less information.

15.2.1 Micromodels

These models can be used at different levels: from a single-component phenomenon investigation to an entire stack model. In the present section, some examples illustrate the possible phenomena models related to some particular components. The modeling of a single cell and a stack, instead, is thoroughly analyzed, in terms of energy, momentum and mass conservation equations.

15.2.1.1 Components and related phenomena modeling

The improvement of single components, such as the electrodes and the electrolyte entails the knowledge of the physical and chemical phenomena related to the components. For the electrodes, for example, great emphasis is given to the polarization phenomenon producing a reduction in cell voltage. When a fuel cell is in operation, in fact, the real voltage is quite different from the open-circuit voltage (OCV). This reduction is due to three main kinds of losses: ohmic, activation and diffusion losses. The cumulative effect of the last two is generally referred to as overpotential and a general mathematical description of it is given in Section 15.2.1.2. All three losses are closely dependent on the materials, on the cell design and on the operating conditions of the components. In order to improve the components performance, and, consequently, the system performance, numerical models are widely used to limit these phenomena and several examples of models can be found in the literature.

The effective contact area between the electrolyte and the electrode, for instance, has strong influence on the effective resistance of an SOFC. If an ionic or electronic flux passes through a material, the resistance is given by:

$$R = \frac{\rho t}{S} \tag{15.1}$$

where ρ is the ionic or electronic resistivity, t is the thickness and S is the surface. This value can be much higher if the current flow is forced to pass only in one area of the surface.

Figure 15.1 shows the resistance change versus the number of discrete contact points. The asymptotic value is the resistance of the material, when a homogenous current flows through it. This result demonstrates how important an accurate cell design ensuring electrode–membrane contact is to reduce the electric resistance.

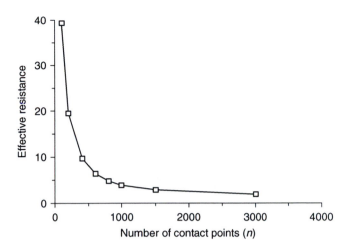

Figure 15.1   Effective resistance variation with respect to the number of discrete points of contact (Herbin 1992).

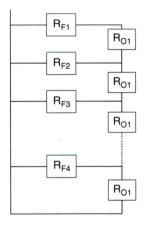

Figure 15.2   Schematization of a porous electrode (Bleise *et al.* 1995). R_F: ionic resistance; R_O: ohmic resistance.

The electrodes models are generally used to investigate the activation and concentration phenomena, or to predict the concentration of one or more chemical species.

Figure 15.2 illustrates a schematization of a porous electrode for the overpotential prediction (Bleise *et al.* 1995). The electrode is schematized by electronic and ionic resistances displaced in a parallel configuration (R_{Fi}), ending on the electronically conductive metal side (left side of the figure) by a Faraday intermediate phase resistance R_{Fi} and on the electrode–electrolyte surface (right on the figure) with ohmic resistance R_O. The values of the R_F resistances are determined by kinetic relations. The results obtained for the current density are reported in Figure 15.3.

Figure 15.4, instead, represents the oxygen concentration variation in a porous PEM cathode and the relative gas channel. The flux of the gas phase is based on

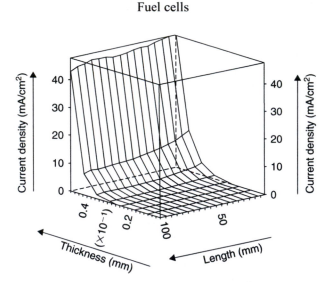

Figure 15.3   Current density distribution for an SOFC anode.

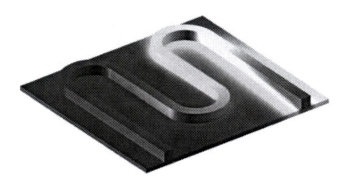

Figure 15.4   Oxygen concentration in a serpentine gas channel and the relative PEM
cathode obtained through FEMLAB software (reproduction under the authorization of
FEMLAB, COMSOL, Inc).

Maxwell–Stefan's equation. As explained under the section Electrochemical performance (in 15.2.1.2), Maxwell–Stefan's equation is just one among the different
relations that can be used to describe the diffusion in a porous medium.

   The previous examples describe just some of the copious mathematic models
used to investigate components-related phenomena. As previously mentioned, there
are many fields of investigation (overpotentials, ohmic resistance, current distribution, temperature gradients, geometry, thickness, etc.) and, for each of them,
specific, detailed equations are used. A complete description of all the models is
not possible and it is beyond the scope of the present book.

15.2.1.2   Single-cell and stack modeling

A stack model starts from a single-cell model, as well as a single-cell model is
defined using models for the single components. The choice of the equations to be

used for describing the phenomena acting in a cell component refers to the accuracy of the results. Scaling from a component to a cell level, some detailed information must be ignored. For example, the model described in Section 15.2.1.1 for the effective ohmic resistance calculation is too complex to be inserted in a cell or stack model. As it will be explained further on, in fact, a single-cell or stack model generally involves chemical, thermal, electrical and mass transport phenomena. If any of these phenomena is treated in a very detailed level, the structure would be too complex, causing problems related to the solution convergence and computational time. It is reasonable, instead, that such models, as for example the one for the effective cell resistance, are not integrated within the cell or stack model. They can be previously implemented and the obtained results can be used as an input for the cell or stack simulation. With reference to the resistance example, the results reported in Figure 15.1 can be used as an input to simulate a specific cell, rather than incorporate the resistance calculation in the model itself.

The choice of the mathematical complexity is a compromise between simplicity, accuracy and number of generable information.

Basic reactions in a fuel cell   Whatever the technology (alkaline fuel cell (AFC), PAFC, PEM, MCFC, SOFC) or the geometry structure (planar, monolithic, tubular), a fuel cell is always composed of a fuel and an oxidant channel, two porous electrodes and, interposed between them, an electrolyte. The electrolyte–electrode complex is also called membrane–electrode assembly (MEA).

When a hydrogen-rich gas is fed to the anode and a gas containing oxygen (typically air) to the cathode, an electrical potential is established between the anode and cathode. The reactions that take place inside the fuel cell depend on the specific technology and on the supplied gas.

An electrochemical reaction can be written in the following generic form (Zhu and Kee 2003):

$$v_f' \left(\sum_{k=1}^{K} n_{f,k} \chi_k \right) + v_o' \left(\sum_{k=1}^{K} n_{o,k} \chi_k \right) \quad \leftrightarrow \quad \sum_{k=1}^{K} v_{f,k}'' \chi_k + \sum_{k=1}^{K} v_{o,k}'' \chi_k \qquad (15.2)$$

where χ_k is the chemical symbol for the kth species (that can participate as an oxidizer, a fuel or both), v_f' and v_o' are the stoichiometric coefficients for the fuel and oxidizer mixtures, and $v_{f,k}''$ and $v_{o,k}''$ are the stoichiometric coefficients for the kth in the anode and oxidizer channels, respectively. Equation (15.2) can be written in a compact form as:

$$\sum_{k=1}^{K} v_k' \chi_k \quad \leftrightarrow \quad \sum_{k=1}^{K} v_k'' \chi_k \qquad (15.3)$$

where $v_k' = v_f' n_{f,k} + v_o' n_{o,k}$ and $v_k'' = v_{f,k}'' + v_{o,k}''$.

The generic formulation provided by Equations (15.2) and (15.3) allows to describe all fuel cells technologies. In the PEM and PAFC, the charge carriers are the H^+ protons that, passing through the electrolyte, move from the anode to the cathode, where they react with the oxygen. For the AFC, MCFC and SOFC, instead, the charge is transferred from the cathode to the anode by OH^-, CO_3^{2-} and O^{2-}, respectively. The result is that for PEM and PAFC, produced water is in the cathode channel, while for the others it is in the anode.

Electrochemical performance   In order to better understand expressions (15.2) and (15.3), consider, as an example, a PEM fuel cell fueled with pure hydrogen and air. The generic reaction formulations given above become:

$$H_2 + \frac{1}{0.42}(0.21O_2 + 0.79N_2) \rightarrow H_2O + \frac{79}{42}N_2 \qquad (15.4)$$

When no electric load is connected, there are no ionic and electronic currents through the MEA, and there are theoretically no losses. The OCV is described by the Nernst's equation:

$$E = E^0 + \frac{RT}{n_e F} \sum_{k=1}^{K} (v'_f n_{f,k} + v'_o n_{o,k} - v''_k) \ln\left(\frac{p_k}{p_o}\right) \qquad (15.5)$$

In expression (15.5), E^0 is Nernst's potential at standard conditions, R is the universal gas constant, n_e is the number of transferred electrons, F is the Faraday's constant, p_k is the partial pressure of the k-species and p_o is the reference pressure. Some experimental results show a deviation between the OCV value obtained applying expression (15.5) and the measured ones (Costamagna and Honegger 1998; Lunghi and Bove 2002). This phenomenon is due to both gas crossover and ionic current that can be present even when no load is connected to the cell (Minh and Takahashi 1995). An empirical parameter is given by Lunghi and Bove (2002) to correct the theoretical value to the actual one, while Baron *et al.* (2004) developed a semi-empirical model to predict the OCV reduction, starting from the results of Gödickemeier *et al.* (1996).

When a load is connected to the cell, an electric current flows through the external electric circuit, while ions (or protons) flow from an electrode to the other. The current, flowing inside the cell, causes a voltage reduction. This performance decrease is due to three different kinds of losses, commonly named as:

(1) ohmic loss,
(2) concentration loss,
(3) activation loss.

The first type is due to the resistance of the electrolyte to the ionic or protonic flow and to the resistance of the electronic flow through the electrodes. The internal resistance can be directly measured or theoretically deducted, as explained in Section 15.2.1.1. The resistivity can be estimated, if the materials and the operating temperature are known. Bossel (1992a) provides the following expression for the ionic (σ_i) and electronic (σ_e) conductivity:

$$\sigma_i = \frac{1}{\rho_i} = A_i e^{\frac{-B_i}{T}} \qquad (15.6)$$

$$\sigma_e = \frac{1}{\rho_e} = \frac{A_e}{T} e^{\frac{-B_e}{T}} \qquad (15.7)$$

where T is the temperature, A_i, A_e, B_i and B_e are specific constants, and ρ_i and ρ_e are the ionic and electronic resistivity, respectively.

When no current flows, the concentration of each gas species is the same in both the bulk channel and the electrode–electrolyte interface. Instead, when an electric current is generated, the gas in the electrode–electrolyte interface reacts, thus generating a concentration gradient. The voltage loss is given by:

$$\eta_{conc} = E_{bulk} - E_{interface} \qquad (15.8)$$

where E_{bulk} and $E_{interface}$ are the Nernst's potentials (expression 15.5) relative to the bulk channel and the electrode–electrolyte interface gas partial pressures. The equation that describes the mass transport inside a porous media is:

$$\frac{\varepsilon}{RT} \frac{\partial(x_i P)}{\partial t} = \nabla N_i + r_i \qquad (15.9)$$

where ε is the porosity, x_i is the molar fraction of the ith species, P is the pressure, N_i is the rate of mass transport and r_i is the rate of reaction inside the media. In the common case of steady-state condition, the left-hand term is equal to zero. Moreover, it is reasonable to assume that the gas reacts at the electrode–electrolyte interface, thus Equation (15.9), inside the porous media is:

$$\nabla N_i = 0 \qquad (15.10)$$

The most commonly used equations to compute N_i are Fick's law, Stefan–Maxwell's equation or the dusty-gas model. Suwanwarangkul et al. (2003) compared the results obtained using the three models for an SOFC anode. The calculations are computed assuming different operating conditions, porosities and gas mixtures. The results show that the dusty-gas model is the most accurate, but at the same time it requires off-line numerical solutions, while Stefan–Maxwell's and Fick's can be directly implemented. The applicability of each model, for any specific condition, is also thoroughly analyzed.

The activation loss is due to the energy barrier that the reactants must overcome. The equation generally used to describe this phenomenon is the Butler–Volmers' equation (Costamagna and Honegger 1998):

$$J = J_0 \left[\exp\left(\frac{\vartheta_a F \eta_{act}}{RT} \right) - \exp\left(\frac{\vartheta_c F \eta_{act}}{RT} \right) \right] \qquad (15.11)$$

where J is the current density, J_0 is the exchange current density and ϑ_a and ϑ_c are two empirical coefficients. Several expressions can be found in the literature to express the exchange current, involving the partial pressure and the temperature of the gas (Mizusoki et al. 1994; Costamagna and Honegger 1998; Iwata et al. 2000; Costamagna et al. 2001; Nagata et al. 2001; Chan et al. 2002; Zhu and Kee 2003).

When one of the two terms of Equation (15.11) can be neglected, the Tafel equation is deducted:

$$\eta_{act} = \frac{RT}{\vartheta F} \ln \frac{J}{J_0} \tag{15.12}$$

Once the losses are estimated, the actual cell voltage can be computed as:

$$V = E - (R + \eta_{conc} + \eta_{act})J \tag{15.13}$$

Figure 15.5 represents the typical voltage variation with respect to the current density (polarization curve). The difference between the OCV and the actual voltage represents the total loss. The curve is composed of three different regions. In the first, when the current density is low, the main contribution to the total loss is represented by the activation polarization. Increasing J, the ohmic loss is dominant with respect to the other losses. The concentration loss also increases with current density and it becomes a noticeable cause of voltage reduction at high current density.

Phenomena acting in a fuel cell   Electrochemical characteristics are generally coupled with thermal and fluid-dynamic phenomena. Temperature profiles inside the cell can be obtained through energy and mass balances, using differential equations that take into account mass transport from the cathode to the anode, reaction enthalpy, conduction, convective and radiation heat transfer. The general equation presents the following form (Bessette *et al.* 1995):

$$\rho c \frac{\partial T}{\partial t} + \rho c \vec{V} \nabla T = \nabla(K \nabla T) + q \tag{15.14}$$

In Equation (15.14), the left side represents the transient and the advection terms; ρ is the gas density, c is the heat capacity, K is the thermal conductivity and q is the heat generation. The term q takes into account the heat fluxes (due to conduction, convention, radiation and mass transport) and the heat generation. Bessette and Wepfer (1996) demonstrate that, for an SOFC stack, the heat transfer due to

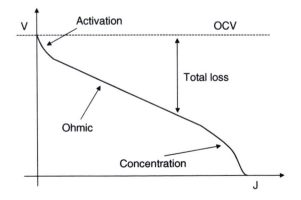

Figure 15.5   Typical fuel cell polarization curve.

radiation yields a consistent contribution to the total exchanged heat, and these results were confirmed by other subsequent studies (Kho *et al.* 2002; Burt *et al.* 2003). Since MCFCs also work at high temperatures, it is reasonable that the contribution of the radiation cannot be neglected even for MCFC stack simulations, while it can be ignored for low-temperature fuel cells.

The mass transport inside the porous media is described by Equation (15.9), while the fluid mechanics of the gas in the channel is described by Navier–Stokes' equation (momentum and mass conservation). A generic expression is reported below (White 1991):

$$\rho \frac{D\vec{V}}{dt} = \rho\vec{g} - \nabla p + \frac{\partial}{\partial x_j}\left[\mu\left(\frac{\partial v_i}{\partial x_j} - \frac{\partial v_j}{\partial x_i}\right) + \delta_{ij}\lambda \operatorname{div}\vec{V}\right] \tag{15.15}$$

$$\frac{\partial \rho}{\partial t} + \operatorname{div}\rho\vec{V} = 0 \tag{15.16}$$

The expression for the voltage prediction is replaced by a differential equation, when 2D or 3D detailed models are implemented, and the detailed distribution needs to be known (e.g. Bernardi and Verbrugge 1991; Ferguson 1992; Springer *et al.* 1993; Gurau *et al.* 1998; Heidebrecht and Sundmacher 2003):

$$\vec{j} = -\sigma\nabla\Phi \tag{15.17}$$

where Φ represents the electric potential.

Figure 15.6 is an example of a detailed model result. The figure shows the temperature variation in a PEM fuel cell, obtained by using the commercial software CFDRC-ACE.

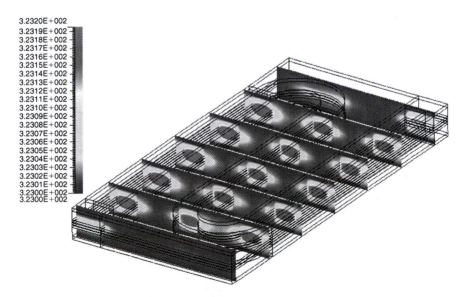

Figure 15.6   Temperature (K) variation in a PEM fuel cell.

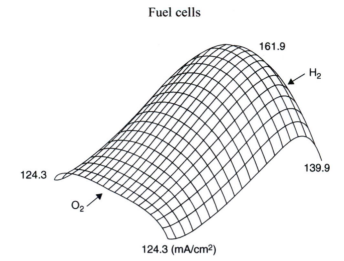

161.9

H₂

124.3

139.9

O₂

124.3 (mA/cm²)

Figure 15.7   Electric current density of a counter-current SOFC (Arato and Costa 1992).

Figure 15.7 represents the simulated current density distribution inside an SOFC, where anodic and cathodic gases flow in a counter-flow configuration (Arato and Costa 1992).

15.2.2   Macromodels

15.2.2.1   The need for macromodels

When fuel cell systems need to be modeled, one very important requirement is the output generation velocity when an input is provided. In order to understand how this requisite is important, consider, a straightforward simulation, where the effect of different fuel utilizations is assessed. If the current provided by the stack is set as an input variable, different fuel utilizations mean that different fuel flow rates to the stack are provided. Furthermore, the oxidant flow rate can now be computed, as a dependent variable, in order to achieve a desired gas temperature at the fuel cell outlet.

Since most of the systems' software (e.g. Aspen Plus™) uses a sequential modular method, every system component is an independent block that must be solved. The results of each block are the input for the next block. For the parametric analysis previously supposed, the fuel cell code must be used several times, as shown in the flow sheet calculation scheme of Figure 15.8.

In Figure 15.8, the anodic and cathodic flow rates are quantified using the coefficient of fuel and oxygen utilization, u_f and u_{ox}:

$$u_f = \frac{I/2F}{m_{H_2} + m_{CO}}$$

(15.18)

$$u_{ox} = \frac{I/4F}{m_{O_2}}$$

(15.19)

where m is the molar flow rate.

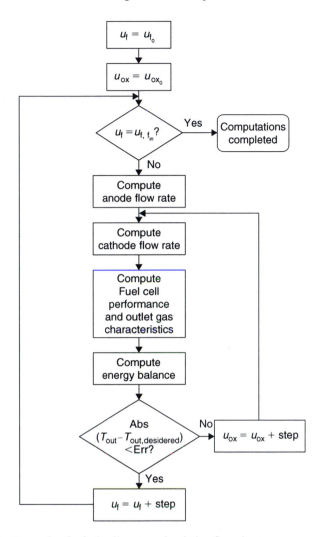

Figure 15.8   Example of a fuel cell system simulation flow sheet.

Once the initial values for u_f and u_{ox} are set, it is possible to compute anode and cathode flow rates (it is supposed that all the other information, such as stack dimension, current density and so on, were previously provided to the simulation software). Using this data, the fuel cell performances are computed, in terms of voltage and electric power. The anodic and cathodic outlet compositions can be obtained considering the electric current provided by the fuel cell, the electrochemical reactions and the initial gas composition. For the energy balance, Equation (15.14) can be replaced by a macrobalance equation (h represents the specific enthalpy):

$$m_{in,anode} h_{in,anode} + m_{in,cathode} h_{in,cathode}$$
$$= m_{out,anode} h_{out,anode} + m_{out,cathode} h_{out,cathode} - VI - Q_{loss} \qquad (15.20)$$

From Equation (15.20) the gas outlet temperature is computed. If the difference between the desired and the computed one is larger than a maximum allowable value, a different oxidant flow rate must be used. As can be observed, there are two loops in the flow sheet of Figure 15.8, each of them can be repeated quite a few times. In complex systems, the parameters involved can be many more than the two supposed in Figure 15.8, and so the fuel cell code can be used several hundred times. Moreover, a system is not only composed of the fuel cell stack, but by several other components, each of which requires time and numerical iterations to get the convergence.

Since analytical models do not require particular complex mathematic methods, and, consequently calculator performance, they were originally used also for both stack and single-part evaluation (Bossel 1992b). Nowadays, thanks to advanced calculators and to commercially available software for differential equations integration, the use of analytical models is restricted to system simulations or preliminary performance analyses.

15.2.2.2   Possible structures of macromodels

Macromodels for system performance evaluation are generally derived from micromodels (refer as an example to Costamagna et al. 2001), thus the Nernst's equation, as well as the equations for ohmic, activation and concentration losses are considered in a finite volume. However, while the ohmic loss depends only on the materials used, OCV, concentration and activation losses are related to gas concentration. In applying Equation (15.13), different results can be obtained if inlet or outlet gas compositions are used. In Hirchenhofer et al. (2002), it is noted that the voltage adjusts to the lowest electrode potential given by the Nernst's value, occurring at the anode and cathode exits. Even if Nernst's equation can be computed at the electrode exits, the current density in that section is, however, not known. What is known as input variable is generally the mean current density.

In order to appreciate the different results obtained using different gas compositions, Bove et al. (2005a) implemented Equation (15.13), using the inlet, the outlet and the average gas composition. The difference is noticeable, especially when the fuel utilization is varied.

For some applications using previous equations, a macrobalance yields high enough accuracy levels. On the contrary, for other applications, numerical models solved in a closed form or with easy numerical methods are needed.

Examples of these models are given by Bove et al. (2005b), Lunghi and Bove (2002) and Standaert et al. (1996). In these models, the gas variation along the cell is considered and the electrochemical performance is analytically estimated. A particularly easy form for the voltage is presented by Standaert et al. (1996):

$$V_{cell}^* = V_{eq}^*(0) - \alpha \left(1 - \exp\left(\frac{-\alpha}{ri_{in}}\right)\right)^{-1} u_1 \qquad (15.21)$$

In expression (15.21), the * symbol indicates that a linearization was conducted, V_{eq} is Nernst's potential (Equation (15.5)), u_1 is the fuel utilization, r is the cell resistance, i_{in} relates to the fuel composition and flow rate, and α is a parameter properly defined for the model.

15.2.3   Systems simulations

The simplicity of the analytical models makes the integration between the fuel cell and the system code fairly easy. Figure 15.9 is the schematic layout of an SOFC system, analyzed by Bove and Sammes (2004).

The system is composed of an internal reforming SOFC section, a pre-reforming reactor, a catalytic after-burner (AB) and three heat exchangers. Methane and water are sent to the pre-reformer reactor, where part of the methane is converted into a hydrogen-rich gas. The remaining methane is then internally reformed in the fuel cell section. Since the fuel utilization is never 100%, the unoxidized fuel coming from the anode is combusted in the AB. The oxygen needed for combustion is provided by the outlet cathodic gas. The heat content of the combusted gas is recycled for the pre-reforming process and for methane, water and air heating. The system is simulated using the commercial software Aspen Plus™, integrated with an SOFC code, based on the results of Bove et al. (2005b). The analysis allows to estimate the system performance under different operating conditions and to compare different fuel pre-processing technologies.

Lunghi et al. (2002) studied the possibility of integrating a dual-bed waste gasifier with an MCFC. Figure 15.10 gives a schematic representation of the simulated and analyzed system, along with some simulation results. Table 15.1 provides detailed results for the streams and the plant performance, obtained through the system simulation (Lunghi et al. 2002).

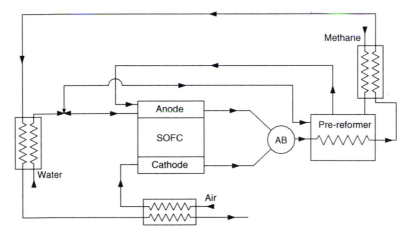

Figure 15.9   Schematic layout of an internal reforming SOFC with pre-reforming (Bove and Sammes 2004).

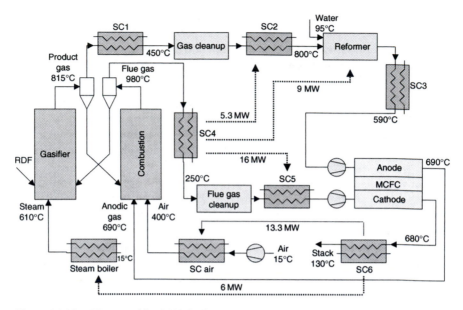

Figure 15.10   Waste gasifier-MCFC plant (Lunghi *et al.* 2002).

Table 15.1   Simulated characteristics of waste gasifier–MCFC power plant.

Compound (kmol/s)	Anode in	Anode out	Cathode in	Cathode out
H_2	0.1972	0.0493	0.0000	0.0000
CO	0.0700	0.0175	0.0000	0.0000
CO_2	0.0556	0.3086	0.3977	0.1973
CH_4	0.0001	0.0001	0.0000	0.0000
C_2H_4	0.0000	0.0000	0.0000	0.0000
C_2H_6	0.0000	0.0000	0.0000	0.0000
H_2O	0.1443	0.2922	0.3497	0.3497
N_2	0.0014	0.0014	0.9184	0.9184
Total	0.4686	0.6690	1.7995	1.4989
Temperature (K)	863.15	963.15	773.15	953.15
Fuel utilization factor	0.75			
Oxygen utilization factor	0.75			
CO_2 utilization factor	0.5			
Cell voltage (V)	0.81			
Cell power (MW)	31.33			

15.3   RELIABILITY AND DURABILITY MODELS

Since fuel cells are still novel technologies and only limited application experiences are available, reliability and durability are often a crucial factor for the economical feasibility of a fuel cell system. These two factors are much more related to the design phase, rather than to operation. Therefore, only an overview of the problem, along with some models, are illustrated in the following sections.

15.3.1 Mechanical characterization

Most of the models available in the literature are focused on the electrochemical and the heat transfer activities related to the cell operation, while only few authors conducted studies aimed at the mechanical characterization. This kind of analysis is of particular importance for some kind of applications where the stack is subject to particular mechanical loads or vibrations, such as automotive applications. When the electrolyte is a liquid (MCFC, PAFC, AFC), the mechanical stresses generated on the solid parts (electrodes, bipolar plates, etc.), are counterbalanced by the electrolyte. On the other hand, the liquid state of the electrolyte represents a limitation for the cell itself, due to possible electrolyte leakages and the inability to create fuel cells with particular shapes.

Due to the high operating temperature and to the thermal cycle that an SOFC is subject to, the stress analysis is very useful for both the production and the operation of the cell itself. The structure of an SOFC, in fact, is made up of different materials, characterized by different thermal expansion coefficients. The result is that, due to the high operating temperature, even small differences of the thermal coefficient can result in a high thermal stress. In general, the problem is mainly due to the anode structure (Minh and Takahashi 1995) since nickel, one of the major components of the anode, has a higher thermal coefficient than the other materials which compose the cell, thus inducing high internal stresses.

Stress is analyzed using the elasticity theory, correlating the temperature variation over the internal stresses.

An example of correlation between the microcrack density (Nb^3) and the heating rate is given by Qu and Haynes (2003):

$$q = \frac{3\pi^2 k r_0}{2\alpha} \left[1 + \frac{16(1 - v^2)Nb^3}{9(1 - 2v)} \right] \sqrt{\frac{G_c \pi (1 - v)}{E_0 b (1 + v)}} \tag{15.22}$$

where E_0 and v are, respectively, Young's modulus and Poisson's ratio of the uncracked material, G_c is the fracture toughness of the material, b is the crack size, N is the number of cracks per unit volume, k is the thermal diffusivity, α is the coefficient of linear thermal expansion and r_0 is a particular parameter, used to characterize the spatial non-uniformity of the heat source. The maximum heating rate, below which the crack density is acceptable, is:

$$q = \frac{3\pi^2 k r_0}{2\alpha} \sqrt{\frac{G_c \pi (1 - v)}{E_0 b (1 + v)}} \tag{15.23}$$

One of the first stress computational studies was conducted by Westinghouse in 1984 where the influence of the thermal expansion on the maximum stress intensity was computed. Majumdar et al. (1986), instead, assessed the influence of the electrodes and electrolyte thickness on the fracture formation, for different thermal expansion coefficients.

When the equation of the elasticity theory is integrated in the fuel cell domain, it is possible to represent mechanical stress distribution. Due to the complexity of

Figure 15.11   Bipolar plate stress analysis (reproduction under the authorization of
FEMLAB, COMSOL, Inc).

the equations and to the complexity of the domain, these equations are usually
integrated, making use of the FEM. Yakabe *et al.* (2001) analyze the interconnect
and electrolyte stresses due to the thermal energy for different SOFC configurations (i.e. co-flow and counter-flow), different operating conditions, and when the
fuel is internally or externally reformed. The results show that the maximum principal stress is low enough in the interconnect, while for the electrolyte it can reach
80 MPa for a counter-flow configuration with internal reforming. In this situation, in
fact, the internal reforming causes a relevant temperature gradient at the fuel inlet
section. The temperature gradient is relatively lower for the co-flow configuration
and about one order of magnitude lower when the fuel is externally reformed.

    Figure 15.11 gives an example of a 3D stress analysis of a bipolar plate.

15.3.2   Durability issues

The factors that influence the durability of a fuel cell stack have different nature
and they mainly depend on the materials used, the construction procedure, the
installation and the operating conditions. For an MCFC, for example, the main
lifetime-limiting phenomena are (Huijsmans *et al.* 2000):

(1) the dissolution of NiO cathode,
(2) electrolyte losses,
(3) corrosion of the separator plate,
(4) electrolyte retention capacity,
(5) catalyst deactivation,
(6) matrix cracking,
(7) high-temperature creep of porous components,
(8) contaminants.

The first phenomenon seems to be the main lifetime limiting (Tanimoto *et al.*
1998; Soler *et al.* 2002). The material widely used for cathode manufacturing (i.e.
porous NiO) is not stable in the MCFC environment and it dissolves, according to
the following reaction:

$$NiO + CO_2 \leftrightarrow Ni^{2+} + CO_3^{2-} \tag{15.24}$$

The Ni^{2+} ions move to the anode, through the electrolyte, and they react with hydrogen, diffusing from the anode:

$$Ni^{2+} + H_2 + CO_3^{2-} \leftrightarrow Ni + CO_2 + H_2O \tag{15.25}$$

The metallic nickel precipitates into the matrix, causing possible short-circuiting and decreasing the relative fuel cell performance. Different solutions have been proposed to reduce this phenomena, such as adding rare particular compounds to the nickel oxide (Soler *et al.* 2002; Colonmer *et al.* 2002) or using alternative materials to prepare the anode (Huijsmans *et al.* 2000).

Tanimoto *et al.* (1998) report the following expression, relating the amount of deposited nickel over the operating time:

$$\frac{M(t_1)}{M(t_2)} = \left(\frac{t_1}{t_2}\right)^{\alpha} \tag{15.26}$$

where M is the amount of the deposited Ni and α is an empirical parameter, estimated to be 0.58 (i.e. close to a parabolic law).

The other major lifetime reduction cause for MCFC is represented by corrosion of the bipolar plate (Colonmer *et al.* 2002). The high polarization that affects the plate, in fact, results in the oxidation of stainless steel components. For more details about durability problems, refer to Colonmer *et al.* (2002) and Huijsmans *et al.* (2000).

For an SOFC, the main causes for performance degradation and lifetime reduction are found to be (Hsiao and Selman 1997):

(1) the increase of contact resistance between electrodes and current collectors;
(2) the instability of electrodes, electrolyte and interconnect materials, when they touch each other;
(3) electrode–electrolyte or electrode–interconnect delamination.

The last phenomenon produces a reduction of the contact surface which then increases the internal electric resistance, as shown in Figure 15.1. The study of Bessette and Wepfer (1996) is a powerful example of a performance degradation model for a SOFC stack, when one of the previous factors occurs. Their results show, in fact, that a contact resistance of only 0.005 Ω causes an output power reduction of about 60%. In the same study, the authors also assess the effect of the failure of one or two single cells, composing a nine tubular cells SOFC stack, and the effect of possible short-circuits. If one single cell fails, the power reduction is between 13% and 17%, while when two cells fail, the output power drops to 46.5%. Finally, they show that the short-circuit failure produces a power reduction between 18% and 24%, according to the position of the cells that fail.

REFERENCES

Arato, E. and Costa, P. (1992) Mathematic Modeling of Monolithic SOFC. In: *Stack Design Tool*. An International Energy Agency SOFC Task Report, Berne, November.

Baron, S., Brandon, N., Atkinson, A., Steele, B. and Rudkin, R. (2004) The impact of wood-derived gasification gases on Ni-CGO anodes in intermediate temperature solid oxide fuel cells. *J. Power Sources* **126**, 58–66.

Bernardi, D.M. and Verbrugge, M.W. (1991) Mathematical model of a gas diffusion electrode bonded to a polymer electrolyte. *AICHE J.* **37**(8), 1151–1163.

Bessette, F.N. and Wepfer, W.J. (1996) Prediction of on-design and off-design for a solid oxide fuel cell power module. *Energ. Convers. Manage.* **37**(3), 281–293.

Bessette, F.N., Wepfer, W.J. and Winnick, J. (1995) A mathematic model of a solid oxide fuel cell. *J. Electrochem. Soc.* **142**(11), 3792–3800.

Bleise, Ch., Divisek, J., de Haart, B., Holtappels, P., Steffen, B. and Stimming, U. (1995) Kinetics modeling of the anodic reaction considering the porous structure. In: *Theory and Measurement of Microscale Processes in Solid Oxide Fuel Cells*. International Energy Agency Report, 1995.

Bossel, U.G. (1992a) *Facts and Figures*. An International Energy Agency SOFC Task Report, Baden, October.

Bossel, U.G. (1992b) *Performance Potential of Solid Oxide Fuel Cell Configuration*. An International Energy Agency SOFC Task Report, Berne, April.

Bove, R. and Sammes, N.M. (2004) Thermodynamic of SOFC systems using different fuel processors. *Proceedings of the 2nd International Conference on Fuel Cell Science, Engineering and Technology*, Rochester, NY, June 14–16.

Bove, R., Lunghi, P. and Sammes, N.M. (2005a) SOFC mathematic model for systems simulation. Part I: From a micro-detailed to a macro-black-box model. *Int. J. Hydrogen Energ.* **30**(2), 181–187.

Bove, R., Lunghi, P. and Sammes, N.M. (2005b) SOFC mathematic model for systems simulation. Part II: Definition of an analytical model. *Int. J. Hydrogen Energ.* **30**(2), 189–200.

Burt, A.C., Celik, I.B., Gemmen, R.S. and Smirvov, A.V. (2003) Influence of radioactive heat transfer on variation of cell voltage within a stack. *Proceedings of the 1st International Conference on Fuel Cell Science, Engineering and Technology*, Rochester, NY, April 21–23.

Chan, S.H., Low, C.F. and Ding, O.L. (2002) Energy and exergy analysis of simple solid-oxide fuel-cell power systems. *J. Power Sources* **103**, 188–200.

Colonmer, H., Ganesan, P., Subramanian, N., Haran, B., White, R.E. and Popov, B. (2002) Optimization of the cathode long-term stability in molten carbonate fuel cells: experimental study and mathematic modeling. US DOE Final Report, DOE Award No. DE-AC26-99FT40714.

Costamagna, P. and Honeger, K. (1998) Modeling of a solid oxide heat exchanger integrated stacks and simulation at high fuel utilization. *J. Electrochem. Soc.* **115**(11), 3995–4007.

Costamagna, P., Magistri, L. and Massardo, A.F. (2001) Design and part-load performance of a hybrid system based on a solid oxide fuel cell reactor and a micro gas turbine. *J. Power Sources* **96**, 352–368.

Cropper, M. (2002) Why is interest in phosphoric acid fuel cells falling? Fuel Cell Today's Knowledge Bank Article.

Ferguson, J.R. (1992) SOFC two dimensional "unit cell" modeling. In: *Stack Design Tool*. International Energy Agency SOFC Task Report, Bern, November.

Gödickemeier, M., Sasaki, K., Gauckler, L.J. and Riess, I. (1996) Perovskite cathodes for solid oxide fuel cells based on ceria electrolytes. *Solid State Ionic.* **86–88**, 691–701.

Gurau, V., Liu, H. and Kakac, S. (1998) Two-dimensional model for proton exchange membrane fuel cells. *AICHE J.* **44**(11), 2410–2422.

Heidebrecht, P. and Sundmacher, K. (2003) Dynamic modeling and simulation of a countercurrent molten carbonate fuel cell (MCFC) with internal reforming. *Fuel Cell. Fundam. Syst.* **2**(3–4), 166–180.

Herbin, R. (1992) On the effective resistance of an electrolyte membrane. In: *SOFC Micromodeling*. International Energy Agency SOFC Task Report, Bern, May.

Hirchenhofer, J.H. *et al.* (2002) *Fuel Cell Handbook*, 6th edn. US Department Energy. Morgantown, West Virginia, USA.

Hsiao, Y.C. and Selman, J.R. (1997) The degradation of SOFC electrodes. *Solid State Ionic.* **98**, 33–38.

Huijsmans, J.P.P., Kraaij, G.J., Makkus, R.C., Rietveld, G., Sitters, E.F. and Reijers, H.Th.J. (2000) An analysis of endurance issues for MCFC. *J. Power Sources* **86**, 117–121.

Iwata, M., Hikosaka, T., Morita, M., Iwanari, T., Ito, K., Onda, K., Esaki, Y., Sakai, Y. and Nagata, S. (2000) Performance analysis of planar-type unit SOFC considering current and temperature distribution. *Solid State Ionic.* **132**, 297–308.

Kho, J.H., Seo, H.K., Yoo, Y.S. and Lim, H.C. (2002) Consideration of numerical simulation parameters and heat transfer models for a molten carbonate fuel cell stack. *Chem. Eng. J.* **87**, 367–379.

Lunghi, P. and Bove, R. (2002) Reliable fuel cell simulation using an experimentally driven numerical model. *Fuel Cell. Fundam. Syst.* **2**, 83–91.

Lunghi, P., Burzacca, R. and Bove, R. (2002) Assessment and analysis of a gasification and fuel cell integration to enhance performances of hybrid plants for electricity production. In: *Fuel Cell Seminar*. Program and Abstracts, Palm Spring, CA, USA.

Majumdar, S., Claar, T. and Flandermeyer, B. (1986) Stress and fracture behavior of monolithic fuel cell tapes. *J. Am. Ceram. Soc.* **69**, 628–633.

Minh, N.Q. and Takahashi, T. (1995) *Science and Technology of Ceramic Fuel Cells.* Elsevier Science, The Netherlands.

Mizusaki, J., Tagawa, H., Saito, T., Kamitani, K., Yamamura, T., Hirano, K., Ehara, S., Takagi, T., Hikita, T., Ippommatsu, M., Nakagawa, S. and Hashimoto, K. (1994) Preparation of nickel pattern electrodes on YSZ and their electrochemical properties in H_2–H_2O atmospheres. *J. Electrochem. Soc.* **114**(8), 2129–2134.

Nagata, S., Momma, A., Kato, T. and Kasuga, Y. (2001) Numerical analysis of output characteristics of tubular SOFC with internal reforming. *J. Power Sources* **101**, 60–71.

Qu, J. and Haynes, C. (2003) An integrated approach to modeling and mitigating SOFC failure. Monthly Project Highlight Report, Agreement No. DE-AC26-02NT41571, Georgia Institute of Technologies – National Energy Technology Laboratory (NETL).

Soler, J., Gonzales, T., Escudero, M.J., Rodrigo, T. and Daza, L. (2002) Endurance test on a single cell of a novel cathode material for MCFC. *J. Power Sources* **106**, 189–195.

Springer, T.E., Wilson, M.S., Gottesfeld, G. (1993) Modeling and Experimental Diagnostics in Polymer Electrolyte Fuel Cells. *J. Electrochem. Soc.* **140**, 3513–3526.

Standaert, F., Hammes, K. and Woudstra, N. (1996) Analytical fuel cell modeling. *J. Power Sources* **63**, 221–234.

Suwanwarangkul, R., Croiset, E., Fowler, M.W., Douglas, P.L., Entchev, E. and Douglas, M.A. (2003) Performance comparison of Fick's, dusty-gas and Stefan–Maxwell models to predict the concentration overpotential of a SOFC anode. *J. Power Sources* **122**, 9–18.

Tanimoto, K., Yanagita, M., Kojima, T., Tamiya, Y., Matsumoto, H. and Miyazaki, Y. (1998) Long-term operation of small-sized single molten carbonate fuel cells. *J. Power Sources* **72**, 77–82.

Westinghouse Electric Corporation (1984) High temperature solid oxide electrolyte fuel cell power generation systems. Quarterly Summary Report, Report No. DOE/ET/17089-2217, Department of Energy (DOE).

White, F.M. (1991) *Viscous Fluid Flow*, 2nd edn. McGraw-Hill Series in Mechanical Engineering, USA.

Yakabe, H., Ogiwara, T., Mishinuma, M. and Yasuda, I. (2001) 3-D model calculation for planar SOFC. *J. Power Sources* **102**, 144–154.

Zhu, H. and Kee, R.J. (2003) A general mathematic model for analyzing the performance of fuel cell membrane–electrode assemblies. *J. Power Sources* **117**, 61–74.

Section IIIB: Low temperature fuel cells

16

Proton exchange membrane fuel cells: description and applications

J. Soler, M.J. Escudero and L. Daza

16.1 INTRODUCTION

A proton exchange membrane fuel cell (PEMFC), also called polymer electrolyte fuel cell (PEFC), is an electrochemical cell that is fed with hydrogen, which is oxidised at the anode and oxygen that is reduced is at the cathode. The protons released during the oxidation of hydrogen are conducted through the proton exchange membrane to the cathode. Since the membrane is not electrically conductive, the electrons released from the hydrogen travel along the electrical detour provided and an electrical current is generated (Figure 16.1).

PEM fuel cells are able to offer an order of magnitude higher power density than any other fuel cell system, with the exception of the advanced aerospace alkaline fuel cell, which has comparable performance. So, at 0.7 V/cell, the Ballard/Dow experimental technology produced 2.15 A/cm^2 on hydrogen/oxygen at 65 psia and over 1.08 A/cm^2 on hydrogen/air at 65 psia.

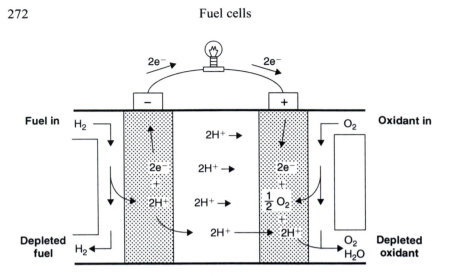

Figure 16.1   Schematic representation of single cell.

A PEM fuel cell delivers a high power density, which offers low weight, cost and volume of the fuel cell stack. The immobilised electrolyte membrane simplifies sealing in the production process, reduces corrosion, and provides for longer cell and stack life. PEM fuel cells operate at low temperature (60–90°C), allowing for faster startups and immediate response to changes in the demand for power (EG&G Services 2000).

16.2   HISTORY

The first application of a PEM in a fuel cell was in the 1960s as an auxiliary power source in the Gemini space flights (Ketelaar 1993). The 1-kW modules, built by General Electric, provided primary power for each of the seven Gemini spacecrafs. Performance and lifetime of these fuel cells was limited due to the polystyrene sulfonic acid membrane used at that time. Later, another General Electric unit was the 350-W module which was powered by to the Biosatellite spacecraft in 1969. An improved Nafion® membrane manufactured by DuPont was used as the electrolyte. Figure 16.2 shows the structural characteristics of Nafion. Performance and lifetime of PEM fuel cells has significantly improved since Nafion's introduction in 1968. In fact, duration has extended the operating time up to 3000 h and power density between 3.8 and 6.5 kW/m^2 (Shah 2003).

General Electric continued working on PEM cells and in the mid-1970s developed PEM water electrolysis technology for undersea life support, leading to the US Navy Oxygen Generating Plant. The British Royal Navy adopted this technology in the early 1980s for their submarine fleet. In the mid-1980s, General Electric technology was transferred both to Siemens AG in Germany and to UTC-Hamilton Standard in the USA. In 1983, Ballard Technologies Corporation in Canada began developing PEM fuel cell technology with funding support from the Canadian Department of National Defense. Subsequently, advances in this technology were

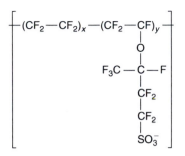

Figure 16.2   Chemical structure of the membrane material Nafion® manufactured by DuPont.

stagnant until the late 1980s when the fundamental design underwent significant modification (Litster and McLean 2004).

Today the PEM fuel cell is widely considered as the most promising fuel cell system, which has widespread applications. Numerous companies try to develop and commercialise PEM fuel cells (Appleby 1996; Evers 2003). The most important are: Ballard Power Systems, H Power, Nuvera Fuel Cells, Plug Power and UTC Fuel Cells.

16.3   CELL COMPONENTS

16.3.1   PEM fuel cell design

The PEM fuel cell is made up of several components. The ion conducting membrane sandwiched between both electrodes (anode and cathode) forms the membrane electrode assembly (MEA). This one is held together by two flow field plates (also called bipolar plates), which distribute the reactant gases. A seal is used to avoid gas leakages (Figure 16.3).

16.3.2   Membrane electrode assembly

The anode and the cathode have a catalyst layer pressed onto one side of each of them. The side containing the catalyst is placed in contact with the proton exchange membrane. The catalyst layer contains a platinum-based catalyst which is supported by carbon support with typical loadings ranging from as low as $0.05\,mg\,Pt/cm^2$ to as high as $0.5\,mg\,Pt/cm^2$. The catalyst layer increases the rate of the electrochemical reaction, allowing it to proceed quickly enough to release electrons that are captured to provide the electrical power, which is then utilised (e.g., by a fuel cell vehicle). Collectively, the membrane and the electrodes form a MEA. The MEA structure is on the order of 725 µm thick (each electrode being 300 µm and the membrane being 125 µm). The state-of-the-art of every component is explained in the sections given below.

Individual cells do not deliver the necessary voltage for normal application. Therefore, the cells are combined into a fuel cell stack (Figure 16.4) of the desired power. In a fuel cell stack, each bipolar plate supports two adjacent cells. The flow

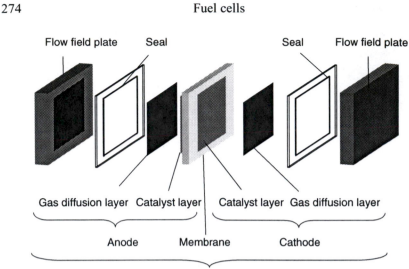

Flow field plate   Seal   Seal   Flow field plate

Gas diffusion layer Catalyst layer | Catalyst layer Gas diffusion layer

Anode   Membrane   Cathode

Membrane-electrode assembly

Figure 16.3   Schematic of a single typical PEMFC.

Figure 16.4   H2E™ PEM fuel cell stack that provides 1–6 kW electric power (courtesy of Nuvera Fuel Cells).

field plates typically have the following functions: to distribute the fuel and oxidant within the cell, to facilitate the water management within the cell, to separate the individual cells in the stack and to carry current away from the cell (Mehta and Cooper 2003).

16.3.3   Electrolyte membrane

The electrolyte membrane allows protons to pass through to the cathode side, but separates hydrogen and oxygen molecules, and therefore prevents direct combustion. The membrane also acts as an electronic insulator between the flow field plates.

Perfluorosulfonic acid (PFSA) is the most commonly used membrane material for PEM fuel cells (Figure 16.2). PFSA consists of three regions:

(1) a polytetrafluoroethylene (PTFE)-like backbone,
(2) side chains of $- O\!-\!CF_2\!-\!CF\!-\!O\!-\!CF_2\!-\!CF_2 -$ which connect the molecular backbone to the third region,
(3) ion clusters consisting of sulfonic acid ions.

The acid molecules are fixed to the polymer and cannot leak out, but the protons on these acid groups are free to migrate through the membrane. The most common membrane material in PEMFC prototypes is Nafion® produced by DuPont.

At the moment, the membrane must remain hydrated in order to be proton conductive (Gottesfeld and Zawodzinski 1997). This limits the operating temperature of PEM fuel cells to under the boiling point of water and makes water management a key issue in PEM fuel cell development. Substantial improvements could be achieved by operating at temperatures above 150°C because the Pt/C electrodes used in PEMFCs are very sensitive to traces of CO present in the reactant gases (normally H_2 and O_2), since CO is adsorbed on the platinum, and thus poisons the electrodes. However, CO desorbs easily from Pt at temperatures above 150°C. Also the oxygen reduction reaction (ORR) on the cathode is faster if temperature increases. With these considerations in mind, it is easy to understand why the most promising trends in the development of PEM fuel cells involve the development of novel membranes for cells working at increased temperatures (150–200°C) using alternative materials to PFSA.

The main membrane development efforts include:

(1) To explore approaches involving polymer synthesis and development, as well as implementation of new "carrier" media to replace the function of water in Nafion.
(2) Study of proton transfer dynamics to inform the synthetic efforts and explore specific possibilities for new acid group types or for acid–base interactions that could lead to progress in proton transfer media (Zawodzinski and Garland 2003).

A summary of proton conducting materials for both low and medium temperatures and the present state of art in the field of solid state protonic conductors has been presented by Alberti and Casciola (2001). The conductivity of the membrane is sensitive to contamination. For example, if the membrane is exposed to metallic impurities, metal ions diffuse into the membrane and displace protons as charge carriers, which lower the membrane conductivity.

Many new materials are presently in preparation and testing. These are:

- Polymers and inorganic materials with controlled pore size, modified with acid groups lining pores or by inclusion of proton conducting subunits.
- Polymer matrices filled with surface-modified inorganic particles.
- Polymeric systems with intrinsically stronger acid groups.
- Polymer systems swollen or embedded with tailored proton acceptors including imidazole and ionic liquids.

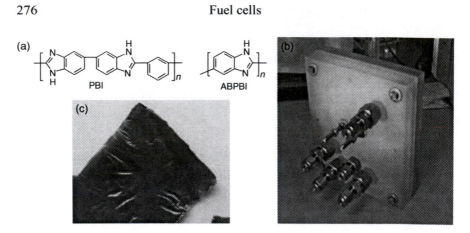

Figure 16.5   New materials for membrane: (a) polibenzimidazole (PBI) and poli 2,5 benzimidazole (ABPBI) chemical formula, (b) single cell for 150–200°C operation and (c) protonic membrane (Asensio *et al.* 2004).

Materials based on polibenzimidazoles (Figure 16.5) have shown a high thermal stability and they constitute a good alternative to PFSAs (Asensio *et al.* 2004).

16.3.4   Electrodes

All electrochemical reactions take place on the electrode surfaces. To speed up fuel cell reactions, electrodes contain catalysts particles, virtually always platinum or an alloy of platinum and other noble metals (Mehta and Cooper 2003). Low operating temperature and pH make the use of catalysts necessary (Srinivasan *et al.* 1993), especially the ORR on the cathode is very slow in the absence of catalyst.

The electrodes consist of a catalyst layer, which are usually made of a porous mixture of carbon supported platinum (Pt/C) and ionomer, and a gas diffusion layer which ensures that reactants effectively diffuse to the catalyst layer. The Pt/C powder, prepared mostly by procedures based on colloid chemistry (Choi *et al.* 2000) and microemulsion methods (Escudero *et al.* 2003), consists of Pt particles of about 2 nm in diameter, supported on carbon particles of about 10 nm in diameter. Platinum is an essential catalyst for the electrochemical conversion of hydrogen and oxygen at the anode and cathode of the fuel cell, respectively, into electric current (electric power). In PEM fuel cells, the processes at the anode and cathode, respectively, are as follows:

$$\text{Anode} \qquad H_2 \longrightarrow 2H^+ + 2e^-$$

$$\text{Cathode} \qquad \tfrac{1}{2}O_2 + 2H^+ + 2e^- \longrightarrow H_2O$$

The Pt/C powder has to be intimately intermixed with recast ionomer to provide sufficient ionic conductivity within the catalyst layer. Thus, the catalyst layer can be described as a Pt/C ionomer composite, where each of the three components are uniformly distributed within the volume of the layer.

An effective electrode is one that correctly balances the transport processes required for an operational fuel cell. The three transport processes required are the transport of:

- protons from the membrane to the catalyst,
- electrons from the current collector to the catalyst through the gas diffusion layer,
- the reactant and product gases to and from the catalyst layer and the gas channels.

16.3.4.1   Catalyst layer

In order to be able to catalyse reactions, catalyst particles must have contact to both protonic and electronic conductors. Furthermore, there must be passages for reactants to reach the catalyst sites and for reaction products to exit. The contacting point of the reactants, catalyst and electrolyte, is conventionally referred to as the three-phase interface. To achieve acceptable reaction rates, the effective area of active catalyst sites must be several times higher than the geometrical area of the electrode. Therefore, the electrodes are made porous to form a three-dimensional network, in which the three-phase interfaces are located. The benefits of the thin-film method include lower price, better utilisation of catalyst and improved mass transport (Wilson and Gottesfeld 1992). The thickness of a catalyst layer is typically 5–15 μm and the catalyst loading is between 0.05 and 0.5 mg/cm^2.

Nowadays, most PEM fuel cells developers have chosen the thin-film approach for their product prototypes, in which the catalyst layers are manufactured directly on the membrane surface.

The other option is to manufacture the catalyst layer on the surface of the porous gas diffusion layer by impregnating with a mixture of carbon supported catalyst and ionomer. In addition to catalyst loading, there are a number of catalyst layer properties that have to be carefully optimised to achieve high utilisation of the catalyst material (i.e. reactant diffusivity and ionic and electrical conductivity). In this way, usually the catalyst is mixed with the binder material (the optimum binder quantity depends on the type of catalyst) and solvents to form a "catalytic ink". A great effort has been made to improve the catalytic ink compositions (i.e. Soler et al. 2002; Sasikumar et al. 2004). The first authors who prepared electrodes using catalytic inks by mixing the solubilised ionomer (Nafion) with the Pt/C catalyst were Wilson and Gottesfeld (1992). They used water and glycerol as solvents and the ink was cast onto the membrane at elevated temperatures (150–190°C) to dry the ink. Other authors (Paganin et al. 1996) used isopropanol as a solvent to form an ink which was quantitatively deposited on the diffusion layer of the electrode by a brushing procedure.

The formation of a good structure of the catalytic layer in the electrode is essential for the suitable performance of the PEMFC in order to increase the Pt utilisation. The preparation of appropriate catalytic inks to create the catalytic layer, where the electrocatalytic reaction takes place, is a key aspect in the improvement of the performance of the membrane electrode assemblies (MEAs).

The influence of PTFE and Nafion on platinum utilisation in the two commonly used catalyst layers – the thin-film catalyst layer and the porous gas diffusion layer electrode impregnate with a mixture of carbon supported catalyst and ionomer – has been investigated by means of SEM and TEM (Cheng *et al.* 1999). In the first method, although some thick Nafion layers and clumps were observed in the Pt/C + Nafion layer, there was an essentially uniform distribution of Nafion in the catalyst layer. In order to avoid the blocking effect of Nafion solids on the electron passageway in the catalyst, the impregnation of the catalyst with Nafion was carried out. For the second one, platinum utilisation was affected mainly by the proton passageway provided by Nafion, while the effect of the PTFE was relatively small. Other researchers (Chun *et al.* 1998) used different catalyst layer preparation procedures and proved that the thin film MEAs manufacturing processes showed a better performance than those prepared using a PTFE-bonded Pt/C electrocatalyst by coating and transfer printing techniques.

Other works (Passalacqua *et al.* 2000) have focused on eliminating the PTFE in the catalytic layer and to substitute it with glycerol or ammonium carbonate as pore former to create an additional porosity. Wilson *et al.* (1995) discovered that Nafion (and other perfluorinated ionomers) could be converted to a thermoplastic form by the ion-exchange inclusion of large, "hydrophobic" counter-ions such as tetrabutylammonium (TBA^+). In this case, the reproducibility was greatly improved and the long-term performance losses were quite low. Another preparation method (Uchida *et al.* 1995), which focused on the colloid formation of the ionomer, was proposed to optimise the network of ionomer in the catalyst layer. They used butylacetate as "poor" solvent to obtain the perfluorosulfonate ionomer colloids. A comparative study between the "solution method", using isopropanol, as solvent and the "colloidal method" with n-butylacetate has been recently published (Shin *et al.* 2002). It was shown that the continuous network of ionomers throughout the catalytic layer increased for the colloidal method, which improved the proton movement from the electrode to the membrane.

16.3.4.2   Gas diffusion layers

In a PEM fuel cell, the MEA is sandwiched between the flow field plates. On each side of the MEA, between the electrode and the flow field plate, there are gas diffusion layers or backings (Figure 16.6). They provide electrical contact between the electrodes and the bipolar plates, and distribute reactants to the electrodes. They also allow the reaction product water to exit the electrode surface and permit passage of water between the electrodes and the flow channels.

Gas diffusion backings are made of a porous, electrically conductive material, usually carbon cloth or carbon paper (Soler *et al.* 2003). The substrate is usually treated with a fluoropolymer and carbon black to improve water management and electrical properties.

16.3.5   Flow field plates

In a fuel cell stack, flow field plates separate the reactant gases of adjacent cells, connect the cells electrically, and act as a support structure (Figure 16.3). Furthermore, flow field plates have reactant flow channels on both sides, forming the anode and cathode compartments of the unit cells on the opposing sides of the

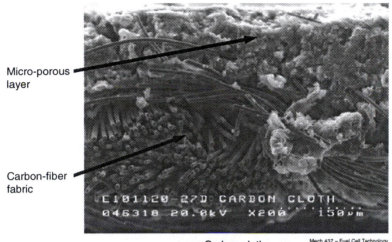

Micro-porous layer

Carbon-fiber fabric

Carbon cloth

Mech 437 – Fuel Cell Technology

Figure 16.6   SEM microphotograph of the structure of the gas diffusion layer (Coudhury 2004).

Figure 16.7   Flow field plates of a PEMFC.

flow field plate. In a unit cell, separator plates have flow channels only on one side and are sometimes called monopolar plates (Figure 16.7).

Flow channel geometry has an effect on reactant flow velocities and mass transfer, and thus on fuel cell performance (Hontañón *et al.* 2000). Flow field plate materials must have high conductivity and be impermeable to gases. Due to the presence of reactant gases and catalyst, the material should be corrosion resistant and chemically inert. For most of the applications, also low weight and high strength are important. For commercial applicability, the material should be cheap and suitable for high-volume manufacturing methods. Most PEMFC flow field plates are made of resin-impregnated graphite (Scholta *et al.* 1999), but also stainless steel has been used (Davies *et al.* 2000). Solid graphite is highly conductive, chemically inert and resistant to corrosion, but expensive and costly to manufacture. Stainless steel is very affordable, but expensive to machine. In addition, stainless steel must often be coated to prevent corrosion and reduce contact resistance.

Flow channels are machined or electrochemically etched to the graphite or stainless steel flow field plate surfaces. These methods are not suitable for mass production and therefore new bipolar materials are being investigated. Best results have been achieved with carbon–polymer composites, which can be moulded (Besmann *et al.* 1998).

16.3.6   MEA manufacturing

The correct performance of a PEM fuel cell depends highly on the quality of the MEAs used. For many years, researchers have looked for the best method to get a MEA with good performance, endurance and reproducibility. For this reason, it is called the "heart" of the PEM fuel cell (Figure 16.8).

In order to manufacture MEAs two different kind of techniques have been developed depending of how the catalyst layer is prepared: (1) application on the gas diffusion layer and (2) application on the membrane (Mehta and Cooper 2003).

For the first one, five methods have been identified:

- *Spreading* (Srinivasan *et al.* 1994): It consists of preparing a catalysed carbon and PTFE dough by mechanical mixing and spreading it on a wet-proofed carbon cloth using a heavy stainless steel cylinder on a flat surface.
- *Spraying* (Srinivasan *et al.* 1994): The electrolyte is suspended in a mixture of water, alcohol and colloidal PTFE. This mixture is then repeatedly sprayed onto wet-proofed carbon cloth.
- *Catalyst power deposition* (Bevers *et al.* 1998): The components of the catalytic layer are mixed in a fast running knife mill under forced cooling. This

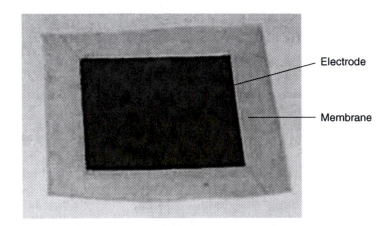

Figure 16.8   Photograph of a MEA used for PEM fuel cells.

mixture is then applied onto a wet-proofed carbon cloth. Also applying a layer of carbon/PTFE mixture flattens out the roughness of the paper, and improves the gas and water transport properties of the MEA.

- *Ionomer impregnation* (Gottesfeld and Zawodzinski 1997): The catalytically active side of the gas diffusion layer is painted with solubilised PFSA in a mixture of lower aliphatic alcohols and water. To improve reproducibility of the gas diffusion layer/catalyst assembly, the catalyst and ionomer are premixed before the catalyst layer is deposited, rather than ionomer impregnation of Pt/C/PTFE layer.
- *Electrodeposition* (Gottesfeld and Zawodzinski 1997): It involves impregnation of the porous carbon structure with ionomer, exchange of the cations in the ionomer by a cationic complex of platinum and electrodeposition of platinum from the complex onto the carbon support. This results in deposition of platinum only at sites that are accessed effectively by both carbon and ionomer.

For the second one, application onto membrane, six different methods have been reported:

- *Impregnation reduction* (Foster *et al.* 1994): In this method the membrane, ion exchanged to the Na^+ form, is equilibrated with an aqueous solution of $(NH_3)_4PtCl_2$ and a co-solvent of H_2O/CH_3OH. Following impregnation, vacuum dried PFSA in the H^+ form is exposed on one face to air and the other to an aqueous reductant $NaBH_4$.
- *Evaporative deposition* (Foster *et al.* 1994): $(NH_3)_4PtCl_2$ is evaporatively deposited onto a membrane from an aqueous solution. After deposition of the salt, metallic platinum is produced by immersion of the entire membrane in a solution of $NaBH_4$.
- *Dry spraying* (Gulzow *et al.* 2000): Reactive materials (Pt/C, PTFE, PFSA powder and/or filler materials) are mixed in a knife mill. The mixture is then atomised and sprayed in a nitrogen stream through a slit nozzle directly onto the membrane. Although adhesion of the catalytic material on the surface is strong, in order to improve the electric and ionic contact, the layer is fixed by hot rolling or pressing.
- *"Novel fabrication method"*, so called by authors (Matsubayashi *et al.* 1994): The PFSA solution is mixed with the catalyst and dried in a vacuum. Then, the PFSA coated catalyst is mixed with a PFSA dispersion, calcium carbonate used to form pores and water. The mixture is passed through a filter and the filtrate is formed into a sheet. The sheet is then dipped in nitric acid to remove any calcium carbonate. The sheet is then dried and PFSA solution is applied to one side of the electrode catalyst layer. Finally the catalyst layer is applied to the membrane.
- *Catalyst decaling* (Gottesfeld and Wilson 1992): Pt ink is prepared by thoroughly mixing the catalyst and solubilised PFSA. The protonated form of PFSA in the ink is subsequently converted to the TBA^+ form by the addition of TBAOH in methanol to the catalyst and PTFA solution. The paintability of the ink and the stability of the suspension can be improved by the addition of glycerol. Membranes are catalysed using a "decal" process in which the ink is cast onto PTFE blanks for transfer to the membrane by hot pressing. When the

PTFE blank is peeled away, a thin casting layer of catalyst is left on the membrane. In the last step, the catalysed membranes are rehydrated and ion-exchanged to the H^+ form by immersing them in lightly boiling sulfuric acid followed by rinsing in deionised water.

- *Painting* (Gottesfeld and Zawodzinski 1995): Pt is prepared as described for the decaling method. A layer of ink is painted directly onto a dry membrane in the Na^+ form and baked to dry the ink.

For both kind of techniques, sputtering can also be used as a single-step option to catalyst preparation and application if catalyst is deposited onto the gas diffusion layer as described by Srinivasan *et al.* (1994) or if it is deposited onto the membrane, which has been reported by Cha and Lee (1999).

After every method, the gas diffusion layer and membrane must be joined. Hot pressing is used in both cases and, during the procedure, the membrane dries out but becomes re-hydrated adequately after insertion in the stack with humidified gases. The primary challenge in the assembly of MEAs is to achieve good contact between the membrane, the gas diffusion layer and the catalyst layer. Good contact maximises catalyst utilisation during cell operation.

16.4   APPLICATIONS OF PEM FUEL CELLS

PEM fuel cell technology developments have been particularly rapid during the 1990s. However, there is scope for further improvements notably concerning the proton conducting membranes presently available. In addition to the cost, existing materials such as Nafion need to be hydrated to be effective proton conductors. In addition, the higher temperature operation would result in an increased CO tolerance of the anode catalyst and extend the range of applications where high-grade heat is a requirement (Acres 2001). These barriers will have to be overcome before the commercialisation stage. Meanwhile, the PEM fuel cell prototypes developed by different companies are being tested for different applications (Costamagna and Srinivasan 2001).

PEM fuel cells found their first application at the beginning of the 1960s, in the NASA's Gemini Space Flights and this was the first application for all types of fuel cells. Since the late 1960s, the PEM fuel cells were displaced by alkaline fuel cells. The main reason for the technology transfer for these applications were the higher attainable efficiency and power density of the AFC compared to the PEM technology during the starting phases of these programs. However, with the major improvements in performance made by PEM since the late 1980s, there is renewed interest for the take-over of the PEM technology in NASA's space station and Mars programs.

Interest in the application of fuel cells as power sources for electric vehicles started in the late 1970s, but received a major boost in the late 1980s and earlier 1990s, firstly because of the California Environmental Legislation to develop ultralow and zero emission vehicles, and secondly because of the Partnership for a New Generation of Vehicles (PNGV) programme to develop cars with three times higher efficiency of the present conventional internal combustion engine powered vehicles.

Two types of PEM fuel cells on-board plants have been proposed: stacks coupled to an electric motor and stacks coupled to an electric motor with back-up batteries (hybrid system). A collaboration between Ballard and Daimler-Chrysler, that after involved Ford as well, has given birth to the fuel cell powered electric cars (NECAR series). The project is now in the fifth phase of development. The production of this vehicle was scheduled for the year 2004 with an entry price around US$ 45,000. Other car manufacturers in the USA (GM, Ford), Japan (Toyota, Mazda, Nissan) and Europe (Renault, Volvo, Fiat) have also entered the arena.

The on-board hydrogen storage system is the main technological obstacle for the development of direct hydrogen fuel vehicles. Hydrogen storage has been demonstrated and includes high-pressure lightweight cylinders, cryogenic liquid systems and solid state metal hydride stores. Although substantial progress has been made with these systems to a level that, for example, compressed hydrogen may be acceptable for use on buses, they are generally considered not to meet volume and weight criteria for light duty vehicles. High levels of hydrogen adsorption in carbon nanotubes are being actively sought (Dillon et al. 1997). While awaiting to overcome this difficulty, an interim solution claimed to enter the market is the so-called processed fuel cell vehicle (PFCV), a vehicle system with fuel cell stacks fed with H_2-rich gas produced on-board by a fuel processor supplied by a primary fuel transported inside the vehicle. The composition of the gas produced by the fuel processor is an essential issue for efficient operation, as the presence of CO causes a rapid decrease of fuel cell performance. An on-board fuel processor usually lowers the overall efficiency of the system and requires high fuel cell catalyst loading, resulting in higher costs and increases in system weight and volume. Nevertheless, it allows the use of different fuels (methanol, propane or gasoline) in a multi-fuel configuration (Cacciola et al. 2001).

Regarding the portable power sources, possible applications are video cameras, electric wheelchairs, portable-power briefcases, laptop computers, intelligent transportation systems (road signs, traffic light, etc.) and military communications. Several companies are close to commercialising fuel cells for portable applications with a power range from 35 to 250 W.

It has been evaluated that commercial buildings, rather than homes, are the segment of market where fuel cells have the best chance to make their entry, at an initial price higher than the target cost where it is affordable. Development of such small units is in progress at a number of US companies, such as American Power Corp., Plug Power-General Electric Power Systems, Avista Laboratories and Northwest Power Systems. Ballard Power Generation Systems is also active in this field (Pokojski 2000), developing a 250 kW unit operating on natural gas (Figure 16.9).

Other versions are planned that will operate on propane, hydrogen or anaerobic digester gas. Demonstration units in the USA, Japan and Europe are now in place having achieved over a year of operation. Cell voltage decay rates are reported to be <0.3% per 1000 h (Acres 2001).

Figure 16.9   Ballard 250 kW system for stationary application of PEMFC.

16.5   CONCLUSIONS

Although PEMFCs have been developed since 1960s, there are still scientific and technical challenges to solve before the era of commercialisation. The necessity to obtain new membranes not depending on water for high temperature operation between 150 and 200°C is a key issue to face in next future. The use of new electrocatalyst with lower noble metal loading or cheaper active compounds would contribute to reach more competitive prices. The development of new manufacturing methods to obtain MEAs would help to both increase the performance and the lifetime.

Meanwhile, many prototypes have been tested for transport, portable and stationary applications, and it is possible to envisage future commercial fields where this technology could have a decisive role to produce electricity with high efficiency and low environmental impact.

REFERENCES

Acres, G.J.K. (2001) Recent advances in fuel cell technology and its applications. *J. Power Sources* **100**, 60–66.

Alberti, G. and Casciola, M. (2001) Solid state protonic conductors, present main applications and future prospects. *Solid State Ionics* **145**, 3–16.

Appleby, A.J. (1996) Issues in fuel cell commercialisation. *J. Power Sources* **69**, 153–176.

Asensio, J.A., Borrós, S. and Gómez-Romero, P. (2004) Polymer electrolyte fuel cells based on phosphoric acid impregnated poly(2,5-benzimidazole) (ABPBI) membranes. *J. Electrochem. Soc.* **151**(2), A304–A310.

Besmann, T.M., Klett, J.W. and Burchell, T.D. (1998) Carbon composite for a PEM fuel cell bipolar plate. *Mat. Res. Soc. Symp. Proc.* **496**, 243–248.

Bevers, D., Wagner, N. and Bradke, M. (1998) Innovative production procedure for low cost PEFC electrodes and electrode membrane structures. *Int. J. Hydrogen Ener.* **23**, 57–63.

Cacciola, G., Antonucci, V. and Freni, S. (2001) Technology up date and new strategies on fuel cells. *J. Power Sources* **100**, 67–79.

Cha, S. and Lee, W. (1999) Performance of proton exchange membrane fuel electrodes prepared by direct deposition of ultra thin platinum on the membrane surface. *J. Electrochem. Soc.* **144**, 3845–3857.

Choi, W.C., Cho, S.I., Sohn, J.M., Kim, M.R. and Woo, S.I. (2000) Preparation of highly dispersed carbon-supported Pt and PtRu alloy electrocatalysts for direct methanol fuel cell. In: *Book of Abstracts of the 12th International Congress on Catalysis*, Granada, Spain, July, 9–14.

Cheng, X., Yi, B., Han, M., Zhang, J., Qiao, Y. and Yu, J. (1999) Investigation on platinum utilization and morphology in catalyst layer of polymer electrolyte fuel cell electrodes. *J. Power Sources* **79**, 75–81.

Chun, Y.G., Kim, C.S, Peck, D.H. and Shin, D.R. (1998) Performance of a polymer electrolyte membrane fuel cell with thin film catalyst electrode *J. Power Sources* **71**, 174–178.

Coudhury, J. (2004) PEM fuel cells. *Lecture in Faculty of Applied Sciences.* Queen's University, California.

Costamagna, P. and Srinivasan, S. (2001) Quantum jumps in the PEMFC science and technology from the 1960s to the year 2000. Part II. Engineering, technology development and application aspects. *J. Power Sources* **102**, 253–269.

Davies, D.P., Adcock, P.L., Turpin, M. and Rowen, S.J. (2000) Stainless steel as a bipolar plate material for solid polymer fuel cells. *J. Power Sources* **86**, 237–242.

Dillon, A.C., Jones, K.M., Bekkedahl, T.A., Kiang, C.H., Bethune, D.S. and Heben, M.J. (1997) Storage of hydrogen in single-walled carbon nanotubes. *Nature* **386**, 377–379.

EG&G Services (2000) *Fuel Cell Handbook*, 5th edn. Parsons, Inc. and Science Applications International Corporation under contract no. DE-AM26-99FT40575 for the US Department of Energy, National Energy Technology Laboratory.

Escudero, M.J., Hontañón, E., Schwartz, S., Boutonnet, M. and Daza, L. (2003) Development and performance characterisation of new electrocatalysts for PEMFC. *J. Power Sources* **106**, 206–214.

Evers, A.A. (2003) Go to where the market is! Challenges and opportunities to bring fuel cells to the international market. *Int. J. Hydrogen Energ* **28**, 725–84.

Foster, S., Mitchel, P. Mortimer, R. (1994) In: *Proceedings of the Fuel Cell Program and Abstracts on the Development of a Novel Electrode Fabrication Technique for Use in Solid Polymer Fuel Cells*, 442–443.

Gottesfeld, S. and Zawodzinski, T. (1997) Polymer electrolyte fuel cells. *Adv. Electrochem. Sci. Eng.* **5**, 195–301.

Gottesfeld, S. and Wilson, M. (1992) Thin film catalyst layers for polymer electrolyte fuel cell electrodes. *J. Appl. Electrochem.* **22**, 1–7.

Gulzow, E., Schulze, M., Wagner, N., Kaz, T., Reissner, R., Steinhilber, G. and Schneider, A. (2000) Dry layer preparation and characterization of polymer electrolyte fuel cell components. *J. Power Sources* **86**, 352–362.

Hontañón, E., Escudero, M.J., Bautista, C., García-Ybarra, P.L. and Daza, L. (2000) Optimisation of flow-field in polymer electrolyte membrane fuel cells using computational fluid dynamics techniques. *J. Power Sources* **86**, 363–368.

Litster, S. and McLean, G. (2004) PEM fuel cell electrodes. *J. Power Sources* **130**, 61–76.

Ketelaar, J.A.A. (1993) *Fuel Cell Systems*. Plenum Press, New York.

Matsubayashi, T., Hamada, A., Taniguchi, S. Miyake, Y. and Saito, T. (1994) In: *Proceedings of the Fuel Cell Program and Abstracts on the Development of the High Performance Electrode for PEFC*, 581–584.

Mehta, V. and Cooper, J.S. (2003) Review and analysis of PEM fuel cell design and manufacturing. *J. Power Sources* **114**, 32–53.

Paganin, V.A., Ticianelli, E.A. and González, E.R. (1996) Development and electrochemical studies of gas diffusion electrodes for polymer electrolyte fuel cells. *J. Appl. Electrochem.* **26**, 297–304.

Passalacqua, E., Gatto, I., Lufrano, F., Patti, A. and Squadrito, G. (2000) Improvements in the catalyst layer structure for low pressure PEFC electrodes. In: *Book of Abstracts of 2000 Fuel Cell Seminar*, Portland, October 30 to November 2.

Pokojski (2000) The first demonstration of the 250-kW polymer electrolyte fuel cell for stationary application (Berlin). *J. Power Sources* **86**, 140–144.

Sasikumar, G., Ihm, J.W. and Ryu, H. (2004) Dependence of optimum Nafion content in catalyst layer on platinum loading. *J. Power Sources* **132**, 11–17.

Scholta J., Rohland, B., Trapp, V. and Focken, U. (1999) Investigations on novel low-cost graphite composite bipolar plates. *J. Power Sources* **84**, 231–234.

Shah, R.K., Desideri, U., Hsueh, K.-L. and Vikar A.V. (2003) Research opportunities and challenges in fuel cell science and engineering. In: *Book of Abstracts 4th Baltic Heat Transfer Conference*, Kaunas, Lithuania, August, 25–27.

Shin, S.J., Lee, J.K., Ha, H.Y., Hong, S.A., Chun, H.S. and Oh, I.H. (2002) Effect of the catalytic ink preparation method on the performance of polymer electrolyte membrane fuel cells. *J. Power Sources* **106**, 146–152.

Soler, J., Benítez, R., Jiménez, S. and Daza, L. (2002) Influence of the solvent in the preparation of catalytic inks for proton exchange membrane fuel cell (PEMFC). In: *Book of Abstracts of 2002 Fuel Cell Seminar*, Palm Springs, November, 18–21.

Soler, J., Hontañón, E. and Daza, L. (2003) Electrode permeability and flow-field configuration: influence on the performance of a PEMFC. *J. Power Sources* **118**, 172–178.

Srinivasan, S., Davé, B.B., Murugesamoorthi, K.A., Parthasarathy, A. and Appleby, A.J. (1993) Overview of fuel cell technology. *Fuel Cell Syst.* 63–67.

Srinivasan, S., Ferreira, A., Mosdale, R., Mukerjee, S., Kim, J., Hirano, S., Lee, S., Buchi, F. and Appleby, A. (1994) In: *Proceedings of the Fuel Cell Program and Abstracts on the PEM Fuel Cells for Space and Electric Vehicle Application*, pp. 424–427.

Uchida, M., Aoyama, Y., Eda, N. and Ohta, A. (1995) New preparation method for polymer-elctrolyte fuel cells. *J. Electrochem. Soc.* **142**, 4143–4149.

Wilson, W. and Gottesfeld, S. (1992) Thin-film catalyst layers for polymer electrolyte fuel cell electrodes. *J. Appl. Electrochem.* **22**, 1–7.

Wilson, M.S., Valerio, J.A. and Gottesfeld, S. (1995) Low platinum loading electrodes for polymer electrolyte fuel cells fabricated using thermoplastic ionomers. *Electrochem. Acta* **40**, 355–363.

Zawodzinski, T. and Garland, N. (2003) High temperature membranes for PEM fuel cells. In: *Book of Abstracts of 2003 Fuel Cell Seminar*, Miami, November 3–7.

17

Advances and perspectives of polymer electrolyte membrane fuel cells

K. Scott and A.K. Shukla

17.1 INTRODUCTION

An important type of fuel cell is the polymer electrolyte membrane fuel cell (PEMFC), also called proton exchange membrane fuel cell, which operates typically in the range 60°C to 100°C and is suitable for transport and portable applications, and for power co-generation in buildings (Larminie and Dicks 2000). PEMFC could also find applications in power generation in providing peak power and avoiding grid-reinforcement and are currently being tested on the 250 kW scale with hydrogen as the fuel (Barbir 2003).

The current, well-developed PEMFC technology, is based on perfluoro-sulfonic acid (PFSA) polymer membranes (e.g. Nafion®) as electrolyte, and has limitations due to the low temperature of operation, namely conductivity and water management issues, slow oxygen reduction reaction (ORR), a low tolerance to fuel impurities;

for example, CO and S, serious cooling problems and poor heat recovery (for residential applications). This chapter considers some of the advances made in the materials for and the performance of PEMFCs since their inception and considers their limitations in relation to the source and type of fuel to be used to generate power.

17.2   CATALYSTS FOR THE PEMFC

Platinum is acknowledged as the best catalyst for the oxidation of hydrogen. Carbon-supported Pt catalysts feature widely as catalyst for oxygen reduction in PEMFC and in direct methanol fuel cells (DMFCs). Although much research has been directed to finding alternative non-precious metal-based catalysts, the performance of Pt has not been matched. As the sluggish kinetics of the ORR, the loading of Pt in the cathode catalyst is significantly greater than in the anode, of the order of 0.4–0.5 mg/cm^2.

Pt-supported catalysts do, however, show a loss of activity with time due to catalyst sintering. To reduce the impact of cathode de-activation alloys of Pt have been researched for the PEMFC largely as a result of the earlier equivalent work on Pt alloys for phosphoric acid fuel cells. Pt alloys using Cr and Fe have been shown to exhibit higher activity than Pt alone and exhibit improved stability (Ralph et al. 1997). The use of Pt alloy catalysts in the DMFC has also been shown to improve cell performance and also offer some methanol tolerance (Thompsett 2003).

In practical applications the hydrogen is generally not pure and can contain traces of CO, NH_3 and sulfur species. Additionally, the hydrogen is typically formed by a reformation process and has a typical "reformate" composition of, ca. 70% H_2, 30% CO_2 with at least 0.5% CO. CO is one of the major poisons for low-temperature fuel cells and poisoning occurs by adsorption of the species onto the active catalyst sites. On Pt, CO adsorbs readily and leads to a surface covered with CO which blocks the sites for oxidation of hydrogen. With Pt catalysts, there is a severe detrimental effect to hydrogen oxidation at levels of CO as low as 5–10 ppm. Avoiding CO contamination in the PEMFC is thus a major issue and can be achieved by a number of methods:

(i)   Gas clean-up in which CO is physically removed from the gas stream to a level which does not affect cell performance.

(ii) Addition of oxygen or peroxide to the fuel stream to chemically oxidise the CO. However, peroxide brings with it the risk of membrane degradation (Gottesfeld and Pafford 1988).

(iii) Improved catalysts. CO-tolerant catalysts based on Pt alloys; for example, Pt–Ru, have been shown to exhibit good performance in the presence of relatively large amounts of CO up to 100 ppm (Gasteiger et al. 1995).

(iv) Bilayer anode. In a bilayer anode, a CO oxidation catalyst is placed next to the Pt-based electrocatalyst (near the feed side) and serves to reduce the CO content with a reduced input of oxygen.

(v)   Higher-temperature operation improves the CO tolerance of Pt. Operation at temperatures above 150°C would enable CO tolerance levels to be increased to several hundred ppm.

17.3   PEMFC WITH BIOFUELS

17.3.1   Water limitations of the PEMFC

To achieve the highest PEMFC performance, the membrane is required to have its greatest conductivity and thus to be fully wet. In operation of the PEMFC there is typically a net flow of water from the anode to the cathode by electro-osmotic flow and hence water is typically added to the anode side to maintain a humidified membrane. This water addition, together with the water produced at the cathode presents a high water load for the fuel cell which, as well as potentially affecting electrochemical performance, say through flooding, can lead to material stability issues. For example, in the use of metallic bipolar plates, the presence of mildly acidic water (with possible trace quantities of peroxide formed via the reduction of water) can lead to the presence of transition metal cations (Cr^{3+}, Ni^{2+} and Fe^{3+}) in the water. Diffusion of these cations to the membrane, or if the water from the cathode is recycled to the anode side, can lead to a reduction in the conductivity of the membrane. Additionally, the presence of the cations with peroxide formed at the cathode can lead to an enhanced oxidation of the membrane which affects it durability and long-term stability.

17.3.1.1   Hydrogen peroxide formation

Production of peroxide at the cathode is a major source of membrane degradation and can be accelerated in regions of low water content. The peroxide can cause loss of base polymer materials as well as ion exchange groups. Degradation can result from the presence of other cell components, for example, bipolar plates (surface films, dissolution, direct contact with the membranes).

In the case of peroxide production, minimising the influence can be achieved by avoiding the accumulation or presence of metal ions; for example, Fe^{2+}, Cu^{2+}, that can accelerate proton exchange membrane (PEM) degradation and by maximising water content (dilute peroxide). The use of free radical inhibitors, stabilisers and sacrificial agents in PEM has also been suggested. One possibility could be the introduction of stable peroxide decomposition catalysts into the membrane (LaConti *et al.* 1977).

17.3.1.2   Ageing of the PEMFC

For long-term operation of fuel cells, ageing is a critical performance issue. Ageing can arise from mechanical and thermal stresses associated with for example, changing temperature, current density and humidity. Such effects are more intensive with cells operating at higher temperatures. The main problems encountered during long-term operation are those of:

- water loss from the electrolyte, causing loss of conductivity;
- flooding of electrodes causing loss of available catalyst surface area;
- delamination of the electrodes;
- poor mechanical strength, especially for very thin electrolytes;
- poisoning of Pt-based anodes by adsorption of CO and other contaminants (e.g. sulfur species);

- severe cell corrosion during electrolyte flooding caused by fuel-deficient conditions;
- fuel starvation leading to carbon corrosion in the catalyst supports and carbon diffusion layers.

Mitigation strategies for many of the aging and degradation problems have been proposed and include the use of water electrolysis catalysts in the anode region to minimise carbon oxidation and the use of gas phase CO_2 reduction catalyst (bilayer anode) to minimise the reverse water gas shift reaction (CO production). Overall, however, systematic studies of ageing of PEMFCs are scarce and urgently needed especially in relation to impurities that could be present in "impure" fuels and atmospheric air.

17.4   HIGHER-TEMPERATURE PEMFC

The operation of PEMFC will be enhanced by elevated temperatures above 100°C by improved kinetics of the cathode and anode reactions, and the reduction of the adsorption of poisoning species such as CO. However, the ionic conductivity of Nafion falls considerably above 80°C due to evaporation loss of water, which is necessary for its conductivity. Composite membranes that exhibit fast proton transport at elevated temperatures are needed for PEMFC operating in the 100°C to 200°C temperature range. Several approaches have been pursued to resolve this issue, such as utilising different proton-conducting ionomer polymers, including polyphenylene sulfide sulfonic acid, sulfonated polyimides the perfluorosulfonyl imide form of the Nafion membrane, sulfonated polyether etherketones (sPEEK) and other novel sulfonated polymeric membranes (Kreuer 2003). The influence of temperature and humidity on the proton conductivity of these systems was frequently similar to that for Nafion membranes, but they often exhibited inferior performance.

Another approach is the use of proton-conducting membranes based on acid-impregnated ionomer polymers such as polystyrene sulfonic acid membranes imbibed with sulfuric acid, Nafion membranes impregnated with 85% phosphoric acid (Savinell et al. 1994) or non-volatile heteropolyacid-impregnated Nafion membranes (Malhotra and Datta 1997). A related strategy is to utilise acid-doped non-ionomeric polymers such as phosphoric acid-doped polybenzimidazol (PBI) (Samms et al. 1996) or trifluoromethane sulfonic acid-doped polyvinylidene fluoride-hexafluoropropylene (PVDF-HFP) copolymers (Peled et al. 1998). With the acid-doped, or impregnated, membranes there are potential problems with acid migration, corrosion of cell components, adsorption of anions on the catalyst and acid volatility.

Ionomer polymer membranes, which derive proton conductivity from fixed acid sites, solvated by water, are unlikely to achieve conductivities greater than 0.1 S/cm, at temperatures far higher than the boiling point of water. However, because the Nafion membrane requires only a dipolar solvent medium for conduction, high ionic conductivity has been achieved in other solvents (not purely aqueous) including water–organic solvent mixtures, alcoholic solvents and some aprotic dipolar solvents.

Consequently a new group of polymer membranes, based on impregnation with ionic liquids, with high ionic conductivity over a wide temperature range and with low solvent volatility, have been investigated. Recently, proton-conducting sulfonated polyether ketone (sPPEK) membranes, imbibed with pyrazole or imidazole liquids under anhydrous conditions, have been produced with conductivities close to 0.01 S/cm at 200°C (Kreuer et al. 1998).

Doyle et al. (2000) produced a high-temperature proton-conducting membrane based on a Nafion-Ionic liquid composite. Ionic conductivities in excess of 0.1 S/cm have been demonstrated using the ionic liquid 1-butyl, 3-methyl imidazolium trifluoromethane sulfonate at temperatures of 180°C. The membranes used in this study were the Nafion (117 and NE105) membrane (1000 equivalent weight (EW)), an 890-EW and also the 805-EW Dow (XUS) membrane. Comparisons between the ionic-liquid-swollen membrane and the neat liquid itself indicate substantial proton mobility in these composites. The demonstration of a perfluorinated ionomer membrane, such as the Nafion membrane, swollen with ionic liquids to give composite free-standing membranes with excellent stability and proton conductivity in fuel cell applications while retaining the low volatility of the ionic liquid has yet to be seen.

17.4.1   PBI membranes

PBI is a relatively low-cost commercially available polymer (150–220 euro/kg) which has excellent stability in reducing and oxidising environments. PBI is a basic polymer (p$Ka \approx 6.0$) which readily sorbs acid and helps to further stabilise the polymer. The PBI membranes are conductive above 100°C even when dry. Acid doping of 50% by weight can be achieved without adverse effects to its mechanical properties.

One of the main attractions of PBI is that the solution form of the polymer can potentially be used to cast membranes and be used as an ionomer ink in the preparation of bonded catalytic electrodes for fuel cells. For example, Samms et al. (1996) prepared PBI membranes containing 15 w/o Pt by casting from a mixture of PBI and fuel cell grade platinum black.

Two procedures for preparation of acid (phosphoric) doped PBI membranes have been researched by Wainwright et al. (1995). First referred to as type I, membranes were cast from a solution of the polymer and doped by immersion in a solution of phosphoric acid, typically 11.0 M in concentration. The membrane films typically contained 5.0 H_3PO_4 molecules per repeat unit in the polymer. The second method casts the membrane films from a solution of polymer in phosphoric acid in a suitable solvent. The membranes, type II, contained 6.0 H_3PO_4 molecules per repeat unit.

In general compared to a Nafion membrane, acid-doped PBI membranes have several advantages. They have good proton conductivity and mechanical flexibility at elevated temperature, and exhibit excellent oxidative and thermal stability.

Table 17.1   Membranes for the PEMFC.

Membrane	Characteristic type	Features
Perfluorinated ionomer	Homogeneous (e.g. Nafion®)	High conductivity and chemical and mechanical stability. High methanol crossover
	Micro-reinforced (e.g. polytetra fluoroethylene (PTFE) filled, GoreSelect®)	Thin, very high conductivity but high crossover of methanol
	Composites with heteropolyacids (e.g. Nafion® containing PWA)	High conductivity
Partially fluorinated ionomer	Grafted membranes (e.g. PVDF-styrene co-polymer)	High conductivity, reduced methanol diffusion Stability unknown
	Based on poly(α,β,β, trifluorostyrene) and co-polymers (e.g. Ballard BAM® membranes)	Lower cost replacement to Nafion for PEM. Excellent PEM performance, good lifetime
Non-fluorinated ionomer	Sulfonated divynyl benzene-cross-linked polystyrene	Poor stability
	Sulfonated styrene/ethylene-butylene/styrene tri-block polymer, DAIS Corp	Good proton conductivity, low cost, sulfonated hydrocarbon polymer ink available for electrode bonding
	Polyphosphazene	High conductivity Low methanol diffusion coefficient
	Homogeneous partially sulfonated (het)arylene main chain polymers (e.g. polyether ketone)	Good chemical and mechanical stability especially when cross-linked
	Covalently cross-linked arylene main chain ionomers and ionomer blends (e.g. Victrex® poly (ethersulfone) with sulfonamide cross-linking)	High-temperature operation. Questions over chemical and material stability
	Ionically cross-linked ionomer networks (e.g. PBI with sulfonated poly(ethersulfone))	High thermal stability and reduced methanol crossover with good conductivity
High-molecular/flow molecular composites	Polymer/inorganic mineral acid composite (e.g. phosphoric acid-doped PBI)	Good conductivity at high temperature. Very low methanol crossover. Liquid operation may leach phosphoric acid
	Acidic polymer/flow-molecular amphoter composites (e.g. sPEEK with imidazole).	Water-free proton conductivity, suitable for high-temperature DMFC
Organic/ inorganic composites	Ionomer/inorganic oxide particle composite (e.g. Nafion/SiO_2)	Good proton conductivity at temperatures $>100°C$. Good DMFC performance

Table 17.1   *(cont'd)*

Membrane	Characteristic type	Features
	Ionomer/inorganic oxide particle composite by sol–gel fabrication (e.g. Nafion/ZrO_2). Organic/inorganic hybrid polymers (e.g. organically modified silane electrolyte (ormolytes))	Good proton conductivity at temperatures >100°C Potential good DMFC performance. Good conductivity.
Inorganic	Nano-porous membrane with immobilised acid (e.g. SiO_2/PVDF binder/ sulfuric acid)	Cheap, high conductivity, low methanol crossover

17.4.2   Alternative membranes

A list of alternative membrane materials that have been used in the PEMFC is presented in Table 17.1.

Ballard developed a range of membranes for PEMFCs which have delivered excellent long-term performance and in particular polymers based on sulfonated trifluorostyrene (BAM3G) (Steck and Stone 1996).

High-performance PEMFC operation has been achieved with a cell based on phosphotungstic acid (PWA) as proton-conducting electrolyte immobilised in a glass fibre matrix (Giordano *et al.* 1996). Excellent power densities for PEMFCs of up to 700 mW/cm^2 have been achieved at 25°C. This was partly attributed to the promotion of the ORR by the PWA electrolyte. An adaptation of the use of this heteropolyacid was its immobilisation into a Nafion membrane (Malhotra and Datta 1997). Preliminary fuel cell data indicated an improved performance with this membrane in comparison with Nafion alone when operating at temperatures of 110°C to 120°C. Nafion membranes immobilised with silicotungstic acid and thiophene were used in PEMFCs to improve the performance. The conductivity of the new membranes was approximately eight times that of equivalent Nafion-117 membranes (Tazi and Savadogo 2000).

Ionomeric membranes made from sulfonated polysulfone (SPSU) filled with 8% phosphatoantimonic acid, which have conductivities of 0.06 S/cm at 80°C have been proposed for high-temperature PEMFCs (Genova-Dimitrova *et al.* 2001).

Interest in the immobilisation of a range of ionic conductors in porous supports for PEMFCs continues with recent reports of immobilisation of sulfuric acid inside glass fibre matrices and polysulfone (PSU) membranes (Haufe and Stimming 2001). Micro-fibre glass fleece with immobilised 5.0 M H_2SO_4 or Nafion-117 exhibited conductivity comparable to Nafion; that is, 0.53 S/cm. PSU-containing immobilised sulfuric acid gave conductivity superior to Nafion. PEMFC performance, using electrodes produced by a "cold-pressing", was as good as that achieved with Nafion membranes.

Kerres *et al.* (2000) prepared acid–base ionomer blends of sPPEK and SPSU with PBI, PSU(NH$_2$)$_2$ and poly(4-vinylpyridine). Polymer membranes were produced, 30 and 60 μm thick, from blends of polyether etherketone (PEEK) and PSU(NH$_2$)$_2$ which had ion exchange capacities of 1.34 and 1.58 with ionic resistances of 0.1 ω cm^2. Hasiotis *et al.* (2001) have tested membranes from blends of PBI and SPSU. Membranes cast from the latter polymer are known to have good mechanical strength and flexibility. On doping with phosphoric acid, the effect of an increase of SPSU was to decrease the doping acid content. The ionic conductivity of the blend membranes increased with temperature. For blends with 44% SPSU and low acid doping of 30 mol.% H$_3$PO$_4$, conductivity increased from 0.002 to 0.01 S/cm with an increase in temperature from 30°C to 180°C. Conductivity of the blend membranes also increased with relative humidity and acid doping level. The blend membranes exhibited superior conductivity to PBI alone under similar test conditions. Typical conductivities of highly doped blend membranes were in the range of 0.01–0.07 S/cm, at temperatures for 20°C to 180°C and were comparable to those achieved with the type II PBI membranes discussed above. The blend membranes have been fabricated into viable membrane electrode assemblies (MEAs) for PEMFCs and exhibited good power densities at temperatures as high as 190°C (Savadogo and Xing 2000).

17.5   DIRECT ALCOHOL FUEL CELLS

17.5.1   Introduction

Although hydrogen-fed PEMFCs are finding several applications as power sources in vehicles, there is a large interest in their application with liquid fuels such as methanol and ethanol. These fuel cells operate with the direct oxidation of the fuel, for example, in the DMFC.

A limitation with the DMFC, in comparison to hydrogen PEMFCs, is the relatively sluggish anode kinetics associated with methanol oxidation. However, progress has been such in recent years that impressive power densities of up to 400 mW/cm^2 have been reported for DMFC systems based on Pt–Ru anode catalysts (Arico *et al.* 2000). Currently, there is a wealth of literature on the potential application of methanol either directly or after reformation in low-temperature fuel cells. However, in order to extend the practical applications of low-temperature fuel cells, and to facilitate their penetration into the commercial market, it is desirable to increase the number of liquid fuels which can be employed in these systems. Among the possible candidates, ethanol offers a number of characteristics that are superior to methanol, including: being far less toxic, having lower vapour pressure, established production from renewable feed-stocks and an established infrastructure for use as a fuel. In addition, ethanol is truly a biomass-renewable fuel, being readily produced from sugar-cane, grain-crops, wood and agricultural waste, and the environmental characteristics of ethanol are comparable to methanol. Thus, a major challenge to science and engineering is developing suitably effective electrocatalysts for ethanol oxidation and thereby increasing the performance of the PEMFC.

17.5.2 DMFCs

The low operating temperature and rapid start-up characteristics together with its robust solid-state construction give the PEMFCs a clear advantage for many applications. The preferred fuel for PEMFCs is hydrogen and while many strategies for providing hydrogen to PEMFCs are being evaluated, the most acceptable option appears to be to generate hydrogen on-demand from liquid hydrocarbons or methanol. The technical challenge, however, lies in modifying large-scale industrial processes like steam-reforming or partial-oxidation reactors to lighter-weight units that can for example, fit inside the car. An elegant solution to the problems associated with the need for gaseous hydrogen fuel lies in operating the PEMFC directly with a liquid fuel. Substantial efforts are therefore being expended on PEMFCs that run on air plus a mixture of methanol and water. A solid–polymer–electrolyte DMFC would be about as efficient as a conventional reformer-based polymer electrolyte fuel cell (PEFC) unit, both in its construction and operation. The realisation of a commercially viable PEMFC that runs on liquid fuel is indeed regarded as the "holy grail" of fuel cell technologies.

17.5.2.1   Operating principle of the DMFC

Historically, direct electro-oxidation of methanol in a fuel cell has been a subject of study far more than three decades. The early cell designs utilised aqueous sulfuric acid electrolyte at about 60°C. About 20 years ago, the Shell Research Centre in the UK and Hitachi Research Laboratories in Japan built DMFC stacks of up to 5 kW but their power densities were only 20–30 m W/cm^2 even with platinum loadings as high as 10 mg/cm^2, which corresponds to specific power densities of 2–3 W/g of platinum catalyst.

In a fuel cell employing an acid electrolyte, methanol can be directly oxidised to carbon dioxide at the anode according to the reaction:

$$CH_3OH + H_2O \rightarrow CO_2 \uparrow 6H^+ + 6e^- \quad E^\circ = 0.03 \text{ V} \qquad (17.1)$$

At the cathode, oxygen gas combines concomitantly with the protons and electrons, and is reduced to water through the reaction:

$$\tfrac{3}{2}O_2 + 6H^+ + 6e^- \rightarrow 3H_2O \qquad (17.2)$$

Accordingly, the net cell reaction is represented as:

$$CH_3OH + 3/2O_2 \rightarrow 2H_2O + CO_2 \qquad (17.3)$$

Accordingly, the standard electromotive force (e.m.f.), is $E^\circ = 1.20$ V. The potential efficiency (ε_e) of a DMFC for an operational cell e.m.f. of 0.5 V is approximately 40% and the specific energy is approximately 6.1 kWh/kg.

The main drawback of such cells is the very sluggish anode reaction, which coupled with the inefficient cathode reaction, gives rise to a very low overall performance, particularly at low temperatures. The performance of the cells utilising sulfuric acid electrolyte is further receded owing to the high internal resistance of the system. In the 1980s, it was realised that a considerable increase in the efficiency might be obtained by using a thin proton-conducting polymer sheet such as

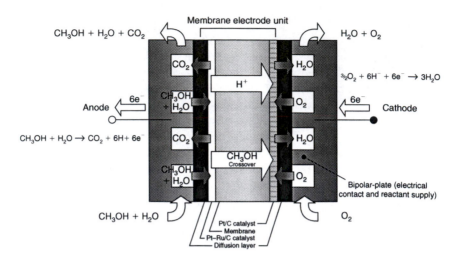

Figure 17.1   Operation of the DMFC.

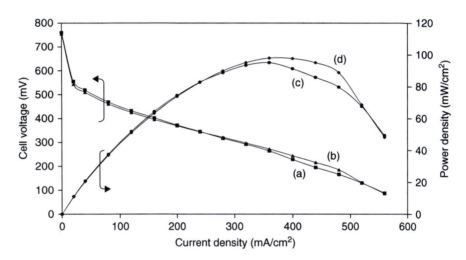

Figure 17.2   Typical power performance of a DMFC (a), (b): cell voltage; (c), (d): power density (Scott *et al.* 1998).

Nafion®: a PFSA polymer. A DMFC with a Nafion® electrolyte membrane is shown schematically in Figure 17.1.

In a DMFC, methanol dissolved in water is supplied to its anode, which has the tendency to pass through the membrane electrolyte and hence affect the performance of the cathode. Therefore, a fundamental limitation in the practical realisation of such a DMFC has been the existence of electrochemical losses at both the anode and the cathode arising mainly due to the electrocatalytic restrictions and methanol crossover though the electrolyte membrane due to osmotic drag.

The typical polarisation curve for a DMFC along with its constituent electrodes is shown schematically in Figure 17.2. Although the thermodynamic potential for reaction (17.1) is 0.03 V vs. the standard hydrogen electrode (SHE), because of there are six electrons involved, the equilibrium value is not readily realisable, even with the best possible electrocatalysts devised so far. Furthermore, because of the high degree of irreversibility of reaction (17.2), even under open-circuit conditions, the overpotential at the oxygen electrode in a PEFC is about 0.2 V which represents a loss of about 20% from the theoretical efficiency of the PEMFCs. The situation is even worse with the DMFC where there is an inherent loss of about 0.1 V at the oxygen electrode owing to the crossover of methanol from anode to the cathode. Consequently, the output cell voltage in a DMFC is much lower than the ideal thermodynamic value and it decreases with increasing load current density as shown in Figure 17.2 (Scott *et al.* 1998).

17.5.2.2 Electrode reaction in the DMFC

Anodic oxidation of methanol   Several schemes have been proffered for the anodic oxidation of methanol. Broadly speaking, the basic mechanism for methanol oxidation can be summarised in two functionalities, namely the electrosorption of methanol onto the substrate, followed by addition of oxygen to adsorbed carbon-containing intermediates to generate carbon dioxide.

In practice, only a few electrode materials are capable of adsorption of methanol. In acidic media, only platinum and platinum-based catalysts have been found to show sensible activity, and almost all mechanistic studies have concentrated on these materials. On platinum itself, adsorption of methanol is believed to take place through a sequence of steps described below.

The first step is dissociative chemisorption of methanol onto the platinum surface, which involves successive donation of electrons to the catalyst as follows:

$$Pt + CH_3OH \rightarrow Pt\!-\!CH_2OH + H^+ + 1e^- \tag{17.4}$$

$$Pt\!-\!CH_2OH \rightarrow Pt_2\!-\!CHOH + H^+ + 1e^- \tag{17.5}$$

$$Pt_2\!-\!CHOH \rightarrow Pt_3\!-\!COH + H^+ + 1e^- \tag{17.6}$$

In reactions 17.4–17.6, Pt_3–COH is the major surface species. A surface rearrangement of the oxidation intermediates generates carbon monoxide, linearly or bridge-bonded to Pt-sites according to the reaction:

$$Pt_3\!-\!COH \rightarrow Pt\!-\!CO + 2Pt + H^+ + 1e^- \tag{17.7}$$

Water discharge occurs at high anodic overpotentials on Pt with the formation of Pt–OH species at the platinum surface as below:

$$Pt + H_2O \rightarrow Pt\!-\!OH + H^+ + 1e^- \tag{17.8}$$

The ultimate step is the reaction of Pt–OH groups with neighbouring methanolic residues to give carbon dioxide according to the following reaction:

$$Pt\!-\!OH + Pt\!-\!CO \rightarrow 2Pt + CO_2\!\uparrow + H^+ + 1e^- \tag{17.9}$$

Accordingly, the overall oxidation of methanol to carbon dioxide proceeds through a six electron donation process.

Platinum alone is not sufficiently active and there is need for a promoter that effectively provides oxygen in some active form to achieve facile oxidation of the chemisorbed CO on platinum. In the literature, various approaches towards platinum promotion have been attempted. The simplest method is to generate more Pt–O species on the platinum surface by incorporating certain metals with platinum to form alloys such as Pt_3Cr and Pt_3Sn, which then dissolve to leave highly reticulated but active surfaces.

A second approach has been the use of surface adatoms produced by underpotential deposition on the platinum surface (Campbell and Parsons 1992). A third type of promotion is the use of alloys of platinum with different metals such as Pt–Ru, Pt–Os, Pt–Ir, etc., where the second metal forms a surface oxide in the potential range for methanol oxidation (Hamnett and Kennedy 1988). The fourth type of promotion described in the literature is a combination of Pt with a base-metal oxide such as Nb, Zr, Ta, etc. (Hamnett 1997). In addition to electrodeposition or reductive deposition of Pt onto an oxide surface such as Pt–WO_3, attempts have also been made to study methanol oxidation on perovskite-based oxides with platinum, such as $SrRu_{0.5}Pt_{0.5}O_3$ (White and Sammells 1993). It also seems that certain amorphous metal alloys, such as Ni–Zr, which form a thick passive oxyhydroxide film, can also facilitate the methanol oxidation reaction (Hays et al. 1993).

Among the various methanol oxidation catalysts described above, perhaps Pt–Ru and Pt–Sn happen to be the most widely studied catalysts (Burnstein et al. 1997). It has been shown that the alloying of Sn and Ru with Pt gives rise to electrocatalysts which strongly promote the oxidation of methanol and related methanolic species. On the platinum surface, at low potentials –CO groups are adsorbed while at high potentials chemisorption of –OH groups takes place during the electro-oxidation of methanol and both the processes are distinctly separated. On a Pt–Ru surface, the chemisorption of –OH groups shifts to lower potentials and overlaps with the region where –CO groups are adsorbed on the catalyst as shown in Figure 17.3. On a Pt–Ru alloy, water discharging occurs on Ru-sites at much lower potentials in relation to pure Pt catalyst, according to the reaction given below:

$$Ru + H_2O \rightarrow Ru\text{–}OH + H^+ + 1e^- \qquad (17.10)$$

The final step is the reaction of Ru–OH groups with the neighbouring methanolic residues adsorbed on Pt to give carbon dioxide according to the reaction:

$$Ru\text{–}OH + Pt\text{–}CO \rightarrow Ru + Pt + CO_2\uparrow + H^+ + 1e^- \qquad (17.11)$$

Cathodic reduction of oxygen   Oxygen reduction can proceed by two different pathways, namely the direct four-electron pathway and the peroxide pathway. The direct four-electron pathway in alkaline medium proceeds as:

$$O_2 + 2H_2O + 4e^- \rightarrow 4OH^- \ (E^\circ = 0.4\,V \text{ vs. SHE}) \qquad (17.12)$$

and in acidic medium, it proceeds as:

$$O_2 + 4H^+ + 4e^- \rightarrow 2H_2O \ (E^\circ = 1.23\,V \text{ vs. SHE}) \qquad (17.13)$$

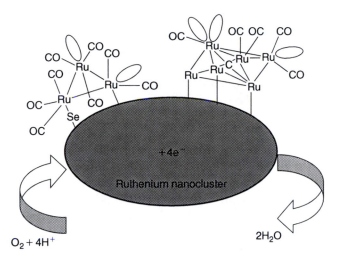

Figure 17.3   Schematic description of –CO oxidation on Pt and Pt–Ru surface by M−OH group.

The peroxide pathway in alkaline medium proceeds as:

$$O_2 + H_2O + 2e^- \rightarrow HO_2^- + OH^- \ (E^\circ = -0.06 \text{ V vs. SHE})   (17.14)$$

followed by peroxide reduction to OH^- ions:

$$HO_2^- + H_2O + 2e^- \rightarrow 3OH^- \ (E^\circ = 0.87 \text{ V vs. SHE})   (17.15)$$

or chemical decomposition of peroxide:

$$2HO_2^- \rightarrow 2OH^- + O_2   (17.16)$$

In an acidic medium, production of dioxygen through the peroxide pathway is possible as follows:

$$O_2 + 2H^+ + 2e^- \rightarrow H_2O_2 \ (E^\circ = 0.67 \text{ V vs. SHE})   (17.17)$$

This is followed by either:

$$H_2O_2 + 2H^+ + 2e^- \rightarrow 2H_2O_2 \ (E^\circ = 1.77 \text{ V vs. SHE})   (17.18)$$

or:

$$2H_2O_2 \rightarrow 2H_2O + O_2   (17.19)$$

The direct four-electron pathway does not involve the peroxide species and hence has a higher Faradaic efficiency in relation to the peroxide pathway. However, it has been difficult to find catalysts which could facilitate the direct four-electron pathway for dioxygen reduction.

As indicated in section "Anodic oxidation of methanol", in addition to irreversible losses, there is an additional overpotential observed on the cathode owing to the methanol crossover in a DMFC from its anode to the cathode. Therefore, cathodes with high loadings of platinum are usually employed in DMFCs.

However, since platinum has the tendency to be poisoned with methanol, it appears mandatory both to develop methanol impermeable membranes as well as methanol-tolerant oxygen reduction catalysts for practical realisation of DMFCs. In recent years, certain Ru-based chalcogenides have shown promise as methanol-tolerant oxygen reduction catalysts. However, these materials have much lower intrinsic specific activity for oxygen reduction than that of platinum.

17.5.2.3 Materials for the DMFC

From the foregoing, it is clear that for the realisation of practical DMFCs, it is mandatory (a) to increase their operational current densities making it desirable to develop catalyst materials which would increase the reaction rates both at the methanol anode and the oxygen cathode, (b) to develop methanol-tolerant oxygen reduction catalysts and (c) to investigate new electrolyte membranes which would have high proton conductivities, would be temperature resistant and would also be methanol impermeable. Various developments in these areas are summarised in the following sections.

Catalyst materials   Pt–Ru is the most commonly used catalyst for the anodic oxidation of methanol in a DMFC (Burnstein *et al.* 1997). The reported difference in activities of Pt–Ru catalyst is usually attributable to the method of catalyst production and resulting varying surface composition. Improvements in catalytic activity can frequently be achieved by dispersing the catalyst on a surface-area graphitic carbon, such as Vulcan XC-72 and KetzenBlack EC-600D carbon. Production of $RuCl_3$ with H_2PtCl_6 in aqueous solution produces relatively poor catalysts. In contrast the use of $Pt(NH_3)_2NO_2$ with $Ru(NO)(NO_3)_x$ in nitric acid or $Na_6Pt(SO_3)_4$ with $Na_6Ru(SO_3)_4$ in sulfuric acid produces effective catalysts for methanol oxidation.

High-surface-area carbon-supported Pt–Ru catalyst has also been prepared by a colloidal dispersion route. Other methods based on thermal decomposition of appropriate high-molecular-weight Pt-precursors and Pt-carbonyl compounds to produce unsupported high-surface-area catalysts have also been documented (Takasu *et al.* 2000). Beside these methods, interest has also been in co-precipitation, sol–gel and sputtering route for producing Pt–Ru catalysts (Cha and Lee 1999).

An effective method of catalyst evaluation is combinatorial synthesis. Through this procedure, both ternary (such as Pt–Ru–Os) and quaternary (such as Pt–Ru–Os–Ir) catalysts have been synthesised which have exhibited superior methanol oxidation capability than the Pt–Ru catalyst (Liu *et al.* 2000).

At present, platinum in particulate form exhibits the highest activity towards ORR in DMFCs; but platinum is not tolerant to methanol. In recent years, alternative cathode catalysts to platinum have been researched for the DMFC, some of which have exhibited tolerance to methanol. In particular, selective oxygen reduction catalysts such as iron tetra-methoxy phenylporphyrin (Fe-TMPP), cobalt tetra-methoxy phenylporphyrin (Co-TMPP), cobalt tetraazaanulene (CoTAA), Ru–Mo–Se and other ruthenium-based cluster catalysts-containing sulfur or selenium have shown substantial promise (Reeve *et al.* 2000). Quite recently, a range of carbon-supported Pt alloys with Co, Cr and Ni have been evaluated with the aim of improving the tolerance of the catalyst to methanol. Among these, the Pt–Co/C binary catalyst has been identified to be more suitable than Pt/C for the DMFC (Shukla *et al.* 2005).

Membrane materials   One of the key problems impeding the development of the DMFC is the properties of the available proton-conducting membrane. Owing to the low reactivity of methanol as well as the low conductivity of commercially available PEMs, namely Nafion, at ambient temperatures, it is necessary to keep an operational temperature near 100°C for DMFCs. Nafion membranes need water inside their skeleton in order to exhibit good proton conductivity at such high temperatures. Although it is quite a difficult requirement for PEFCs, it is of little concern in DMFCs, operating at low temperatures, since water along with methanol is constantly circulated in DMFCs during their operation. But the problem of methanol crossover associated with the Nafion membranes is serious. Methanol crossover is detrimental to the performance of a DMFC since it reduces both the coulombic efficiency of the fuel cell and the cell voltage. A crossover of methanol equivalent to $80 \, mA/cm^2$ can occur at a load current density of $150 \, mA/cm^2$ using a Nafion membrane electrolyte in a liquid-feed SPE–DMFC at 80°C (Hikita *et al.* 2001). The effects of methanol crossover can be controlled to a certain extent by strictly correlating the methanol-feed concentration with the actual demand of the cell. It is therefore mandatory to find electrolyte membranes which would reduce the methanol and water crossover sufficiently. Efforts are therefore expended to develop methanol impermeable PEMs either by modifying the available Nafion membranes or by developing novel PEMs.

Membranes comprising metallic blocking layers were proposed (Verbetsky *et al.* 1998). These membranes are composite electrolytes where a film of a methanol impermeable proton conductor, such as metal hybrid, is sandwiched between the Nafion sheets. Such composite electrolyte membranes were found to exhibit lower methanol crossover than Nafion with the best performance obtained for Nafion-115/Pt/Pd/Pt/Nafion-115 system.

Among the organic–inorganic composite membranes-containing zirconium-phosphonates, tin-doped morderites, zeolites or silica investigated for their methanol permeability, the results of an $80 \, \mu m$ thick membrane cast from a composite of Nafion ionomer and silica were particularly encouraging (Antonucci *et al.* 1999). DMFCs tests conducted with this membrane, and commercial Pt–Ru anode and Pt cathode, both with $2 \, mg/cm^2$ Pt loadings, demonstrated peak power densities of 240 and $150 \, mW/cm^2$ with oxygen and air cathodes, respectively. The open-circuit voltage at 145°C was as high as 0.95 V and methanol crossover as low as $4 \times 10^{-6} \, mol/cm^2 min$, which is equivalent to a methanol crossover current density of $40 \, mA/cm^2$.

Plasma etching of Nafion and palladium sputtering were also investigated to reduce methanol crossover in DMFCs (Choi *et al.* 2001). Plasma etching increases the roughness of the membrane and decreases the pore size, thereby decreasing its methanol permeability. But plasma etching can potentially remove sulfonic acid groups from the membrane surface and hence reduce its perform-ance. By contrast, palladium sputtering, whilst not affecting proton-transfer rate, has been reported to improve performance in comparison to that with an unmodified membrane. Ion-beam radiation has also been successfully deployed as a technique to increase the roughness of Nafion membranes and enhance cell performance (Oh *et al.* 2000).

Oxidation of methanol can be promoted substantially by operating the DMFC at temperatures above 100°C. The kinetics of oxidation will be accelerated and the influence of poisoning species, such as adsorbed CO species, will be reduced. However, the ionic conductivity of Nafion falls considerably above 100°C due to loss of water, which is necessary for its conductivity, by evaporation. Composite membranes that exhibit fast-ion proton transport at elevated temperatures are needed for PEMFCs operating between 100°C to 120°C. Several approaches have been pursued to resolve this issue, such as utilising different proton-conducting ionomer polymers including polyphenylene sulfide sulfonic acid (Miyatake *et al.* 1996), sulfonated polymides (Faure 1997), the perfluorosulfonyl imide form of Nafion (Kobayashi *et al.* 1998), sPPEK (Sumner *et al.* 1998) and other novel sulfonated polymeric membranes (Savadogo 2001). Although the influence of temperature and humidity on the proton conductivity of these systems was frequently similar to that for the Nafion membrane, they often exhibited inferior performance.

Another approach is the use of proton-conducting membranes based on acid-impregnated ionomer polymers, such as polystyrene sulfonic acid membranes imbibed with sulfuric acid, Nafion impregnated with 85% phosphoric acid (Savinell *et al.* 1994) or non-volatile heteropolyacid-impregnated Nafion membranes (Malhotra and Datta 1997). A related strategy is to utilise acid-doped non-isomeric polymers, such as phosphoric acid-doped polybenzimidazole (PBI) (Samms *et al.* 1996) or tri-fluromethane sulfonic acid-doped PVDF-HFP (Peled *et al.* 1998). Materials, such as polyvinyl alcohol (PVA), poly-aniline (PANI) and PVDF, showed very low methanol permeability but, however, even when acid impregnated, showed much lower proton conductivity. Of all the materials, only the acid-doped PBI showed characteristics suitable for the DMFC.

The use of acid-doped PBI as a polymer electrolyte is attractive in terms of thermal and oxidative stability but in order to attain the necessary mechanical flexibility and proton conductivity, the SPE–DMFC has to be operated at temperatures in excess of 200°C. Proton conductivity in a Nafion membrane is assigned to the vehicle mechanism by which protons are transported through membrane accompanied by another molecular species such as H_3O^+, $H_5O_2^+$ or $CH_3OH_2^+$. Materials assigned to this mechanism generally show a high degree of methanol crossover and high proton conductivity, which accurately describes the properties of a Nafion membrane. Conversely, PBI exhibits both low proton conductivity and low methanol permeability (2.4% to that of Nafion-117 membrane), a behaviour described by the Grottus "jump" mechanism. Theoretically, this mechanism can facilitate high proton conductivity but in practice it is an anomaly.

Another promising alternative is composite membranes made from blends of acid–base polymers. These membranes are produced by blending acid polymers, such as SPSUs, sPPEK, sPEEK, with basic polymers, such as poly-4-vinylpyridine (P4VP), PBI or a basically substituted PSU (bPSU). In these materials, electrostatic

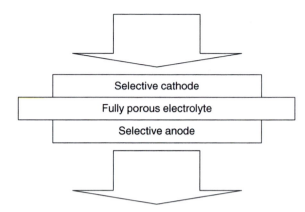

Figure 17.4   Schematic presentation of a mixed-reactant (MR) fuel cell.

forces from salt formation between acidic and basic groups achieve reversible cross linking of the polymer (Kerres *et al.* 2000). Recently, the treatment of Nafion-117 with dilute PBI/dimethyl acetamide (DMAC) solutions can produce films which exhibit a significantly higher DMFC performance than the parent materials. But, even to-date, the challenge of combining inherent conductivity with a low permeability to methanol remains.

A solution to these problems lies in the use of mixed-reactant (MR) DMFC, where unlike the conventional DMFC, the fuel and the oxidant are allowed to mix (Shukla *et al.* 2003b). Such a fuel cell relies on the selectivity of anode and cathode electrocatalysts to separate the electrochemical oxidation of fuel and electrochemical reduction of the oxidant. In such a fuel cell system (Figure 17.4) with no physical separation of fuel and oxidant, there is no longer any need for a gas-tight structure inside the stack and hence considerable relaxation of sealing, manifold and reactant delivery structures is possible.

17.5.2.4   Conventional vs. MR-DMFC

During the last decade, significant advances have been made in the direct methanol single cell development. Maximum power densities of 450 and 300 mW/cm^2 under oxygen and air-feed operation, respectively, and 200 mW/cm^2 at a cell potential of 0.5 V have been achieved for cells operating at temperatures close to or above 100°C under pressurised conditions, with Pt loadings of 1–2 mg/cm^2. The best results achieved with DMFC stacks for electro-traction are 1 kW/l power density with an overall efficiency of 37% at the design point of 0.5 V/cell (Shukla *et al.* 2003a).

Quite recently, researchers from International fuel cell (IFC) and University Austin have demonstrated that the performance of a DMFC in MRs mode with selective electrodes could exceed that of the conventional DMFCs when the fuel and the oxidant are supplied at identical rates to the anode and the cathode, respectively (Barton *et al.* 2001). They also conducted a design study in which the dimensions of a series of MRs, strip cell DMFC were optimised. In the single cell tests, a two-phase reactant mixture of 1 M methanol (3 ml/min) and air (3 l/min) was supplied to both sides of a conventional geometry MEA at 80°C. The 32 cm^2 MEA was a Nafion-117 membrane coated on one side with a hydrophobic Pt–Ru/C (5 mg/cm^2) anode and on the other side with Fe-TMPP, a methanol-tolerant cathode material. The half-cell experiments demonstrated that, in the system, there was little reaction between oxygen and methanol at the anode and that the main effect of the entrained air or nitrogen in the mixed-feed was to impede mass-transport of the feed to the anode at current densities above 100 mA/cm^2. At the cathode, half-cell measurements again showed little difference while operating in MRs and conventional separated-reactants modes. Consequently, performance of the DMFCs in MR mode was identical to the performance of DMFCs in conventional mode. In the MR-DMFCs, one may argue that methanol crossover offers a performance advantage, a situation quite opposite to a conventional SPE–DMFC. In the latter, methanol leaks constantly from the anode to the cathode causing both a lowering in its potential and wastage of fuel through direct chemical oxidation. This has its adverse effect on fuel efficiency in a conventional DMFC and is one of the primary reasons for various selective cathode materials, such as Ru–Se, being investigated for conventional DMFCs.

   In the DMFC, cathodes selectivity is paramount and is accomplished by using oxygen reduction catalysts, which in addition to being tolerant to methanol, do not oxidise it. The performance characteristics of certain MR-DMFCs, employing various methanol-resistant oxygen reduction catalysts at the cathode and a Pt–Ru/C catalyst at the anode are shown in Figures 17.5 and 17.6 (Shukla *et al.* 2003). The performance curves at 90°C for the DMFCs employing 1.0 mg/cm^2 of Fe-TMPP/C, Co-TMPP/C, FeCo-TMPP/C and RuSe/C at the cathode are shown in Figure 17.5. Among these, the best performance, with a maximum power output of ~30 mW/cm^2, is observed for the MR-DMFC employing 1 mg/cm^2 of RuSe/C catalyst at the cathode. The performance curves at 90°C for the MR-DMFCs employing varying amounts of RuSe/C at the cathode are shown in Figure 17.6. It is found that the best performance at the maximum output power of about 50 mW/cm^2 is obtained for the MR-DMFC with the RuSe/C loading of 2.5 mg/cm^2, while operating the MR-DMFC with methanol plus oxygen. A maximum output power of ~20 mW/cm^2 is obtained when operating the cell with methanol plus air.

17.5.3   Direct ethanol fuel cells

In many respects the development of the direct ethanol fuel cell (DEFC) is similar to that of the DMFC as many of the cell materials investigated are similar. However, the major limitation of the DEFC is due to the relatively poor kinetics of ethanol oxidation, even compared to methanol, requiring removal of 12 electrons:

$$C_2H_5OH + 3H_2O \rightarrow 12H^+ + 12e^- + 2CO_2 \qquad (17.20)$$

   Up until recently, a major problem of the DEFC was perceived to be the significant production of intermediates such as ethanal and ethanoic acid. It is still the case that a key problem in the rapid development of the DEFC is the facile and controlled cleavage of the C–C bond. In addition, in contrast to DMFC, chemisorption of ethanol itself is relatively slow, with the result that overall current

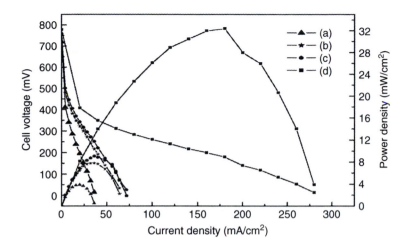

Figure 17.5   Performance data for DMFCs with Pt–Ru/C anode and cathode comprising: (a) Co-TMPP/C, (b) FeCo-TMPP/C, (c) Fe-TMPP/C and (d) RuSe/C.

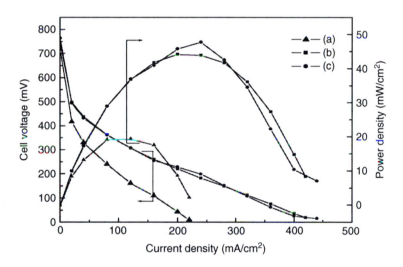

Figure17.6   Performance data for DMFCs with Pt–Ru/C anode and cathode with (a) 1 mg/cm^2, (b) 2 mg/cm^2 and (c) 2.5 mg/cm^2 of RuSe/C.

densities are significantly below those observed with methanol under similar conditions. The oxidation of ethanol on Pt and its alloys is a complex process that takes place through two different intermediates, one strongly adsorbed and one that is significantly less strongly adsorbed.

Some work has also been done on promoters for ethanol oxidation. As expected, Ru offers very specific advantages in the provision of active oxygen species at low potentials. Both Wang *et al.* (1995) and Arico *et al.* (1998) have reported the use of ethanol in solid PEMFCs using Pt–Ru as the anode electrocatalyst. Wang *et al.* (1995) reported a power density of 100 mW/cm^2 *ca.* 150 mA/cm^2 and 160°C to

180°C and Arico *et al.* (1998) reported a maximum power density of *ca.* 110 mW/cm^2 at 0.32 V and 145°C. This latter data may be contrasted with an optimum power density of *ca.* 240 mW/cm^2 at 0.4 V using methanol as the fuel in the same cell. Optimisation of the conditions for Pt–Ru operation in practical cells also leads to regimes in which ethanol formation is much reduced, though conditions have not been identified in which it is wholly eliminated. Recently, studies have also been reported on the use of ethanol as a fuel in the PEMFC after steam reforming or partial oxidation (Ioannides and Neophytides 2000) as well as the direct application of other hydrocarbons in which C–C bond scission takes place such as ethylene glycol (Peled *et al.* 2001) and dimethyl oxalate and 1-butanol (Chen 2000).

It is clear from the literature that the problem of incomplete oxidation of ethanol in a DEFC is still a factor and the performance of the DEFC is, at present, significantly below that of the DMFC. Overall, however, there appears to be considerable unrealised potential in the use of ethanol as a fuel in SPE fuel cells.

17.6   CONCLUSIONS

PEMFC are currently based on Nafion® or similar membranes and operate at low temperatures of less than 80°C. The main challenges for technology development are:

(1) high materials cost (noble metal catalysts, polymer membrane, etc.);
(2) complex system construction and operation with respect to water and thermal management;
(3) fuel supply, that is, on-board storage and refuelling of hydrogen or reformer-purification units for hydrocarbons/alcohols;
(4) low value of heat energy, low overall efficiency (~30%) and limited co-generation of heat and power for stationary applications.

The technical problems to developing improved PEMFC technology can be solved by the research of new PEMFCs for operation at temperatures significantly higher than 100°C. In terms of applications, there appear two temperature ranges for higher-temperature PEMFC operation:

(i)   110°C to 130°C for automotive operation;
(ii) 150°C to 200°C for stationary operation.

At these higher temperatures, the oxidation of fuels such as ethanol comes much more attractive and the use of such biofuels could become an attractive option for fuel cell derived power.

REFERENCES

Antonucci, P.L., Arico, A.S., Creti, P., Ramunni, E. and Antonucci, V. (1999) Investigation of a direct methanol fuel cell based on a composite Nafion (R)-silica electrolyte for high temperature operation. *Solid State Ionics* **125**, 431.

Arico, A.S., Creti, P., Antonucci, P.S. and Antonucci, V. (1998) Comparison of ethanol and methanol oxidation in a liquid-feed solid polymer electrolyte fuel cell at high temperature. *Electrochem. Solid State Lett.* **1**, 66.

Arico, A.S., Creti, P., Modica, E., Monforte, G., Baglio, V. and Antonucci, V. (2000) Investigation of direct methanol fuel cells based on unsupported Pt-Ru anode catalysts with different chemical properties. *Electrochim. Acta* **45**, 4319.

Barbir, F. (2003) System design for stationary power generation. *Handbook of Fuel Cells – Fundamentals, Technology and Applications* (eds. W. Vielstich, A. Gasteiger and A. Lamm), Chapter 51. John Wiley and Sons Ltd. Chichester.

Barton, S.C., Patterson, T., Wang, E., Fuller, T.F. and West, A.C. (2001) Mixed-reactant, strip-cell direct methanol fuel cells. *J. Power Sources* **96**, 329.

Burnstein, G.T., Barnett, C.J., Kucernak, A.R. and Williams, K.R. (1997) Aspects of the anodic oxidation of methanol. *Catal. Today* **38**, 425.

Campbell, S.A. and Parsons, R.J. (1992) Effect of Bi and Sn adatoms on formic acid and methanol oxidation at well defined platinum surfaces. *Chem. Soc. Farad. Tran.* **88**, 833.

Cha, S.Y. and Lee, W.M. (1999) Performance of proton exchange membrane fuel cell electrodes prepared by direct deposition of ultrathin platinum on the membrane surface. *J. Electrochem. Soc.* **146**, 4055.

Chen, S.L. and Schell, M. (2000) Excitability and multistability in the electrochemical oxidation of primary alcohols. *Electrochim. Acta* **45**, 3069.

Choi, W.C., Kim, J.D. and Woo, S.I. (2001) Modification of proton conducting membrane for reducing methanol crossover in a direct-methanol fuel cell. *J. Power Sources* **96**, 411.

Doyle, M., Choi, S.K. and Proulx, G. (2000) High-temperature proton conducting membranes based on perfluorinated ionomer membrane-ionic liquid composites. *J. Electrochem. Soc.* **147**, 34.

Faure, S., Cornet, N., Gebel, G., Mercier, R., Pineri, M. and Sillion, B. (1997) Sulfonated polyimides as novel proton exchange membranes for H_2/O_2 fuel cells. *Proceedings of the Second International Symposium New Materials for Fuel Cell and Modern Battery Systems*, Montreal, Canada, July 6–10, 818.

Gasteiger, H.A., Markovic, N.M. and Ross, P.N. (1995) Hydrogen and CO electrooxidation on well defined Pt, Ru and Pt–Ru. 2. Rotating disk electrode studies of CO/H_2 mixtures at 62°C. *J. Phys. Chem.* **99**, 318.

Genova-Dimitrova, P., Baradie, B., Foscallo, D., Poinsignon, C. and Sanchez, J.Y. (2001) Ionomeric membranes for proton exchange membrane fuel cell (PEMFC): sulfonated polysulfone associated with phosphatoantimonic acid. *J. Membr. Sci.* **185**, 59.

Giordano, N., Staiti, P., Hocevar, S. and Arico, A.S. (1996) High performance fuel cell based on phosphotungstic acid as proton conducting electrolyte. *Electrochim. Acta* **41**, 397.

Gottesfeld, S. and Pafford, J. (1988) A new approach to the problem of carbon monoxide poisoning in fuel cell operating at low temperatures. *J. Electrochem. Soc.* **135**, 2651.

Hamnett, A. (1997) Bimettallic carbon supported anodes for the direct methanol air fuel cell. *Catal. Today* **38**, 445.

Hamnett, A. and Kennedy, B.J. (1988) Bimettallic carbon supported anodes for the direct methanol air fuel cell. *Electochim. Acta* **33**, 1613.

Hasiotis, C., Qingfeng, L., Deimede, V., Kallitsis, J.K., Kontoyannis, C.G. and Bjerrum, N.J. (2001) Development and characterization of acid-doped polybenzimidazole/sulfonated polysulfone blend polymer electrolytes for fuel cells. *J. Electrochem. Soc.* **148**, A513.

Haufe, S. and Stimming, U. (2001) Proton conducting membranes based on electrolyte filled microporous matrices. *J. Membr. Sci.* **185**, 95.

Hays, C.C., Manoharan, R. and Goodenough, J.B. (1993) Methanol oxidation and hydrogen reactions on NiZr in acid solution. *J. Power Sources* **45**, 291.

Hikita, S., Yamane, K. and Nakajima, Y. (2001) Measurement of methanol crossover in direct methanol fuel cell. *JSAE review* **22**, 151.

Ioannides, T. and Neophytides, S. (2000) Efficiency of a solid polymer fuel cell operating on ethanol. *J. Power Sources* **91**, 150.

Kerres, J., Ullrich, A., Haring, Th., Baldauf, M., Gebhardt, U. and Preidel, W. (2000) Preparation, characterization and fuel cell application of new acid–base blend membranes. *J. New Mater. Electrochem. Syst.* **3**, 129.

Kobayashi, T., Rikukawa, M., Sanui, K. and Ogata, N. (1998) Proton-conducting polymers derived from poly(ether-etherketone) and poly(4-phenoxybenzoyl-1,4-phenylene). *Solid State Ionics* **106**, 21.

Kreuer, K.D. (2003) Hydrocarbon membranes. *Handbook of Fuel Cells – Fundamentals, Technology and Applications* (eds. W. Vielstich A. Gasteiger and A. Lamm), Chapter 33. John Wiley and Sons Ltd. Chichester.

Kreuer, K.D., Fuchs, A., Ise, M., Spaelh, M. and Maier, J. (1998) Imidazole and pyrazole-based proton conducting polymers and liquids. *Electrochim. Acta* **43**, 1281.

LaConti, A.B., Fragala, A.R. and Boyack, J.R. (1977) Solid polymer electrolyte electrochemical cells: electrode and other material considerations. *Proceedings of the Symposium on Electrode Materials and Process for Energy Conversion and Storage* (eds. J.D.E. McIntyre, S. Srinivasan and F.G. Wills), The Electrochem. Soc. Inc., Princeton, NJ, Vol. 77(6), 354.

Larminie, J. and Dicks, A. (2000) *Fuel Cell Systems Explained*. John Wiley and Sons, New York.

Liu, R., Iddir, H., Fan, Q., Hou, G., Bo, A., Ley, K., Smotkin, E., Sung, Y., Kim, H., Thomas, S. and Wieckowski, A. (2000) Potential-dependent infrared absorption spectroscopy of adsorbed CO and X-ray photoelectron spectroscopy of arc-melted single-phase Pt, PtRu, PtOs, PtRuOs, and Ru electrodes. *J. Phys. Chem. B.* **104**, 3518.

Malhotra, S. and Datta, R. (1997) Membrane-supported nonvolatile acidic electrolytes allow higher temperature operation of proton-exchange membrane fuel cells. *J. Electrochem. Soc.* **144**, L23.

Miyatake, K., Iyotani, H., Yamamoto, K. and Tsuchida, E. (1996) Synthesis of poly(phenylene sulfide sulfonic acid) via poly(sulfonium cation) as a thermostable proton-conducting polymer. *Macromolecules* **29**, 6969.

Oh, I-H., Cho, S-A., Shim, J.P., Ha, H.Y. and Cha, S.Y. (2000) Characteristics of ion beam treated nafion membranes and reinforced composite membranes for PEMFC. *Fuel Cell Seminar*, Portland, Oregon.

Peled, E., Duvdevani, T. and Melman, A.J. (1998) A novel proton conducting membrane. *Electrochem. Solid State Lett.* **1**, 210.

Peled, E., Duvdevani, T., Aharon, A. and Melman, A. (2001) New fuels as alternatives to methanol for direct oxidation fuel cells. *Electrochem. Solid State Lett.* **4**, A38.

Ralph, T.R., Keating, J.E., Collins, N.J. and Hyde, T.I. (1997) *Catalysis for Low Temperature Fuel Cells*. ETSU Contract Report F/02/00038.

Reeve, R.W., Christensen, P.A., Dickinson, A.J., Hamnett, A. and Scott, K. (2000) Methanol-tolerant oxygen reduction catalysts based on transition metal sulfides and their application to the study of methanol permeation. *Electrochim. Acta* **45**, 4237.

Samms, S.R., Wasmus, S. and Savinell, R.F. (1996) Thermal stability of proton conducting acid doped polybenzimidazole in simulated fuel cell environments. *J. Electrochem. Soc.* **143**, 1225.

Savadogo, O. (2001) The effect of heteropolyacid mechanism and electrocatalysis in the direct methanol fuel cells and isopolyacids on the properties of chemically bath deposited CdS thin films. *Sol. Energ. Mat. Sol.* **70**, 71.

Savadogo, O. and Xing, B. (2000) Hydrogen/oxygen polymer electrolyte membrane fuel cell (PEMFC) based on acid-doped polybenzimidazole (PBI). *J. New Mater. Electrochem. Syst.* **3**, 345.

Savinell, R., Yeager, E., Tryk, D., Landau, U., Wainright, J., Weng, D., Lux, K., Li, M. and Rogers, C. (1994) A polymer electrolyte for operation at temperatures up to 200°C. *J. Electrochem. Soc.* **141**, L46.

Scott, K., Taama, W.M. and Argyropoulos, P. (1998) Material aspects of the liquid feed direct methanol fuel cell. *J. Appl. Electrochem. Soc.* **28**, 40.

Shukla, A.K., Scott, K. and Jackson, C. (2003a) The promise of fuel cell based automobiles. *Bull. Mater. Res.* **26**, 207.

Shukla, A.K., Scott, K., Jackson, C. and Meuleman, W.R. (2003b) A mixed reactant solid polymer electrolyte direct methanol fuel cell. *J. Electrochem. Soc.* **150**, A12131.

Shukla, A.K., Shukla, R.K. and Scott, K. (2005) Advances in mixed reactant fuel cells. *Fuel Cells* (in press).

Steck, A.E. and Stone, C. (1996) Development of the BAM membrane for electrochemical applications. *10th International Forum on Electrolysis in the Chemical Industry – the Power of Electrochemistry*, November 10–14, 62–69.

Sumner, J., Creager, S.E., Ma, J.J. and Desmarteau, D.D. (1998) Proton conductivity in Nafion (R) 117 and in a novel is[(perfluoroalkyl)sulfonyl]imide ionomer membrane. *J. Electrochem. Soc.* **145**, 107.

Takasu, Y., Fujiwara, T., Murakami, Y., Sasaki, K., Oguri, M., Asaki, T. and Sugimoto, W. (2000) Effect of structure of carbon-supported PtRu electrocatalysts on the electrochemical oxidation of methanol. *J. Electrochem. Soc.* **147**, 4427.

Tazi, B. and Savadogo, O. (2000) Parameters of PEM fuel-cells based on new membranes fabricated from Nafion (R), silicotungstic acid and thiophene. *Electrochim. Acta* **45**, 4329.

Thompsett, D. (2003) Pt alloys as oxygen reduction catalysts. *Handbook of Fuel Cells – Fundamentals, Technology and Applications* (eds. W. Vielstich, A. Gasteiger and A. Lamm), Chapter 37. John Wiley and Sons Ltd. Chichester.

Verbetsky, V.N., Malyshenko, S.P., Mitrokhin, S.V., Solovei, V.V. and Shmalko, Y.F. (1998) Metal hydrides: properties and practical applications. Review of the works in CIS-countries. *Int. J. Hydrogen Energy* **23**, 1165.

Wainright, J.S., Wang, J-T., Weng, D., Savinell, R.F. and Litt, M. (1995) Acid doped polybenzimidazoles – A new polymer electrolyte. *J. Electrochem. Soc.* **142**, L121.

Wang, J.T., Wasmua, S. and Savinell, R.F. (1995) Evaluation of ethanol, 1-propanol, and 5-propanol in a direct oxidation polymer-electrolyte fuel cell – A real-time mass spectrometry study. *J. Electrochem. Soc.* **142**, 4218.

White, J.H. and Sammells, A.F. (1993) Pervoskite anode electrocatalysis for direct methanol fuel cells. *J. Electrochem. Soc.* **140**, 2167.

Section IIIC: High-temperature fuel cells

18

High-temperature fuel cells

*A. Moreno, R. Bove, P. Lunghi and
N.M. Sammes*

18.1 INTRODUCTION

A variety of fuel cells (FC) are in different stages of development. The materials used for the cell are strictly connected to their resulting physicochemical and thermomechanical properties. In particular, the operating temperature plays an important role in assessing the ionic or protonic conductivity of the electrolyte itself. While for molten carbonate fuel cells (MCFC), the alkali carbonates of the electrolyte form a highly conductivity molten salt at about 600–700°C, solid oxide fuel cells (SOFC) can operate at a wider range of operating temperatures (550–1000°C), depending on the materials used for the electrolyte.

The high-temperature range of both MCFC and SOFC provides these technologies with common peculiarities compared to low-temperature fuel cells.

First of all, low-temperature fuel cells use expensive and carbon monoxide (CO)-sensitive catalysts, thus restricting, in most practical applications, the fuel to be pure hydrogen (H_2). On the contrary, high-temperature fuel cells can use Ni-based catalysts, thus tolerating the presence of carbon dioxide (CO_2) and using CO, that accompanies methane (CH_4) in biogas and reformed gas, as a fuel.

Moreover, nickel is much less expensive than platinum, that is generally used as the catalyst in low-temperature fuel cells.

Secondly, the high thermal energy content due to the high temperature allows one to define combined heat and power (CHP) applications as well as to recover the outward energy in a bottoming (thermal) cycle, made of a gas turbine (GT) or a steam turbine (ST).

Moreover, the fuel consumed in a fuel cell must be reformed in some way to create a hydrogen-rich gas to feed to the stack. Since the reforming process is an endothermic reaction, while the oxidation of H_2 and CO are both exothermic, a high system efficiency can be obtained if the heat generated through water production is recovered for the reforming process. As shown in Section 18.2.1 this practice can be conducted internally or externally, and each of those solutions presents advantages and disadvantages.

At present, FCs have been developed mainly considering natural gas or other hydrocarbons as fuels. Fuel cell requirements towards the quality of biofuels are extremely high as low concentrations of certain detrimental gases can damage irreparably the fuel cell. Current research deals with balance of plant (BoP) and increasing the lifespan of the fuel cell stack materials, aiming at cost reduction. The development of less contaminant-sensitive fuel cells is of major interest for feeding biofuel into fuel cells. Due to the particular sensitivity of FCs to trace gases such as H_2S, NH_3, siloxane, chlorine and fluorine compounds, biofuel utilization requires the reduction and control of these accompanying traces of detrimental gases. For upgrading fuels derived from biomass fermentation (BF), several physical, chemical and biological gas cleaning methods are already known. However, research is needed to adapt available technologies to the specific needs for the combination of BF and fuel cell and to improve their effectiveness in terms of performance and economical feasibility.

18.2 COMMON CHARACTERISTICS OF MCFC AND SOFC

MCFC and SOFC present some common characteristics, mainly due to the high operating temperature. In the following, these main features are analyzed, showing the related advantages and disadvantages of each option.

18.2.1 Internal and external reforming

18.2.1.1 Hydrocarbon reforming methods

If the energy source of FCs is represented by conventional hydrocarbons, like natural gas, propane, gasoline, etc., a reaction that transforms these into a hydrogen-rich gas mixture is required. There are three main practices that are commonly used:

(1) Steam reforming
(2) Partial oxidation
(3) Autothermal reforming

The general hydrocarbon conversion reaction can be written in the following form (Hirchenhofer *et al.* 2002):

$$C_nH_mO_p + x(O_2 + 3.76N_2) + (2n - 2x - p)H_2O$$
$$\rightarrow nCO_2 + (2n - 2x - p + m/2)H_2 + 3.76N_2 \qquad (18.1)$$

The amount of air used in the reaction, denoted with the x symbol, determines the minimum mole number of the required water, that is $2n-2x-p$. In practice, the reaction is conducted with excess water to ensure the reaction and to avoid carbon deposition. When no air is used for the fuel conversion ($x = 0$), the process is steam reforming (SR), and is strongly endothermic. By increasing x, the reaction becomes less endothermic and, according to the selected hydrocarbon, there is a value of x that makes the reaction thermoneutral. In this case, the conversion process is commonly called autothermal reforming (ATR). When $x = 1$, no water is needed for the reaction and the reaction is called partial oxidation (POX).

A straightforward thermodynamic consideration allows one to estimate which of the three processes can lead to the highest system efficiency. According to the first thermodynamic law (energy conservation) and ignoring the thermal losses (adiabatic reactor), in fact, if heat is provided (i.e. the reaction is endothermic), the reformed gas presents an energy content that is higher than the unprocessed fuel. Since in a high-temperature fuel cell system, the heat required for SR is generally recycled from the fuel cell section, no additional fuel is required for the reforming reaction. This means that the more endothermic the reaction is, the higher the energy content in the produced gas is, thus enhancing the system efficiency. When the fuel cell operates at low temperature, or when the fuel is externally processed and then delivered to the fuel cell system, the enhanced energy content of the reformed gas is provided by the combustion of additional fuel and so a system efficiency reduction is possible.

In the case of the ATR and POX, instead, no external heat is provided for the reaction and, therefore, according to the first law, the system efficiency is expected to be lower than that of the SR.

For the reason explained above and considering that for high-temperature fuel cell the required heat can be recovered from the cell, only SR is considered as the fuel processing reaction in this section.

18.2.1.2 Methane reforming

When the shift reaction of CO is considered (Equation 18.3), Equation (18.1) becomes, for methane:

$$CH_4 + H_2O \rightarrow CO + 3H_2 \qquad (18.2)$$

$$CO + H_2O \leftrightarrow CO_2 + H_2 \qquad (18.3)$$

Reactions (18.2) and (18.3) can be considered intermediate of the following reaction (18.4), even though CO is always present in the final gas mixture.

$$CH_4 + 2H_2O \leftrightarrow CO_2 + 4H_2 \tag{18.4}$$

The enthalpy of reaction (18.2) under standard condition is 206 kJ/mol, while for reaction (18.4) it is 165 kJ/mol (Heinzl *et al.* 2002). The enthalpy of the exothermic reaction (18.5), that occurs inside the fuel cell is 285.8 kJ/mol:

$$H_2 + \frac{1}{2}O_2 \rightarrow H_2O \tag{18.5}$$

Obviously, this energy is not totally released as heat, but a part is converted into electric energy and the rest into heat. Considering that for every CH_4 mole, a number between 3 and 4 moles of H_2 are produced (depending on the reaction conditions of (18.2) and (18.3)), it is reasonable to suppose that the heat produced by the fuel cell is enough for the SR. The problem is related to the heat transfer from the fuel cell section to the reformer. If high-temperature fuel cells are considered, as in the present situation, thermal energy can easily be transferred from the outlet anodic and cathodic gases to the reformer section. Another possible solution for heat recycling is the use of the so-called internal reforming fuel cells. In this case, reactions (18.2), (18.3) and (18.5) occur inside the cell itself, thus solving the heat transfer problem.

18.2.1.3   Internal reforming of methane

When internal reforming is performed, no external devices are needed and heat transfer can take place with minimum losses. Moreover, reaction (18.5) removes H_2 and produces H_2O, thus promoting reaction (18.4).

Internal reforming can be conducted in a direct or indirect configuration. As illustrated in Figure 18.1, in the case of direct internal reforming (DIR), methane is converted into hydrogen inside the anode section, together with hydrogen oxidation. For indirect internal reforming (IIR), instead, the reforming section is adjacent to the anode, but reactions (18.2) and (18.3) do not take place simultaneously with reaction (18.5). This last solution is an intermediate situation between external and internal reforming.

18.2.1.4   Modular integrated reformer

Based on the same principle of the IIR, the modular integrated reformer (MIR) (Figure 18.2) allows the system to obtain a high conversion efficiency. In this solution, the reformer and the catalytic combustor (that is used to oxidize the anodic outlet gas) are contiguous, so that, not only is the heat surplus from the cell recovered, but also the chemical energy of the unoxidized anodic gas. The consequence is that the reforming process can be performed at a temperature even higher than the fuel cell operating temperature.

The reason why some companies chose to develop external reforming is because, besides the advantages illustrated above, internal reforming presents also some disadvantages. First of all, if, for some reason, for example an excess of

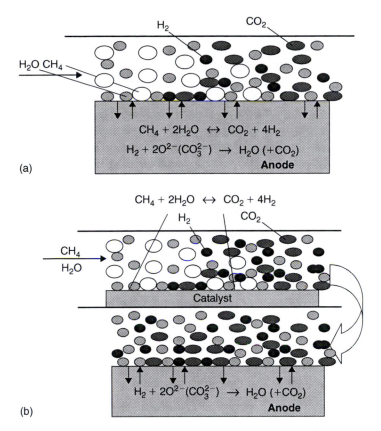

Figure 18.1 (a) Direct internal reforming and (b) Indirect internal reforming.

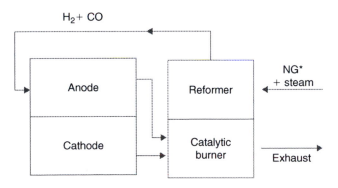

Figure 18.2 Modular integrated reformer. *NG = Natural gas.

impurities in the fuel, the catalyst used for the reforming reaction is deactivated, the whole stack must be replaced, even if the fuel cell components are still in good condition. In the case of external reforming it is possible to replace only the fuel processing section.

Secondly, fuel cell modules that are developed and optimized to run on natural gas need particular modification when fed with syngas derived from alternative fuels, such as biomasses, wastes, wastewater, etc. The BoP modification is much simpler if the gas processing unit (GPU) is external of the fuel cell.

In the case of MCFC, moreover, the carbonate vapors of the electrolyte can influence the reforming catalysts.

Finally, the internal reforming creates temperature gradients inside the fuel cell and carbon deposition problems. While carbon deposition can be avoided using an adequate steam to carbon ratio, temperature gradients reduce the efficiency and, in the case of SOFC, can cause cell cracking (Meusinger *et al.* 1998). In order to avoid these problems, and for reforming complex hydrocarbons, part of the fuel is usually pre-converted in an external reactor, before entering the anode, even for internal reforming fuel cell.

18.2.2   CHP systems and combined cycles

18.2.2.1   *Electrical efficiency and heat production*

As stated by the second law of thermodynamic, a thermal energy system cannot convert all the thermal energy of a heat source into useful work, but a part of it must be released to a heat sink, that is at a lower temperature than the heat source. As a consequence, the maximum allowable efficiency obtainable by an energy conversion system operating between a thermal energy source at temperature T_a and a sink at temperature T_b is the Carnot efficiency:

$$\eta_{Carnot} = 1 - \frac{T_b}{T_a} \tag{18.6}$$

The efficiency of real processes are lower because of the losses involved. The concepts of combined cycles (CC) and cogeneration systems are both based on this principle, that is, to recover the energy content that is always present in thermal systems exhausts. If heat is readily used, the system produces power and heat simultaneously (CHP), while if another thermal cycle, operating at a lower-temperature range, is used to recover the outward energy, a CC is realized.

If a fuel cell is considered, the expression (18.6) is not valid, because no combustion occurs inside the fuel cell. Nevertheless, the electric efficiency of a fuel cell can be expressed as:

$$\eta = \frac{\text{Useful Work}}{\text{Energy Input}} = \frac{\text{Useful Work}}{\Delta H} \tag{18.7}$$

If a reversible reaction is supposed, then expression (18.7) can be written as:

$$\eta = \frac{\Delta G}{\Delta H} \tag{18.8}$$

In expressions (18.7) and (18.8), ΔH is the formation enthalpy of H_2O, that is, the energy supplied to the fuel cell and ΔG represents the change in free energy associated with the same reaction. Bearing in mind that, for a chemical reaction with pressure and temperature constant, the following expression is valid:

$$\Delta G = \Delta H - T\Delta S \qquad (18.9)$$

Expression (18.8) can be written as:

$$\eta = 1 - T\frac{\Delta S}{\Delta H} \qquad (18.10)$$

Relation (18.10) shows that even fuel cells have significant limitations for electric efficiency because of thermodynamic reasons. For example, if pure hydrogen is considered as fuel and standard conditions are supposed, ΔG is 237.1 kJ/mol and ΔH is 285.8 kJ/mol. Substituting these values in relation (18.8) an ideal electric efficiency of 83% is obtained (Hirchenhofer *et al.* 2002).

The result is that, as for conventional thermal energy systems, even for fuel cells an efficiency limit exists, thus part of the inward energy is converted into heat. Due to the analogy with the thermal systems, it is reasonable to recover the energy content of the outlet gases for CHP applications or CCs.

18.2.2.2   CHP applications

Several studies of CHP solutions, using MCFCs (Lunghi *et al.* 2001; Silvera *et al.* 2001; Sugiura and Naruse 2002) or SOFCs (Entchev 2003; Van herle *et al.* 2003; Bove and Sammes 2004) can be found in the literature. All the studies show that a high conversion efficiency can be achieved. Nevertheless, for residential applications, fuel cells do not fit with the heat and electricity requirements. Fuel cells, in fact, present a high electric efficiency, that is, they present a very high electricity/ thermal energy ratio, while for residential applications the heat requirement is much higher than the electricity. Moreover the electrical requirements in residential applications is characterized by very high peaks, for just few hours, that are well above the average requirements. In order to satisfy both thermal and electric load needs, the fuel cell needs to be over-dimensioned, with consequently high, and not justified, investment costs.

The solution is to operate the fuel cell in a parallel configuration with the national electric grid and purchase the electricity during the few peak hours, or to provide the fuel cell systems with a battery that is charged during the off-peak hours (Torrero and McClelland 2002; Gunes and Ellis 2003; Bove and Sammes 2004). As an example, the results for an SOFC cogeneration system operating for a typical residential application is depicted in Figure 18.3 (Bove and Sammes 2004). In the Figure 18.3, the variation of the electrical and total efficiency during the hours of the day are reported, as well as the amount of heat recovered from the fuel cell and the electric load ratio. The electric load ratio is defined as the ratio between the provided and the nominal power. In this study, the fuel cell system operates in parallel with the electric grid, and electricity is provided by the grid in the hours corresponding to the load ratio equal to 1.

18.2.2.3   Combined cycles

Another way to recover thermal energy from fuel cells is to use the outlet gas for a thermal cycle. The temperature range of both MCFC and SOFC suggests the use

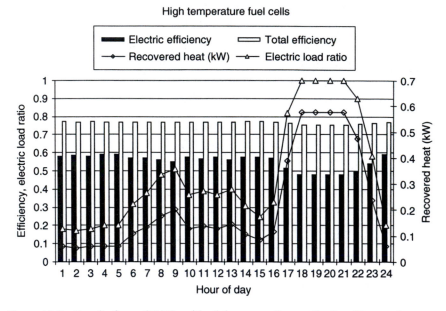

Figure 18.3   Results for an SOFC residential cogeneration application (Bove and Sammes 2004).

of a GT or a ST as a bottoming cycle. Fuel cell and GT can be combined in two different configurations: direct or indirect. In the first situation (Figure 18.4a), the outlet gas from the fuel cell directly evolves into a GT, thus the combustion chamber of the GT is replaced by the fuel cell section (Lunghi and Ubertini 2001).

The main drawback of this configuration is that the operating pressure of the SOFC must be high in order to achieve an acceptable efficiency of the GT. The development of micro-turbines and the improvements in fuel cell materials will probably allow the problem to be overcome. If the bottoming cycle is arranged in an indirect configuration (Figure 18.4b), fuel cell and the GT operate with two different fluids and the GT can work at its optimal pressure, without interfering with the fuel cell operations. On the other hand, a high-temperature heat exchanger is needed, thus increasing system cost and complexity.

Thermodynamic analysis of FCGT plants has been fully investigated and the computed electric efficiency can reach 60%, while the total (thermal and electrical) efficiency can be up to 85% (Gemmen et al. 2000; Riensche et al. 2000; Costamagna et al. 2001; Chan et al. 2003; Lunghi et al. 2003).

Since ST plants are more complex than GTs, few studies have been conducted on FCST combined plants (Hirchenhofer et al. 2002; Bove and Lunghi 2003; Lunghi and Bove 2003a).

18.2.3   High-temperature fuel cells and alternative fuels

As previously mentioned, high-temperature fuel cells are manufactured without using precious metals, but nickel is usually used as a catalyst. For this reason CO

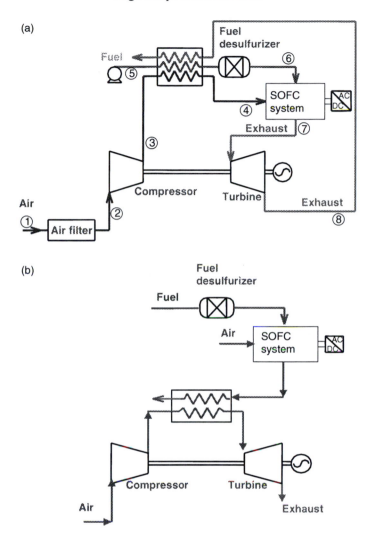

Figure 18.4   Combination of fuel cell/GT in a direct (a) and indirect (b) configuration.

and CO_2, that are usually contained in alternative fuels, are not harmful for the cell. Furthermore, CO is not just harmless, but can be directly oxidized, or shifted into hydrogen, thus representing additional fuel.

18.2.3.1   Life cycle assessment

When renewable energy is converted through a fuel cell, the intrinsic environmental advantages of its usage (i.e. fossil preservation and CO_2 emission reduction) are further enhanced by the high energy conversion efficiency and the low operating emissions related to FCs. As shown in Figure 18.5 the life-cycle of a molten carbonate fuel cell, fuelled with landfill gas (LFG), for example, produces

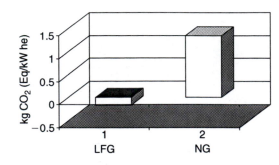

Figure 18.5   Greenhouse gases emissions (CO_2 equivalents) related to the life cycle of an MCFC fuelled with LFG and NG.

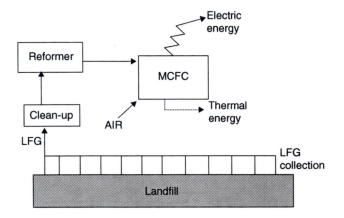

Figure 18.6   Schematic representation of an MCFC system fuelled with LFG.

much less greenhouse gas than the same fuel cell powered by steam reformed natural gas (Lunghi *et al.* 2004). The life-cycle comprises stack production, fuel purification, operating emissions and stack system disposal.

Even if specific life cycle analyses on other alternative fuels usage in fuel cells have not been performed yet, it is reasonable to assume that greenhouse gas reduction is achievable using any renewable energy source. Hydrogen production via SR, in fact, releases $10.621 \, g \, CO_2/g \, H_2$ (Spath and Mann 2001).

18.2.3.2   *Alternative fuels*

Alternative gases usable for fuelling fuel cells are by-products of processes (gas from landfill, anaerobic digester, industrial by products) or are produced on purpose (waste and biomass gasification, pyrolysis, fermentation, etc.).

Figure 18.6 is a schematic representation of an MCFC system fuelled with LFG, which is collected through wells and then sent to a clean-up system; impurities embedded in the LFG are reduced to a desired level. The purified gas is then sent to the reforming section, where methane is reformed into hydrogen, carbon monoxide and carbon dioxide. The resulting gas is finally used as anodic gas.

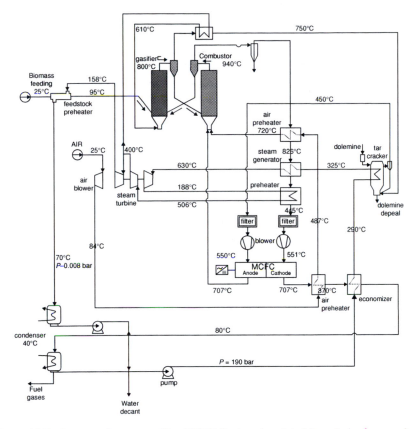

Figure 18.7   Integrated steam gasifier-MCFC System (reprinted from Lobachyov and Richter (1998) with permission of Elsevier).

Figure 18.7, instead, depicts an MCFC system using a dual bed steam gasifier, fuelled with biomass, as proposed by Lobachyov and Richter (1998). The peculiarity of this gasifier is that there are two reactors, named combustor and gasifier, operating at two different temperatures.

The heat needed for the gasification process is transferred from the combustor to the gasifier through sand re-circulation from one bed to the other. The main advantage of conducting gasification in a separate reactor is that the process results independent from combustion, in terms of materials mixing and operating parameters, such as temperature and pressure. The main drawback of this configuration is that a large amount of heat is required, thus leading to poor energy conversion efficiency for stand-alone gasifier systems. Nevertheless, MCFC outlet gas contains a high thermal energy content at high temperature, that can be used for the process itself. The high performance of the process integration is quantified by Lobachyov and Richter (1998), who computed an efficiency of 53%.

Operations of a Sulzer–Hexis SOFC module on agricultural biogas are reported by Van herle et al. (2004) as well as investigations on operational parameters (Van herle et al. 2003). The farm where the 1 kW SOFC module has been installed is located in Lully (Switzerland). The biogas production rate is $70 \, m^3$ per day, corresponding to 0.55 TJ per year (Van herle *et al.* 2004). The stack, operating in a cogeneration configuration, showed an electric efficiency of 38.2% and an overall efficiency of 84.2%. Prior to the SOFC installation, the biogas was converted with a gas engine, whose performance was noticeably poorer, that is, 18% for the electric efficiency and 73% for the overall efficiency.

18.3   SOLID OXIDE FUEL CELLS

18.3.1   General features

An SOFC is an all-solid-state high-temperature electrochemical device. The system came about after the work of Nernst in the 19th century. Nernst published a patent (Nernst 1899) proposing that a solid electrolyte could be made to electrically conduct, using a heater. The system then "glowed" by the passage of an electric current. The systems originally studied by Nernst were based on simple-oxides. However, the later work of Nernst and others including Baur and Preis (1937) on the so-called "Nernst Mass" (85% zirconia and 15% yttria), and other zirconia-based materials showed that ionic conduction could occur at high temperature (600–1000°C).

This work was the prelude to the modern SOFC. The electrolyte is based on yttria-stabilized zirconia (YSZ), traditionally fully cubic, having a relatively high ionic conductivity at temperatures above 700°C, with negligible electronic conductivity (required in an SOFC). The material is stable in both oxidizing and reducing environments and can be fabricated in many forms, using ceramic processing and thin-film technologies. The requirement for a dense YSZ is critical as it is imperative that the fuel and air gases are not physically (and thus chemically) mixed. The porous cathode (traditionally doped $LaMnO_3$) must be electronically conducting, must have good catalytic activity towards oxygen reduction, and must be stable under oxidizing environments. The anode must also be electronically conducting, must be stable under reducing environments and must be catalytically active towards the oxidation reaction of the fuel of choice. Traditionally hydrogen is used as the fuel and thus Ni is used as the electrocatalyst. However, Ni has a very high thermal expansion coefficient and tends to sinter (thus lowering the active sites available for oxidation, and thus lowering the overall efficiency of the fuel cell) with time at temperature and under electrical load. Thus, YSZ is added to form a Ni/YSZ composite to reduce the sinterability and also to lower the coefficient of thermal expansion closer to that of the electrolyte. YSZ in the composite matrix can also act to increase the number of active sites (the triple phase boundary) by allowing some oxygen ion conduction within the anode structure (Singhal and Kendall 2004).

At the cathode, the reduction of oxygen occurs via:

$$O_2 + 4e^- \rightarrow 2O^{2-} \tag{18.11}$$

The oxygen ion reacts with an oxygen vacancy present within the structure of the YSZ (Carter and Roth 1968; Hohnkle 1968), and moves from the high oxygen partial pressure side (cathode) to the low partial pressure side (anode) via the well-known vacancy hopping mechanism (Kilner and Brook 1982).

At the anode, the following reaction takes place, whereby the oxide ion is oxidized and the electron is released to flow back round the external circuit:

$$2O^{2-} \rightarrow O_2 + 4e^- \tag{18.12}$$

Hence, the SOFC can be considered as a concentration cell. Now, consider the oxidation reaction, not of an oxygen ion at the anode, but of a fuel such as hydrogen, viz:

$$H_2 + \frac{1}{2}O_2 \rightarrow H_2O \tag{18.13}$$

The oxygen partial pressure at the anode is, therefore, given by:

$$P_{O_2} = \left[\frac{P_{H_2O}}{P_{H_2} K_3} \right] \tag{18.14}$$

where K_3 is the equilibrium constant for reaction (18.13). The EMF of the cell is, thus, given by:

$$E = E^O + \frac{RT}{4F} \ln P_{O_2} + \frac{RT}{2F} \ln \frac{P_{H_2}}{P_{H_2O}} \tag{18.15}$$

Where E^O is the reversible voltage at the standard-state, given by:

$$E^O = \frac{RT}{2F} \ln K_3 = \frac{\Delta G^0}{2F} \tag{18.16}$$

18.3.2   State of the art

18.3.2.1   Materials

The traditional materials used in the SOFC, described above, are summarized in Table 18.1.

Due to the use of doped-zirconia in the traditional SOFC, a relatively high operating temperature (up to 1000°C) is required to allow for a reasonable oxygen ion flux. This causes a number of problems in the SOFC stack, including:

- lack of chemical stability of individual components and interdiffusion between cell components;
- densification of porous electrodes;
- delamination of cell components due to thermal expansion mis-match and thermal cycling;
- expensive interconnect and manifolding materials required;
- heat-up time of the order of several hours;
- expensive materials are required for the BoP;

Table 18.1   Traditional materials for SOFC.

Component composition	Specific conductivity (S/m) at SOFC running temperature	Conductivity depends upon
Anode		
Ni/YSZ cermet	400–1000	Particle size ratio Ni content
Cathode		
$Sr_xLa_{1-x}MnO_{3-\delta}$	6–60	Cathode Porosity Sr content
Electrolyte		
Y_2O_3–ZrO_2	10–15	Electrolyte density

- order of magnitude more expensive than the existing stationary power generation systems.

Hence, there is a drive nowadays towards using alternative lower temperature materials, and also towards using anode-systems that can tolerate (and potentially internally reform) real hydrocarbon fuels and biogas without causing cracking of the hydrocarbon into carbon. There are a number of advantages to a lower temperature of operation for the SOFC stack, including:

- longer lifetime of cell (decreases material interactions between electrode and electrolyte);
- cheaper manifolding materials;
- less time for startup of SOFC generator.

Operation at less than 700°C means that low-cost metallic materials such as ferritic stainless steels can be used as the interconnect and construction materials. This makes both the stack and balance of the plant cheaper and more robust. Using ferritic materials also significantly reduces the problems associated with chromium evaporation from the interconnect to the cathode (Yamaji *et al.* 1998).

The use of alternative electrolytes or much thinner YSZ electrolytes on anode-supports or cathode-supports, has been the main driving force in reducing the SOFC operation temperature. Alternative zirconia-based systems (such as scandia stabilized zirconia (ScSZ)), and other oxides including doped ceria and doped perovskite materials (e.g. lanthanum gallate) have been studied. ScSZ has the highest ionic conductivity of all the doped zirconium oxide systems (approximately twice that of YSZ). The cubic phase exists above 8.5 mol% Sc_2O_3 and above 10 mol% a rhombohedral (β-phase) co-exists with the cubic. However, the material tends to show a large deterioration in performance with time (formation of an ordered β-phase), but alumina additions have been shown to have some effect on the long-term stability of the electrolyte, although they have been shown to reduce the ionic conductivity of the electrolyte. Doped-CeO_2 has been found to have a very high ionic conductivity, but is prone to reduction in reducing atmospheres (Ce^{4+} becomes Ce^{3+}), as described in Figure 18.8 (Godickemeier and Gaucker 1998).

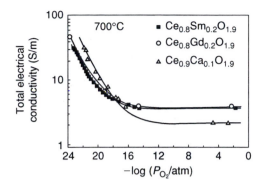

Figure 18.8   Electrical conductivity as a function of the O_2 partial pressure.

Doped $LaGaO_3$ has been shown to be an excellent oxygen ion conductor. It is stable in both reducing and oxidizing atmospheres, although recent work has shown the potential for Ga depletion in hydrogen atmospheres (Singhal and Kendall 2004).

18.3.2.2   SOFC configurations

Although materials are very important for SOFC performance and life-time, alone, they are not enough to realize high-performance fuel cells. There are a variety of options to be considered when fuel cells are realized. Two important factors are the cell supporting elements and the geometry of the cell.

According to the element that acts as structural part of the cell, an SOFC can have a self-supporting or a supported structure. In the first situation one of the fuel cell elements supports the cell structure, thus it is possible to realize anode-, electrolyte- or cathode-supported cells. In the second solution, instead, the structural support is provided by an additional element. The main advantages of supported SOFC is that the electrodes and the electrolyte can be very thin, thus reducing ohmic losses. On the other hand, a supporting structure usually interferes with the operations, in particular it can limit gas diffusion activity.

According to the geometry, SOFC can be constructed in a tubular or planar configuration. Although both tubular and planar SOFC can be constructed in different configurations, macro differences between the two solutions occur (Table 18.2).

Table 18.2   Main characteristics of tubular and planar SOFC.

	Tubular	Planar
Power density	Low	High
Volumetric power density	Low	High
High temperature sealing	Not necessary	Required
Start-up cool down	Faster	Slower
Interconnect	Difficult	High cost
Manufacturing cost	High	Low

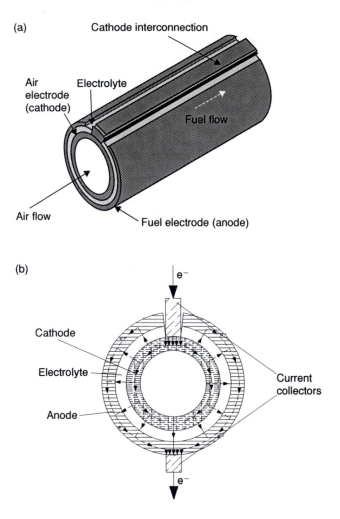

Figure 18.9   Single cell design (a) and current path and (b) in the Siemens–Westinghouse.

The most advanced tubular SOFC design is the sealless, realized by Siemens–Westinghouse. The support of the cell is guaranteed by a supporting porous tube. The electrolyte and the anode cover the cathode (Figure 18.9a), with the exception of a small strip, where a current collector is placed (Minh and Takahashi 1995). The cell is closed at one end, thus air is provided through an injector tube. Single cells are then assembled to form power modules, as explained in Section 18.3.3.

The sealless design is currently the most advanced SOFC design, but, at the same time, it presents high ohmic losses due to its design. The electrons and ions flow around all the cylindrical surface, thus reducing the cell performance (Figure 18.9b).

Another solution for tubular SOFC is to displace current collectors at the two ends of the tube, one in contact with the anode and another with the cathode. Figure 18.10

Figure 18.10   Tubular SOFC with terminal current collectors.

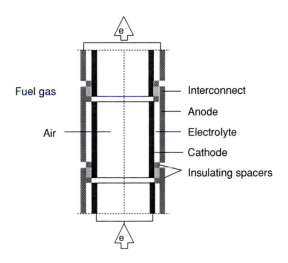

Figure 18.11   Example of segmented SOFC (Bossel 1992).

shows this case for an anode-supported fuel cell. In this situation, however, the current path is all along the tube surface, thus producing high ohmic resistance.

The effect of the in-plane resistance, of this configuration can be reduced, fabricating small length tubes connected in gas flow and electric series configuration. This configuration is usually referred to as "segmented" cells, and an example is given in Figure 18.11. Better performance, compared to that of traditional tubular SOFC, is due to the fact that the first fuel cell units work with a reduced fuel utilization, thus producing high voltage.

The design of planar SOFC can be flat or monolithic. A schematic representation of a flat plane design is shown in Figure 18.12a. The main advantage of this solution is the low in-plane ohmic losses, thus performances are mostly independent of cell size. Moreover, the manufacturing process is simple and inexpensive. On the other hand the sealing system, often located at the edges, can create non-uniform mechanical tension distribution and, consequently, cracking problems. Another disadvantage is the high contact resistance between electrolyte and electrodes, due to the geometry configuration and the solid state of all the components.

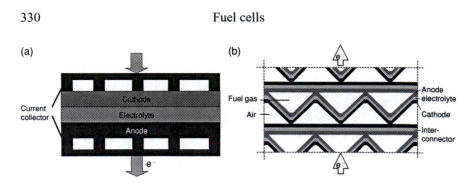

Figure 18.12   Schematic cross section of (a) a flat planar and (b) monolithic (Bossel 1992) SOFC.

Table 18.3   Leading SOFC developers (Hirchenhofer *et al.* 2002).

US	Europe	Japan	Other
General electric[a]	Forschungszentrum Julich GmbH	Chubu Electric Power Company (Mitsubishi Heavy Industries, MHI)	Ceramic Fuel Cells Limited (CFCL) Australia
Delphi and Battelle	Sulzer–Hexis Ltd.	Tokyo Gas Co.	
Cummins Power Generation (CPG) and SOFCo-EFS[b]	Rolls-Royce	Osaka Gas Co. (Murata Mfg. Co.)	
Fuel cell energy (FCE)[c]		Engineering & Shipbuilding Co.	
Siemens–Westinghouse			
Department of Energy			
Department of Defense NIST			

[a]Formerly Honeywell, [b]Formerly McDermott Technology, [c]Acquired Global Thermoelectric Inc, Canada.

The monolithic SOFC is the most evolved flat planar configuration (Figure 18.12b). It consists of corrugated thin cell components. The result is that the volumetric power density is very high. On the other hand, the structure is more complex to be realized, as well as the manifold system compared to the flat planar.

18.3.3   Demonstration projects

Worldwide, a number of developers are involved in SOFC technology. As for the other fuel cell technologies, most of the developers are located in US, Europe and Japan. Table 18.3 reports the most important players involved in SOFC development and commercialization.

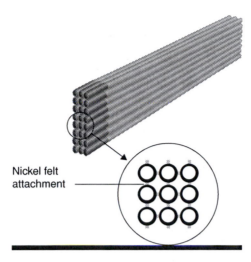

Figure 18.13   Connection concept of Siemens–Westinghouse tubular cells.

An overview of SOFC patents (Kozhukharov *et al.* 2001) shows that most of the recent inventions are focused on increasing power density and efficiency, enhancing the life-time, finding simple and low-cost BoP solutions.

18.3.3.1 USA

There are several companies, national laboratories and universities that are currently developing SOFC technology in US. In 1999, some of the most relevant of them formed an alliance, called the Solid State Energy Conversion Alliance (SECA). The main goal of SECA is to develop an SOFC module in the 3–10 kW range. Targets are: achieving a stack cost of 800$/kW by 2005 and 400$/kW by 2010. The stack life-time target is 40,000 h. These targets will allow to develop, by 2015, very-high-efficiency SOFC-GT systems and integrated gasification fuel cell systems (IGFCC), also called Vision 21 Power Plants (Utz 2004; NETL 2004).

In the following, important achievements of the SECA industrial partners are illustrated. *Siemens–Westinghouse* is currently the world leader in SOFC. The SOFC is made up of an electrolyte and two electrode layers in an unique tubular design, as explained in Section 18.3.2.2.

To generate commercially meaningful quantities of electricity, many cells must be connected together into a generator module or stack. They are connected into bundles using a nickel felt which makes the electrical connection between cells, as shown in Figure 18.13.

Being compliant, the nickel felts allow each cell to expand or contract independently of the adjacent cells as the temperature in the stack changes. In a typical stack design, the tubular cells are oriented with the axis vertical and the closed end down.

The cell is nominally 2.2 cm in diameter by 150 cm in active length with one closed end. To generate electricity efficiently, the cell must be maintained at an operating temperature of about 1000°C, air must be supplied to the cell interior using an air delivery tube, and fuel is delivered to the cell exterior. At atmospheric pressure, a uniform temperature of 1000°C, 85% fuel utilization and 25% air utilization, a single tubular SOFC generates power of up to 210 W.

A pressurized 220 kW hybrid SOFC-GT system has been installed at the Fuel Cell National Research Center, University of California, at Irvine, and has run for 1700 h (Figure 18.14).

The main drawback of the Siemens–Westinghouse design is that the current path along the cylinder generates relevant in-plane electric resistance, that, together with the unavoidable cross-plane resistance, result in low power density. For this reason, Siemens–Westinghouse has realized a new single designed, called high power density (HPD). This configuration presents all the advantages of tubular fuel cells (i.e. internal stress reduction and fast start-up), while this configuration allows the cell to operate with low ohmic losses. Cross sections of the standard tubular cell, the HPD5R0 and HPD5R1 (five channels) and HPD10 (10 channels) are depicted in Figure 18.15 (Vora 2003).

While in the past Siemens-Westinghouse focused on large stationary SOFCs, a 10 kW system for residential application and a 3–5 kW auxiliary power generation unit (APU) system are now being developed within the SECA program.

Figure 18.14   220 kW hybrid system with a SOFC generator and a down-stream micro hot-GT.

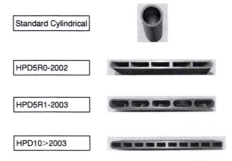

Figure 18.15   Siemens–Westinghouse tubular and HPD configurations.

Fuelcell Energy Inc acquired Global Thermal Electric Technology in 2003, which has developed SOFC systems since 1997 (Borglum 2003). The target is to develop an SOFC system in the range of 1–10 kW. A picture of an FCE stack is shown in Figure 18.16.

Each single cell is composed of a 1-mm thick nickel–zirconia cermet anode, an electrolyte made of YSZ of about 5 μm and a cathode made of conducting ceramic and about 50 μm thick. The stack has internal manifolds, that ensure a cross-flow configuration. The operation temperature is in the range of 700–750°C. The low cost of the stack is potentially ensured, since interconnects, current collectors and end plates are not made of expensive materials, but of stainless steel. The stack electric power is estimated to be 1.6 kW (Borglum 2003). Figure 18.17 depicts a 2 kW natural gas system, realized in 2001 at Global Thermoelectric.

In the *Acumentrics* design, the fuel reformation process occurs inside the tubular cell itself. The process is nearly co-mingled with the generation process itself. With the introduction of a small amount of air with the fuel, the inherent high temperature of the process reforms the fuel, producing the needed hydrogen as well as CO. The small tube fuel cell design avoids one of the biggest problems in many fuel cell concepts, that is, catastrophic damage due to temperature gradients. Gradients occur during fuel cell thermal cycling from repeated start-up and

Figure 18.16   Fuel cell energy SOFC stack.

Figure 18.17   2 kW natural gas SOFC system.

shut-down over the lifetime of the unit. Thermal gradients also occur due to the endothermic and exothermic operation of hydrocarbon fuels within the stack. This is especially an issue with planar fuel cells; the thin flat plates are arranged in dense layers of contiguous interconnecting surfaces that have difficulty withstanding temperature changes.

The radial geometry of the Acumentrics tubular design allows fast radial-to-axial heat transfer that severely limits the effects of unequal expansion caused by temperature changes. Further, as discrete elements, tubes react independently to temperature changes. There are no dense layers of material to differentially warp under temperature stress, as found in a planar fuel cell stack.

Sealing is another design issue with most fuel cell systems. However in the Acumentrics small tube design, only the round tube ends require a seal at the manifold. The area to be sealed is a small fraction of the required sealing area in flat plane designs. In this small tube design, the failure of a tube seal is not significant; the unit continues to function. In planar designs, a seal failure could be catastrophic. This small tube design also permits a rapid fuel cell start, particularly useful in back-up power applications.

Acumentrics has recently demonstrated stable cell performance for over 6000 h, with a degradation rate of 0.25% for 500 h. This achievement is close to the Phase III SECA goal of 0.1% (Bessette 2004).

Delphi Corporation, in partnership with *Battelle,* is developing a 5 kW SOFC system, capable of running on a variety of fuels, including gasoline, diesel, natural gas and coal-derived gas. The system is designed to satisfy several applications, including small residential, Auxiliary Power Unit (APU) for cars and heavy-duty trucks, as well as military applications. The fuel cell design is planar, the operating temperature is 750°C and a power density of 330 W/cm^2, when running on hydrogen, has been demonstrated (Shaffer 2004). Figure 18.18 represents the compact APU realized in 2002; the volume is 44 l and the weight is 70 kg (Zizelman 2003).

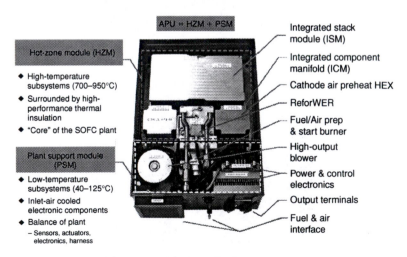

Figure 18.18   APU Delphi system (Zizelman 2003).

General Electric recently acquired the *Honeywell* SOFC technology and is currently developing an anode-supported SOFC. The system under development is targeted for residential applications and presents a net power of 5 kW; 3000 h has been demonstrated. The operating temperature is about 800°C. The fuel processor is a catalytic reactor that has the function of pre-reformer. Depending on the type of fuel and operating conditions, the reaction can be an POX or an ATR (Minh 2004). Stack stable performance on unreformed methane, at power density of 0.163 W/cm^2 and a voltage of 0.6 V has been demonstrated (Minh 2004).

Together with the stack and BoP development for residential applications, *GE* is also performing studies on hybrid SOFC-GT combined systems for power plant realizations.

Cummins Power Generation has a strong background in designing, manufacturing and servicing of power generation solutions. *SOFCo-EFS Holdings LL* (formerly McDermott technology), instead, developed SOFC and fuel processing technologies. The two companies have teamed together to develop a 10 kW power system.

The stack is based on planar configuration, and is "all ceramic". Two stack prototypes have been realized: one is composed of fifty-five 10 × 10 cm single cells (C1 Prototype), and another one, named C2 Prototype, composed of 55 cells, 15 × 15 cm.

18.3.3.2   Europe

Sulzer–Hexis was founded in 1997 as a division of the Sulzer Group. Although the theoretical Hexis "idea" was born in 1989, the first stack run during 1997–1998 for more than 12,000 h on hydrogen. The Hexis stack has a unique configuration: it is characterized by a disk shape and current collectors, that also act as heat exchangers and fuel distribution channels (Figure 18.19a). The fuel gas enters the channels of the current collector from inside and flows through them in a radial direction, passing over the anode. Atmospheric oxygen is fed from the outside, heats up inside the current collector and is diverted back over the cathode.

The system based on this stack configuration is the HXS 1000 PREMIER (Figure 18.19b). The system has a nominal power of 1 kW$_e$ and 3 kW$_t$, and, consequently, the electric efficiency is about 33%. Although natural gas is the primary fuel, Sulzer demonstrated the use of biofuel by installing an HXS Premier on a farm in Switzerland (see also Section 18.2.3).

Rolls-Royce   Between 1987 and 1992, theoretical studies on PEM and SOFC were occasionally undertaken at *Rolls-Royce*. In 1992 a program started at the Corporate Applied Science Laboratory, now included in the Corporate Strategic Research Center (Gardner *et al.* 2000). The technology is based on planar electrolyte-supported rectangular cells, displaced to form the Integrated Planar-SOFC (IP-SOFC). The development stage, that took place in 2003, had the main aim of producing small stacks in the range of 2–10 kW. Using these blocks, Rolls-Royce plans to realize a 10 kW system and later on a 60 kW. The final step is to realize a 1 MW hybrid SOFC-GT system, that combines a 800 kW SOFC system and a 200 kW Rolls-Royce GT.

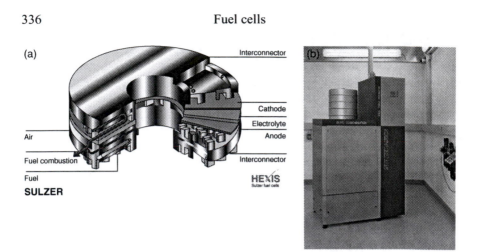

Figure 18.19   (a) Sulzer–Hexis stack and (b) HEXIS 1000 PREMIER 1 kW system.

***Energy Research Center of the Netherlands* (ECN)**   principally works on SOFC fuel cells development both on materials, cells, stack and systems design. During the past years, several SOFC solutions have been developed, including electrolyte-supported cells (specifically developed for Sulzer–Hexis); anode-supported cells, operating at intermediate temperature (680–800°C). An interconnect-supported fuel cell with ferritic stainless steel is currently under development.

In 1999, ECN created a spin off company called InDEC whose sole mission was commercializing their own technologies. InDEC manufactures the cells by tape casting (the ticker part that can be either the electrolyte or the anode) and by screen printing (the thinner parts). In 2003, the German company H.C. Starck bought the majority of InDEC stocks from ECN.

***Forschungszentrum Julich GmbH* (FZJ)**   is one of the largest SOFC Research Center in Europe. Their main objectives are to develop reliable, low cost, durable and tolerant to pollutant materials; to produce high power and mechanically strong cells; to develop high efficient, low weight, low volume and thermal cycle resistant stacks; to manufacture in a cost-effective way whole systems. They have also developed numerical models for understanding SOFC thermodynamics and behavior.

Their present research efforts are mainly on reducing steel corrosion, developing protective layers, enhancing resistance to thermal cycling and testing fuel cells under long-term operation. One of the main goals of the SOFC group at FZJ is to demonstrate a 20 kW SOFC for industrial CHP, whose cost is lower than 1000€/kW, and durability is >40,000 h.

18.3.3.3   Japan

Japan, together with USA and Europe, is a leader developer of SOFC technology. All the basic SOFC designs (planar, tubular and monolithic) are being developed. Mitsubishi Heavy Industries (MHI) has realized a multi kW stack and is currently considered the developer leader. For example, a stack made of 414 tubular cells, each 1.5 cm in diameter and 70 cm active length, achieved a maximum output of

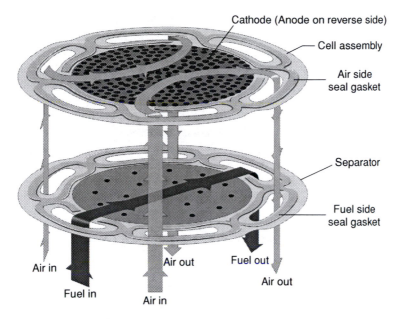

Figure 18.20 Ceramic fuel cell single cell.

21 kW (Iritani and Kougami 2001). Toto and Nippon Steel (Nakayama and Suzuki 2001) have developed a 36-cell stack, made of single cells 2.2 cm in diameter and 90 cm in length. The system runs on natural gas, operates at 1 atm, 70% fuel utilization, and the electric power is 3 kW.

18.3.3.4 Other countries

Ceramic Fuel Cell Ltd (CFCL) has been developing SOFC technology in Australia since 1992. In 2000, CFCL switched from metal ceramic composite stacks to all ceramic stacks, in order to overcome reliability and thermal cycling problems (Lawrence and Barker 2002). A scheme of a so-called all ceramic "layer set" is depicted in Figure 18.20.

Stacks are designed and manufactured in a highly modular approach. The CFCL stack consists essentially of four main components:

  (i) Solid oxide cell plate with anode and cathode coatings.
 (ii) Solid oxide separator plate with conductive coatings (interconnect).
(iii) Glass-ceramic air seal.
(iv) Glass-ceramic fuel seal.

These four components form a layered set. The layered sets, assembled together, form a stack.

The first reported tests on a stack (Godfrey *et al.* 2000) were conducted in a 4 × 2 array, made up of 50 layers, electrolyte supported, with metallic interconnects. The power was 5.5 kW, and the tests run for 400 h. After that, a 10 layers 2 × 2 array stack was tested for 3500 h at 820°C, generating about 1.3 MWh, with a degradation rate of 1–2% per 1000 h (Godfrey *et al.* 2000). Lately a 2 kW, 22 layer 2 × 2 anode supported stack operating at 760°C has been tested (Godfrey *et al.* 2000).

In June 2000, tests on a 25 kW test bed ended, providing feedback to realize a 40 kW system designed to run on natural gas (Foger and Godfrey 2002). This stack is composed of electrolyte-supported cells, and the structure is all-ceramic. The operating temperature is 850°C. The electrolyte is made of 10 YSZ, the cathode of La–Sr Manganite and the anode is a Ni-10YSZ cermet, modified to allow internal reforming. The power section of the system is composed of six modules, each composed of four stacks with a nominal power of 2.1 kW. The total DC power is about 50 kW, while the AC power of the whole system is 40 kW. Important achievements of the 40 kW stack, compared to the 25 kW previously realized, are a substantial cost reduction, high reliability and low degradation (Foger and Godfrey 2002).

18.4   MOLTEN CARBONATE FUEL CELLS

18.4.1   General features

Molten carbonate fuel cells originate from the work conducted in the 1960s to develop a fuel cell running on coal (DoD 2004). Nowadays, the prospective of using coal directly seems to be far away, while the use of a variety of fuels has been demonstrated or are in the demonstration phase (DoE 1998; Bove *et al.* 2004; EPA 2004).

The typical structure of an MCFC is schematically illustrated in Figure 18.21. The electrolyte is in the form of liquid and is embedded in a matrix. Ionic transfer inside the electrolyte is conducted via CO_3^{2-} ions that migrate from the cathode to the anode side.

The chemical reactions that govern the operations are:

$$CO_2 + \frac{1}{2}O_2 + 2e^- \rightarrow CO_3^{2-} \tag{18.17}$$

on the cathode side, while, on the anode:

$$H_2 + CO_3^{2-} \rightarrow H_2O + CO_2 + 2e^- \tag{18.18}$$

$$CO + H_2O \leftrightarrow H_2 + CO_2 \tag{18.19}$$

Expression (18.19) is commonly called a shift reaction and converts carbon monoxide and water into hydrogen. As a consequence of Equations (18.18) and (18.19), water is formed in the anode side and CO_2 is needed in the cathode side. Since the CO_2 required for reaction (18.17) is the same formed as a consequence of reaction (18.18), anodic gas is generally recycled from the anode to the cathode.

The partial pressure of CO_2 is not necessarily the same in the cathode and in the anode, thus the Nernst equation, providing the ideal voltage, is the following:

$$E = E^0 + \frac{RT}{2F} \ln \frac{P_{H_2} P_{O_2}^{0.5} P_{CO_2,cathode}}{P_{H_2O} P_{CO_2,anode}} \tag{18.20}$$

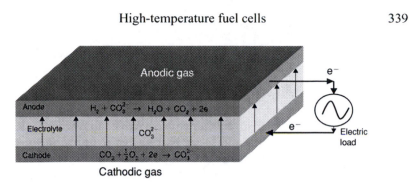

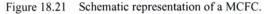

Cathodic gas

Figure 18.21   Schematic representation of a MCFC.

where E^0 is the voltage at standard pressure, R, T, F are, respectively, the universal gas constant, the temperature and the Faraday constant, while P_i is the partial pressure of the ith chemical species.

The stable electrolyte/gas interface in the electrodes is based on a capillary pressure balance (Larinne and Dicks 2000; Hirchenhofer *et al.* 2002). At thermodynamic equilibrium, the diameter of the largest pores that are flooded, is regulated by the following equation:

$$\frac{\gamma_c \cos \vartheta_c}{D_c} = \frac{\gamma_a \cos \vartheta_a}{D_a} = \frac{\gamma_e \cos \vartheta_e}{D_e} \qquad (18.21)$$

where γ is the interfacial surface tension, ϑ is the contact angle of the electrolyte and D is the diameter of the pores. The subscriptions c, a, e refer, respectively, to the cathode, anode and electrolyte matrix. All the pores with a diameter smaller than D are filled with the electrolyte, while the pores with a larger diameter remain empty. The matrix pores of the smallest diameters are totally filled with the electrolyte, while the electrodes are partially filled, according to the pores diameter distribution.

18.4.2   State of the art

18.4.2.1   Materials

The materials typically used for MCFC manufacturing are: nickel–chromium or nickel–aluminum for the anode, NiO lithiate for the cathode, Li_2CO_3/K_2CO_3 for the electrolyte, and α-LiAlO$_2$ or γ-LiAlO$_2$ for the matrix (Carrette *et al.* 2001; Hirchenhofer *et al.* 2002; Lunghi and Bove 2003b). In order to improve the cell performance and durability, as well the tolerance of some chemical substances present in most of the fuels, alternative materials or particular treatment can be adopted. As an example, $LiNi_xCo_{1-x}O_2$ or coated nickel cathodes can be considered as alternatives to the typical NiO Lithiate (Colonmer *et al.* 2002).

As reported in Section 15.3.2, one of the most important problems that reduces MCFC longevity, is the dissolution of the cathode in the electrolyte. NiO in fact reacts with CO_2 in the cathode, according to the following reaction:

$$NiO + CO_2 \leftrightarrow Ni^{2+} + CO_3^{2-} \qquad (18.22)$$

Nickel ions migrate through the matrix toward the anode, where they react with the incoming H_2:

$$Ni^{2+} + H_2 + CO_3^{2-} \rightarrow Ni + H_2O + CO_2 \qquad (18.23)$$

Besides the cathode dissolution, another problem related to reactions (18.22) and (18.23) is that the resulting metallic nickel precipitates in the matrix, thus leading to short circuiting across the matrix. As can be noted from expression (18.22), a way to reduce cathode solubility consists of decreasing the CO_2 partial pressure. The CO_2 partial pressure depends on the cathode operating pressure and cathodic gas composition:

$$P_{CO_2} = P_{cathode} \cdot X_{CO_{2,cathode}} \qquad (18.24)$$

where X represents the molar fraction, and so less durability is expected when the stack operates under pressurized conditions. Several studies have been conducted to assess NiO solubility, considering different electrolytes and cathodic gas compositions (Baumgartner 1984; Peelen et al. 1997; Yazici and Selman 1998).

Various materials are also considered to replace NiO for cathode manufacturing; among them $LiFeO_2$, Li_2MnO_3 and $LiCoO_2$ (Veldhuis et al. 1992; Giorgi et al. 1994; Lundblad et al. 2000) were found to be more stable than NiO, but their relative performances are noticeably lower. Other possibilities are to reduce the electrolyte acidity, using particular additives, the performance of the fuel cell is approximately the same for small percentages of additives such as $CaCO_3$, $SrCO_3$, $BaCO_3$ (Hirchenhofer et al. 2002) or by substituting Li/K electrolyte mixtures with the Li/Na one, with the aim to find an acceptable compromise between low NiO solubility, ionic conductivity and low chemical aggressive behavior.

18.4.2.2   Impurities in fuel

As stated before, MCFCs can operate on a variety of different fuels, such as coal-derived fuel, natural gas, gasified biomass, gasified waste and LFG. While fuel flexibility is a dramatic advantage for MCFCs, on the other hand, the poisoning effect of some substances contained in these fuels becomes a primary issue.

Since the most used fuel is currently natural gas, several investigations have been performed on the effect of sulfur compounds on the anode and, consequently on the entire fuel cell performance. Other harmful substances are NH_3, siloxane, chlorine and fluorine. Table 18.4 gives some typical limit values for these compounds. Moreover, since the anodic gas is generally recycled to the cathode after catalytic combustion, the presence of NOx in the cathodic gas must also be considered (Kawase et al. 2002). At the present time, the effects of these impurities on the MCFC performance are not totally clear. For this reason, very small concentration levels are taken when the fuel processing unit is designed. The consequence is that the fuel processor is often oversized, conducting a fuel purification that is not really needed and leading to an increased cost of the BoP. A review of the effects of several fuel impurities has been recently presented (Desideri et al. 2003).

Table 18.4   Summary of MCFC tolerance to impurities (Desideri *et al.* 2003).

Contaminants	Tolerance limits	Reference
Sulfur (H$_2$S)	0.1 ppm	Steinfeld and Sanderson (1998)
	0.5 ppm	Hirchenhofer *et al.* (2002)
	5 ppm	Kawase *et al.* (2001)
	0.3 ppm (suppression of shift reaction)	Takahashi *et al.* (2002)
Nitrogen compounds		
• NH$_3$	1 vol%	Hirchenhofer *et al.* (2002)
	no effects with 500 ppm	Kawase *et al.* (2001)
• NO$_x$	20 ppm	Kawase *et al.* (1999)
Halogens (HCl)	0.1 ppm	Steinfeld and Sanderson (1998)
	0.5–1 ppm	Hirchenhofer *et al.* (2002)
Alkali metals	1–10 ppm	Lobachyov *et al.* (1998)
Particulates (>3 μm)	100 ppm	Lobachyov *et al.* (1998)

18.4.3   Demonstration projects

There are at least six major players, that are currently developing MCFC technology:

- Fuel Cell Energy (FCE, Danbury, CT, USA)
- GenCell Corporation (Southbury, CT, USA)
- MTU CFC Solutions GmbH (Munich, Germany)
- Ansaldo Fuel Cell S.p.A. (AFCo, Genoa, Italy)
- Technology Research Association for MCFC power generation system/Hitachi/HIH under the coordination of NEDO (Japan)
- Korean Electric Power Corporation (KEPCO) – Korean Electric Power Research Institute (KEPRI, Korea).

During the last few years they have all demonstrated the feasibility of the technology and they are now focusing their activities on developing commercial products. Possible applications are stand-alone power plants, CHP plants, grid support and marine application (APU).

18.4.3.1   USA

Fuel Cell Energy (formerly Energy Research Corporation) started developing MCFC technology about 30 years ago. The product developed, the direct fuelcell® (DFC®), is based on an internal reforming technology. The first power plant realized in 1992 was the 2 MW Santa Clara installation, in California. FCE has developed three classes of products, named DFC 300, DFC 1500 and DFC 3000. The relative output electric power is, respectively, 250 kW, 1 MW and 2 MW, with an efficiency of about 47–50%. A hybrid MCFC-GT system has also been successfully demonstrated. The GT is arranged in an indirect configuration (Figure 18.4b, Section 18.2.2.2). The turbine adds 10–15% points to the efficiency of the system and is totally unfired. The company is also working under a US Navy contract to equip a ship with a fuel cell, running on diesel, as an APU.

Figure 18.22   FCE DFC®300 installed at Ocean County Community College, NJ.

FCE units have been installed in several sites in the US, such as universities, hospitals, wastewater treatment plants, telecom, data center, commercial and industrial facilities. Moreover, the company is receiving a growing number of orders from both Marubeni and MTU. Figure 18.22 shows the DFC®300 recently installed at the Ocean County Community College, in Tom Rivers, NJ.

GenCell Corporation is located in Southbury (Connecticut) and was founded in 2002, through the fusion of Allen Engineering (founded in 1997) and Fuel Cell Technology (founded in 1996). Despite the world trend, GenCell is now focusing its products on the 40–100 kW market. A prototype has been realized and operated for 1200 h.

18.4.3.2   Europe

MTU CFC Solutions GmbH is a DaimlerChrysler company that developed, using the FCE technology, the HotModule unit. A complete description of the HotModule is given in Chapter 27. Several HotModules have been installed in Europe, in hospitals, universities and industrial facilities. Figure 18.23 illustrates the installation at EnBW Michelin in Karlsruhe (Germany).

Ansaldo Fuel Cells S.p.A., part of the Finmeccanica Group, was funded in December 2001, to continue the work carried out by Ansaldo Ricerche on fuel cells. AFCo is primarily involved in the commercialization of MCFC power plants. The modules currently produced are the "Series 100" and "Series 500", whose nominal power are, respectively, 100 and 500 kW. The Series 100 capability has been fully demonstrated in two installations in Guadalix (Spain) and in Milan (Italy) during 1998–1999. Series 500 is planned to be scaled up for realizing multi-MW power plants.

The Series 500 is based on the TwinStack® configuration (Figure 18.24), that is composed of two electrochemical stacks displaced in a symmetric configuration. AFCo modules are equipped with the MIR. This device is constructed according

Figure 18.23   MTU HotModule at EnBW Michelin, Karlsruhe (Germany).

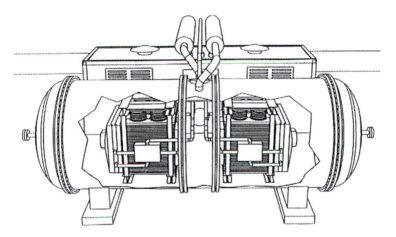

Figure 18.24   TwinStack® configuration.

to the principle of Figure 18.2 (Section 18.2.1.4), so that it presents both the advantages of internal and external reforming.

AFCo has targeted a select number of projects to demonstrate fuel flexibility. Gases tested include natural gas, LFG, gas from biomass gasification, digester gas and reformed diesel.

18.4.3.3   Japan

The Japanese *New Energy and Industrial Technology Development Organization (NEDO)* was established in October 1980, immediately after the second oil crisis, as a Japanese semi-governmental organization under the Ministry of International Trade. NEDO is an unique organization that works to coordinate the funds, personnel and technological strengths of both public and private sectors in Japan.

The activity of NEDO on MCFC started in 1984 testing a 10 kW stack. Later, a 100 kW MCFC was successfully tested from 1987 to 1992. The results provided the feedback to realize the first Japanese 1 MW power plant, in Kawagoe, that operated for about 5000 h, producing 2.103 MWh. Since 2000, NEDO has installed a compact 300 kW system in Kawagoe. The power section operates at 0.4 MPa and a current density of about 200 mA/cm^2.

18.4.3.4   Other countries

The *Korean Electric Power Research Institute (KEPRI)* is a research department of the Korean Electric Power Corporation (KEPCO), the monopoly utility of the country. Together with the Korean Institute of Science and technology (KIST), KEPRI and KEPCO are currently involved in MCFC development. The first activity started in 1993, when a 100 W stack was realized and tested. This was followed by 1.5 and 12 kW stacks which were tested from 1994 to 1996. After demonstrating the stacks capabilities, KEPRI developed and operated small size systems, that is, a 6 kW internal reforming system in 1998 and a 25 kW, external reforming system, in 1999.

18.5   MAIN FACTORS AFFECTING HIGH-TEMPERATURE FUEL CELLS COMMERCIALIZATION

As shown in Figure 18.25, there are several factors that affect high-temperature fuel cells (HTFC) commercialization. A successful commercialization does not depend just on the technological status, but is also affected by the market, and the political regulations.

Market   As for most of the industrial products, a large volume production will consistently reduce the production price, as estimated by most of the fuel cell companies.

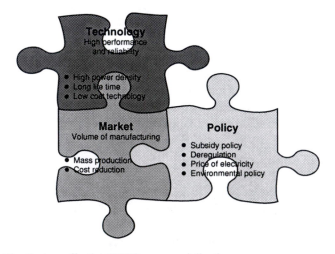

Figure 18.25   Factors affecting HTFC commercialization.

Policy Political regulation of different countries is also an important key for HTFC market penetration. This includes economical incentives to support demonstration projects, as well as continuous R&D for technological improvement. Economical incentives, however, must be considered as a transitional form of support that is required during the pre-commercial phase, but not as a long-term form of support. The deregulation and, consequently, the need for small stationary systems with high efficiency, instead, will conspicuously favor HTFC. Moreover, at the present time, one issue of HTFC systems is the high capital cost, compared to traditional systems. The high efficiency, however, is translated into a fuel consumption reduction, and consequently, a reduced operating cost. If the fuel price is elevated, the operating cost can play a role determinant over the capital cost. Moreover, the use of alternative fuels can be economically attractive, thus favoring HTFC usage.

Finally, US, Europe and Japan are introducing taxation related to the amount of pollutants emitted, thus sustaining clean technologies, such as fuel cells.

Technology Most of the technological barriers are currently represented by the stack life time and system production cost. As previously explained, MCFC life time is mainly affected by cathode stability and electrolyte management problems. The performance of the stack, instead, has already reached a high level, even if further improvements will allow the system to operate with lower operating costs.

Current SOFC issues are mainly related to internal mechanical stress that originates from thermal cycling and different coefficients of thermal expansion. As a consequence, planar single cells can currently be manufactured only in small dimension. Large size SOFC power plants (MW size) can only be constructed using Siemens–Westinghouse tubular technology, that however, results high cost and with low power density. Other issues of SOFC are material degradation and sealing problems.

All these limiting factors make SOFC and MCFC investment cost high. Not only because of the manufacturing costs, but also due to the short life-time of these fuel cell system.

All the HTFC producers, however, agree that the high manufacturing cost is mainly due to the BoP, that represents about two-thirds of the total capital cost. For this reason, research is needed for reducing the manufacturing cost of the components, as well as reducing the complexity of the BoP.

Among the three main factors (market, policy and technology), fuel cell companies and developers can deal only with the technological improvements. However, realization of systems that demonstrate the maturity of the technology can strongly stimulate political decisions and market dynamics.

REFERENCES

Baumgartner, C.E. (1984) NiO solubility in molten Li/K carbonate under molten carbonate fuel cell cathode environment. *J. Electrochem. Soc.* **131**, 1850.

Baur, E. and Preis, H. (1937) Uber Brennsto.-Ketten mit Festleitern. *Z. Elektrochem.* **43**(9), 727–732.

Bessette, N. (2004) Development of a low cost 10 kW tubular SOFC power system. *SECA Annual Workshop and Core Technology Program Peer Review*, May 11, 2004, Boston, MA.

Borglum, B. (2003) *From Cells To Systems: Global Thermoelectric's Critical Path Approach To Planar SOFC Technology And Product Development.* Eighth Grove Fuel Cell Symposium, London, UK, September 25, 2003.

Bossel, U.G. (1992) *Facts & Figures,* an International Energy Agency SOFC Task Report, Baden, October 1992.

Bove, R. and Lunghi, P. (2003) Comparison between MCFC/Gas Turbine and MCFC/Steam Turbine combined power plants. *Proceedings of ASME-IMECE*, Washington DC, USA, November 15–21, 2003.

Bove, R. and Sammes, N.M. (2004) Behavior of an SOFC system for small stationary applications under full and partial load operation. *Proceedings of 6th European Solid Oxide Fuel Cell Forum*, 28 June–2 July 2004, Lucerne, Switzerland.

Bove, R., Lunghi, P., Lutazi, A. and Sammes, N.M. (2004) Biogas as fuel for a fuel cell system: investigation and first experimental results for a molten carbonate fuel cell. *ASME Fuel Cell Science and Technology* **1**, 21–24.

Carrette, L., Friedrich, K.A. and Stimming, U. (2001) Fuel Cells-fundamentals and applications. *Fuel Cells from Fundamental to Systems* **1**, 5–39.

Carter, R.E. and Roth, W.L. (1968) *Electromotive Force Measurements in High Temperature Systems* (eds. C.L. Alchok), p 653. I.M.M., London.

Chan, S.H., Ho, H.K. and Tian, Y. (2003) Modeling for part-load operation of solid oxide fuel cell-gas turbine hybrid power plant. *J. Power Sources* **114**, 213–227.

Colonmer, H., Ganesan, P. and Subramanian (2002) Optimization of the cathode long-term stability in molten carbonate fuel cells: experimental study and mathematical modeling. *Final Report submitted to US DOE.* DOE Award number DE-AC26-99FT40714.

Costamagna, P., Magistri, L. and Massardo, A.F. (2001) Design and part-load performance of a hybrid system based on a solid oxide fuel cell reactor and a micro gas turbine. *J. Power Sources* **96**, 352–368.

Desideri, U., Lunghi, P. and Burzacca, R. (2002) State of the art about the effects of impurities on MCFCs and pointing out of additional research for alternative fuel utilization. *Proceedings of 1st International Conference of Fuel Cell Science, Engineering and Technology*, April 21–23, 2003, Rochester, NY, USA.

DoD (2004) Department of Defense/Fuel cell/ERDC-CERL Programs, web site: http://www.dodfuelcell.com

DoE (1998) Landfill gas clean up for molten carbonate fuel power generation. *US Department of Energy and Electric Power Research Institute,* Final Report CRADAMC95-031.

EPA (2004) Database of Landfills and Energy Projects, downloadable on http://www.epa.gov/lmop/index.htm

Entchev, E. (2003) Residential fuel cell energy systems performance optimization using "soft computing" techniques. *J. Power Sources* **118**, 212–217.

Foger, K. and Godfrey. (2002) SOFC product development at Ceramic Fuel Cells Ltd. *Proceedings of the 5th European Solid Oxide Fuel Cell Forum*, Lucerne, Switzerland.

Gardner, F.J., Day, M.J., Brandon, N.P., Pashley, M.N. and Cassidy, M. (2000) SOFC technology development at Rolls-Royce. *J. Power Sources* **86**, 122–129.

Gemmen, R.S., Liese, E., Rivera, J.G. and Brouwer, J. (2000) Development of dynamic modeling tools for solid oxide and molten carbonate hybrid fuel cell gas turbine systems. *Proceedings of 2000 ASME Turbo Expo*, May 8–11, Munich, Germany.

Giorgi, L., Carewska, M., Patriarca, M., Scaccia, S., Simonetti, E. and Di Bartolomeo, A. (1994) Development and characterization of novel cathode materials for molten carbonate fuel cell. *J. Power Sources* **49**, 227–243.

Godfrey, B., Foger, K., Gillespie, R., Bolden, R. and Badwal, S.P.S. (2000) Planar solid oxide fuel cells: the Australian experience and outlook. *J. Power Sources* **86**, 68–73.

Godickemeier, M. and Gaucker, L.J. (1998) Engineering of solid oxide fuel cells with ceria-based electrolytes. *J. Electrochem. Soc.* **145**, 414–420.

Gunes, M.B. and Ellis, M.W. (2003) Evaluation of energy, environmental, and economic characteristics of fuel cell combined heat and power systems for residential applications. *J. Energ. Resour. Tech.* **125**, 208–220.

Heinzl, A., Vogel, B. and Hübner, P. (2002) Reforming of natural gas-hydrogen generation for small scale stationary fuel cell systems. *J. Power Sources* **105**, 202–207.

Hirchenhofer, J.H. *et al.* (2002) *Fuel Cell Handbook,* 6th edn, US Department of Energy.

Hohnkle, D.K. (1968) *Fast Ion Transport in Solid* (eds. P. Vashista, J.N. Mundy and G.K. Shenoy), p 669. North Holland, Amsterdam.

Iritani, J. and Kougami, K. (2001) Pressurized 10 kW class module of SOFC. *Proceedings of the Seventh International Symposium on Solid Oxide Fuel Cells* (SOFC VII), Tsukuba, Japan.

Kawase, M., Nakatsu, T., Ujihara, T. (1999) The effect of coal gas on the performance of molten carbonate fuel cells. *Proceedings of the 3rd International Fuel Cell Conference, Nagoya Congress Center,* November 30–December 3, 1999.

Kawase, M., Mugikura, Y., Watanabe, T., Hiraga, Y. and Hujihara, T. (2002) Effects of NH_3 and NOx on the performance of MCFCs. *J. Power Sources* **104**, 265–271.

Kawase, M., Mugikura, Y. and Watanabe, T. (2001) Electrochemical studies on the high performance electrodes and the effect of impurity gas in molten carbonate fuel cells. *NEDO Final Report.*

Kilner, J.A. and Brook, R.J. (1982) A study of oxygen ion conductivity in doped nonstoichiometric oxides. *Solid State Ionics* **6**, 237–252.

Kozhukharov, V., Machkova, M., Ivanova, M. and Brashkova, N. (2001) Patents state of the art in SOFCs application. *Proceedings of Seventh International Symposium on Solid Oxide Fuel Cells (SOFC VII),* Tsukuba, Japan.

Larinne, J. and Dicks, A. (2000) *Fuel Cell Systems Explained.* Wiley and Sons, UK.

Lawrence, J. and Barker, J. (2002) Design of an SOFC Stack. *Proceedings of the 5th European Fuel Cell Forum,* Lucerne, Switzerland.

Lobachyov, K.V. and Richter, H.J. (1998) An advanced integrated biomass gasification and molten fuel cell power system. *Energ. Convers. Manage.* **39**, 1931–1943.

Lundblad, A., Schwartz, S. and Bergman, B. (2000) Effects of sintering procedures in development of $LiCoO_2$-cathodes for the molten carbonate fuel cell. *J. Power Sources* **90**, 224–230.

Lunghi, P. and Bove, R. (2003a) Life cycle assessment of a molten carbonate fuel cell stack. *Fuel Cells From Fundamental to Systems* **3**, 224–230.

Lunghi, P. and Bove, R. (2003b) Analysis and optimization of a hybrid MCFC-steam turbine plant. *Proceedings of 1st International Conference of Fuel Cell Science, Engineering and Technology,* April 21–23, 2003, Rochester, NY, USA.

Lunghi, P., Bove, R. and Desideri, U. (2004) Life cycle assessment of fuel cells based landfill gas energy conversion technology. *J. Power Sources* **131**, 120–126.

Lunghi, P., Bove, R. and Desideri, U. (2003) Analysis and optimization of hybrid MCFC Gas Turbines plants. *J. Power Sources* **118**, 108–117.

Lunghi, P., Burzacca, R. and Desideri, U. (2001) Celle a combustibile gassificazione per la trigenerazione da scarti industriali: criteri di progettazione, prestazioni e test case in un'industria alimentare. *Proceedings of the 28th National Congress of ANIMP-OICE-UAMI,* 25–26 October, Spoleto, Italy (in Italian).

Lunghi, P. and Ubertini, S. (2001) Solid oxide fuel cells and regenerated gas turbines hybrid systems: a feasible solution for future ultra high efficiency power plants. *Proceedings of the Seventh International Symposium on Solid Oxide Fuel Cells (SOFC VII),* June 3–8, 2001, Tokyo, Japan.

Meusinger, J., Riensche, E. and Stimming, U. (1998) Reforming of natural gas in solid oxide fuel cell system. *J. Power Sources* **71**, 315–320.

Minh, N. and Takahashi, T. (1995) *Science and Technology of Ceramic Fuel Cells.* Elsevier Science, the Netherlands.

Minh, N. (2004) GE Power Systems, Hybrid Power Generation Systems. *SECA Annual Workshop and Core Technology Program Peer Review*, May 11, 2004, Boston, MA.

Nakayama, T. and Suzuki, M. (2001) Current status of SOFC R&D program at NEDO. *Proceedings of the Seventh International Symposium on Solid Oxide Fuel Cells (SOFC VII)*, Tsukuba, Japan.

Nernst, W. (1899) Uber die elektrolytische Leitung fester Korper bei sehr hohen Temperaturen. *Z. Elektrochem.* **6**, 41–43.

NETL (2004) *Fuel Cell Annual Report 2003*. US Department of Energy, Office of Fossil Energy, National Energy Technology Laboratory, Final Report.

Ota, K., Takeishi, Y., Shibata, S., Yoshitake, H., Kamiya, N. and Yamazaki, N. (1995) Solubility of cobalt oxide in molten carbonate. *J. Electrochem. Soc.* **142**, 3322.

Peelen, W.H., Hemmes, K. and de Wit J.H. (1997) Diffusion constants and solubility values of CO^{2+} and Ni^{2+} in Li/Na and Li/K carbonate melts. *Electrochem. Acta* **42**, 2389–2397.

Riensche, E., Achenbach, E., Froning, D., Haines, M.R., Heidug, W.K., Lokurlu, A. and von Andrian, S. (2000) Clean combined-cycle SOFC power plant-cell modeling and process analysis. *J. Power Sources* **86**, 404–410.

Shaffer, S. (2004) Solid state energy conversion alliance – Delphi SECA Industry. *SECA Annual Workshop and Core Technology Program Peer Review*, May 11, 2004, Boston, MA.

Silveira, J.L., Leal, E.M. and Ragonha, L.F. (2001) Analysis of a molten carbonate fuel cell: cogeneration to produce electricity and cold water. *Energy* **26**, 891–904.

Singhal, S.C. and Kendall, K. (2004) *High-temperature Solid Oxide Fuel Cells: Fundamentals, Design and Applications*. Elsevier Science, London.

Spath, P.L. and Mann, M.K. (2001) Life cycle assessment of hydrogen production via natural gas steam reforming, *National Renewable Energy Laboratory Technical Report*.

Steinfeld, G. and Sanderson, R. (1998) Landfill gas cleanup for carbonate fuel cell power generation. *ERC Final Report for NREL*.

Sugiura, K. and Naruse, I. (2002) Feasibility study of the co-generation system with direct internal reforming-molten carbonate fuel cell (DIR-MCFC) for residential use. *J. Power Sources* **106**, 51–59.

Takahashi, S., Mizukami, T. and Kahara, T. (2002) Effects of H_2S and HCl in coal gas on MCFC performance. *Proceedings of Fuel Cell Seminar, Palm Spring*, CA, November 18–21, 2002.

Torrero, E. and McClealland, R. (2002) Residential Fuel Cell Demonstration Handbook, *National Renewable Energy Laboratory Technical Report No NREL/SR-560-32455*.

Utz, B.R. (2004) NETL Distributed Generation Program. *Program and Abstracts of the 1st International Conference on Fuel Cell Development and Deployment*, March 7–10 2004, Storrs, CT.

Van herle, J., Membrez, Y. and Bucheli, O. (2004) Biogas as fuel source for SOFC co-generators. *J. Power Sources* **127**, 300–312.

Van herle, J., Maréchal, F. and Favrat, D. (2003) Energy balance model of a SOFC co-generator operated with biogas. *J. Power Sources* **118**, 375–383.

Veldhuis, J.B.J., Eckes, F.C. and Plomp, L. (1992) The dissolution properties of LiCoO2 in molten 62 : 38 mole percent LI : K carbonate. *J. Electrochem. Soc.* **139**, 1, L6.

Vora, S.D. (2003) SECA Program at Siemens–Westinghouse. *4th SECA Workshop*, April 15–16, Seattle, WA.

Yamaji, K., Horita, T., Sakai, N., Ishikawa, M. and Yokokawa, H. (1998) Reaction between $(La_{0.9}Sr_{0.1})(Ca_{0.8}Mg_{0.2})O_{2.85}$ and Platinum. *Proceedings of the 3rd European Solid Oxide Fuel Cell Forum*, 2–5 June 1998, Nantes, France, pp. 373–384.

Yazici, M.S. and Selman, J.R. (1998) Dissolution of partially immersed nickel during in situ oxidation in molten carbonate: cycling, stripping and square wave voltammetry measurements. *J. Electroanal. Chem.* **457**, 89–97.

Zizelman, J. (2003) Development update on Delphi's solid oxide fuel cell systems: from gasoline to electric power. *4th SECA Workshop*, April 15–16, Seattle, WA.

19

Metallic construction materials for solid oxide fuel cell interconnectors

W.J. Quadakkers, J. Pirón Abellán, P. Huczkowski, V. Shemet and L. Singheiser

19.1 INTRODUCTION

The most important properties required for the interconnector material of a solid oxide fuel cell (SOFC) are high electronic conductivity, thermal stability in the cathode and anode side gas at the high service temperatures (650–1000°C), and a thermal expansion coefficient similar to that of the ceramic, electro-active components. Most SOFC designs use yttria stabilized zirconia (YSZ) as electrolyte, (La, Sr)MnO$_3$ as cathode, and Ni/ZrO$_2$-cermet as anode (Minh 1993; Steele 2000; Stöver *et al.* 2002). Ceramic, perowskite type materials on the basis of La-chromite have been shown to possess the property combination required (Singnal 1997;

Steele 2000). However, especially in planar cell designs, the interconnector acts as the mechanical support for the thin ceramic parts and it is the gas-proof separation between fuel gas and oxidant. Although La-chromites are being considered as construction materials for interconnectors in planar SOFC designs (Stolten *et al.* 1997), metals have gained more and more attention as possible replacement for the La-chromite-based ceramics (Greiner *et al.* 1995; Köck *et al.* 1995; Buchkremer *et al.* 1997; Stöver *et al.* 1999) in recent years.

Metallic construction materials offer a number of advantages over ceramic materials: they are easier, and therefore cheaper to fabricate than ceramics, they are less brittle, easier to machine, and they can be joined with a number of standard welding and brazing techniques. Additionally, they possess higher electrical and thermal conductivities than most ceramics.

19.2   HIGH-TEMPERATURE ALLOYS

The main drawback for SOFC application of metals and alloys is their reactivity with the anode and cathode side service environments at the high operating temperatures. The resulting high-temperature corrosion phenomena not only lead to dimensional changes and loss in load bearing cross sections of the components, but also to formation of oxide scales on the component surface which mostly possess poor electrical conductivities (Kofstad 1988; Lai 1990). Therefore, noble metals have in a few cases been considered as interconnector materials. However, because of limited availability and high cost, this solution has been abandoned for large scale, commercial application, and conventional high-temperature alloys (England and Virkar 1999) have received by far the most attention as possible candidate metallic interconnector materials. A major property of these high-temperature alloys is their ability to form protective surface oxide scales which possess sufficiently slow growth rates to keep the oxidation attack below an acceptable level from the viewpoint of dimensional changes and loss of load bearing cross section.

If one would consider oxidation resistance at high temperatures as the first selection criterion for interconnector application, alloys of the type NiCrAl, CoCrAl and especially FeCrAl would be the materials to be chosen (Lai 1990). These types of high-temperature alloys rely for their resistance against oxidation/corrosion attack on the formation of an extremely slowly growing alumina scale which forms on the material surface upon high-temperature exposure. Also alumina forming intermetallics on the basis of NiAl have been considered as interconnector materials (Yamamoto 2000) because Ni-aluminides possesses not only superior oxidation resistance, but also a low thermal expansion coefficient. The excellent oxidation resistance of alumina forming metallic materials is, however, accompanied by an extremely low electronic conductivity of the surface scales (Kofstad and Bredesen 1992; Kofstad 1996) and the use of such materials would thus require measures to overcome this problem by developing special stack designs to assure suitable long-term stable electrical connections of the interconnector with anode and cathode.

19.3   CHROMIA-FORMING ALLOYS

Most studies on qualification and/or development of metallic high-temperature materials for interconnector application relate to chromia-forming alloys. Chromia provides less oxidation/corrosion protection than alumina (Kofstad 1988), however, its electronic conductivity in the envisaged SOFC operation temperature range of approximately 600–1000°C is orders of magnitude larger than that of alumina. The vast majority of the commercial chromia-forming alloys is based on the systems NiCr, NiFeCr or FeCr. The two first mentioned alloy systems seem less promising for SOFC application because their thermal expansion coefficient is substantially higher than that of the commonly used electro-active ceramic cell materials. Nickel-based alloys, for example, of the type INCONEL600 have been used as interconnector (Matsuzaki and Yasuda 2000; Yasuda *et al.* 2001), however, application of such a material would require a special cell design to overcome the stresses expected to be generated during thermal cycling in contact and joining areas of the interconnector with other cell components.

For these reasons, the materials most frequently studied for SOFC interconnector application are based on the binary alloy system FeCr. As the thermal expansion coefficient in FeCr alloys decreases with increasing chromium content (Malkow *et al.* 1997), chromium-based materials have in most cases been considered for application in zirconia-electrolyte-based SOFC concepts (Quadakkers *et al.* 1994; Greiner *et al.* 1995; Schmidt *et al.* 1995; Thierfelder *et al.* 1997). If a slightly higher thermal expansion coefficient is required, as is, for example, the case in Ni-cermet anode-based cell concepts or in cells using ceria as electrolyte, iron-rich alloys, that is ferritic steels, are being considered (Quadakkers *et al.* 1996). Therefore, the following discussion will concentrate mainly on chromium-based alloys and chromium containing ferritic steels.

19.4   CHROMIUM AND CHROMIUM-BASED ALLOYS

Tracer studies using ^{18}O isotopes revealed, that in most cases outward chromium diffusion is the dominating transport mechanism in chromia scales formed at high temperatures on Cr surfaces (Cotell *et al.* 1987; Graham and Hussey 1995). The outward chromium transport results in formation of voids and cavities at the scale/metal interface frequently leading to poor scale adherence (Quadakkers *et al.* 1988; Graham and Hussey 1995). The scales tend to be buckled which has been attributed by many authors to the parallel diffusion of chromium and oxygen along chromia grain boundaries (Kofstad 1988; Graham and Hussey 1995).

The buckling and poor adhesion of the scale can be minimized by the addition of so-called reactive elements (REs), either in metallic form, in form of an oxide dispersion or by surface modification (Hou and Stringer 1992). Apart from improved oxide scale adhesion the RE addition (e.g. Y, La, Ce) leads to a substantial decrease in oxide growth rate and in the case of low-Cr alloys it promotes the selective oxidation of chromium (Whittle and Stringer 1980). Several authors proposed that the presence of the RE tends to decrease the scale growth rate by the suppression of outward chromium transport (Golightly *et al.* 1976; Pint 1996; Quadakkers *et al.* 1996).

Extensive research work at Plansee recently lead to the development of chromium which was dispersion strengthened (ODS) by RE-oxides (Figure 19.1). The material showed substantially improved oxidation and mechanical properties compared to conventional chromium alloys (Greiner et al. 1995; Köck et al. 1995). The ODS materials are manufactured by elemental mixing of the starting powders (abbreviated as MIX alloy) or by mechanical alloying (MA), that is by high energy ball milling of chromium and oxide powders in a non-oxidizing environment (Quadakkers et al. 1994, 1996). The latter process can produce a very finely distributed oxide dispersion (typical particle diameter $\approx$ 15 nm) embedded in the chromium matrix, compared to the coarser dispersion resulting from the elemental mixing process (Greiner et al. 1995; Köck et al. 1995). The improvement in ductility which can be obtained by the material modification depends on a number of factors, for example dispersion type and distribution. Very recently a substantial ductility increase of Cr was demonstrated by using MgO dispersions (Brady et al. 2001), however, no further data on other mechanical properties have yet been reported. The commercial ODS Cr alloys specifically developed for SOFC application mostly contain a few per cent of iron to adjust the thermal expansion coefficient as close as possible to that of the electrolyte material YSZ (Greiner et al. 1995; Köck et al. 1995).

In using Cr (alloys) in air or oxygen, it has to be considered that the chromia scales tend to form volatile species (Ebbinghaus 1993; Quadakkers et al. 1994) such as CrO_3 and/or $CrO_2(OH)_2$. The rate of volatilization increases with increasing temperature and gas flow rate (Kofstad 1988). Several authors have demonstrated that the existence of volatile chromium species in the gas phase can cause serious cell degradation (Badwal et al. 1997; Günther et al. 1996). Based on thermodynamic considerations and experimental results, this deterioration was found to be mainly related to poisoning of the cathode/electrolyte interface (Das et al. 1994; Hilpert et al. 1996). Other authors have claimed that the driving force for cell degradation is related to spinel formation in the cathode. The extent of degradation should thus depend on the cathode composition (Jiang et al. 2001). Several protection methods have been proposed to minimize evaporation of volatile Cr-species, such as coating of the interconnector with La-chromite (Gindorf et al.

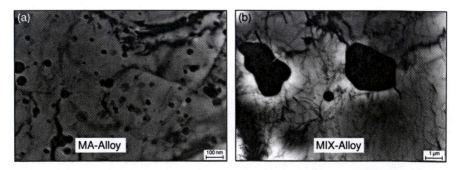

Figure 19.1   TEM pictures showing distribution of Y_2O_3 dispersions in Cr-based ODS alloys: (a) manufactured by MA and (b) MIX of the metal and oxide powders.

2000) or – manganite (Batawi *et al.* 1999), metallic layers (Quadakkers 1995a), oxide layers or aluminium surface enrichment to promote alumina surface scale formation on interconnector areas, where electrical conductivity is not a major issue (Quadakkers 1995a, b).

19.5   CHROMIUM-BASED ALLOYS IN ANODE SIDE GAS

The vast majority of publications on oxidation of chromium and chromium alloys relates to exposures in pure oxygen or air (Linderoth *et al.* 1996; Linderoth 1998). Several reviews on this subject are available and substantial information can be found in textbooks (Kofstad 1988; Lai 1990). Concerning the oxidation behaviour of Cr and Cr-alloys in SOFC relevant anode side H_2/H_2O-based gases far fewer data are available. The oxidation rate of Cr-based ODS alloys at 950–1050°C in an $Ar/H_2/H_2O$-mixture (equilibrium oxygen partial pressure at 1000°C approximately 10^{-15} bar) were found to be higher than in high-pO_2 environments (air, Ar/O_2), whereby the difference could only partly be explained by formation of volatile oxides and hydroxides in the high-pO_2 gases (Hänsel *et al.* 1998). The actual oxidation rate strongly depended on the manufacturing method (Figure 19.2). In the $Ar/H_2/H_2O$-environment the alloys tended to form whisker type oxide morphologies, the extend of whisker formation being decreased by an "optimum" addition of a reactive element (oxide dispersion).

An important observation, which has hardly been discussed in literature, is that the chromia scales formed on Cr-based ODS alloys in $Ar/H_2/H_2O$ exhibit far better adherence to the metallic substrate than those formed in air or oxygen (Quadakkers *et al.* 1996; Hänsel *et al.* 1998). This effect of atmosphere composition on scale adhesion is even more dramatic in case of non-RE-doped alloys and elemental chromium (Figure 19.3). A possible explanation for the improved chromia scale adherence in H_2/H_2O-based gases is a mechanism related to

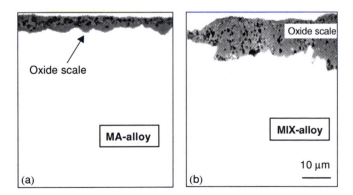

Figure 19.2   Cross sections of oxide scales on Cr-based ODS alloys, manufactured by different processing methods (cf. Figure 19.1) after 1000 h exposure in an $Ar/H_2/H_2O$ mixture at 950°.

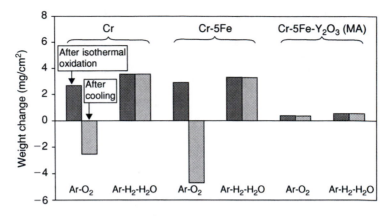

Figure 19.3   Weight changes of Cr, Cr-5Fe and the ODS alloy Cr-5Fe-Y₂O₃ after isothermal oxidation in Ar-20%O$_2$ and Ar-4%H$_2$-2%H$_2$O at 1000°C and after subsequent cooling to room temperature. The results illustrate, especially for the RE-free materials, the importance of the test atmosphere for oxide adherence (derived from results in reference Hänsel *et al.* 1998).

dissociation processes occurring within scale voids, as proposed for porous iron oxide scales by Rahmel and Tobolski (1965) as well as Gil *et al.* (1996).

During air or oxygen oxidation of non-doped chromium (alloys) the outward scale growth is correlated with inward cation vacancy transport (Kofstad 1988). The vacancies tend to coalesce at the scale/metal interface, eventually leading to formation of large voids which result in buckled and poorly adherent scales. Incorporation of water vapour or hydrogen into the scale can cause a rapid re-oxidation of the exposed metal by a dissociation mechanism involving "H$_2$/H$_2$O-bridges" within the voids (Rahmel and Tobolski 1965). In this way a crack/pore healing occurs, thus preventing growth of the voids at the metal/scale interface.

This repetitive process of void formation and subsequent re-oxidation should result in oxide layers containing substantial in-scale micro-voidage, as was indeed observed experimentally (Gil *et al.* 1996; Hänsel *et al.* 1998).

A strong indication for the fact that the formation of the micro-porous scales in H$_2$/H$_2$O-mixtures is indeed a result of a repeated formation/healing process of voids which in turn are correlated with outward scale growth, was provided by implantation experiments (Hänsel *et al.* 1998; Quadakkers *et al.* 1998). Exposure of a Y-implanted Cr-alloy in an Ar/H$_2$/H$_2$O- mixture at 1000°C initially resulted in formation of an extremely thin, dense scale, which exclusively grew by inward oxygen transport and was completely free of micro-porosity. After longer exposure time, that is after the implanted yttrium became depleted, the scale locally started to change into the well-known micro-porous, more rapidly growing scale as known from non-implanted material. After around 1000 h exposure the effect of the implanted yttrium on the oxide scale morphology had nearly completely vanished and the more rapidly growing, micro-porous scale covered virtually the whole specimen surface. However, also in this case the relatively thick oxide scale formed in the H_2/H_2O atmosphere exhibited excellent adherence to the metallic substrate (Hänsel *et al.* 1998).

19.6   MIXED GAS CORROSION OF CHROMIUM-BASED ALLOYS

In an actual SOFC cathode side environment, oxygen will not be the only gaseous specie which reacts with the Cr-based interconnector alloys. The main reactive gases to be considered if air is used as oxidant are nitrogen and water vapour. Especially for non-RE-doped chromium and chromium alloys, reaction with nitrogen is of major concern because it leads to environmentally induced embrittlement during high-temperature exposure. The sensitivity for nitrogen uptake depends on the gas-tightness of the chromia layer because nitrogen transport between gas atmosphere and metal or alloy is known to occur via gas molecules rather than via solid-state diffusion through the scale. Generally, it can be said that the thick chromium oxide scales formed on non-RE-doped Cr are more permeable for molecules of, for example, nitrogen than the thinner, that is more protective, scales on RE- or RE-oxide-doped metals or alloys. This observation in fact indicates that the protective properties of the chromium oxide scales are to a large extend determined by their ability to prevent molecular transport of oxidizing species rather than by the type and concentration of point defects in the chromia lattice.

The second reactive gaseous specie which is of importance during air oxidation of Cr-based materials is water vapour. The first well-known effect of water vapour is that it enhances the volatilization of chromia due to formation of volatile hydroxides and oxyhydroxides (Ebbinghaus 1993; Das *et al.* 1994; Hilpert *et al.* 1996). A second, less apparent effect was observed by Hänsel *et al.* (1998). They found the scale growth rate in the temperature range 950–1050°C to be enhanced if water vapour was present in air or Ar-O_2, and this effect was related to the enhanced oxidation of cracks which occurred in the chromia layers even during isothermal oxidation. The enhanced oxidation may therefore be related to a dissociation mechanism occurring in voids and cracks (Rahmel and Tobolski 1965), similar to that described in the previous section.

Depending on the cell-operating conditions mixed gas corrosion of Cr-based alloys also has to be considered in the anode side gas. If carbon-containing gas species such as CO and/or CH_4 are present, all Cr-based alloys tend to form a sub-scale layer containing Cr-carbide, that is, mainly $Cr_{23}C_6$ sometimes in combination with Cr_7C_3. In case of a CH_4-based test gas the carbides exist as a near-continuous layer (Thierfelder *et al.* 1997), whereas in the case of a CO-based gas the sub-scale layer tends to consist of an oxide/carbide mixture (Crone *et al.* 1997; Quadakkers *et al.* 1998). As already discussed for nitridation, the extent of carbon attack, which occurs by molecular gas transport through the scale, also decreases with decreasing growth rate of the oxide layer (Quadakkers *et al.* 1998). This again is a strong indication that in the case of chromia scales, protectiveness means to a large extent resistance against molecular transport from the gaseous atmosphere to the metal or alloy (Figures 19.2 and 19.4).

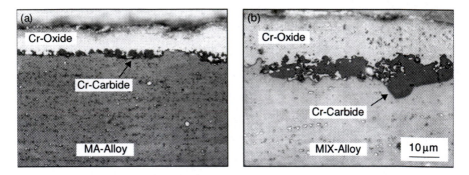

Figure 19.4   Cross sections of oxide scales on Cr-based ODS alloys, manufactured by different processing methods (cf. Figure 19.1) after 1000 h exposure in a CO containing H_2/H_2O mixture at 950°C.

19.7   SCALE GROWTH KINETICS

The growth rate of the oxide scales formed on the interconnector surfaces is an important parameter to be considered in SOFC applications because of dimensional constraints in a multi-layered cell stack. Additionally, fracture mechanics concepts describing scale integrity during thermal cycling (Schütze 1997) predict that the onset for scale spalling depends on scale thickness.

If chromia growth proceeded by diffusion of point defects in the oxygen and/or chromium sub-lattice, scaling rates could be calculated by Wagner theory (Kofstad 1988), provided that the self-diffusion coefficients of chromium and oxygen were known. Lattice self-diffusion coefficients of both chromium and oxygen in chromia are reported to be extremely small, but the literature data shows considerable scatter. At 1100°C the self-diffusion coefficients are in the range 10^{-17}–10^{-18} cm²/s (Atkinson and Taylor 1984; Sockel et al. 1988; Sabioni et al. 1992).

The growth rates of scales on Cr-based ODS alloys is substantially slower than the rates of scales formed on pure Cr or Cr alloys (Linderoth et al. 1996; Hänsel et al. 1998; Linderoth 1998) without addition of a reactive element (oxide). It is difficult to give a quantitative figure of this reduction in oxidation rate by the presence of the RE because Cr as well as Cr-based alloys do not exhibit the classical parabolic time dependence for oxide scale growth, frequently assumed in literature. Classical Wagner oxidation theory (Kofstad 1988) predicts that the scale growth will obey a parabolic time dependence if oxidation is controlled by diffusion of metal and/or oxygen ions through the oxide lattice:

$$x^2 = k \cdot t \tag{19.1}$$

in which x is the scale thickness, t the time and k the parabolic oxidation rate constant. Expressed as oxygen-uptake per unit area (Δm), Equation (19.1) is mostly written in the following form:

$$(\Delta m)^2 = K_p \cdot t \tag{19.2}$$

For Cr, Cr-alloys and especially Cr-based ODS alloys, Equation (19.2) is hardly ever obeyed (Rahmel and Tobolski 1965; Gil et al. 1996; Hänsel et al. 1998) because the assumptions made to describe scale growth by Equation (19.2) are mostly not fulfilled (Quadakkers 1990, 1993; Bongartz *et al.* 1993; Quadakkers and Singheiser 2001). Chromia scale growth proceeds via rapid diffusion paths, such as oxide grain boundaries (Cotell *et al.* 1987; Quadakkers *et al.* 1988; Sabioni *et al.* 1992), the density of which in most cases is not time independent. The scale growth rate (k) can for such cases be written as (Quadakkers 1990):

$$k \sim \frac{D_o \cdot \delta}{r} \cdot \frac{\Delta \mu_o}{RT} \tag{19.3}$$

in which D_o is the oxygen grain boundary diffusion coefficient, δ the grain boundary width, r the grain size and $\Delta\mu_o$ the oxygen potential gradient across the scale. In many cases r increases in the scale growth direction, that is initially small oxide grains are formed whereas the grains formed after longer times are larger in size (Quadakkers 1990; Bongartz *et al.* 1993). This leads to a k-value which decreases with increasing oxidation time.

A further factor which has to be considered in evaluating growth rates is that the scales are not completely gas-tight due to formation of micro-voids and micro-cracks and consequently molecular gas transport contributes to the scale growth process. Furthermore, in oxygen-rich environments (e.g. air) formation of volatile oxides and/or hydroxides can affect chromia growth, especially at high temperatures (Kofstad 1988). The combination of all these factors makes the description of the scaling kinetics quite complex and a (near) fit of a measured curve to a parabolic rate law is frequently more or less accidental (Quadakkers 1990; Bongartz *et al.* 1993).

It has been demonstrated unequivocally that a reduction in scale growth rate in a RE (oxide) containing Cr-alloy requires the RE to become incorporated into the scale (Hou *et al.* 1991; Quadakkers and Singheiser 2001). In ODS alloys the rate of RE incorporation in the scale is therefore strongly affected by size and distribution of the oxide dispersion (cf. Figures 19.2 and 19.4). In respect to the oxidation properties of chromia-forming alloys in general, several authors have tried to explain the RE-induced reduction in scale growth rate and improvement in scale adherence by a single mechanism. However, implantation experiments (Hou *et al.* 1991) revealed that the presence of the RE yttrium in the oxide scale decreased scale growth rate but it did not necessarily decrease scale adherence. Apparently, the improvements of the two major oxide scale properties are imparted by two fundamentally different mechanisms. Similar conclusions can be drawn from TG studies on Cr-based ODS alloys: the reduction in scale growth rate strongly depends on dispersion size and distribution, whereas scale adherence does not (Quadakkers 1995b; Hänsel *et al.* 1998).

19.8   ELECTRONIC CONDUCTIVITY OF CHROMIA-BASED OXIDE SCALES

Chromia is an electronic conductor and at high temperatures ($>1000°C$) the electrical conductivity is claimed to be independent of the oxygen partial pressure

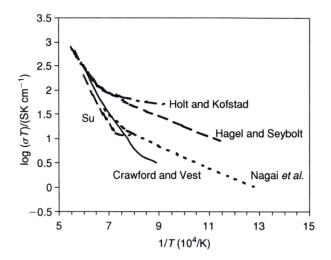

Figure 19.5   Electrical conductivity of Cr_2O_3 as a function of reciprocal temperature measured by different authors (derived from Hagel and Seybolt 1961; Crawford and Vest 1964; Nagai *et al.* 1983; Nagai and Ohbayashi 1989; Holt and Kofstad 1997, 1999; Liu *et al.* 1998).

(Crawford and Vest 1964; Hagel and Seybolt 1961; Park and Natesan 1990). At lower temperatures the concentration of electronic defects due to the intrinsic electronic equilibrium becomes so small that chromia changes into an extrinsic electronic conductor, the electronic conductivity being dominated by the presence of impurities or dopants (Holt and Kofstad 1997,1999). For chromia-based scales on a Ni–20%Cr alloy in a H_2/H_2O mixture, the change from intrinsic to extrinsic behaviour appeared to occur at lower temperatures, that is 700–900°C (Liu *et al.* 1998). The electrical conductivity of bulk chromia samples at 800–1000°C (Figure 19.5) has been reported to be in the range 1.10^{-2}–5.10^{-2}S/cm (Hagel and Seybolt 1961; Crawford and Vest 1964; Holt and Kofstad 1997), although lower values also have been found (Nagai *et al.* 1983). Nagai *et al.* found doping by Y_2O_3 or La_2O_3 to increase the electronic conductivity of chromia (Nagai *et al.* 1983; Nagai and Ohbayashi 1989). The strongest increase of the conductivity was observed for doping by NiO, whereby in this case no oxygen partial pressure dependence was found.

19.9   COMMERCIALLY AVAILABLE FERRITIC STEELS

In commercially available ferritic steels for high-temperature application, the Cr-content varies between approximately 7 and 28 mass% (Table 19.1), whereby, the oxidation resistance generally increases with increasing Cr content (Malkow *et al.* 1998). During air exposure, for example in the temperature range 700–1000°C, the scale composition strongly varies with Cr content. On low-Cr steels (>5% Cr) the scales consist of nearly pure Fe-oxide accompanied by internal

Table 19.1   Typical concentrations of main alloying elements (mass%) in commercially available high-Cr ferritic steels.

DIN-designation	Commercial name	DIN-Nr.	Concentration in mass%			
			Cr	Al	Si	Mn
X10CrAl7	Ferrotherm 4713	1.4713	6–8	0.5–1	0.5–1	–
X10Cr13	Nirosta 4006	1.4006	11.9	<0.02	0.49	0.30
	–	1.4509	15–16	<1.25	<0.5	<1
X10CrAl18	Ferrotherm 4742	1.4742	17.3	1.04	0.93	0.31
X10CrAl24	Ferrotherm 4762	1.4762	23.5	1.82	1.01	0.38
			24.4	1.38	1.14	0.54
Fe–25Cr–Mn	RA 446	–	24.2	–	0.43	0.67
Fe–26Cr–1Mo	E Brite	–	25.8	–	0.24	0.02
X18CrN28	Sandvik 4C54	1.4749	26.5	<0.01	0.47	0.70
Fe–29Cr–4Mo–Ti	Al29-4C	–	27.3	–	0.26	0.28

oxide precipitates of Cr_2O_3 and/or $FeCr_2O_4$-spinels. With increasing Cr content the scales become richer in spinel and chromia which is accompanied by a decrease of the scale growth rate. Formation of a protective, single-phase chromia layer requires a chromium content of approximately 17–20%, depending on temperature, surface treatment, minor alloying additions and impurities (Figure 19.6).

With respect to oxidation behaviour, the most important minor alloying additions in commercial ferritic steels are Mn, Ti, Si and Al, their concentrations commonly being a few tenths of a per cent. Upon high-temperature oxidation, the two first mentioned elements tend to become incorporated in the scale (Ennis and Quadakkers 1987), although titanium can also be present in the form of internal titania precipitates in the sub-surface region. Titanium can become dissolved in chromia at low oxygen partial pressures, that is in the inner part of the scale. However at higher oxygen partial pressures, thus in the outer part of the scale, it tends to become re-precipitated. Mn possesses only a very limited solubility in chromia. It is mostly found in the form of a Cr–Mn–spinel layer in the outer part of the scale (Malkow et al. 1998). Si and Al form oxides which are thermodynamically far more stable than chromia. Therefore, these elements mostly prevail in internal oxides rather than being incorporated into the scale.

If the Al and/or Si contents are increased to approximately 0.5–1%, the internal silica and/or alumina precipitation can change into protective, external scale formation or formation of a silica sub-layer in direct contact with the inner part of the chromia scale. In some cases this leads to spallation of the chromia-based oxide scale.

The sub-scale silica and especially the external alumina scale formation is accompanied by a strong increase in oxidation resistance (Malkow et al. 1998). The external alumina formation depends on Cr-content, concentrations of minor alloying additions and component surface treatment. The latter effect was, for example, clearly demonstrated during oxidation of an 18Cr–1Al steel (DIN designation 1.4742) at 800°C in air (Buchkremer et al. 1996; de Haart et al. 2001). Diamond polishing prior to oxidation appeared to promote external alumina formation (Malkow et al. 1998). It is important to mention that the exact surface scale

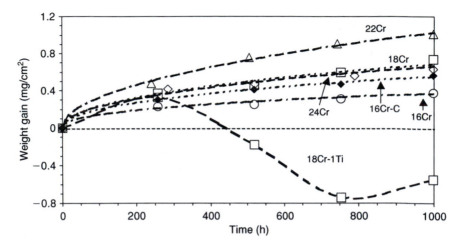

Figure 19.6   Weight changes of selected, commercially available ferritic steels of
different chromium contents (cf. Table 19.1) during discontinuous oxidation at 800°C
in air. Data illustrate that scale growth rates and spallation resistance are not solely
determined by Cr content but are substantially affected by commonly available minor
alloying additions.

composition on the 18Cr–1Al steel is strongly time dependant (Figure 19.7). After
an exposure times of 25 h at 800°C the scale on the polished sample surface is
Fe- and Cr-rich and longer exposure times of around 100 h are required for the
external alumina scale to become stabilized (Malkow et al. 1997, 1998).

A change of the test atmosphere from air to an anode gas-simulating H_2/H_2O
mixture has similar effects for high-Cr ferritic steels as described for Cr-based
alloys. In high-Cr alloys the morphology of the chromia- and spinel-rich surface
layers (Sakai et al. in press) is slightly modified and the adherence of the scale is
improved (Malkow et al. 1998; Horita et al. 2002). In alloys with substantially
lower Cr contents than approximately 20%, the anode-side environments can sup-
press the formation of oxides with low thermodynamic stability, especially in the
transient oxidation stage (Malkow et al. 1998). This may result in a type and morphology of external and internal oxide formation which completely differs from
that occurring during air or oxygen exposure (Malkow et al. 1998). The effect can
be even more pronounced if the anode-side gas contains carbonaceous gas species
such as CO and/or CH_4. As described for the Cr-based alloys, carbon transfer
from the gas to the alloy can occur, however, in the case of ferritic steels this does
generally not result in a continuous sub-scale carbide layer, but in the formation
of finely-distributed carbide precipitates in the alloy matrix. The formation of
these Cr-rich carbides ($Cr_{23}C_6$ and/or Cr_7C_3) depletes Cr from the ferritic matrix
in the near-surface region of the alloy, resulting in reduced availability to the surface of Cr required for protective scale formation. The presence of C-containing
gas species can thus substantially degrade the oxidation resistance of ferritic
steels, especially if the alloy Cr-content is lower than approximately 20%. In this
case the extent of C-uptake is also correlated with the protective properties of the

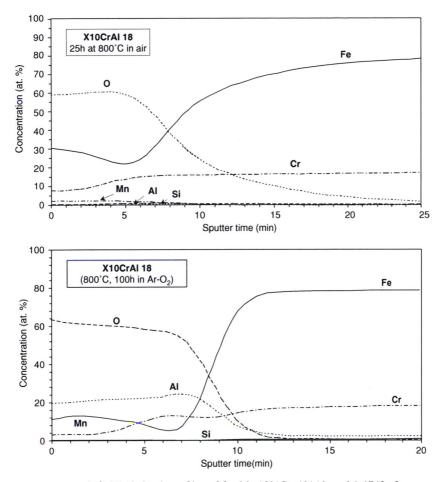

Figure 19.7   MCs$^+$-SIMS depth profiles of ferritic 18%Cr–1%Al steel 1.4742 after 25 and 100 h oxidation in Ar-O$_2$ at 800°C showing time dependence of external scale formation.

external scale, that is alloys with smaller oxidation rates generally possess better resistance against carburization.

19.10   FERRITIC STEELS DESIGNED FOR SOFC APPLICATION

The optimum steel composition for an SOFC interconnector will depend on the cell design (e.g. electrolyte or anode supported concept) and the service conditions (temperature, thermal cycling, fuel gas composition, required service life) (Dulieu et al. 1998). If an YS-ZrO$_2$ electrolyte-supported design is considered, a crucial requirement for the interconnector is a low thermal expansion coefficient of approximately 10.10^{-6}/K. Using binary Fe–Cr alloys with Cr contents commonly present in ferritic steels (see Table 19.1) such a low value cannot be achieved (Malkow *et al.* 1997). Some authors, therefore, developed alloys on the basis of Fe–20% Cr with large amounts of refractory elements such as Mo or W (Ueda and Taimatsu 2000; Ueda *et al.* 2000) to decrease the CTE. The oxidation resistance of these alloys could be improved by adding small amounts of REs. However, it has to be proved that the scaling rates are sufficiently low for the materials to be used in SOFCs with the envisaged high application temperature of 1000°C.

Substantially lower service temperatures (between 600 and 800°C) seem to be more appropriate for ferritic steels if they are to be used as interconnectors. In this temperature range, for example, ceria should be used as an electrolyte (Steele 2000) or an anode-supported cell design, allowing an extremely thin YS-ZrO$_2$ electrolyte to be applied (Buchkremer *et al.* 1996; de Haart *et al.* 2001). The choice of such concepts would require interconnectors with CTEs in the range $11–12 \times 10^{-6}$/K. Such CTE-values can be achieved by ferritic steels with Cr contents of approximately 20%.

The suitability for interconnector applications of model steels with Cr contents of 16–25% were studied mainly aiming at the effect of various RE additions to achieve an optimum combination of low-scale growth rate (Figure 19.8) and excellent scale adherence (Quadakkers *et al.* 2000; Piron-Abellan *et al.* 2001). Additions of La appeared to be most beneficial, mainly because, contrary to other commonly used REs, it does not form intermetallic compounds with iron. This allows La to become evenly distributed in the alloy rather than being precipitated in the form of intermetallics on the alloy grain boundaries as is, for example, observed if Y or Ce additions (Bredesen and Kofstad 1996) are being used. The actual scale growth rate strongly depends on commonly present minor alloying additions/impurities such as Mn and Ti (Figure 19.8). Although in RE-doped

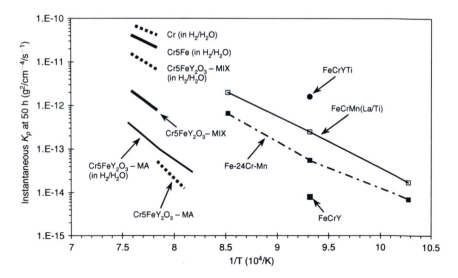

Figure 19.8 Instantaneous oxidation rates at 50 h of FeCr-RE and FeCrMn(La/Ti) alloys during isothermal oxidation at various temperatures compared with data of various Cr alloys. If not otherwise indicated, data relate to exposures in Ar-20%O$_2$ (compiled from Quadakkers *et al.* 1996; Hänsel *et al.* 1998).

steels very protective behaviour can be obtained if the Cr-content is only 16–18%, a substantially higher level was chosen to obtain a larger "Cr-reservoir" to counteract the danger of increased growth rates due to presence of high water vapour contents in the anode gas (Quadakkers *et al.* 1997), or Cr-loss due to interaction with contacting materials (Huang *et al.* 2000). Also in very thin interconnects a sufficiently high Cr-reservoir is of great importance (Huczkowski *et al.* 2004a). The significance of a high reservoir if long-term SOFC operation is considered was recently demonstrated. In components with a thickness of say 0.1 mm breakaway oxidation can occur due to the scale growth induced exhaustion of the Cr- and Mn-reservoir in the bulk alloy (Figures 19.9 and 19.10).

The newly developed steels contain small amounts of Mn and Ti to obtain external spinel formation which is expected to decrease the formation of volatile Cr-species (Piron-Abellan *et al.* 2001; Gindorf *et al.* 2000). Minor Ti additions were added to obtain fine internal oxide precipitates of titania (Figure 19.11) which results in strengthening of the near-surface region, thus reducing the tendency for occurrence of surface wrinkling caused by relaxation of oxidation-related stresses during isothermal exposure and thermal cycling. The electrical conductivity of the oxide scales on these steels is generally higher than that of oxide formed on Cr and Cr-alloys (Figure 19.12).

Further indications for the development of steels which are especially designed for SOFC applications are described by other authors (Kung *et al.* 2000; Ghosh *et al.* 2001; Krumpelt *et al.* 2002). However, no detailed informations on steel composition were given. Honegger *et al.* developed ferritic materials for SOFC application with approximately 20% Cr which were produced by powder metallurgical techniques, thus allowing incorporation of oxide dispersions to achieve dispersion strengthening (Honegger *et al.* 2001).

Very recently, developments on the basis of 22% Cr-containing ferritic steels with systematic variations of minor alloying additions were described (Uehara *et al.* 2002).

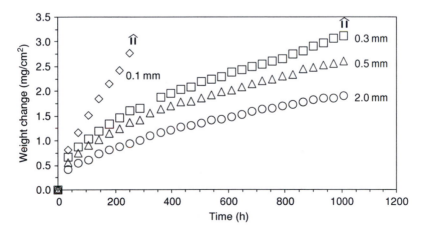

Figure 19.9   Weight change data during oxidation of FeCrMn(La/Ti) steel specimens at 900°C in air showing thickness dependence of oxidation rate and earlier occurrence of breakaway oxidation (arrows) in thinner specimens.

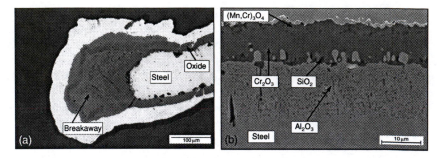

Figure 19.10   Metallographic cross sections of ferritic steel specimens of different
thickness (0.1 and 2 mm) after exposure in air at 900°C ((a) 250 h, (b) 1000 h).

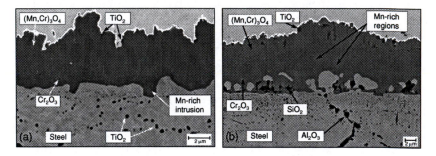

Figure 19.11   Cross sections of oxide scales on type FeCrMn (La/Ti) steels after
6000 h exposure in air at 800°C: (a) without further additions and (b) with minor
additions (≈0.1%) of manufacturing related impurities of Si and Al.

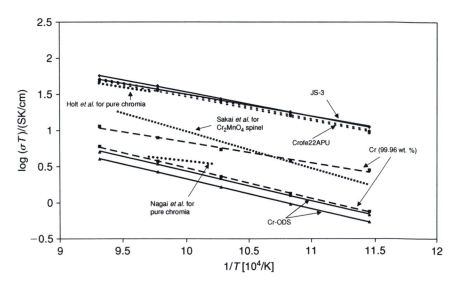

Figure 19.12   Temperature dependence of oxide conductivity for FeCrMn (La/Ti) type
ferritic steels (JS3 and Crofer 22APU) as well as Cr and Cr-based ODS alloy
during exposure at 800°C in air, compared with literature data for bulk chromia and
Cr/Mn-spinel (see Figure 19.5 as well as Huczkowski (2004); Sakai et al. (2005)).

19.11   INTERACTION OF INTERCONNECT WITH CATHODE SIDE CONTACT MATERIALS

Most studies concerning compatibility of metallic interconnectors with electrodes or contact materials relate to the cathode side of the cell. Here the interconnector is in most designs in direct contact with La-based perowskites. Due to interaction of the interconnector with the perowskite, two important processes can occur, that is a change of the oxide composition on the interconnector surface and a change in the alloy composition in the surface near region, mainly due to Cr transport into the perowskite (Tietz *et al.* 1999; Uehara *et al.* 2002). These processes can change the contact resistance, as well as the oxidation resistance of the interconnector (Tietz *et al.* 1999; Teller *et al.* 2001). With perowskite type materials commonly used in SOFCs, formation of spinel layers (e.g. $CoCr_2O_4$, $MnCr_2O_4$) is frequently observed at the interface between surface oxide film and perowskite (Quadakkers *et al.* 1993; Urbanek *et al.* 1996; Batfalsky *et al.* 1999; Tietz *et al.* 1999; Teller *et al.* 2001).

The changes in interface composition can affect the contact resistance in a positive as well as a negative way. The relatively pure chromia layers formed on Cr-based (ODS) alloys are frequently reported to exhibit high contact resistances which can even at 1000°C be in the range of several $\Omega \ cm^2$ (Huczkowski *et al.* 2004b), although recently substantially lower values were reported (Huczkowski 2004). Larring and Norby (2000) found extrapolated contact resistances after 10,000 h of 24 m$\Omega \ cm^2$ for LSM + LSC coatings on a Cr-ODS alloy at 900°C . It was claimed that the formation of CoMn-spinel at the interface reduced the transport of Cr from the interconnector into the perowskite. Due to the significance of this effect for cell performance, some authors tried to select the perowskite-based intermediate layer in such a way that it not only acted as the contacting material but also as a "gettering agent" for the evaporating chromium species. Other authors proposed that a dense and gas-tight, Cr-free spinel layer could be formed on top of a protective chromia scale by reaction of manganese, which outwardly diffuses through the chromia scale on the ferritic steels with contact layers containing large amounts of spinel-forming elements such as Co or Ni (Figure 19.13). The dense, Cr-free spinel layer, formed by interdiffusion, acts as barrier against vapour-phase transport of volatile chromium oxides and hydroxides (Larring and Norby 2000; Gindorf *et al.* 2001).

Several authors found that Sr, frequently present as a dopant in La-manganites, -chromites and -cobaltites, is easily transported from the cathode or contact layer in the direction of the Cr-rich or Cr-based interconnector surface (Kadowaki *et al.* 1993; Quadakkers 1996). This effect leads to the formation of compounds of the type $SrCrO_4$ and/or $Sr_3Cr_2O_8$ in the chromia layer (Figure 19.14) and was claimed to be beneficial for the electrical conductivity of the scale.

In the case of ferritic steels, the chemical compatibility between alloy and perowskite mainly depends on the relative stability of the mixed oxides (e.g. spinels) which can form due to reaction of Cr from the alloy with the B-type elements (Co, Mn, Cr, Fe, Cu etc.) in the perowskite (Malkow *et al.* 1998; Batfalski *et al.* 1999; Tietz *et al.* 1999; Teller *et al.* 2001). From this viewpoint it seems obvious that manganites, and especially chromites, provide the best compatibility with the

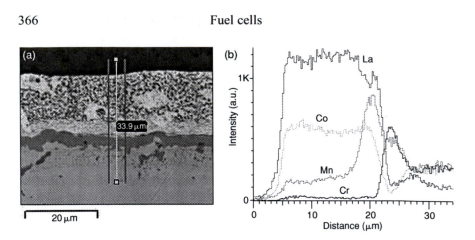

Figure 19.13   Interface between steel of the type FeCrMn (La/Ti) and LaCoO₃ contact layer after 1000 h exposure at 800°C in air: (a) SEM cross section and (b) element profiles measured along trace indicated in figure (a).

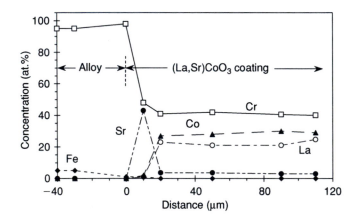

Figure 19.14   Element profiles measured near interface between Cr-ODS-alloy and (La, Sr)CoO₃ contact layer after 500 h exposure at 950°C in air (Quadakkers 1996).

ferritic steels, whereas cobaltites and ferrites tend to result in extensive reaction products with the ferritic steel, thereby decreasing its oxidation resistance (Maruyama *et al.* 1995; Malkow 1998; Hou *et al.* 1999; Godfrey *et al.* 2000; Oishi *et al.* 2000; Yoo and Dauga 2001). The effect of these interface reactions on the contact resistance cannot simply be correlated with the thickness of the reaction layer. The resistance depends not only on the type of perowskite used, but also on perowskite and oxide layer morphology. The results also seem to be affected by the way in which the perowskite is applied, that is to a bare or to a pre-oxidized metal surface. Indications were found that the dense inner chromia layer frequently present at the interface with the alloy, possesses a higher specific resistivity than the outer mixed oxide layers of the spinel type, formed either due to oxidation of the alloying elements in the steel or to reaction of the chromia layer with perowskite constituents. However, recent data on electrical conductivities of

Cr/Mn-spinel revealed values which were not too different from those of chromia (Sakai *et al.* in press). A substantial increase in spinel conductivity was found to occur if Fe was added.

19.12   INTERACTION WITH ANODE SIDE CONTACT MATERIALS

Far less information is available concerning the interaction of metallic interconnectors with Ni–ZrO$_2$–cermet or Ni-based contacting materials at the anode side. The type and extent of interaction will depend on the contacting method actually used. If, for example, a Ni-mesh is spot-welded to the interconnector, excellent contact resistance values are initially observed. In the case of a Cr-based interconnector, however, interdiffusion can lead to excessive Kirkendall-void formation, eventually resulting in deterioration of the contact. Ni-plating of the interconnector has been applied to reduce this problem. Interdiffusion is of course suppressed if the Ni-mesh is not directly welded to the interconnector. In that case the Ni will, in fact, be in contact with an ever-present surface oxide on the Cr-based alloy and the contact resistance will mainly be determined by the resistivity of the intermediate chromia layer in the low-pO_2, anode-side environment. For Cr-based alloys, as well as ferritic steels, indications were found that this resistivity is generally higher than that in cathode side gases (Teller *et al.* 2001). In the case of a ferritic steel, interdiffusion between interconnector and the spot-welded nickel wire mesh will lead to Ni transport into the steel, resulting in local austenite formation and thus to the related changes in oxidation resistance as well as thermal expansion coefficient. Vice versa, transport of Fe, Cr and other steel constituents into the wire mesh converts the latter into an alloy which will form surface oxide layers in the anode-side environment.

In C-containing anode gases a spot-welded Ni-wire or – mesh might lead to an additional problem. As described before, the metallic interconnect (Cr alloy or ferritic steel) is protected against rapid carburization by the surface oxide layer. If Ni is spot welded to the interconnector surface, it may be a pathway for carbon transfer from the environment into the alloy. Direct contacting of the metallic interconnect to a Ni-wire or -mesh may in this way deteriorate the "intrinsic" carburization resistance of the interconnect material.

REFERENCES

Atkinson, A. and Taylor, R. (1984) *Transport in Non-Stoichiometric Compounds.* (eds. G. Simkovich and G. Stubican). NATO ASI Series, Series B: Physics Col. 129. Plenum Press, New York, USA.

Badwal, S.P.S., Deller, R., Foger, K., Ramprakash, Y. and Zhang, J.P. (1997) Interaction between chromia forming alloy interconnects and air electrode of solid oxide fuel cells. *Solid State Ionics* **99**(3–4), 297–310.

Batawi, E., Plas, A., Straub, W., Honegger, K. and Diethelm, R. (1999) New cost-effective ceramic oxide phases used as protective coatings for chromium based interconnects. In: *Solid Oxide Fuel Cells* (*SOFC VI*) (eds. S.C. Singhal and M. Dokiya), pp. 767–773. The Electrochemical Society Proceedings Series. Pennington, USA.

Batfalsky, P., Buchkremer, H.P., Froning, D., Meschke, F., Nabielek, H., Steinbrech, R.W. and Tietz, F. (1999) Operation and analysis of planar SOFC stacks. In: *Proceedings of the 3rd International Fuel Cell Conference*, p. 349. New Energy and Industrial Technology Development Organization (NEDO) and Fuel Cell Development Information Centre (FCDIC), Japan.

Bongartz, K., Quadakkers, W.J., Pfeifer, J.P. and Becker, J.S. (1993) Mathematical modelling of oxide growth mechanisms measured by [18]O tracer experiments. *Surf. Sci.* **292**, 196–208.

Brady, M.P., Wright, I.G., Anderson, I.M., Sikka, V.K., Ohriner, E.K., Walls, C., Westmoreland, G. and Weaver, M.L. (2001) Ductilization of Cr via oxide dispersions. In: *15th International Plansee Symposium* (eds. G. Kneringer, P. Rodhammer and H. Wildner), pp. 108-1-108-13. Plansee AG, Reutte, Austria.

Bredesen, R. and Kofstad, P. (1996) *17th Risø International Symposium on Materials Science: High Temperature Electrochemistry, Ceramics and Metals* (eds. F. Poulsen, N. Bonanos, S. Linderoth, M. Mogenson and B. Zachau-Christiansen), pp. 187–192, Riskilde, Denmark.

Buchkremer, H.P., Diekmann, U. and Stöver, D. (1996) Components manufacturing and stack integration of an anode supported planar SOFC system. In: *Preprints 2nd European Solid Oxide Fuel Cell Forum*, 2 (ed. B. Thorstensen), pp. 221–222, Oberrohrdorf, Switzerland.

Buchkremer, H.P., Diekmann, U., de Haart, L., Kabs, H., Stimming, U. and Stöver, D. (1997) Advances in the anode supported planar SOFC technology. In: *Solid Oxide Fuel Cell (SOFC-V)* (eds. U. Stimming, S.C. Singhal, H. Tagawa and W. Lehnert), pp.160–170. The Electrochemical Society Proceedings Series, Pennington, USA.

Cotell, C., Yurek, G., Hussey, R., Mitchel, D. and Graham, M. (1987) The influence of implanted yttrium on the mechanism of growth of Cr_2O_3 on Cr. *J. Electrochem. Soc.* **134**(7), 1871–1872.

Crawford, J.A. and Vest, R.W. (1964) Electrical conductivity of single-crystal Cr_2O_3. *J. Appl. Phys.* **35**(8), 2413–2418.

Das, D., Miller, M., Nickel, H. and Hilpert, K. (1994) Chromium evaporation from SOFC interconnector alloys and degradation process by chromium transport. In: *First European Solid Oxide Fuel Cell Forum*, 2 (ed. U. Bossel), pp. 703–713, Oberrohrdorf, Switzerland.

de Haart, L., Vinke, I., Janke, A., Ringel, H. and Tietz, F. (2001) New developments in stack technology for anode substrate based SOFC. In: *Solid Oxide Fuel Cells (SOFC VII)* (eds. H. Yokokawa and S.C. Singhal), pp. 111–119. The Electrochemical Society Proceedings Series, Pennington, USA.

Dulieu, D., Cotton, J., Greiner, H., Honegger, K., Scholten, A., Christie, M., Sedguelong Th. (1998). Development of interconnect materials for intermediate temperature SOFC's. In: *3rd European Solid Oxide Fuel Cell Forum* (ed. U. Bossel), pp. 417–458, Nantes, France.

Ebbinghaus, B.B. (1993) Thermodynamics of gas phase chromium species – the chromium oxides, the chromium oxyhydrides, and volatility calculations in waste incineration processes. *Combust. Flame* **93**(1–2), 119–137.

England, D. and Virkar, A.V. (1999) Oxidation kinetics of some nickel-based superalloy foils and electronic resistance of the oxide scale formed in air. Part I. *J. Electrochem. Soc.* **146**(9), 3196–3202.

Ennis, P.J. and Quadakkers, W.J. (1987) Corrosion and creep of nickel-base alloys in steam reforming gas. In: *Preprints of the Conference High Temperature Alloys – Their Exploitable Potential* (eds. J.B. Marriot, M. Merz, J. Nihoul and J.O. Ward), pp. 465–474. Elsevier Applied Science, London, New York.

Gil, A., Penkalla, H.J., Hänsel, M., Norton, J., Köck, W. and Quadakkers, W.J. (1996) The oxidation behaviour of Cr based ODS alloys in H_2/H_2O at 1000°C. In: *IX Conference on Electron Microscopy of Solids*, pp. 441–446, Krakow-Zakopane, Poland.

Gindorf, C., Hilpert, K., Nabielek, H., Singheiser, L., Ruckdäschel, R. and Schiller, G. (2000) Chromium release from metallic interconnects with and without coatings. In:

Fourth European Solid Oxide Fuel Cell Forum, 2 (ed. J. McEvoy), pp. 845–854, Oberrohrdorf, Switzerland.

Ghosh, D., Tang, E., Perry, M., Prediger, D., Pastula, M. and Boersma, R. (2001) Status of SOFC developments at Global Thermoelectric. In: *Solid Oxide Fuel Cells* (SOFC VII) (eds. H. Yokokawa and S.C. Singhal), pp. 101–110, The Electrochemical Society Proceedings Series, Pennington, USA.

Gindorf, C., Singheiser, L. and Hilpert, K. (2001) Determination of chromium vaporisation from Fe, Cr base alloys used as interconnect in fuel cells (SOFC). *Steel Res.* **72**, 528–533.

Godfrey, B., Föger, K., Gillespie, R., Bolden, R. and Badwal, S.P.S. (2000) Planar solid oxide fuel cells: the Australian experience and outlook. *J. Power Sources* **86**, 68–73.

Golightly, F., Stott, H. and Wood, G. (1976) Influence of yttrium additions on oxide-scale adhesion to an iron–chromium–aluminium alloy. *Oxid. Metal.* **10**(3), 163–187.

Graham, M. and Hussey, R. (1995) Analytical techniques in high-temperature corrosion. *Oxid. Metal.* **44**, 339–374.

Greiner, H., Grögler, T., Köck, W. and Singer, R. (1995) Chromium based alloys for high temperature SOFC applications. In: *Solid Oxide Fuel Cells* (*SOFC IV*) (eds. M. Dokiya, O. Yamamoto, H. Tagawa and S.C. Singhal), pp. 879–888. The Electrochemical Society Proceedings Series, Pennington, USA.

Günther, C., Beie, H.-J., Greil, P. and Richter, F. (1996) Parameters influencing the long-term stability of the SOFC. In: *2nd European Solid Oxide Fuel Cell Forum*, 2 (ed. B. Thorstensen). pp. 491–502, Oberrohrdorf, Switzerland.

Hagel, W.C. and Seybolt, A.U. (1961) Cation diffusion in Cr_2O_3. *J. Electrochem. Soc.* **108**(12), 1146–1152.

Hänsel, M., Quadakkers, W.J., Singheiser, L. and Nickel, H. (1998) *Korrosions- und Kompatibilitätsstudien an Cr-Basislegierungen für den metallischen Interkonnektor der Hochtemperaturbrennstoffzelle (SOFC)*. Report No. Jül-3583, Forschungszentrum Jülich, Germany.

Hilpert, K., Das, D., Miller, M., Peck, D.H. and Weiß, R. (1996) Chromium vapor species over solid oxide fuel cell interconnect materials and their potential for degradation processes. *J. Electrochem. Soc.* **143**(11), 3642–3647.

Holt, A. and Kofstad. P. (1997) Electrical conductivity and defect structure of Mg-doped Cr_2O_3. *Solid State Ionics* **100**(3–4), 201–209.

Holt, A. and Kofstad. P. (1999) Electrical conductivity of Cr_2O_3 doped with TiO_2. *Solid State Ionics* **117**(1–2), 21–25.

Honegger, K., Plas, A., Diethelm, R. and Glatz, W. (2001) Evaluation of ferritic steel inter-connects for SOFC stacks. In: *Solid Oxide Fuel Cells* (*SOFC VII*) (eds. H. Yokokawa and S.C. Singhal), pp. 803–810. The Electrochemical Society Proceedings Series, Pennington, USA.

Horita, T., Xiong, Y., Yamaji, K., Sakai, N. and Yokokawa, H. (2002) Stability of Fe-Cr based alloys in H_2-H_2O atmosphere for SOFC Interconnector. In: *Preprints of the 5th European Solid Oxide Fuel Cell Forum*, 1 (ed. J. Huijsmans), pp. 401–408, Oberrohrdorf, Switzerland.

Hou, P. and Stringer, J. (1992) Oxide scale adhesion and impurity segregation at the scale metal interface. *Oxid. Metal.* **38**(5–6), 323–345.

Hou, P., Brown, I.G. and Stringer, J. (1991) Study of the effect of reactive-element addition by implanting metal ions in a preformed oxide layer. *Nucl. Instrum. Meth. Phys. Res. B. BeamInterac. Mater. Atoms* 59–60, Part 2, 1345–1349.

Hou, P., Huang, K. and Bakker, W. (1999) Promises and problems with metallic intercon-nects for reduced temperature solid oxide fuel cells. In: *Solid Oxide Fuel Cells* (*SOFC VI*) (eds. S.C. Singhal and M. Dokiya), pp. 737–748. The Electrochemical Society Proceedings Series, Pennington, USA.

Huang, K., Hou, P.Y. and Goodenough, J. (2000) Characterization of iron based alloy inter-connects for reduced temperature solid oxide fuel cells. *Solid State Ionics* **129**(1–4), 237–242.

Huczkowski, P., Christiansen, N., Shemet, V., Piron-Abellan, J., Singheiser, L. and Quadakkers, W.J. (2004a) Oxidation limited life times of chromia forming ferritic steels. *Mater. Corros.* **55**(11), 825–830.

Huczkowski, P., Christiansen, N., Shemet, V., Singheiser, L. and Quadakkers, W.J. (2004a) Growth rates and electrical conductivity of oxide scales on ferritic steels proposed as interconnect materials for SOFCs. In: *Proceedings 6th European Solid Oxide Fuel Cell Forum*, 3 (ed. M. Mogensen), pp. 1594–1601, Lucerne, Switzerland.

Jiang, S.P., Zhang, J.P., Apateanu, L. and Froger, K. (2001) Deposition of chromium species at Sr-doped LaMnO₃ electrodes in solid oxide fuel cells. III. Effect of air flow. *J. Electrochem. Soc.* **148**(7), 447–455.

Kadowaki, T., Shiomitsu, T., Matsuda, E., Nakagawa, H., Tsuneizumi, H. and Maruyama, T. (1993) Applicability of heat resisting alloys to the separator of the planar type solid oxide fuel cell. *Solid State Ionics* **67**, 65–69.

Köck, W., Martinz, H., Greiner, H. and Janousek, M. (1995) Development and processing of metallic Cr based materials for SOFC parts. In: *Solid Oxide Fuel Cells (SOFC IV)* (eds. M. Dokiya, O. Yamamoto, H. Tagawa and S.C. Singhal), pp. 841–849. The Electrochemical Society Proceedings Series, Pennington, USA.

Kofstad, P. (1988) *High Temperature Corrosion*. Elsevier Applied Sciences/Chapman and Hall, London, UK.

Kofstad, P. (1996) High temperature oxidation of chromium and chromia-forming alloys. In: *Proceedings of the 2nd European Solid Oxide Fuel Cell Forum*, 2 (ed. B. Thorstensen), pp. 479–490, Oberrohrdorf, Switzerland.

Kofstad, P. and Bredesen, R. (1992) High-temperature corrosion in SOFC environments. *Solid State Ionics* **52**, 69–75.

Krumpelt, M., Ralph, J., Cruse, T. and Bae, J.M. (2002) Materials for low temperature SOFCs. In: *Proceedings of the 5th European Solid Oxide Fuel Cell Forum*, 1 (ed. J. Huijsmans), pp. 215–224, Oberrohrdorf, Switzerland.

Kung, S., Cal, T., Moris, T., Barringer, E., Elangovan, S. and Hartvigsen, J. (2000) Performance of metallic interconnect in solid oxide fuel cells. In: *Proceedings of the Fuel Cell Seminar*, pp. 585–588, Portland, OR, USA.

Lai, G. (1990) *High Temperature Corrosion of Engineering Alloys*. ASM International, Materials Park, OH, USA.

Larring, Y. and Norby, T. (2000) Spinel and perovskite functional layers between Plansee metallic interconnect (Cr-5 wt% Fe-1 wt% Y₂O₃) and ceramic (La $_{0.85}$ Sr $_{0.15)0.91}$ MnO₃ cathode materials for solid oxide fuel cells. *J. Electrochem. Soc.* **147**(9), 3251–3256.

Linderoth, S. (1998) Oxidation of Cr94Fe5(Y₂O₃) at high temperatures. *High Temp. Mat. Process.* **17**(4), 217–222.

Linderoth, S., Hendriksen, P., Mogensen, M. and Langvad, N. (1996) Investigations of metallic alloys for use as interconnects in solid oxide fuel cell stacks. *J. Mat. Sci.* **31**(19), 5077–5082.

Liu, H., Stack, M. M. and Lyon, S. B. (1998) Reactive element effects on the ionic trans-port processes in Cr₂O₃ scales. *Solid State Ionics* **109**(3–4), 247–257.

Malkow, T., v.d. Crone, U., Laptev, A.M., Koppitz, T., Breuer, U. and Quadakkers, W.J. (1997) Thermal expansion characteristics and corrosion behaviour of ferritic steels for SOFC interconnects. In: *Solid Oxide Fuel Cells (SOFC V)* (eds. U. Stimming, S.C. Singhal, H. Tagawa and W. Lehnert), pp. 1245–1253. The Electrochemical Society Proceedings Series, Pennington, USA.

Malkow, Th., Quadakkers, W.J., Singheiser, L. and Nickel, H. (1998) *Untersuchungen zum Langzeitverhalten von metallischen Interkonnektor Werstoffen der Hochtemperatur-Brennstoffzelle (SOFC) im Hinblick auf die Kompatibilität mit kathodenseitigen Kontaktschichten*. Report No. Jül-3589, Forschungszentrum Jülich, Germany.

Maruyama, T., Ioue, T. and Nagata, K. (1995) Electrical conductivity of SrCrO₄ and Sr₃Cr₂O₈ at elevated temperatures in relation to the highly conductive chromia scale formed on an alloy separator in SOFC. In: *Solid Oxide Fuel Cells (SOFC IV)* (eds. M. Dokiya, O. Yamamoto, H. Tagawa and S.C. Singhal), pp. 889–894. The Electrochemical Society Proceedings Series, Pennington, USA.

Matsuzaki, Y. and Yasuda, I. (2000) Electrochemical properties of an SOFC cathode in contact with a chromium-containing alloy separator. *Solid State Ionics* **132**, 271–278.

Minh, N.Q. (1993) Ceramic fuel-cells. *J. Am. Ceram. Soc.* **176**(3), 563–588.

Nagai, H. and Ohbayashi, K. (1989) Effect of TIO_2 on the sintering and the electrical-conductivity of Cr_2O_3. *J. Am. Ceram. Soc.* **72**(3), 400–403.

Nagai, H., Fujikawa, T. and Shoji, K.-I. (1983) Electrical conductivity of Cr_2O_3 doped with La_2O_3, Y_2O_3 and NiO. *Trans. Jpn. Inst Metal.* **24**(8), 581–588.

Oishi, N., Namikawa, T. and Yamazaki, Y. (2000) Oxidation behaviour of an La-coated chromia-forming alloy and the electrical property of oxide scales. *Surf. Coat. Tech.* **132**(1), 58–64.

Park, J.H. and Natesan, K. (1990). Electronic transport in thermally grown Cr_2O_3. *Oxid. Metal.* **33**(1/2), 31–54.

Pint, B. (1996) Experimental observations in support of the dynamic-segregation theory to explain the reactive-element effect. *Oxid. Metal.* **45**(1–2), 1–37.

Piron-Abellan, J., Shemet, V., Tietz, F., Singheiser, L. and Quadakkers, W.J. (2001) Ferritic steel interconnect for reduced temperature SOFC. In: *Solid Oxide Fuel Cells (SOFC VII)* (eds. H. Yokokawa and S.C. Singhal), pp. 811–819. The Electrochemical Society Proceedings Series, Pennington, USA.

Quadakkers, W.J. (1990) Growth mechanisms of oxide scales on ODS alloys in the temperature range 1000–1100°C. *Werkst. Korros.* **41**, 659–668.

Quadakkers, W.J. (1993) Oxidation of ODS alloys. *J. Phys. IV*, **3**(1), 177–186.

Quadakkers, W.J. (1995a) Bipolare platte mit selektiver Beschichtung. *German Patent*, DE 195 47 699 C2, 20.12.

Quadakkers, W.J. (1995b) *German Patent*, DE 195 46 614 C2, 13.12.

Quadakkers, W.J. and Singheiser, L. (2001) Practical aspects of the reactive element effect. *Mat. Sci. Forum* **77**, 369–372.

Quadakkers, W.J., Holzbrecher, H., Briefs, K.G. and Beske, H. (1988) The effect of yttria dispersions on the growth mechanisms and morphology of chromia and alumina scales. In: *European College on the Role of Active Elements in the Oxidation Behaviour of High Temperature Alloys* (ed. E. Lang), pp. 155–173. European Communities, Petten, The Netherlands.

Quadakkers, W.J., Mallener, W., Grübmeier, H. and Wallura, E. (1993) Corrosion and compatibility studies on metallic interconnector materials for SOFCs. In: *Preprints of the 5th IEA Workshop SOFC; Materials, Process Engineering and Electrochemistry*, pp. 87–99, Forschungszentrum Jülich, Germany.

Quadakkers, W.J., Greiner, H. and Köck, W. (1994) Metals and alloys for high temperature SOFC applications. In: *First European Solid Oxide Fuel Cell Forum*, 2 (ed. U. Bossel), pp. 525–541, Oberrohrdorf, Switzerland.

Quadakkers, W.J., Baumanns, F., Nickel, H. (1995) Metalische bipolare Platte für HT-Brennstoffzellen und Verfahren zur Herstellung derselben. *German Patent* DE 44 10 711, 7.9.

Quadakkers, W.J., Greiner, H., Köck, W., Buchkremer, H.P., Hilpert, K. and Stöver, D. (1996) The chromium based metallic bipolar plate- fabrication, corrosion and Cr evaporation. In: *2nd European Solid Oxide Fuel Cell Forum*, 2 (ed. B. Thorstensen), pp. 297–306, Oberrohrdorf, Switzerland.

Quadakkers, W.J., Greiner, H., Hänsel, M., Pattaniak, A., Khanna, A.S. and Mallener, W. (1996) Compatibility of perovskite contact layers between cathode and metallic interconnector plates of SOFCs. *Solid State Ionics* **91**, 55–67.

Quadakkers, W.J., Norton, J.F., Penkalla H.J., Breuer, U., Gil, A., Rieck, T. and Hänsel, M. (1996) SNMS and TEM-studies concerning the oxidation of Cr-based ODS alloys in SOFC relevant environments. In: *3rd International Conference on 'Microscopy of Oxidation' Cambridge* (eds. S.B. Newcomb and J.A. Little), pp. 221–230. The Institute of Materials, USA.

Quadakkers, W.J., Thiele, M., Ennis, P.J., Teichmann, H. and Schwarz, W. (1997) Application limits of ferritic and austenitic materials in steam and simulated combustion gas of advanced fossil fuelled power plants. In: *Proceedings of the European Federation of Corrosion* II, pp. 35–40, Trondheim, Norway.

Quadakkers, W.J., Hänsel, M. and Riech, T. (1998) Carburization of Cr-based ODS alloys in SOFC relevant environments. *Materials and Corrosion* **49**, 252–257.

Quadakkers, W.J., Malkow, T., Piron-Abellan, J., Flesch, U., Shemet, V. and Singheiser, L. (2000) Suitability of ferritic steels for application as construction material for SOFC interconnects. In: *Proceedings of the Fourth European Solid Oxide Fuel Cell Forum*, 2 (ed. J. McEvoy), pp. 827–836, Oberrohrdorf, Switzerland.

Rahmel, A. and Tobolski, J. (1965) Einfluss von Wasserdampf und Kohlendioxid auf die Oxidation von Eisen in Sauerstoff bei hohen Temperaturen. *Corros. Sci.* **5**, 333–342.

Sabioni, A.C.S., Huntz, A.M., Millot, F. and Monty, C. (1992) Self-diffusion in Cr_2O_3. II: oxygen diffusion in single crystals. *Philos. Mag. A* **66**, 351–360.

Sakai, N., Horita, T., Ping Xiong, Y., Yamaji, K., Kishimoto, H., Brito, M., Yokokawa, H. and Maruyama, T. (2005) Structure and transport property of manganese–chromium–iron-oxide as a main compound in oxide scales of alloy interconnects for SOFCs. *Solid State Ionics* **16**, 681–686.

Schmidt, H., Brückner, B. and Fischer, K. (1995) Interfacial functional layers between the metallic bipolar plate and the ceramic electrodes in the high temperature solid oxide fuel cell. In: *Solid Oxide Fuel Cells (SOFC IV)* (eds. M. Dokiya, O. Yamamoto, H. Tagawa and S.C. Singhal), pp. 869–878. The Electrochemical Society Proceedings Series, Pennington, USA.

Schütze, M. (1997) *Protective Oxide Scales and Their Breakdown* (ed. D.R. Holmes). Wiley, New York, USA.

Singhal, S.C. (1997) Recent progress in tubular solid oxide fuel cell technology. In: *Solid Oxide Fuel Cell (SOFC-V)* (eds. U. Stimming, S.C. Singhal, H. Tagawa and W. Lehnert), pp. 37–44. The Electrochemical Society Proceedings Series, Pennington, USA.

Sockel, H.G., Saal, B. and Heilmaier, M. (1988) Determination of the grain boundary diffusion coefficient of oxygen in Cr_2O_3. *Surf. Interface Anal.* **12**, 531–533.

Steele, B.C.H. (2000) Materials for IT-SOFC stacks 35 years R&D: the inevitability of gradualness. *Solid State Ionics* **134**, 3–20.

Stolten, D., Späh, R. and Schamm, R. (1997) Status of SOFC development at Daimler-Bentz/Dornier. In: *Solid Oxide Fuel Cell (SOFC-V)* (eds. U. Stimming, S.C. Singhal, H. Tagawa and W. Lehnert), pp. 88–93. The Electrochemical Society Proceedings Series, Pennington, USA.

Stöver, D., Diekmann, U., Flesch, U., Kabs, H., Quadakkers, W.J., Tietz, F. and Vinke, I. (1999) Recent developments in anode supported thin film SOFC at Research Centre Jülich. In: *Solid Oxide Fuel Cells (SOFC VI)* (eds. S.C. Singhal and M. Dokiya), pp. 812–821. The Electrochemical Society Proceedings Series, Pennington, USA.

Stöver, D., Buchkremer, H.P., Tietz, F. and Menzler, N.H (2002) Trends in processing of SOFC components. In: *Proceedings of the 5th European Solid Oxide Fuel Cell Forum*, 1 (ed. J. Huijsmans), pp. 1–9, Oberrohrdorf, Switzerland.

Teller, O., Meulenberg, W., Tietz, F., Wessel, E. and Quadakkers, W.J. (2001) Improved material combinations for stacking of solid oxide fuel cells. In: *Solid Oxide Fuel Cells (SOFC VII)* (eds. H. Yokokawa and S.C. Singhal), pp. 895–903. The Electrochemical Society Proceedings Series, Pennington, USA.

Thierfelder, W., Greiner, H. and Köck, W. (1997) High-temperature corrosion behaviour of chromium based alloys for high temperature SOFC. In: *Solid Oxide Fuel Cells (SOFC V)* (eds. U. Stimming, S.C. Singhal, H. Tagawa and W. Lehnert), pp. 1306–1315. The Electrochemical Society Proceedings Series, Pennington, USA.

Tietz, F., Simwonis, D., Batfalsky, P., Diekmann, U. and Stöver, D. (1999) Degradation phenomena during operation of solid oxide fuel cells. In: *Proceedings of the 12th IEA Workshop on SOFCs: Materials and Mechanisms* (ed. K. Nisancioglu), pp. 3–11. International Energy Agency, Trondheim, Norway.

Ueda, M. and Taimatsu, H. (2000) Thermal expansivity and high-temperature oxidation resistance of Fe–Cr–W alloys developed for a metallic separator of SOFC. In: *Preprints of the Fourth European Solid Oxide Fuel Cell Forum*, 2 (ed. J. McEvoy), pp. 837–844, Oberrohrdorf, Switzerland.

Ueda, M., Kadowaki, M. and Taimatsu, H. (2000) Effect of La and Ce on the reactivity of Fe–18 mass%Cr–7 mass%W alloy with $La_{0.85}Sr_{0.15}MnO_3$. *Mat. Trans. JIM* **41**(2), 317–322.

Uehara, T., Ohno, T. and Toji, A. (2002) Development of ferritic Fe–Cr alloy for SOFC separator. In: *Proceedings of the 5th European Solid Oxide Fuel Cell Forum*, 1 (ed. J. Huijsmans), pp. 281–288, Oberrohrdorf, Switzerland.

Urbanek, J., Miller, M., Schmidt, H. and Hilpert, K. (1996) Reduction of the chromium vaporization from the metallic interconnect by perovkite coatings. In: *Proceedings of the 2nd European Solid Oxide Fuel Cell Forum*, 2 (ed. B. Thorstensen), pp. 503–512, Oberrohrdorf, Switzerland.

v.d. Crone, U., Hänsel, M., Quadakkers, W.J. and Vaßen, R. (1997) Oxidation behaviour of mechanically alloyed chromium based alloys. *J. Anal. Chem.* **358**, 230–232.

Whittle, D. and Stringer, J. (1980) Improvements in high-temperature oxidation resistance by additions of reactive elements or oxide dispersions. *Phil. Trans. Roy. Soc. London A* **295**(1413), 309–326.

Yamamoto, O. (2000) Solid oxide fuel cells: fundamental aspects and prospects. *Electrochim. Acta*, **45**, 2423–2445.

Yang, Z., Weil, K.S., Paxton, D.M. and Stevenson, J.W. (2003) Selection and evaluation of heat-resistant alloys for SOFC interconnect applications. *J. Electrochem. Soc.* **150**(9), 1188–1201.

Yasuda, I., Baba, Y., Ogiwara, T., Yakabe, H. and Matsuzaki, Y. (2001) Development of anode supported SOFC for reduced temperature operation. In: *Solid Oxide Fuel Cells* (*SOFC VII*) (ed. H. Yokokawa and S.C. Singhal), pp. 131–139. The Electrochemical Society Proceedings Series, Pennington, USA.

Yoo, Y. and Dauga, M. (2001) The effect of protective layers formed by electrophoretic deposition on oxidation and performance of metallic interconnects. In: *Solid Oxide Fuel Cells* (*SOFC VII*) (ed. H. Yokokawa and S.C. Singhal), p. 837. The Electrochemical Society Proceedings Series, Pennington, USA.

Section IIID: Microbiological fuel cells

20

Microbial fuel cells: performances and perspectives

K. Rabaey, G. Lissens and W. Verstraete

20.1 INTRODUCTION

Bacteria gain energy by transferring electrons from an electron donor, such as glucose or acetate, to an electron acceptor, such as oxygen. The larger the difference in potential between donor and acceptor, the larger the energetic gain for the bacterium, and generally the higher the growth yield. In a microbial fuel cell (MFC), bacteria do not directly transfer their electrons to their characteristic terminal electron acceptor, but these electrons are diverted towards an electrode (i.e. an anode). The electrons are subsequently conducted over a resistance or power user towards a cathode and thus, bacterial energy is directly converted to electrical energy (Rao *et al.* 1976). To close the cycle, protons migrate through a proton-exchange membrane (PEM) (Figure 20.1).

Simplified, an MFC can be compared to a car battery, in which the anode contains bacteria as catalyst to liberate electrons. MFCs could be applied for the treatment of liquid waste streams. Indeed, the conventional aerobic sewage treatment faces several problems:

- *Energy cost*: aeration and recirculation require considerable amounts of electricity.
- *Footprint*: the sequence of different reactors requires a considerable surface.
- *Investment costs*: two technologies need to be installed, that is, a system to handle the water as such, and a complex additional system to separate and handle the bio-solids.
- *Sludge treatment costs*: the large amount of sludge formed during aerobic conversion represents a significant part of overall treatment costs.
- *Reliability*: nitrification and phosphorus removal are sensitive processes (Verstraete and Philips 1998), sedimentation of the sludge is often a critical factor (Seka *et al.* 2001).
- *Nutrient removal efficiency*: influents containing distorted ratios (chemical oxygen demand/nitrogen/phosphorus (COD/N/P)) often show limited potential for adequate nutrient removal.

As an alternative strategy, anaerobic digestion of wastewater with concomitant methane formation has been developed (Van Lier *et al.* 2001), but up till now this strategy is not widely applied, mainly due to the low COD concentrations in the domestic sewage, which require a reactor dimension that is economically not feasible.

MFCs might provide an answer to several of the problems traditional wastewater treatment faces, as further discussed in Section 20.4. Moreover, their configuration

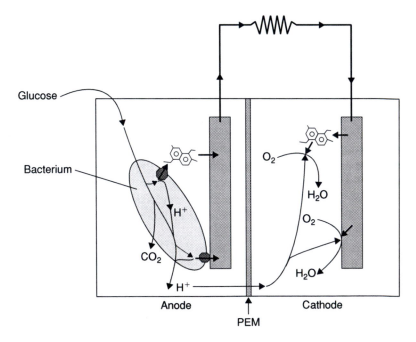

Figure 20.1   Schematic drawing of an MFC. Electron transfer can occur through either soluble redox mediators ( ) or membrane bound complexes ( ).

opens up possibilities for other applications, such as small-scale power plants, batteries and sensors. However, for an extended period since their conception, MFCs have remained a scientific curiosity because of their limited efficiency. Several bottlenecks still exist in this technology, each needing further attention. The growing success of catalytic fuel cells has given rise to a wide variety of technological solutions for most electrochemical bottlenecks. Therefore, the MFC research focuses mainly onto the optimization of the biological compartment at this moment.

In this chapter, the technological aspects of MFCs will be assessed in relation to the current bottlenecks and advantages of the concept. Furthermore, some perspectives of MFCs in other domains of interest will be discussed.

20.2 CONVERSION OF ORGANIC MATTER TO BIOFUELS

20.2.1 Stoichiometric reactions

By means of microbiological fermentation, a whole range of biofuels and related bioproducts can be produced from organic biomass present in solid waste and wastewater (Chapters 7–13). Considering glucose as the principal building block of biomass, one can compare the stoichiometric reactions for the production of bio-ethanol, biogas (CO_2 and CH_4) and hydrogen gas with the overall reaction taking place in a MFC:

$$C_6H_{12}O_6 \xrightarrow{\text{Bio-ethanol}} 2\ C_2H_5OH + 2\ CO_2$$

$$C_6H_{12}O_6 \xrightarrow{\text{Biogas}} 3\ CH_4 + 3\ CO_2$$

$$C_6H_{12}O_6 + 6\ H_2O \xrightarrow{\text{Hydrogen gas}} 12\ H_2 + 6\ CO_2$$

$$C_6H_{12}O_6 + 6\ O_2 \xrightarrow{\text{Microbial fuel cell}} 6\ H_2O + 6\ CO_2$$

Evidently, the total energetic yield that can be reached in practice for a particular biofuel is smaller than the theoretical yield of the above-mentioned stoichiometric reactions. Based on mass weight, the theoretical and practical yield for bio-ethanol production is clearly the highest (Table 20.1). The practical yields shown in Table 20.1 are based on average conversion efficiencies, namely 50–80% for methane (Liu *et al.* 2002), 50–90% for bio-ethanol (Bjerre *et al.* 1996) and 15–33% for hydrogen gas (Logan 2004).

Table 20.1   Energetic yields of conventional biofuels (after Lay *et al.* 1999; Logan 2004).

	Methane	Bio-ethanol	Hydrogen gas
Theoretical yield (g/g glucose)	0.27	0.51	0.13
Yield in practice (g/g glucose)	0.14–0.22	0.3–0.46	0.02–0.04
Energetic yield (kJ/g glucose)	7.8–12.2	8–12.3	2.4–4.9

Based on the calorific energy content of biogas (55.5 kJ/g methane), bio-ethanol (26.7 kJ/g ethanol) and hydrogen gas (122 kJ/g hydrogen), the energetic yields for each biofuel from glucose (15.6 kJ/g glucose) can be computed (Table 20.1). Despite the much higher calorific energy content of hydrogen gas, the practical energetic yield is about three times lower for hydrogen gas production from glucose (Table 20.1) than for methane (biogas) and bio-ethanol production.

In a conventional biofuel process, the biochemically bound energy contained in biomass and organic waste is recovered under the form of an energy carrier that can be either a liquid fuel (bio-ethanol) or a gas (biogas and hydrogen gas). In order to effectively use the energy from the carrier liquid or gas, further physico-chemical incineration processes after the fermentation step are needed to release the energy (e.g. for electricity production) from the carrier.

In an MFC, the biochemical energy contained in the organic matter is directly converted into electricity in what can be called a microbiologically mediated "incineration" reaction. This implies that the overall conversion efficiencies that can be reached are potentially higher for MFCs compared to other biofuel processes.

20.2.2   Conversion efficiencies: benchmarking technology

20.2.2.1   Bio-ethanol

Compared to biogas, bio-ethanol has the main advantage that it is a liquid fuel with high performance in internal combustion engines. Despite its decreasing production cost over recent years due to the development of more efficient pre-treatment methods and enzymes, bio-ethanol production even from refined materials (e.g. sugar cane or pure starch) is still not cost competitive with gasoline production. The main cost involved in the process is the need for high enzyme loadings during the simultaneous saccharification and fermentation process (Sheehan and Himmel 2001). The distillations costs have already decreased considerably and will probably further decrease in the future because of increased heat recovery (Seeman 2003).

The overall conversion efficiency from organic waste to electricity via ethanol is low (10–25%) and involves the use of an energy consuming distillation step. As a result, bio-ethanol is expected to primordially play a role in the powering of combustion engines by blending with gasoline.

20.2.2.2   Biogas

The biogas production process is a well-established technology and can undoubtedly handle the widest variety of all kinds of heterogeneous wastes. The energy content of biogas, with a calorific value of 17–25 MJ/m^3 (about 10% lower than natural gas), is usually recovered by burning the biogas in diesel-stationary engines or dual-fuel engines with a thermal efficiency in the range of 30–38% (Henham and Makkar 1998; Bilcan *et al.* 2003). Smaller engines (<200 kW-h) generally have an electrical conversion efficiency of less than 25% while larger engines (>600 kW-h) can reach efficiencies up to 38%. In case hot water and steam from the engine's exhaust and cooling systems are recovered, an overall conversion efficiency of more than 80% can be reached of which 35% under the form of electricity and about 50% under the form of heat (Ross *et al.* 1996). Overall, from every kilogram of biodegradable waste approximately 1 kW-h electricity and 2 kW-h heat are produced during biogas production.

20.2.2.3 Hydrogen gas

The volumetrically based calorific yield of hydrogen gas is much lower than the calorific biogas yield because of the nearly 10-fold lower gas density of hydrogen gas compared to methane gas. Furthermore, the microbiological conversion of organics into hydrogen is relatively inefficient (15–30% efficiency) and the produced hydrogen gas is rapidly consumed by hydrogen consuming bacteria (Segers and Verstraete 1985; Liessens and Verstraete 1986). This also causes the need for much larger storage volumes of hydrogen gas compared to other biofuels. However, contrary to the low biogas to electricity conversion rate, the main advantage of hydrogen is that it can be very efficiently (90%) converted into electricity by means of chemical fuel cells. Therefore, hydrogen production from wastewater is a feasible option because the market value of hydrogen gas is nearly 20 times higher per kilogram of hydrogen gas compared to methane gas (Logan 2004) and because of the "negative cost" of waste and wastewater treatment. In all cases, the overall efficiency from waste to electricity via hydrogen gas remains relatively low (1 kW-h for every kilogram of biodegradable waste) mainly due to the low efficiency in the fermentation step.

20.2.2.4 MFC

Based on the calorific content of glucose, an MFC can theoretically (at 100% efficiency during fermentation) deliver 3 kW-h for every kilogram of organic matter (dry weight) in one single fermentative step (instead of 1 kW-h of electricity and 2 kW-h of heat per kilogram in hydrogen and biogas production by employing several process steps). This means that during fermentation in MFCs, hardly any energy is released under the form of external heat, and that all the biochemical energy in the waste can potentially be converted into electricity. Recent work shows that depending on the experimental conditions employed, overall efficiencies up to 80% can be reached in practice (see Section 20.3).

20.2.3 Current bottlenecks in conventional biofuel production

20.2.3.1 Biodegradation kinetics

Hydrolysis of complex polymers (mainly lignocellulose, proteins and lipids) by hydrolytic organisms is the first and one of the most important steps in the bioconversion of organic waste (Table 20.2).

Despite the hydrolytic capabilities of many anaerobic bacteria by secretion of exocellular enzymes or attachment of the bacteria to the solid substrate, this step is considered to be most rate limiting in the fermentation of organic matter and is mostly also yield-limiting in biological conversion processes (Mata-Alvarez *et al.* 2000). This is particularly true for lignocellulosic substrates due to their inherent rigid structures. As a result, the hydrolysis of most organic matter and hence their

Table 20.2   First order kinetic rates for hydrolysis (Mata-Alvarez *et al.* 2000).

Component	Hydrolysis rates k values (d^{-1})
Lipids	0.005–0.010
Proteins	0.015–0.075
Carbohydrates	0.025–0.200
Food wastes	0.4
Solid wastes	0.006–0.078 (pH from 4 to 10)
Biowaste components	0.03–0.15 (20°C), 0.24–0.47 (40°C)

bioconversion into biofuels is normally never complete (Sanders *et al.* 2000). The lower efficiencies of anaerobic digestion in practice are a result of the presence of slowly biodegradable compounds (e.g. lignocellulose derived from plant waste) in mixed waste streams (see Chapters 7 and 10). Efficiencies higher than 80% can be reached with high-quality biomasses such as cellulolytic crops or carbohydrate-rich wastewaters from the food industry.

20.2.3.2   Gas treatment

One of the major drawbacks of biogas and hydrogen gas production compared to MFCs is the need for further gas treatment. In fact, biogas can contain considerable amounts of H_2S (up to 2–3%) whereas hydrogen production may generate gas contaminated with volatile fatty acids (VFAs) and other compounds that can interfere with catalysts in chemical fuel cells. Both energy carriers therefore need further treatment before they can be valorized.

20.2.3.3   Overall conversion efficiencies

The overall conversion efficiency for waste-to-electricity conversion in conventional biofuel processes is generally of the order of 10–25% (see Section 20.2.2). This is largely due to high thermal losses in biofuel burners after fermentation.

20.2.3.4   Post-treatment of solid residues

In conventional biofuel processes, such as anaerobic digestion, post-treatment of the digested material is required for nutrient removal and further COD removal from the wastewater. These post-treatments require extra energy and thus lower the efficiency of the overall biofuel process.

20.3   MFC: STATE OF THE ART

20.3.1   MFC reactor configurations

Four basic reactor configurations using bacteria can be distinguished in biofuel cells:

- *The uncoupled bioreactor MFC*: microbiological hydrogen or methane production (with subsequent reforming) in a separate bioreactor followed by a

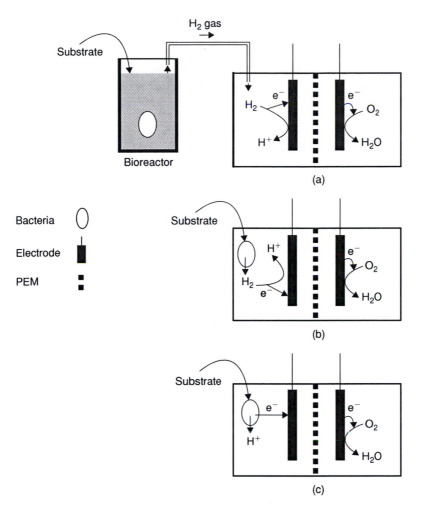

Figure 20.2   Three typical MFC configurations. (a) Uncoupled bioreactor MFC
in a bioreactor followed by a chemical fuel cell. (b) Integrated bioreactor MFC.
(c) MFC with bacteria–anode interaction (after Katz *et al.* 2003).

chemical fuel cell (generally high-temperature solid oxide fuel cells, SOFC) to
convert hydrogen gas into electricity (Figure 20.2a).

- *The integrated bioreactor MFC*: microbiological hydrogen production and
  hydrogen to electricity conversion in a single cell (Figure 20.2b).
- *The MFC with direct electron transfer*: microbiological electricity generation
  and direct electron transfer to the anode (Figure 20.2c).
- *The MFC with mediated electron transfer*: microbiological electricity generation
  and electron transport towards the anode by means of electron shuttling mediators.

The first two systems cannot be regarded as a real MFC since they use a chemical
fuel cell to oxidize the generated gas.

In an uncoupled MFC, a biofuel (e.g. hydrogen or methane gas) is produced in a bioreactor prior to a chemical fuel cell. One of the major drawbacks of this configuration is that the conventional bottlenecks of both separate systems remain valid, mainly the low efficiencies of biological substrate to hydrogen conversion and the requirement of high fuel cell temperatures to obtain sufficient hydrogen oxidation. Moreover, the produced biofuel gas is not sufficiently pure for direct use in a fuel cell, due to contamination caused by CO, H_2S and (poly)siloxanes (see Part IV). The second configuration (Figure 20.2b) is basically the same as the first one, apart from the fact that the fermentation (mostly to hydrogen gas) now takes place in the fuel cell itself (Cooney et al. 1996, Scholz and Schroder 2003). This type of MFC commonly involves the use of expensive catalysts such as Pt to provide the most optimal environment for the conversion of hydrogen gas to electricity. The third configuration (Figure 20.2c) is what can be called the real MFC whereby electrons are transferred from the bacteria to the anode without the intervention of an intermediate fermentation product (Tender et al. 2002).

Since micro-organisms act as a catalyst in the transfer of electrons from the substrate to the anode, the selection of a high performing microbial consortium (either pure or mixed culture) is of crucial importance in the "real" MFCs. The electron transfer from the bacterium to the anode can proceed in a direct way from the bacterial membrane to the anode surface or indirectly by means of a mediator.

Alternatively, biofuel cells exist that use enzymes, which are immobilized on the anode surface. These fuel cells cannot be regarded as MFCs *sensu strictu*. Due to the high costs involved in enzyme production, enzymatic fuel cells are only suitable for miniaturized small-scale applications. As a result, dimensions and corresponding power outputs are significantly smaller. Commonly used enzymes are glucose-oxidase and dehydrogenases (Palmore et al. 1998; Katz et al. 1999; Pizzariello et al. 2002; Kim et al. 2003).

20.3.2   Mediating compounds

Both transfer through bacterial contact with the electrode and through soluble shuttles can be regarded as mediated. In the first case, a bacterial redox enzyme, immobilized in the cell wall, provides the electron transfer. Examples of such bacteria are *Geobacter sulfurreducens* (Bond and Lovley 2003) and *Rhodoferax ferrireducens* (Chaudhuri and Lovley 2003). These organisms have been reported to form biofilms onto the electrode surface.

When a soluble mediator is used, the electrons are shuttled by mediator molecules between the redox enzyme(s) of the bacteria and the electrode surface, thereby facilitating electron transport (Figure 20.3) (Roller et al. 1984). Mediators are typically redox molecules (e.g. ubiquinones, dyes and metal complexes) that can form reversible redox couples, are stable in both oxidized and reduced form, are not biologically degraded and are not toxic towards the microbial consortium (Willner et al. 1998; Park and Zeikus 2003).

Although externally supplied mediators can considerably enhance the electron transfer efficiency, they are generally too expensive to apply in practice and they can exhibit toxic effects or can be degraded over longer time periods (Delaney et al.

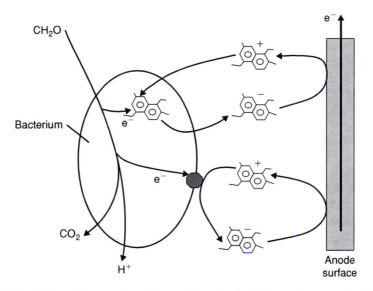

Figure 20.3 Schematic overview of the working principle of a mediator in an MFC. Soluble redox mediators ( ) can be transported into the bacteria for reduction, or reduction can occur at the bacterial surface through membrane bound complexes ( ).

Table 20.3 Overview of metal-reducing bacteria applied in MFCs.

Organism	References
S. putrefaciens	Bond and Lovley (2003); Kim *et al.* (1999a, b, 2002); Schroder *et al.* (2003)
G. sulfurreducens	Bond and Lovley (2003)
G. metallireducens	Bond *et al.* (2002)
D. acetoxidans	Bond *et al.* (2002)
R. ferrireducens	Chaudhuri and Lovley (2003)

1984; Gil *et al.* 2003). Alternatively, it has been demonstrated that electrochemically active bacteria enriched in an MFC can produce mediator compounds *in situ* (Rabaey *et al.* 2004). Relative to the feasibility of practical applications of MFCs, this overview only focuses on non-mediator-amended MFCs.

20.3.3 Micro-organisms

20.3.3.1 Axenic bacterial cultures

Some bacterial species in MFCs, of which metal-reducing bacteria are the most important, have recently been reported to directly transfer electrons to the anode (Table 20.3). Metal-reducing bacteria are commonly found in sediments, where they use insoluble electron acceptors such as Fe (III) and Mn (IV). Specific cytochromes at the outside of the cell membrane make *Shewanella putrefaciens* electrochemically active (Kim *et al.* 2002) in case it is grown under anaerobic conditions. The same holds true for bacteria of the family *Geobacteraceae*, which have been reported to form a biofilm on the anode surface in MFCs and to transfer the electrons from acetate with high efficiency (Bond and Lovley 2003).

Rhodoferax species isolated from an anoxic sediment were able to efficiently transfer electrons to a graphite anode using glucose as a sole carbon source (Chaudhuri and Lovley 2003). Remarkably, this bacterium is the first reported strain that can completely mineralize glucose to CO_2 while concomitantly generating electricity at 90% efficiency (Chaudhuri and Lovley 2003). In terms of performance, current densities in the order of 0.2–0.6 mA and a total power density of 1–17 mW/m^2 graphite surface have been reported for *S. putrefaciens*, *G. sulfurreducens* and *R. ferrireducens* at conventional (woven) graphite electrodes (Kim *et al.* 2002; Bond and Lovley 2003; Chaudhuri and Lovley 2003) (Table 20.4). However, in case woven graphite in the *Rhodoferax* study was replaced by highly porous graphite electrodes, the current and power output was increased up to 74 mA/m^2 and 33 mW/m^2, respectively.

Although these bacteria generally show high electron transfer efficiency, they have a slow growth rate, a high substrate specificity (mostly acetate or lactate) and relatively low energy transfer efficiency compared to mixed cultures (Rabaey *et al.* 2003). Furthermore, the use of a pure culture implies a continuous risk of contamination of the MFCs with undesired bacteria.

20.3.3.2   *Mixed bacterial cultures*

MFCs that make use of mixed bacterial cultures have some important advantages over MFCs driven by axenic cultures: higher resistance against process disturbances, higher substrate consumption rates, smaller substrate specificity and higher power output (Rabaey *et al.* 2004, 2005). Mostly, the electrochemically active mixed cultures are enriched either from sediment (both marine and lake sediment) (Tender *et al.* 2002; Bond and Lovley 2003) or activated sludge from wastewater treatment plants (WWTPs) (Park *et al.* 2001; Tender *et al.* 2002; Bond and Lovley 2003; Lee *et al.* 2003; Kim *et al.* 2004; Rabaey *et al.* 2004). By means of molecular analysis, electrochemically active species of *Geobacteraceae*, *Desulfuromonas*, *Alcaligenes faecalis*, *Enterococcus faecium*, *Pseudomonas aeruginosa*, *Proteobacteria*, *Clostridia*, *Bacteroides* and *Aeromonas* species were detected in the before-mentioned studies. Most remarkably, the study of Kim *et al.* (2004) also showed the presence of nitrogen-fixing bacteria (e.g. *Azoarcus* and *Azospirillum*) amongst the electrochemically active bacterial populations. The study of Rabaey *et al.* (2004) showed that by starting from methanogenic sludge and by continuously harvesting the anodic populations over a 5-month period using glucose as carbon source, an electrochemically active consortium can be obtained that mainly consists of facultative anaerobic bacteria (e.g. *Alcaligenes*, *Enterococcus* and *Pseudomonas* species). In this particular study, very high glucose-to-power efficiencies could be reached in the order of 80% (Table 20.4) (Rabaey *et al.* 2004).

So far, only the study of Tender *et al.* (2002), Kim *et al.* (2004) and Liu *et al.* (2004) reported the use of complex substrates, respectively, sediments on the seafloor and organics present in wastewater (latter two) to generate electricity in

Table 20.4 Overview of the power output delivered by MFCs without mediator addition.

Micro-organism	Substrate	Anode	Current (mA)	Power (mW/m^2)	Reference
S. putrefaciens	Lactate	Woven graphite	0.031	0.19	Kim *et al.* 2002
G. sulfurreducens	Acetate	Graphite	0.40	13	Bond and Lovley 2003
R. ferrireducens	Glucose	Graphite	0.2	8	Chaudhuri and Lovley 2003
	Glucose	Woven graphite	0.57	17.4	Chaudhuri and Lovley 2003
	Glucose	Porous graphite	74	33	Chaudhuri and Lovley 2003
Mixed seawater culture	Acetate	Graphite	0.23	10	Bond *et al.* 2002
	Sulfide/acetate	Graphite	60	32	Tender *et al.* 2002
Mixed active sludge culture	Acetate	Graphite	5	–	Lee *et al.* 2003
	Glucose	Graphite	30	3600	Rabaey *et al.* 2003
	sewage	Woven graphite	0.2	8	Kim *et al.* 2004

MFCs. Table 20.4 summarizes the most relevant facts of the main MFC studies reported so far. It should be remarked that in order to maximize the power output, experiments with varying external resistance should be performed. However, from the data derived from the references given in Table 20.4, it appears that this optimization was not always performed.

To estimate the power per unit surface to putative power output per unit reactor volume, one can take into account that at present some 100–500 m^2 of anode surface can be installed per m^3 anodic reactor volume. Hence, the state of the art power supply ranges from approximately 1 to 1800 W per m^3 anode reactor volume installed.

To render the anode more susceptible for receiving electrons from the bacteria, electrochemically active compounds can be incorporated in the electrode material. This approach has been investigated by Park and Zeikus (2003), who incorporated dyes such as neutral red and metals such as Mn^{4+} into Fe^{3+} containing graphite anodes. In this way, the main disadvantages of mediators in solution, namely toxicity and degradation, can thus be circumvented since the mediator is not released from the electrode material and thus has a longer life time. Moreover, bacteria are still able to form a biofilm on the modified anode surface.

20.4 ADVANTAGES OF MFC

MFCs present several advantages, both operational and functional, in comparison to the currently used technologies for generation of energy out of organic matter or treatment of waste streams.

20.4.1   Generation of energy out of biowaste or organic matter

This feature is certainly the most "green" aspect of MFCs. Electricity is being generated in a direct way from biowastes and organic matter. This energy can be used for operation of the waste treatment plant, or sold to the energy market. Furthermore, the generated current can be used to produce hydrogen gas. Since waste flows are often variable, a temporary storage of the energy in the form of hydrogen, as a buffer, can be desirable.

20.4.2   Direct conversion of substrate energy to electricity

As previously reported, in anaerobic processes the yield of high value electrical energy is only one third of the input energy during the thermal combustion of the biogas. While recuperation of energy can be obtained by heat exchange, the overall effective yield still remains of the order of 30%.

An MFC has no substantial intermediary processes. This means that if the efficiency of the MFC equals best 30% conversion, it is the most efficient biological electricity producing process at this moment. However, this power comes at potentials of approximately 0.5 V per biofuel cell. Hence, significant amounts of MFCs will be needed, either in stack or separated in series, in order to reach acceptable voltages. If this is not possible, transformation will be needed, entailing additional investments and an energy loss of approximately 5%.

Another important aspect is the fact that a fuel cell does not – as is the case for a conventional battery – need to be charged during several hours before being operational, but can operate within a very short time after feeding, unless the starvation period before use was too long to sustain active biomass.

20.4.3   Sludge production

In an aerobic bioconversion process, the growth yield is generally estimated to be about 0.4 g CDW/g COD removed where CDW is cell dry weight. Due to the harvesting of electrical energy, the bacterial growth yield in an MFC is considerably lower than the yield of an aerobic process. The actual growth yield, however, depends on several parameters:

–   The amount of electrons diverted towards the anode and the energy they represent. This energy (J) can be calculated as $E = P \times t = V \times I \times t$, with E energy (J), P power (W), t time (s), V voltage (V) and I current (A).
–   *The amount of substrate converted to VFAs that are not further converted*: often, the effluent of an MFC still contains considerable amounts of VFA (Rabaey *et al.* 2003) that need removal during post-treatment. These VFA represent an additional loss in energetic efficiency, and will yield additional sludge if the effluent is post-treated aerobically.
–   *The amount of hydrogen formed*: per equivalent of bio-hydrogen formed, two equivalents of electrons are not diverted to the anode. Hydrogen formation appears to be in competition with anodic electron transfer (Rabaey *et al.*

2004). Normally, bio-hydrogen formation can be completely suppressed in MFCs, indicating that the anode is a more energetically feasible electron acceptor than protons, due to a higher overall redox potential.

20.4.4 Omission of gas treatment

Generally, off-gases of anaerobic processes contain high concentrations of nitrogen gas, hydrogen sulfide and carbon dioxide next to the desired hydrogen or methane gas. The off-gases of MFCs have generally no economic value, since the energy contained in the substrate was prior directed towards the anode. The separation has been done by the bacteria, draining off the energy of the compounds towards the anode in the form of electrons. The gas generated by the anode compartment can hence be discharged, provided that no large quantities of H$_2$S or other odorous compounds are present in the gas, and no aerosols with undesired bacteria are liberated into the environment.

20.4.5 Aeration

The cathode can be installed as a "membrane electrode assembly", in which the cathode is precipitated on top of the PEM or conductive support, and is exposed to the open air. This omits the necessity for aeration, thereby largely decreasing electricity costs. However, from a technical point of view, several aspects need additional consideration when open-air cathodes are used.

First, the cathode needs to remain sufficiently moist to ensure electrical contact. Preliminary experiments by Rabaey *et al.* (unpublished data) indicated that the water formation through oxygen reduction is insufficient to keep the cathode moist. Therefore, a water recirculation needs to be installed, possibly entailing extra energy costs. Secondly, the cathode needs to contain a non-soluble redox mediator to efficiently transfer the electrons from the electrode to oxygen. Generally, platinum is being used as a catalyst, at concentrations up to 40% w/w, representing considerable costs. However, new catalysts need to be developed, which would compensate their possible lower efficiency by a significantly reduced cost and higher sustainability.

20.5 BOTTLENECKS OF MFC

20.5.1 Anode compartment: potential losses decrease MFC voltage

The direct transfer of electrons from the bacteria towards the electrode is hampered by so-called overpotentials, which can be described as transfer resistances (Larminie and Dicks 2000). These overpotentials lower the potential attained over the MFC and hence decrease the energetic efficiency (Figure. 20.4). The losses can be categorized as activation overpotentials, ohmic losses and concentration polarization.

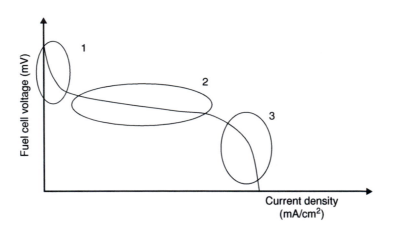

Figure 20.4   The voltage over a fuel cell as a function of the current density. Three zones can be distinguished, in which losses mainly occur through one form of the overpotentials: 1. zone of activation losses, 2. zone of ohmic losses, 3. zone of concentration polarization (after Larminie and Dicks 2000).

20.5.1.1 Activation overpotential

Either oxidizing a compound at the anode surface, or reducing a compound at the bacterial surface or in the bacterial interior surface requires certain energy to activate the oxidation reaction. This activation incurs a potential loss generally described as activation overpotential (Larminie and Dicks 2000).

The extent of the activation losses can be described by the Tafel equation (Bockris *et al.* 2000):

$$V = A \log\left(\frac{i}{i_0}\right)$$

with *V* as the overvoltage (V), *A* the correlation coefficient determined by the reaction, *i* the current density (A) and i_0 the limit current density, the current density at which the overpotential is zero. Clearly, the activation overpotential depends on the current density and a correlation coefficient.

Activation potentials are mainly important in the zone of 0 to 1 mA/cm^2 and are hence most important for MFCs. The possible solutions mentioned below for activation losses are of crucial importance for MFCs. In order to decrease this activation overpotential, several solutions are possible for MFCs. They can be grouped as follows.

Increasing the operation temperature   For catalytic fuel cells, raising the temperature up to several hundreds degrees celsius (°C) can decrease the activation losses significantly. However, the temperature range of the current MFCs does not allow the temperature increases needed (above 70°C). Only for indirect systems, for example hydrogen production in a bioreactor with subsequent catalytic oxidation in a fuel cell, higher temperatures in the fuel cell are possible.

Decreasing the activation losses at the electrode surface
– *Addition of a catalyst to the electrode, which decreases the activation energy*:
  inserting platinum into the electrode has been previously described to be successful. However, Pt is being polluted rapidly by the bacterial suspension. This
  can be counteracted by covering the electrode with a more robust conductive
  layer (Schroder *et al.* 2003). Also organic compounds can be immobilized onto
  the electrode surface, such as neutral red (Park and Zeikus 2000). It has to be
  noted that neutral red can also be used as soluble redox mediator (so it can
  function both as mediator and catalyst). In addition, mixtures of the catalyst
  into the electrode matrix have been described, such as manganese oxide in
  graphite-kaolin electrode matrices (Park and Zeikus 2003). The addition of
  these catalysts can significantly increase the MFC power output on short term.
– *Increasing the roughness and specific surface of the electrode*: an increase of
  the specific surface decreases the current density and hence the activation
  losses. Carbon felt mats can be a solution for short-term operations (experiments of several hours). However, on the longer term (days and weeks), clog-
  ging has been repeatedly observed (Rabaey *et al.* unpublished data), actually
  transforming the electrode surface to a plate-shaped surface. Larger structures like graphite granules are more suited for the purposes of MFC anode
  for several reasons (see further).

Decreasing the activation losses at the bacteria   A redox mediator can be added
to the anode compartment to facilitate effective electron transfer. Two effects are
caused by mediators in solution: (1) increasing physical transport of electrons
from the bacteria to the bulk solution; (2) by selecting a suitable mediator, one
also decreases the activation losses occurring at the level of the bacterial cells.

Hence, a good redox mediator needs to meet several requirements (Park and
Zeikus 2002; Katz *et al.* 2003):

– The oxidized state of the mediator should easily access the membrane bound
  redox active complexes to be reduced.
– The oxidized state of the mediator should easily penetrate the bacterial membrane to reach the reductive species inside the bacterium.
– The redox potential of the mediator should fit the potential of the reductive
  metabolite (the mediator potential should be positive enough to provide fast
  electron transfer from the metabolite, but it should not be so positive as to
  prevent significant loss of potential).
– Neither oxidation state of the mediator should interfere with other metabolic
  processes (should not inhibit them or be decomposed by them).
– The reduced state of the mediator should easily escape from the cell through
  the bacterial membrane.
– Both oxidation states of the mediator should be chemically stable in the electrolyte solution. They should be well soluble and they should not adsorb on
  the bacterial cells or electrode surface.
– The electrochemical kinetics of the oxidation process of the mediator-
  reduced state at the electrode should be fast (electrochemically reversible).

20.5.1.2   Ohmic overpotential

Ohmic losses are caused by electrical resistances of the electrodes, the electrolyte and the membrane. They are important at higher current levels if the sum of the resistances is limited. However, the resistance over the MFC can increase rapidly by suboptimal contacts or limited conductivity and turbulence of the electrolyte. A resistance of only 15 Ω causes a potential loss of 150 mV at a current of 10 mA, a loss not to be neglected.

20.5.1.3   Concentration polarization

Concentration polarization occurs when, due to the large oxidative force of the anode, compounds are being oxidized faster at the anode than they can be transported to the surface. However, this is a problem only occurring at higher current densities and therefore not important for MFCs. For example, only in cases where diffusion is seriously hampered by a thick non-conductive biofilm, concentration polarization would be a problem.

20.5.2   Transport of charge and ions in the electrolyte: the influence of turbulence

For a good operation of the MFC, both protons and electrons need to migrate from the anode to the cathode, be it through another medium, at the highest possible rate. Diffusion is not sufficient to reach acceptable levels of current and cell potential. Moreover, lacking proton transport could decrease the pH of the anode to undesired levels for bacteria (e.g. below 5).

Therefore, turbulent conditions need to be introduced to the anode and the cathode. Both shaking and recirculation of the influent can be applied to solve this problem. Both methods cost energy, depleting the overall efficiency of the MFC system. It has to be noted that in a realistic MFC system, this will be the largest energy cost by far.

20.5.3   Membrane resistance, selectivity and O_2 permeability

The selection of a membrane, separating anode and cathode, represents a choice between two opposing interests:

- *High selectivity*: the higher the selectivity for protons, the better the biofuel cell will operate and the lower the resistance of the membrane.
- *High stability*: membranes need to be robust in a colloidal and nutrient-rich environment, which the bacterial suspension generally is.

Nafion™ has been widely used as PEM for fuel cells and MFCs (Park and Zeikus 2000; Bond and Lovley 2003, Liu *et al.* 2004), and has the large advantage of being very selective for protons. However, this membrane contains sulfonic acid groups, that are binding with ammonia present in the bacterial solution. Hence, at this moment, this membrane type scores high for selectivity but low for stability.

A second approach is the use of a more general cation exchange membrane (CEM), such as Ultrex™ (Rabaey *et al.* 2003). This type of membrane has a larger resistance and is less selective but generally shows larger stability. These membranes have been reported to perform adequately for over 3 months (Rabaey *et al.* 2005).

One can also omit the selective membrane and use a rigid carbon electrode as separator and open-air cathode (Liu and Logan 2004). This approach provides satisfying power output, although conversion efficiencies are lowered due to the increased oxygen diffusion.

The approach used will depend on the application foreseen. When clean influents will be used, the first approach can be a possibility if no ammonia is present. Probably, Nafion™ will be the material of choice for battery MFCs. When wastewater will be used, approach two and three are more feasible due to their robustness. The difference between the two approaches can be made on the basis of two determinants: efficiency (approach two provides higher coulombic and energetic efficiency) and cost (approach two is more expensive).

20.5.4   The structure of the anode

Several anode configurations are currently used, such as plate-shaped plain graphite (Bond and Lovley 2003), graphite felt (Bond and Lovley 2003, Liu *et al.* 2004) and graphite granules (Rabaey *et al.* 2005). These three possible configurations have already shown reasonable to good results, depending on the application desired. However, fine tuning still needs to be performed, based on several requirements:

- Free flow of influent and effluent through the electrode matrix, in order to supply sufficient feed, and to remove biodegradation products.
- Adequate surface for growth of a biofilm, which will perform most of the electron transfer.
- Sufficient support and conductive surface.
- Sufficient turbulence for adequate proton diffusion towards the membrane and the cathode.

20.5.5   The role of the cathode performance

As in the anode compartment, losses occur in the cathode compartment due to overpotentials. Although small current densities flow through the electrode surface, these losses need consideration. To decrease the activation overpotential, catalysts need to be added to the electrode, or a suitable mediator is needed to transfer the electrons from the cathode to oxygen. Generally, Pt is used as a catalyst in the electrode (Schroder *et al.* 2003), at concentrations up to 45% w/w, entailing a considerable cost. Another option is adding $K_3Fe(CN)_6$ to the liquid catholyte (Park *et al.* 2000), which is aerated. The use of Pt enables an open-air cathode, decreasing aeration cost (operation cost – opex), but the investment cost (capex) is considerable. Less expensive catalysts are currently being developed, which may render the open-air cathodes more feasible in terms of capex.

Theory dictates that, when the reactant concentration (oxygen in this case) is increased, this will increase the transfer efficiency (Larminie and Dicks 2000). However, in most cases, the use of pure oxygen seems not feasible for practical design reasons.

20.5.6   Upscaling problems

Several aspects needed for an efficient MFC are hampering upscaling:

–   The influent needs to reach the whole anode matrix sufficiently.
–   Protons need rapid diffusion towards the membrane.
–   Sufficient electrical contact needs to be established between bacteria in suspension and the anode.
–   Sufficient voltage needs to be reached over the MFC to have a useful power.
–   Instatement of an aeration device should be avoided.

To overcome these bottlenecks, several techniques can be copied from conventional fuel cells, such as the use of fuel cell stacks, constructed from several plate-shaped fuel cells (Larminie and Dicks 2000).

20.6   FUTURE APPLICATIONS OF MFC

MFCs have still some way to go to reach large scale commercialization. The largest reactors reported thus far had an anode internal volume of 0.388 l (Liu *et al.* 2004). Two types of implementation into practice offer perspectives within a reasonable time scale: MFCs for treatment of wastewater and MFCs converting renewable biomass in batteries.

20.6.1   MFCs for wastewater treatment

Consider a conventional WWTP designed for 30,000 IE, receiving a daily influent flow of 5400 m^3. At a biodegradable chemical oxygen demand (bCOD) concentration of 500 mg/l, this represents an influx of organic matter of 2700 kg dry weight per day. The amount of sludge formed at a nominal yield of 0.4 g CDW per g bCOD converted (Verstraete and Van Vaerenberg 1986) will be 1080 kg per day. This needs to be disposed off at a cost which can rise up to €500 per ton dry matter (Weemaes and Verstraete 2001). The other costs contained in the operational costs are the aeration costs and pump costs for recirculation and processing.

If an MFC is used with an open-air cathode, no aeration is needed. The putative energy of the input organic matter amounts to 8950 kW-h per day. The costs for sludge processing will be lower, since no aerobic cell yields can be attained. For methanogenesis, the cell yield is about 0.05 g CDW/g substrate; for MFC the yield can be estimated somewhere in between aerobic and methanogenic conditions. At an energetic efficiency of 35%, which should be attainable on large scale, approximately 3150 kW-h per day of useful energy will be produced. This comparison does not take into account the capital cost of both systems. However, if the capital cost is of the same order, the comparison illustrates a significant difference in operational costs. Hence, if large scale MFCs can be built at an acceptable price, this will be a viable technology.

20.6.2   Renewable biomass conversion

Conventional batteries have several drawbacks:

– They need to be charged for several hours in order to be used.
– They are environmentally unfriendly due to the heavy metal content.
– One needs electricity to power them up.

Therefore, much research is performed with respect to the development of fuel cells that can use hydrogen, methanol or ethanol to power portable applications. At this moment, several fuel cell types based on hydrogen and methanol work appropriately, and applications already exist, for example in portable computers. However, the question can be raised whether this energy generation is really sustainable. Furthermore, the customer may not like to carry hydrogen gas (even captured within a metal hydride matrix) or methanol.

MFCs can operate on a large variety of substrates that are readily available, even in any supermarket. Substrates such as plain sugar and starch are easy to store, contain more energy than any other feed type per unit of volume, and are easy to dose. Furthermore, they have a more "green" image than, for example, methanol. Moreover, MFCs can be developed that are environmentally friendly in terms of material composition.

If the development of MFCs leads to a product that has a reasonable (read: usable) power output per unit of MFC volume, it will be a viable product. A customer will accept a larger battery, and a larger feeding tank, provided the feeding is easy to perform and has a green and safe label.

20.7   EMERGING OPPORTUNITIES

20.7.1   Body fluid batteries

In the future, the amount of low-power devices implanted in the human body will significantly expand. These devices need long-term, stable power provision. To provide this power, an MFC can be used. Two possibilities exist: enzymatic and MFCs. In enzymatic fuel cells, the potential difference is created by the use of two electrodes with different enzymatic reactions, creating a potential difference based on the reaction redox potential (Pizzariello *et al.* 2002).

Also micro-organisms can be used, such as *Saccharomyces cerevisiae* (Chiao *et al.* 2003). Micro-organisms have the advantage of providing a more long-term stability than enzymes immobilized onto a surface.

20.7.2   Electricity from photosynthesis

Plants produce, as a product of photosynthesis, sucrose and other low-molecular carbohydrates. These carbohydrates are transported through the stem during certain periods of the year. These plant saps can be harvested and the possibility exists to use this flow as a feed for MFCs that would be installed in a stationary way. Some plant saps such as maple syrup have already been tested and yielded a conversion efficiency of up to 50% (Rabaey *et al.* 2005). The minerals remaining after the MFC could be recycled to the trees or the plants. This way, a forest could function as a continuous, green power providing system that converts directly light energy into electricity.

20.7.3   Bio-sensors

Bacteria shows lower metabolic activity when inhibited by toxic compounds. This will cause a lower electron transfer towards an electrode. Bio-sensors could be constructed, in which bacteria are immobilized onto an electrode and protected behind a membrane. If a toxic component diffuses through the membrane, this can be measured by the change in potential over the sensor. Such sensors could be extremely useful as indicators of toxicants in rivers, at the entrance of WWTPs, to detect pollution or illegal dumping, or to perform research on polluted sites (Meyer *et al.* 2002, Chang *et al.* 2004).

20.7.4   Sediment electricity

MFCs can be used to generate electricity based on the potential difference generated by bacteria between a sediment and the aqueous phase above it (Tender *et al.* 2002). Two anode reactions appear to occur: *oxidation of sulfide* present in the sediment, which is formed through bacterial oxidation of organic carbon, and *oxidation of organic matter* by bacteria growing onto the anode. The potential for energy generation from the seafloor is large, although the accessibility will often pose a problem.

20.8   CONCLUSIONS

MFCs do hold promise towards sustainable energy generation in the near future. Many bottlenecks yet exist, which pose a challenge that will take a multidisciplinary approach and intensive research. As shown in this chapter, a multitude of solutions exists to answer these (mainly) technical problems. Aside from the main goal of bioreactors producing electricity in an elegant way, several serendipities of this technology emerge, applicable within a broad range of life sciences.

REFERENCES

Bilcan, A., Le Corre, O. and Delebarre, A. (2003) Thermal efficiency and environmental performances of a biogas-diesel stationary engine. *Environ. Technol.* **24**, 1165–1173.

Bjerre, A.B., Olesen, A.B., Fernqvist, T., Plöger, A. and Schmidt, A.S. (1996) Pretreatment of wheat straw using combined wet oxidation and alkaline hydrolysis resulting in convertible cellulose and hemicellulose. *Biotechnol. Bioeng.* **49**, 568–577.

Bockris, J.O'M., Reddy, A.K.N. and Gamboa-Aldeco, M. (2000) *Modern Electrochemistry: Fundamentals of Electrodics*, 2nd edn. Kluwer Academic Publishers, New York.

Bond, D.R. and Lovley, D.R. (2003) Electricity production by *Geobacter sulfurreducens* attached to electrodes. *Appl. Environ. Microbiol.* **69**, 1548–1555.

Bond, D.R., Holmes, D.E., Tender, L.M. and Lovley, D.R. (2002) Electrode-reducing microorganisms that harvest energy from marine sediments. *Science* **295**, 483–485.

Chaudhuri, S.K. and Lovley, D.R. (2003) Electricity generation by direct oxidation of glucose in mediatorless microbial fuel cells. *Nat. Biotechnol.* **21**, 1229–1232.

Chang, I.S., Jang, J.K., Gil, G.C., Kim, M., Kim, H.J., Cho, B.W. and Kim, B.H. (2004) Continuous determination of biochemical oxygen demand using microbial fuel cell type biosensor. *Biosens. Bioelectron.* **19**, 607–613.

Chiao, M., Kien, B., Lam, Y.C.S. and Lin, L. (2003) A miniaturized microbial fuel cell as a power source for MEMS. Berkeley: UCB, 2 p.

Cooney, M.J., Roschi, E., Marison, I.W., Comninellis, C. and von Stockar, U. (1996) Physiologic studies with the sulfate-reducing bacterium *Desulfovibrio desulfuricans*: evaluation for use in a biofuel cell. *Enzyme Microb. Technol.* **18**, 358–365.

Delaney, G.M., Bennetto, H.P., Mason, J.R., Roller, S.D., Stirling, J.L. and Thurston, C.F. (1984) Electron-transfer coupling in microbial fuel-cells. 2. Performance of fuel-cells containing selected microorganism mediator substrate combinations. *J. Chem. Technol. Biotechnol. Biotechnology* **34**, 13–27.

Gil, G.C., Chang, I.S., Kim, B.H., Kim, M., Jang, J.K., Park, H.S. and Kim, H.J. (2003) Operational parameters affecting the performance of a mediator-less microbial fuel cell. *Biosens. Bioelectron.* **18**, 327–334.

Grommen, R. and Verstraete, W. (2002) Environmental biotechnology: the ongoing quest. *J. Biotechnol.* **98**, 113–123.

Henham, A. and Makkar, M.K. (1998) Combustion of simulated biogas in a dual-fuel diesel engine. *Energ. Convers. Manage.* **39**, 2001–2009.

Katz, E., Filanovsky, B. and Willner, I. (1999) A biofuel cell based on two immiscible solvents and glucose oxidase and microperoxidase-11 monolayer-functionalized electrodes. *New J. Chem.* **23**, 481–487.

Katz, E., Shipway, A.N. and Willner, I. (2003) Biochemical fuel cells. In: Handbook of Fuel Cells – Fundamentals, Technology and Applications (eds. W. Vielstich, H.A. Gasteiger and A. Lamm), pp. 355–381. John Wiley & Sons, New York, USA.

Kim, B.H., Ikeda, T., Park, H.S., Kim, H.J., Hyun, M.S., Kano, K., Takagi, K. and Tatsumi, H. (1999a) Electrochemical activity of an Fe(III)-reducing bacterium, *Shewanella putrefaciens* IR-1, in the presence of alternative electron acceptors. *Biotechnol. Tech.* **13**, 475–478.

Kim, B.H., Kim, H.J., Hyun, M.S. and Park, D.H. (1999b) Direct electrode reaction of Fe(III)-reducing bacterium, *Shewanella putrefaciens*. *J. Microbiol. Biotechnol.* **9**, 127–131.

Kim, H.J., Park, H.S., Hyun, M.S., Chang, I.S., Kim, M. and Kim, B.H. (2002) A mediator-less microbial fuel cell using a metal reducing bacterium, *Shewanella putrefaciens*. *Enzyme Microb. Technol.* **30**, 145–152.

Kim, H.H., Mano, N., Zhang, X.C. and Heller, A. (2003) A miniature membrane-less biofuel cell operating under physiological conditions at 0.5 V. *J. Electrochem. Soc.* **150**, A209–A213.

Kim, B.H., Park, H.S., Kim, H.J., Kim, G.T., Chang, I.S., Lee, J. and Phung, N.T. (2004) Enrichment of microbial community generating electricity using a fuel-cell-type electrochemical cell. *Appl. Microbiol. Biotechnol.* **63**, 672–681.

Larminie, J. and Dicks, A. (2000) *Fuel Cell Systems Explained*, 308 pp. John Wiley & Sons, Chichester.

Lay, J.-J.L., Lee, Y.-J. and Noike, T. (1999) Feasibility of biological hydrogen production from organic fraction of municipal solid waste. *Water Res.* **33**, 2579–2586.

Lee, J., Phung, N.T., Chang, I.S., Kim, B.H. and Sung, H.C. (2003) Use of acetate for enrichment of electrochemically active microorganisms and their 16S rDNA analyses. *FEMS Microbiol. Lett.* **223**, 185–191.

Liessens, J. and Verstraete, W. (1986) Selective inhibitors for continuous non-axenic hydrogen-production by *Rhodobacter Capsulatus*. *J. Appl. Bacteriol.* **61**, 547–557.

Liu, H. and Logan, B.E. (2004) Electricity generation using an air-cathode single chamber microbial fuel cell in the presence and absence of a proton exchange membrane. *Environ. Sci. Technol.* **38**, 4040–4046.

Liu, H.W., Walter, H.K., Vogt, G.M. and Vogt, H.S. (2002) Steam pressure disruption of municipal solid waste enhances anaerobic digestion kinetics and biogas yield. *Biotechnol. Bioeng.* **77**, 121–130.

Liu, H., Ramnarayanan, R. and Logan, B.E. (2004) Production of electricity during wastewater treatment using a single chamber microbial fuel cell. *Environ. Sci. Technol.* **38**, 2281–2285.

Logan, B.E. (2004) Extracting hydrogen electricity from renewable resources. *Environ. Sci. Technol.* **38**, 160A–167A.

Mata-Alvarez, J., Macé, S. and Llabrs, P. (2000) Anaerobic digestion of organic solid wastes: an overview of research achievements and perspectives. *Bioresour. Technol.* **74**, 3–16.

Meyer, R.L., Larsen, L.H. and Revsbech, N.P. (2002) Microscale biosensor for measurement of volatile fatty acids in anoxic environments. *Appl. Environ. Microbiol.* **68**, 1204–1210.

Palmore, G.T.R., Bertschy, H., Bergens, S.H. and Whitesides, G.M. (1998) A methanol/dioxygen biofuel cell that uses NAD(+)-dependent dehydrogenases as catalysts: application of an electro-enzymatic method to regenerate nicotinamide adenine dinucleotide at low overpotentials. *J. Electroanal. Chem.* **443**, 155–161.

Park, D.H. and Zeikus, J.G. (2000) Electricity generation in microbial fuel cells using neutral red as an electronophore. *Appl. Environ. Microbiol.* **66**, 1292–1297.

Park, D.H. and Zeikus, J.G. (2002) Impact of electrode composition on electricity generation in a single-compartment fuel cell using *Shewanella putrefaciens*. *Appl. Microbiol. Biotechnol.* **59**, 58–61.

Park, D.H. and Zeikus, J.G. (2003) Improved fuel cell and electrode designs for producing electricity from microbial degradation. *Biotechnol. Bioeng.* **81**, 348–355.

Park, D.H., Kim, S.K., Shin, I.H. and Jeong, Y.J. (2000) Electricity production in biofuel cell using modified graphite electrode with neutral red. *Biotechnol. Lett.* **22**, 1301–1304.

Park, H.S., Kim, B.H., Kim, H.S., Kim, H.J., Kim, G.T., Kim, M., Chang, I.S., Park, Y.K. and Chang, H.I. (2001) A novel electrochemically active and Fe(III)-reducing bacterium phylogenetically related to *Clostridium butyricum* isolated from a microbial fuel cell. *Anaerobe* **7**, 297–306.

Pizzariello, A., Stred'ansky M. and Miertus S. (2002) A glucose/hydrogen peroxide biofuel cell that uses oxidase and peroxidase as catalysts by composite bulk-modified bioelectrodes based on a solid binding matrix. *Bioelectrochemistry* **56**, 99–105.

Rabaey, K., Lissens, G., Siciliano, S.D. and Verstraete, W. (2003) A microbial fuel cell capable of converting glucose to electricity at high rate and efficiency. *Biotechnol. Lett.* **25**, 1531–1535.

Rabaey, K., Boon, N., Siciliano, S.D., Verhaege, M. and Verstraete, W. (2004) Biofuel cells select for microbial consortia that self-mediate electron transfer. *Appl. Environ. Microbiol.* **70**, 5373–5382.

Rabaey, K., Ossieur, W., Verhaege, M. and Verstraete, W. (2005) Continuous microbial fuel cells convert carbohydrates to electricity, *Wat. Sci. Tech.* (in Press).

Rao, J.R., Richter, G.J., Vonsturm, F. and Weidlich, E. (1976) Performance of glucose electrodes and characteristics of different biofuel cell constructions. *Bioelectrochem. Bioenerg.* **3**, 139–150.

Roller, S.D., Bennetto, H.P., Delaney, G.M., Mason, J.R., Stirling, J.L. and Thurston, C.F. (1984) Electron-transfer coupling in microbial fuel-cells. 1. Comparison of redox-mediator reduction rates and respiratory rates of bacteria. *J. Chem. Technol. Biotechnol.* **34**, 3–12.

Ross, C., Charles, C., Drake, T.J. and Walsh, J.L. (1996) *Handbook on Biogas Utilization*, 2nd edn. Muscle Shoals, AL, Southeastern Regional Biomass Energy Program, Tennessee Valley Authority.

Sanders, W.T.M., Geerink, M., Zeeman, G. and Lettinga, G. (2000) Anaerobic hydrolysis kinetics of particulate substrates. *Water Sci. Technol.* **41**(3), 17–24.

Scholz, F. and Schröder, U. (2003) Bacterial batteries. *Nat. Biotechnol.* **21**, 1151–1152.

Schröder, U., Niessen, J. and Scholz, F. (2003) A generation of microbial fuel cells with current outputs boosted by more than one order of magnitude. *Angew. Chem. Int. Edit.* **42**, 2880–2883.

Seeman, F. (2003) Energy reduction in distillation for bioethanol plants. *Int. Sugar J.* **105**, 420–427.

Segers, L. and Verstraete, W. (1985) Ammonium as an alternative nitrogen-source for hydrogen producing photobacteria. *J. Appl. Bacteriol.* **58**, 7–11.

Seka, M., Van de Wiele, T. and Verstraete, W. (2001) Feasibility of a multi-component additive for efficient control of activated sludge filamentous bulking. *Water Res.* **35**, 2995–3003.

Sheehan, J.S. and Himmel, M.E. (2001) Outlook for bioethanol production from lignocellulosic feedstocks: technology hurtles. *Agro Food Ind. Hi Tech* **12**, 54–57.

Tender, L.M., Reimers, C.E., Stecher, H.A., Holmes, D.E., Bond, D.R., Lowy, D.A., Pilobello, K., Fertig, S.J. and Lovley, D.R. (2002) Harnessing microbially generated power on the seafloor. *Nat. Biotechnol.* **20**, 821–825.

Van Lier, J.B., Tilche, A., Ahring, B.K., Macarie, H., Moletta, R., Dohanyos, M., Hulshoff, L.W., Lens, P. and Verstraete, W. (2001) New perspectives in anaerobic digestion. *Water Sci. Technol.* **43**, 1–18.

Verstraete, W. and Van Vaerenberg, E. (1986) In: *Aerobic Activated Sludge,*Biotechnology, Vol. 8, Chapter 2. VCH Verlagsgesellschaft, Weinheim.

Verstraete, W. and Philips, S. (1998) Nitrification–denitrification processes and technologies in new contexts. *Environ. Pollut.* **102**, 717–726.

Weemaes, M. and Verstraete, W. (2001) Other treatment techniques. In: '*Sludge into Biosolids*'. IWA Publishing, London.

Willner, I., Katz, E., Patolsky, F. and Buckmann, A.F. (1998) Biofuel cell based on glucose oxidase and microperoxidase-11 monolayer-functionalized electrodes. *J. Chem. Soc.-Perkin Trans.* **2**, 1817–1822.

PART FOUR: Upgrading of fermentation biofuels to fuel cell quality

21

Biofuel quality for fuel cell applications

M. Haberbauer

21.1 INTRODUCTION

If fuel cells are chosen as a tool for achieving a sustainable and clean energy future, then it is essential to know from where the prime energy hydrogen comes from. Fuel cells have very important advantages but if the energy source they transform into electricity is not sustainable, no environmental advantages will be gained. Therefore the used energy source has to be a renewable one and be as far as possible directly suitable for its energy transformation in fuel cells. If the prime energy undergoes several processes in order to adapt it to the fuel cells, the process will lead to high costs. Therefore, biofuel upgrading has to be a cost-competitive process in order to avoid a neutralization of the fuel cell and biofuel advantages.

Further, biofuels can be upgraded to vehicle fuel quality for incorporating it into the mobile fuel cell application sector (transport). The transportation sector is a very interesting emerging market for biofuels where not only new drive propulsion systems are being developed, but also the necessary fuels are needed. However, issues as the high costs for the upgrading of biofuels with conventional technologies to the demanded vehicle fuel quality or the still too expensive fuel cells slow down a widespread implementation of biofuel products in both the transportation and the electricity sector. Therefore, cost competitive and sustainable upgrading technologies are the key issues for a broad use of biofuels in different applications.

21.2   WHY BIOFUEL USAGE IN FUEL CELLS?

The prime energy carrier is currently electricity, which can be distributed easily via the electricity grid. The conversion from the original energy source to electricity always leads to a loss of part of the energy content. However, this has to be taken into account if the produced renewable energy is to reach the different end-users in a user-friendly form. Therefore, it is important to use conversion technologies that have a maximum electricity output and low losses, meaning devices with high efficiencies.

A possible alternative to conventional gas motors is the use of fuel cells. Although the overwhelming majority of existing fuel cells use natural gas as their source of hydrogen, other fuels including biofuels can be used. In combination with biofuels, the fuel cells can have substantial advantages over traditional systems. Fuel cells enable an increased electrical output out of biofuels while reducing drastically the usual particle emissions in the exhaust gas. If using a biofuel, these advantages are joined by the fact that the system would potentially reduce the CO_2 emissions. So the drive for using biofuels in fuel cells is mainly environmental.

Several European Directives encourage biofuel and/or fuel cell utilization. The White Paper for a Community Strategy and Action Plan "Energy for the Future: Renewable Sources of Energy" (Commission of the European Communities 1997) as well as the Green Paper on renewable sources of energy (Commission of the European Communities 2000) encourage the widespread and efficient use of renewable energy. The European landfill directive is of relevance as it aims at reducing the amount of disposed organic waste in landfills. One alternative way of treating part of this organic waste is anaerobic digestion (see Chapter 7), which produces the biofuel methane.

The European Commission proposed also a new European Union (EU) policy to promote the use of biofuels in the transport sector. One directive aims at promoting the use of biofuels for transport by setting biofuel sales targets for the EU Member States (Commission of the European Communities 2003). The actual directive obliges the Member States to establish a minimum of 2% by December 31, 2005 by volume of biofuels to be sold in their respective national markets. This amount is to increase to 5.75% in the year 2010. The biofuels that are taken into consideration as such are defined as: bioethanol, biodiesel, biogas, biomethanol, biohydrogen, biodimethylether, bio-ethyltertiobutylether, bio-methyltertiobutylether, synthetic biofuels and pure vegetable oil (Directive 2003/30/EC).

Biofuel and fuel cell technology contribute to the quality of life: public perceptions are changing and consumers are increasingly interested for ecological problems from the use of fossil energy sources like greenhouse gases, emissions, such as SO_2, NO_x or volatile organic compounds (VOCs) from power stations. The combination of biofuel, a CO_2 neutral energy source, as a fuel and fuel cell technology – a clean energy technology with lowest emissions – will help to improve the quality of life because the environmental conditions, which themselves to a large extent depend on air quality, strongly influence the public health. Thus, fuel cells as a clean and efficient technology will have a great acceptance in the population. Moreover, if energy of renewable sources will be competitive to other energy sources, the consumers have more freedom for choice of the energy source.

21.3 THE ROLE OF THE BIOFUEL QUALITY

A pre-condition for the use of biofuels is the elimination or at least the reduction of accompanying traces of detrimental gases in a cost-competitive way. Biofuel upgrading is therefore an essential issue for coupling biofuel–fuel cell systems. A cost-competitive treatment unit has a fundamental importance in the process integration of biofuels and fuel cell systems since the increase of the use of biofuels for power generation can be pursued only if the costs associated with using this resource are lower than the purchase of traditional fuels.

21.3.1 Biogas

The composition of biofuels varies according to the feedstock and conversion techniques used. Table 21.1 summarizes the composition of biogases from different sources.

Anaerobic digestion is an established technology for environmental protection through treatment of organic substrates (De Baere 2000; Converti *et al.* 1999). Wastewater treatment facilities all over the world use anaerobic digestion to reduce the organic content of sewage sludge. Anaerobic digestion is also being considered to treat animal wastes (Francese *et al.* 2000; Weiland 2000; Salminen and

Table 21.1   Composition of biogas from different sources (Spiegel *et al.* 1999a; Spiegel and Preston 2000; De Mes *et al.* 2003).

Component	Agricultural biogas	Sewage gas	Landfill gas
Methane	55–75%	55–65%	40–45%
CO_2	25–45%	30–40%	35–50%
N_2	0–10%	0–10%	0–20%
H_2S	0–1.5%	Up to 200 ppm	~200 ppm
Water	Saturated	Saturated	Saturated
Halogens	Trace amount	Up to 4 ppm	Always, amount depending on the landfill
Higher hydrocarbons	Trace amount	Trace amount	Up to 200 ppm

Table 21.2   Siloxane concentrations in landfill gas and sewage gas (Asten, Austria).

Siloxanes	Landfill gas (mg/m^3)	Sewage gas (mg/m^3)
Hexamethyldisiloxane L2	6.07	0.02
Hexamethylcyclotrisiloxane D3	0.49	0.04
Octamethyltrisiloxane L3	0.32	0.02
Octamethylcyclotetrasiloxane D4	12.53	0.93
Decamethylcyclopentasiloxane D5	4.73	6.03
Total	24.15	7.04

Rintala 2002). Biogas resulting from anaerobic digestion is a renewable energy source with a high potential for the reduction of greenhouse gas emissions. In addition to the main components (methane and carbon dioxide), biogas can also contain a variety of contaminants and impurities such as sulfur compounds (H_2S, mercaptans, etc.), halogens, dust, ammonia, etc.

The composition of landfill gases collected at individual sites can be found in the literature (He *et al.* 1997; Spiegel *et al.* 1999b) and, while the bulk gas compositions are similar, the levels of the trace compounds as chlorine content can vary significantly. Landfill gas is usually richer in siloxanes than sewage gas (Table 21.2). These volatile methyl siloxanes concentrations were found in landfill and sewage gas in Austria.

21.3.2   Bioethanol

Ethanol (CH_3CH_2OH) is obtained by the fermentation of the sugar components of biomass or it can also be produced from cellulosic biomass such as crop wastes (sawdust, etc.) or municipal solid waste. Cellulosic materials first need to be processed to form sugar that can then be fermented. Ethanol is a liquid fuel. Compared with biogas, ethanol is a relatively clean fuel and does not require complex clean-up processes.

21.3.3   Hydrogen

In the bioprocess for hydrogen from biomass, CO_2 is one of the products. Besides CO_2, formed in a ratio of 1:2 (CO_2:H_2), no other volatile products are expected from the dark fermentation (de Vrije and Claassen 2003) or they were not measured till yet. Regarding the photoheterotrophic H_2 production, the product gas composition consists of hydrogen as well as carbon dioxide and the composition of the product gas at the photoautotrophic H_2 production is hydrogen and oxygen. More research is necessary for the determination of contaminating trace gases in these H_2 producing processes.

21.4   THE REQUIREMENTS ON FUEL QUALITY

Fuel cell systems can utilize a variety of fuels for conversion to electricity. However, the common reductant in all systems is H_2. Unless the H_2 is supplied directly to the fuel cell system, hydrogen needs to be generated by reforming other fuels and processed to meet the requirements of a given fuel cell system.

Prior to its use in a fuel cell (as well as in other power generation systems), biofuels must be essentially free of harmful compounds like:

- particulates;
- water (although a harmless product for the fuel cell itself, it damages for example the flow controls and it becomes corrosive in combination with NH_3 and especially H_2S in the biogas);
- ammonia;
- carbon dioxide, which is only harmful for the alkaline fuel cells (AFCs);
- hydrogen sulfide;
- halogenated hydrocarbons;
- volatile methyl siloxanes.

Particulates can be removed by passing the gas through a filter pad made of stainless steel wire or through a ceramic filter pack or alternatively using cyclone separator (Crane *et al.* 1992; Alvin 1998; Environmental Agency 2002). Water removal, which is needed for the prevention of accumulation of condensate and corrosion in the fuel line, is generally carried out by refrigeration in shell and tube heat exchangers or by means of absorbent materials which physically absorb moisture (Fuel Cell Handbook 2000). Sulfur removal systems will be required in all hydrocarbon-reforming fuel cell systems because both fuel cells and steam-reforming catalysts do not tolerate sulfur (CADDET 1999). They will be necessary even for natural gas, the simplest hydrocarbon fuel. Halogen guard beds will likely be required for all fuel cell power systems as well to minimize long-term, high-temperature corrosion. Due to their complexity, biofuel clean-up technologies for removal of H_2S, halogenated hydrocarbons, siloxanes and CO_2 will be discussed in more details in the forthcoming chapters.

21.5 FUEL CELL TOLERANCES

Any type of fuel cell could, in theory, be used with biofuels (Table 21.3). However, the economic and technological implications of low- and high-temperature fuel cells are different. The need for increased fuel flexibility and greater resistance to impurities was a motivation for developing high-temperature molten carbonate fuel cell (MCFC) and solid oxide fuel cell (SOFC) systems. These systems are currently in various stages of development and demonstration, although MCFC systems are nearer to commercialization (Eichenberger 1998; Figueroa and Otahal 1998; Steinfeld *et al.* 2000). Among all fuel cells, high-temperature fuel cells seem to be the most promising ones because of their flexibility towards the gas quality and the increased systems efficiency (e.g. internal reforming of biogas). Especially MCFCs seem to be ideally suitable for the usage of biogas as the CO_2 in the biogas can be used as a reactant in the fuel cell, increasing the electrical efficiency of the system by approximately 2%. Low temperature fuel cells, such as AFCs, proton exchange membrane fuel cells (PEMFCs) and phosphoric

Table 21.3   Compatibility of three types of biofuels (bioethanol, biogas and biohydrogen) with four types of fuel cells.

Biofuel	PEMFC	PAFC	MCFC	SOFC
Biogas	~	+	+ +	+ +
Bioethanol	+	+	+ +	+ +
Biohydrogen (R)	+ +	+ +	+ +	+ +

+ +: strong long-term potential; +: some compatibility; ~: not very compatible; R: more research necessary.

acid fuel cells (PAFCs), require relatively pure hydrogen, the biofuel needs first to be treated to eliminate impurities and contaminants, and also reformed into usable hydrogen. This involves the use of two additional processes: gas clean-up and fuel reforming (Baron 2004).

This section will summarize the effects and impact of impurities and the tolerances the individual fuel cell systems have towards these impurities. The impurity tolerances for the fuel cell systems are summarized in Table 21.4.

21.5.1   Alkaline fuel cells

AFCs are extremely sensitive to impurities. The presence of N_2 and impurities in the gas streams substantially reduce the cell efficiency. The presence of even small amounts of CO_2 in air ($\sim$300 ppm) is detrimental because CO_2 will react with the KOH to form K_2CO_3 and inhibits gas diffusion through the carbon electrodes. Only pure H_2 and O_2 can be used in these systems and therefore this fuel cell is not suited for utilization of biofuels. This restriction limits the use of these fuel cell systems to applications such as space and military programs, where the high cost of providing pure H_2 and O_2 is permissible.

21.5.2   Proton exchange membrane fuel cells

PEMFCs are tolerant to CO_2, so air can be used as the oxidant and reforming-hydrocarbon fuels can produce the H_2 for the reductant. The PEMFC tolerance for CO is in the low ppm level because CO blocks the active sites in the Pt catalyst (Wagner and Gülzow 2004). Therefore, if a hydrocarbon reformer is used to produce H_2, the CO content of the fuel gas needs to be greatly reduced. In PEMFC systems, the CO content of the input gas stream should be reduced to less than 10 ppm (Ledjeff-Hey et al. 2000). This is usually accomplished by oxidation of CO to CO_2, using a water gas shift reactor, or using pressure swing adsorption to purify the hydrogen. Sulfur gases (mainly H_2S) can also poison the Pt catalyst in the electrodes (Daza et al. 2004). Sulfur also deactivates the Ni-based fuel-reforming catalysts. Recent developments in PEMFC systems are also going to direct alcohol fuel cells, in which alcohol is used directly as the fuel (Lamy et al. 2002). Except methanol, the direct oxidation of which is now widely studied in a PEMFC very few other alcohols have been investigated. Ethanol seems to be the most convenient and the more reactive alcohol after methanol. However, the relatively complex reaction mechanism, leading to a low electroreactivity of most

Table 21.4 Summary of fuel cell tolerances (Fuel Cell Handbook 2000; Dayton *et al.*
2001; Daza *et al.* 2004).

	PEMFC	AFC	PAFC	MCFC	SOFC
CO_2	Diluent	Poison 300 ppm	Diluent	Consumed in the cathode	Diluent
CO	Poison <10 ppm	Poison	Poison <1%	Fuel: with water shifted to make hydrogen	Fuel: with water shifted to make hydrogen
CH_4	Inert, fuel with reformer	Poison	Inert, fuel with reformer	Fuel: reformed internally or externally	Fuel: reformed
$C_2–C_6$	Poison	Poison	Poison <0.5% olefins	Fuel: reformed, saturated HC: 12 vol.%, Olefins: 0.2 vol.%, Aromatics: 0.5 vol.%, Cyclics: 0.5 vol.%	Fuel similar to MCFC
Sulfur	Poison <1 ppm H_2S, poisoning is cumulative and not reversible	Poison	Poison <4 ppm H_2S	Poison <10 ppm H_2S in fuel <1 ppm SO_2 in oxidant	Poison <0.1 ppm H_2S
NH_3	Poison	Poison	Poison <0.2 mol.% ammonium phosphate <1 ppm NH_3	Fuel (?), inert <1 vol.%	Fuel <5000 ppm
Halides (HCl)	Poison	Poison	Poison <4 ppm	Poison <1 ppm	Poison <1 ppm
Alkali metals	No studies to date	No studies to date	No studies to date	Electrolyte loss 1–10 ppm	
SiO_2	No studies to date	No studies to date	No studies to date	No studies to date	Deposition <1 mg/m^3

alcohols, impose the need to investigate new platinum-based electrocatalysts as
well as new oxygen reduction electrocatalysts insensitive to the presence of alcohols (Lamy *et al.* 2002).

21.5.3 Phosphoric acid fuel cells

Air is used as the oxidant. In contrast to the AFC, PAFCs are tolerant of CO_2
because concentrated phosphoric acid (H_3PO_4) is used as the electrolyte and CO_2
and the air does not react with the electrolyte. PAFC still requires hydrocarbon fuels
to be externally reformed into a H_2-rich gas. PAFCs use high-cost precious metal

catalysts such as platinum. CO poisoning of the Pt electrodes is slower at PAFC operating temperatures ($\sim 200°C$) than at lower temperatures so up to 1 vol.% CO in the fuel gas produced during the reforming process can be tolerated. At lower temperatures, CO poisoning of the Pt in the anode is more severe. Aside from the CO produced during hydrocarbon reforming, additional impurities in the fuel gas are also an issue with PAFC systems. Sulfur gases (mainly H_2S, mercaptans and COS) that originate from the fuel gas can poison the anode by blocking active sites for H_2 oxidation on the Pt surface and also the catalysts in the fuel reformer. So the H_2S level should be below 4 ppm and halogens should be kept to similar levels to avoid corrosion in the fuel processing system (Spiegel et al. 1999a, b). Molecular nitrogen acts as a diluent but other nitrogen compounds like NH_3, HCN and NO_x are potentially harmful impurities. NH_3 acts as a fuel, however, NH_3 reacts with H_3PO_4 to form ammonium phosphate salts $(NH_4)H_2PO_4$, which decreases the rate of oxygen reduction. A concentration of less than 0.2 mol.% $(NH_4)H_2PO_4$ must be maintained to avoid unacceptable performance (Fuel Cell Handbook 2000). Therefore, ammonia in the fuel gas should be kept below 1 ppm.

21.5.4   Molten carbonate fuel cells

Unlike PAFC systems, MCFCs can tolerate the high concentrations of CO that are produced in hydrocarbon-reforming processes (Huijsmans et al. 2000; Steinfeld et al. 2000; Fuel Cell Handbook 2000). This eliminates the need for water gas shifting and selective CO oxidation. MCFCs typically operate at temperatures between 600°C and 700°C and are therefore much more fuel flexible. Internal reforming of CH_4 and other hydrocarbons at the anode can be achieved at these operating temperatures (Craig and Mann 1996; Lobachyov and Richter 1998). Consequently, MCFC systems are very tolerant for hydrocarbons in the fuel gas. According to Bossart et al. (1990) the input fuel gas to a MCFC system can contain 12 vol.% saturated hydrocarbons (methane included), 0.2 vol.% olefins, 0.5 vol.% aromatics and 0.5 vol.% cyclics. Excessive levels of higher molecular weight hydrocarbons can cause typical problems like plugging and fouling of pipes, fuel transfer lines, heat exchangers and particle filters.

A higher operating temperature also means that less expensive materials can be used as catalysts in the electrodes and so Ni is used as material. Another advantage of the MCFC is that it operates efficiently with CO_2-containing fuels such as biogas. CO_2 is actually a necessary component in maintaining the carbonate content in the electrolyte and is expected to be supplying by recycling the anode gas exhaust to the cathode. CO is also a directly usable fuel. Catalyst poisoning (both in the reformer as well in the fuel cell) is also an issue if the sulfur content of the reagent gases is greater than 10 ppm, similar to all Ni-based fuel-reforming systems. H_2S will deactivate the Ni catalysts decreasing the fuel-reforming efficiency. In addition, the tolerant limit in the oxidant stream is 1 ppm SO_2. Coke formation on the anode from fuel reforming can also be an issue. HCl levels should be reduced below 1 ppm in the fuel gas (Fuel Cell Handbook 2000) and HF levels should be kept below 2.3 ppm in the fuel gas (Kawase et al. 2002), because HCl and HF react with molten carbonate to form the respective volatile alkali halides, which cause more rapid vaporization and accelerated electrolyte loss, which decreases the long-term performance of the fuel cell. The introduction of alkaline earth metals into the electrolyte also leads to reduced cell performance (Kordesch and Simader 1996).

The contaminants of solid particulates can be originated from a variety of sources, and their presence can block gas passages and/or the anode surface. The tolerance limit of MCFC to particulates larger than 3 micrometer in diameter is below 0.1 g/l. Compounds such as ammonia and HCN are discharged from the anode, and do not affect the cell voltage (Kawase *et al*. 2002).

21.5.5   Solid oxide fuel cells

SOFC systems operate between 800°C and 1000°C, higher than any other fuel cell systems. The high operating temperatures of the SOFCs provides fuel flexibility and no need for expensive catalysts in the electrodes. Overall, SOFC systems can tolerate impurities because of their high operating temperatures. This fuel cell accepts hydrocarbons and carbon dioxide in the fuel. CO can be a fuel in a SOFC or with water vapor it can be shifted to form H_2 that is consumed. The main problem impurity is sulfur, which will poison the catalytic nickel surface of the cell anode (Staniforth and Kendall 1998). Therefore, energy efficient sulfur removal methods are necessary to lower the sulfur content of the fuel to less than 10 ppm. Hydrocarbon fuels are usually reformed in an external-reforming unit and coking of the Ni-based catalysts is always a concern. By careful consideration of the materials used in SOFC anodes, it is possible to make fuel cells that are resistant to coke formation and reasonably sulfur tolerant. For example, Gorte *et al*. (2002) have shown that the Cu-based anodes for a SOFC can perform without difficulty in fuels containing at least 100 ppm of sulfur. SOFC systems require very stringent removal systems because they can only tolerate 1 ppm halides in the fuel gases and ammonia can be tolerated up to 0.5 vol.% (Fuel Cell Handbook 2000).

21.6   CONCLUSIONS

The removal of impurities from biofuels is always needed if they will be used in fuel cells. The technical challenge will be to design a biofuel-cleaning system to pre-condition the product gases before they enter the fuel cell stack. Sulfur removal systems will be required in all hydrocarbon-reforming fuel cell systems because both fuel cells and steam-reforming catalysts do not tolerate sulfur. They will be necessary even for systems using natural gas, the simplest hydrocarbon fuel. Halogen guard beds will likely be required for all fuel cell power systems as well to minimize long-term, high-temperature corrosion. The main challenge involves the fact that biomass feedstocks have incredibly variable compositions. This would suggest that fuel-cleaning systems, while generic in one sense, will have to be customized for individual situations. Among all fuel cells, the MCFC and SOFC high temperature systems seem to offer the best thermal integration opportunities with biofuel systems because of greater resistance to impurities.

REFERENCES

Alvin, M.A. (1998) Impact of char and ash fines on porous ceramic filter life. *Fuel Process. Technol.* **56**, 143–168.

Baron, S. (2004) Biofuels and their use in fuel cells. *Fuel Cell Today*, www.fuelcelltoday.com

Bossart, S.J., Cicero, D.C., Zeh, C.M. and Bedick, R.C. (1990) *Gas Stream Cleanup, Technical Status Report*. Morgantown Energy Technology Center Report No. DOE-METC-91/0273.

CADDET Japanese National Team (1999) Fuel cell CHP using biogas from brewery effluent. *CADDET Renew. Energ. Newslett.* July 1999.

Commission of the European Communities (1997) *White Paper for a Community Strategy and Action Plan: Energy for the Future: Renewable Sources of Energy.* COM(97)599 final 26 November 1997, Brussels.

Commission of the European Communities (2000) *Green Paper: Towards a European Strategy for the Security of Energy Supply.* COM(2000) 769 final, Brussels.

Commission of the European Communities (2003) *Directive 2003/30/EC of the European Parliament and of the Council of 8 May 2003 on the Promotion of the Use of Biofuels or Other Renewable Fuels for Transport*, Brussels.

Converti, A., Del Borghi, A., Zilli, M., Arni, S. and Del Borghi, M. (1999) Anaerobic digestion of the vegetable fraction of municipal refuses: mesophilic versus thermophilic conditions. *Bioproc. Eng.* **21**, 371–376.

Craig, K.R. and Mann, M.K. (1996) *Cost and Performance Analysis of Biomass-Based Integrated Gasification Combined-Cycle (BIGCC) Power Systems.* NREL Technical Report, NICH Report No. TP-430-21657, October 1996.

Crane, R.I., Barbaris, L.N. and Behrouzi, P. (1992) Particulate behaviour in cyclone separators with secondary gas extraction. *J. Aerosol Sci.* **23**, 765–768.

Dayton, D.C., Ratcliff, M. and Bain, R. (2001) *Fuel cell integration: A Study of the Impacts of Gas Quality and Impurities.* Milestone Completion Report, NREL/MP-510-30298, June 2001.

Daza, L., Escudero, M.J and Soler, J. (2004) Fuel cell technology. In: *Biogas Powered Fuel Cells.* Trauner Verlag Linz, Austria.

De Baere, L. (2000) Anaerobic digestion of solid waste: state-of-the-art. *Water Sci. Technol.* **41**, 283–290.

De Mes, T.Z.D., Stams, A.J.M., Reith, J.H. and Zeeman, G. (2003) Methane production by anaerobic digestion of wastewater and solid wastes. In: *Bio-methane and Bio-hydrogen.* The Hague, Dutch Biological Hydrogen Foundation.

De Vrije, T. and Claassen, P.A.M. (2003) Dark hydrogen fermentations. In: *Bio-methane and Bio-hydrogen.* The Hague, Dutch Biological Hydrogen Foundation.

Eichenberger, P.H. (1998) The 2 MW Santa Clara Project. *J. Power Sources* **71**, 95–99.

Environmental Agency (2002) *Guidance on Gas Treatment Technologies for Landfill Gas Engines.* December 2002. Bristol, www.environment-agency.gov.uk

Figueroa, R.A. and Otahal, J. (1998) Utility experience with a 250-kW molten carbonate fuel cell cogeneration power plant at NAS Miramar, San Diego. *J. Power Sources* **71**, 100–104.

Fuel Cell Handbook, 5th edn (2000) Report prepared by EG&G Services, Parsons, Inc. and Science, Applications International Corporation under Contract No. DE-AM26-99FT40575 for the US Department of Energy, National Energy Technology Laboratory, October.

Francese, A.P., Aboagye-Mathiesen, G., Olesen, T., Córdoba, P.R. and Sineriz, F. (2000) Feeding approaches for biogas production from animal wastes and industrial effluents. *World J. Microbiol. Biotechnol.* **16**, 147–150.

Gorte, R.J., Kim, H., and Vohs, J.M. (2002) Novel SOFC anodes for the direct electrochemical oxidation of hydrocarbon. *J. Power Sources* **106**, 10–15.

He, C., Herman, D.J., Minet, R.G. and Tsotsis, T.T. (1997) A catalytic/sorption hybrid process for landfill gas cleanup. *Ind. Eng. Chem. Res.* **36**, 4100–4107.

Huijsmans, J.P.P., Kraaij, G.J., Makkus, R., Rietveld, G., Sitters, E.F. and Reijers, H.T.J. (2000) An analysis of endurance issues for MCFC. *J. Power Sources* **86**, 117–121.

Kawase, M., Mugikura, Y., Watanabe, T., Hiraga, Y. and Ujihara, T. (2002) Effects of NH3 and NOx on the performance of MCFCs. *J. Power Sources* **104**, 265–271.

Kordesch, K. and Simader, G. (1996) *Fuel Cells and Their Applications*. VCH Verlagsgesellschaft, Weinheim, Germany.

Lamy, C., Lima, A., LeRhun, V., Delime, F., Coutanceau, Ch. and Léger, J.-M. (2002) Recent advances in the development of direct alcohol fuel cells (DAFC). *J. Power Sources* **105**, 283–296.

Ledjeff-Hey, K., Roes, J. and Wolters, R. (2000) CO_2-scrubbing and methanation as purification system for PEFC. *J. Power Sources* **86**, 556–561.

Lobachyov, K.V. and Richter, H.J. (1998) An advanced integrated biomass gasification and molten fuel cell power system. *Energ. Convers. Manage.* **39**, 1931–1943.

Salminen, E. and Rintala, J. (2002) Anaerobic digestion of organic solid poultry slaughterhouse waste – a review. *Biores. Technol.* **83**, 13–26.

Spiegel, R.J. and Preston, J.L. (2000) Test results for fuel cell operation on anaerobic digester gas. *J. Power Sources* **86**, 283–288.

Spiegel, R.J., Preston, J.L. and Trocciola, J.C. (1999a) Fuel cell operation on landfill gas at Penrose Power Station. *Energy* **24**, 723–742.

Spiegel, R.J., Thorneloe, S.A., Trocciola, J.C. and Preston, J.L. (1999b) Fuel cell operation on anaerobic digester gas: Conceptual design and assessment. *Waste Manage.* **19**, 389–399.

Staniforth, J. and Kendall, K. (1998) Biogas powering a small tubular solid oxide fuel cell. *J. Power Sources* **71**, 275–277.

Steinfeld, G., Ghezel-Ayagh, H., Sanderson, R. and Abens, S. (2000) Integrated gasification fuel cell (IGFC) demonstration test. In: *Proceedings of the 25th International Technical Conference on Coal Utilization and Fuel Systems*, March 6–9 in Clearwater, FL.

Wagner, N. and Gülzow, E. (2004) Change of electrochemical impedance spectra (EIS) with time during CO-poisoning of the Pt-anode in a membrane fuel cell. *J. Power Sources* **127**, 341–347.

Weiland, P. (2000) Anaerobic waste digestion in Germany – status and recent developments. *Biodegradation* **11**, 415–421.

22

Biogas upgrading with pressure swing adsorption versus biogas reforming

A. Schulte-Schulze Berndt

22.1 INTRODUCTION

The common method for utilisation of biogas from, for example, fermentation plants, landfill sites or waste water treatment plants in Europe is, by far, the production of heat and electric energy with cogeneration plants (Figure 22.1) or in gas engines, exclusively for the production of electric power and/or heat with boilers.

The disadvantage of using gas engines or cogeneration plants is the reduced recovery of the energy content in the biogas (OTTI-KOLLEG 2002; Weiland 2003). Using gas engines, only 30–40% of the potentially usable energy can be recovered as electric power whereas the off-heat has to be regarded as wasted. By using cogeneration plants the loss heat can be recovered, but there still is a problem

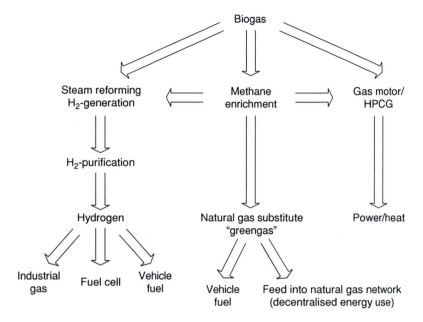

Figure 22.1   Utilisation of biogas.

related to the site location. A profitable use of the loss heat is very often either not or only possible with technical difficulties, due to the peripheral location of most biogas production plants.

A different option to utilise biogas is the production of hydrogen with a steam reformer followed up by a gas purification system. Applications for this hydrogen are industrial raw gas, car fuel or fuel for the production of electric energy with fuel cells. This alternative is typified by its complex and not (yet) industrial-scale tested process and thus not to be regarded as a profitable option – neither in the short nor in the medium term. In addition, also an infrastructure of hydrogen hand-ling to the consumer is – at the moment – not at hand.

Thus, the most promising alternative to utilise the biogas remains the purifica-tion up to natural gas grade and the use of it as car fuel or natural gas substitute. Hereunder, this type of purified biogas shall be called "bio-natural gas".

In Germany, this technology has been applied for the first time in 2002 on a demonstration plant in Albersdorf/Schleswig-Holstein, operated by the company Farmatic Biotech Energy AG (Intercompany communications). Considering also long-term experiences of similar plants in Switzerland, Sweden, Norway and the USA during the last years, it has turned out that this technology is profitable not only from an economical and ecological view but has also been developed up to the point for industrial-scale use.

This chapter gives a short introduction to both technologies of biogas upgrading by purification to "bio-natural gas" as well as to the conversion to hydrogen.

22.2 COMPOSITION AND QUALITY OF BIOGAS FOR THE USE AS CAR FUEL OR SUBSTITUTE FOR NATURAL GAS

Table 22.1 shows the usual composition of biogas from fermentation processes (Bayerisches Landesamt für Umweltschutz 1997). The main components are methane with a volumetric content of approximately 55–70 vol.% and carbon dioxide with approximately 30–45 vol.%. In most biogases, also small percentages of oxygen and nitrogen are present, both are inevitably residuals from biological desulfurisation processes (Rieger *et al.* 2003).

The biogas is saturated with water and also contains some hundred parts per million of hydrogen sulfide as well as usually some higher hydrocarbons.

The utilisable components are methane (CH_4), as fuel or as educt for the hydrogen conversion and carbon dioxide (CO_2), which can be, after further purification, applied as industrial raw gas for the production of dry ice or for example to increase the CO_2 content in greenhouse atmospheres (Table 22.1).

Compared to natural gas, raw biogas is a "heavy" gas due to its high CO_2 content with a density of approximately 1.05–1.2 kg/Nm³. Also related to the high CO_2 content, the upper heating value of only 20–24 MJ/Nm³ is approximately 30–40% lower as of natural gas.

The requirements on purified "bio-natural gas", also known as "greengas", derive from the current regulations of the Deutsche Vereinigung des Gas- und Wasserfachs e.V. (DVGW 2000). Additionally also long-term experiences with the operation of cars running on greengas were taken into account.

Referring to Table 22.1, the process of upgrading the raw biogas bases on the one hand on the purification from the impurities hydrogen sulfide, higher hydrocarbons, oxygen and water and on the other hand on shifting the heating value. Shifting the upper heating value is mainly based on the removal of carbon dioxide down to values of approximately 1 vol.%.

Especially with the aim of a later use as car fuel or natural gas substitute, the critical compounds are hydrogen sulfide and water. Not only that hydrogen sulfide is

Table 22.1 Specification of raw and upgraded ("greengas") biogas.

Component	Symbol	Raw biogas	"Greengas"	DVGW260
Methane[a]	CH_4	55–70%	>97%	No minimum values
Carbon dioxide[b]	CO_2	30–45%	<1%	No minimum values
Nitrogen	N_2	<2%	<2%	No maximum values
Oxygen	O_2	<0.5%	<0.5%	<0.5%
Hydrogen sulfide	H_2S	<500 ppm v	<5 mg/Nm³	<5 mg/Nm³
Hydrocarbons	C_xH_y	<100 ppm v	<10 ppm v	<Condensation point
Water	H_2O	Saturated	<0.03 g/m³	<Condensation point
Calorific value	$H_{S,M}$	6–7.5 kWh/m³	Maximum 11 kWh/m³	8.4–13.1 kWh/m³

Usable and/or marketable components: [a]due to the high calorific value and as hydrogen carrier; [b]as industrial gas, dry ice products, greenhouses.

a strong poison, but particularly the corrosive properties of H_2S in water are problematic. This automatically leads to the necessity to mainly reduce the content of water, hydrogen sulfide and at least carbon dioxide in the biogas to ensure a trouble-free utilisation of biogas. Only in case of higher nitrogen and/or oxygen concentrations in the biogas the removal of these components also has to be considered to avoid further negative influences on the heating value.

22.3 BIOGAS UPGRADING PROCESSES

22.3.1 Upgrading technologies

Due to the demands on quality for "bio-natural gas", there is the need to upgrade raw biogas with a safe, stable and economical advantageous process to remove carbon dioxide as well as the other impurities.

Table 22.2 shows the different biogas upgrading technologies for CO_2 removal, which are currently at hand and in use. The classical procedures are gas scrubbing, adsorption and CO_2-liquefaction, but since some years also wet and dry membrane separation processes have become available.

While gas scrubbing with, e.g. monoethanolamine (MEA), wet or dry membrane separation processes and CO_2-liquefaction can mainly remove only the carbon dioxide from biogas, the advantage of the adsorption process on activated carbon is characterised by the simultaneous removal of water, hydrogen sulfide and other impurities like silicon compounds.

In this chapter, only the adsorption process shall therefore be presented in detail.

22.3.2 Basic principle of biogas upgrading by means of adsorption

The centre of a biogas upgrading plant is the adsorber, a vessel filled with a carbon molecular sieve (CMS, Figure 22.2). From a biogas stream through this adsorber from bottom to top preferably carbon dioxide will adsorb on the inner surface of

Table 22.2   Process for methane enrichment from biogas.

Process	Description
Gas scrubbing	CO_2 is adsorbed by means of washing liquid (e.g. water, caustic soda solution, MEA washing)
Adsorption	CO_2 is bound at an adsorbent over electrostatic forces, adsorbed
Membrane process, wet	CO_2 is separated due to different permeation rates at a membrane and afterwards adsorbed by a washing liquid (MEA)
Membrane process, dry	CO_2 is separated due to different permeation rates at a membrane
CO_2-liquefaction	Phase separation of liquid CO_2 and gaseous methane

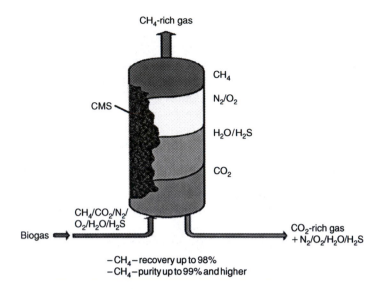

Figure 22.2   Methane enrichment by PSA by means of CMS.

the adsorbing agent. Besides CO_2 also the other biogas components, such as nitrogen, oxygen and residual humidity (water) will be adsorbed.

In relation to the superficial velocity of the flow response the mean dwell time of the biogas in the adsorber, a product gas with a purity of almost 100% CH_4 can be obtained on the adsorber top nozzle. Inside the vessel, concentration profiles of the impurities will be formed. After a while, when the adsorption capacity of the CMS is reached, impurities such as $N_2/O_2/CO_2/H_2O/H_2S$ will start to break through. At this point of the process the adsorption phase has to be stopped and regeneration must be started.

For the regeneration, the adsorber (pressurised to approximately 5–10 bar g for the adsorption) has to be depressurised and afterwards evacuated with a vacuum pump. During this evacuation all adsorbed components will be removed from the molecular sieve and thus a totally regenerated adsorber will be obtained.

As already mentioned, the advantage of the adsorption process is the simultaneous removal of all biogas impurities in addition to CO_2. Unfortunately one will find all these impurities in the evacuation gas. Utilisation of this CO_2-rich gas (e.g. in a green house) is thus not possible without further conditioning of the off-gas.

22.3.3   Industrial process of adsorptive biogas upgrading

22.3.3.1   Pressure swing adsorption

In principle it is possible to upgrade biogas in a plant with only one adsorber. To obtain a continuous product gas flow as well as to increase the gas recovery and efficiency of the plant, it is advantageous to operate with more than one adsorber.

Figure 22.3 Biogas upgrading to "greengas", 150 Nm³/h biogas – 100 Nm³/h product gas, purity: 97 ± 1 vol.% CH₄ (Courtesy: RÜTGERS CarboTech Engineering GmbH).

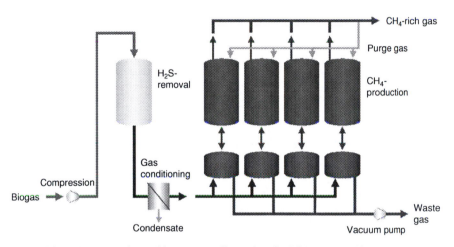

Figure 22.4 Process scheme biogas upgrading using the PSA process (Courtesy: RÜTGERS CarboTech Engineering GmbH).

Considering a high plant efficiency and recovery rate compared with the investment costs, plants with four adsorbers have over the years turned out as an optimal choice.

The adsorption process shifts the problem of the impurity H₂S from the raw gas side to the off-gas side and even increases the concentration of this component. This resulted in the requirement of either a difficult off-gas treatment system or in the installation of an additional H₂S prefilter system. Figure 22.3 gives an example of full-scale applications of biogas upgrading systems to "greengas".

These requirements resulted in the design of a continuous operating biogas upgrading plant by means of pressure swing adsorption (PSA) as shown in Figure 22.4. The main steps of this process are (Figure 22.5):

(1) raw gas compression,
(2) H_2S removal stage,
(3) biogas conditioning,
(4) methane production.

Together with product gas monitoring with gas analysers (for CH_4, CO_2, H_2O and H_2S), the plant can be controlled by a central programmable logic controller (PLC)

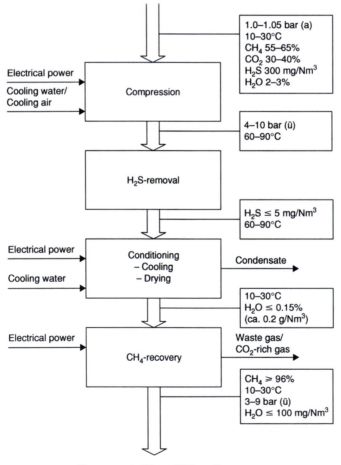

Figure 22.5   Process flow sheet biogas upgrading (courtesy: RÜTGERS CarboTech Engineering GmbH).

and can be operated fully automatically. Permanent attendance of on-site operators is not necessary.

22.3.3.2   H₂S removal

The raw biogas is at first compressed (oil-free compression by a piston-type reciprocating compressor to approximately 4–10 bar g) and lead with a temperature of approximately 60–90°C into the H₂S-removal reactor.

The H₂S removal (Figure 22.6) is based on the principle of cracking the H₂S molecule on a activated carbon surface at temperatures of 60–90°C. The resulting sulfur is subsequently adsorbed on the surface of the activated carbon filling of the removal stage. By using a special type of impregnated activated carbon, an adsorption capacity of more than 100% of weight is obtainable, thus guaranteeing life cycles of the removal adsorbent of more than 1 year. The resulting H₂S content in the biogas is 5 mg/Nm³ and lower.

In the following conditioning system, the biogas temperature is reduced to approximately 20–30°C and also a dew point of approximately 3–5°C is obtained by means of cold drying. The drying serves as protection against corrosion of the

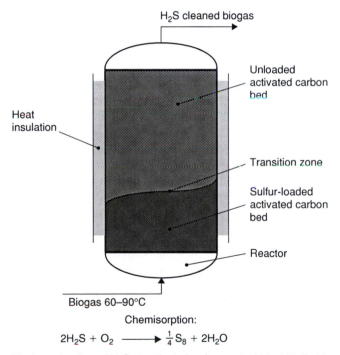

Chemisorption:

$$2H_2S + O_2 \longrightarrow \tfrac{1}{4}S_8 + 2H_2O$$

Maximum loading of H₂S at activated carbon up to 100 wt.%, that is per 1 kg activated carbon can be taken up to 1 kg H₂S and more!

Figure 22.6   H₂S removal: Conversion of H₂S to elementary sulfur and water by catalytic oxidation at activated carbon.

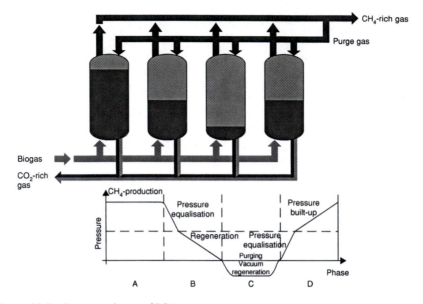

Figure 22.7   Process scheme of PSA.

following parts of the plant due to condensation as well as to decrease the necessary size of the adsorbers and the vacuum pump.

The almost H_2S-free and dry biogas is then lead into a four-bed PSA plant to
purify the methane. Every adsorber of the plant is operated in a four-step-cycle of
adsorption, depressurisation, regeneration and repressurisation (Figure 22.7).
After adsorption under high pressure (Phase A), a pressure exchange with a regenerated adsorber of low pressure is performed. After a further depressurisation down
to atmospheric pressure, Phase B is completed. During Phase C the adsorbent in
the adsorber is totally regenerated by evacuation. In Phase D, the regenerated adsorber is repressurised by pressure exchange with another adsorber and by repressurisation with product gas and thus prepared for the next cycle.

Due to this four-bed and four-cycle method a continuous product gas flow can
be obtained.

22.3.3.3   Additional components of the plant

An air injection system prior to the gas compression can be necessary in case the
raw biogas contains absolutely no oxygen. The smallest amounts of oxygen are
necessary to ensure a complete conversion of the H_2S in the H_2S removal stage
according to the formula:

$$2H_2S + O_2 \rightarrow 2H_2O + \tfrac{1}{4}S_8.$$

Odourless methane used as fuel or as natural gas substitute has to be odorised
with THT (tetrahydrothiophen). Legislation requires this odourisation prior to the
use as fuel or induction into public pipelines to enable fast detection of potential
leakages due to the strong smell of THT.

Table 22.3   Advantages of adsorptive biogas upgrading.

- Dry process, that is:
  - No process water
  - No waste water handling
  - No chemicals
  - No corrosion problems
- Co-adsorption minimises/reduces besides CO_2 also other components:
  - Removal of water up to a dew point of $-40°C$, therefore no need of further drying for feed into natural gas network
  - Partial removal of nitrogen and oxygen, therefore no enrichment in product gas, high calorific value
  - Removal of chlorofluorocarbons (CFCs), volatile organic components and silicon compounds
- Low utility consumption
- Compact and space-saving design
- Simple and automatic operation

In addition, all adsorbers are equipped with small prefiltering adsorbers. These prefilters, filled with an ordinary low-priced activated carbon, serve as safety filters and adsorb higher hydrocarbons as well as other impurities in the raw gas that normally cannot be detected. Using these prefilters, the lifetime of the CMS in the main adsorbers can be extended to more than 10 years.

22.3.4   Advantages of the adsorptive biogas upgrading

Compared to other processes of biogas upgrading the adsorptive process features some evident advantages as shown in Table 22.3. These advantages mainly result from the fact that this is a dry process.

Thus, there is no requirement for a process water system, a wastewater treatment, any additional reactants, and additional protection against corrosion.

Another benefit is the automatic co-adsorption of impurities other than carbon dioxide.

The PSA biogas plants have a very low utility consumption and a compact and space-saving set up. Resulting from the simple process also the simple and easy-to-learn usability of these plants should be mentioned.

22.3.5   Operating and cost efficiency of biogas upgrading with PSA

Drawing for example the energy balance of a medium-size PSA biogas upgrading plant with a raw gas intake of approximately 400 Nm³/h biogas, it is evident, that this is a highly efficient gas separation process (Figure 22.8). An energy efficiency of over 94% is possible.

The high degree of effectiveness will also be reflected in the economic calculations for such a plant.

Based on the general conditions of today, the overall specific costs of upgrading biogas to natural gas quality have been compared for different sizes of plants (Figure 22.9). It is evident that for plants with a capacity of >250 Nm³/h product gas specific total upgrading charges of down to 0.05 €/Nm³ can be considered

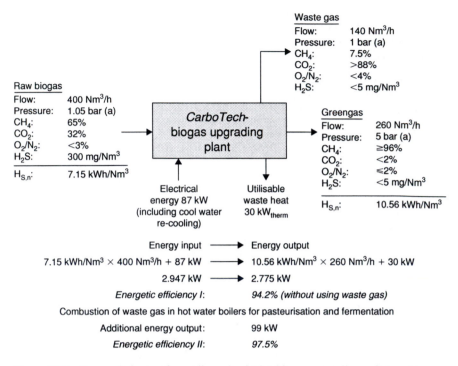

Energetic efficiency I: 94.2% (without using waste gas)

Combustion of waste gas in hot water boilers for pasteurisation and fermentation

Additional energy output: 99 kW

Energetic efficiency II: 97.5%

Figure 22.8   Energy balance of a medium-sized PSA biogas upgrading unit (raw biogas flow 400 Nm^3/h).

Cost:
• Investment cost including buildings/foundations/analytics
• Utilities (electrical power, cooling water)
• Service/maintenance/spare parts

Assumption:
• Depreciation   15 years linear
• Interest rate   6%
• Power cost   5 Cent/kWh
• Operation time   8.600 h/year
• Gas specification   CH$_4$: raw gas 65%/product gas >96%
                                    H$_2$S: raw gas 300 mg/Nm^3/product gas <5 mg/Nm^3

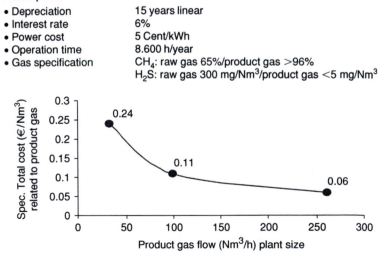

Figure 22.9   Overall specific costs of biogas upgrading for different plant sizes.

possible. This corresponds with approximately 0.5 €cent/kWh. Compared with a selling price of approximately 3.6 cent/kWh for ordinary natural gas this is a low value. Taking further into account a petroleum tax exemption for bio-natural gas, a selling price on gas stations slightly below the natural gas price could be possible (Figure 22.9).

22.4   PROCESS OF CONVERSION OF BIOGAS TO HYDROGEN: BIOGAS REFORMING

22.4.1   Basics on CH_4-steam reforming

Methane, as the main component of natural gas and also of biogas is usually catalytically cracked at high temperatures and pressures in a special type of reactor, called "reformer". The cracking products, a gas containing namely hydrogen, carbon monoxide, carbon dioxide, water and residual methane is known as "reformate".

Simplified, steam reforming is chemically a combination of two reactions, the so-called "reforming reaction", an endothermic reaction that converts hydrocarbons to hydrogen and carbon monoxide and the "shift reaction" that converts carbon monoxide to carbon dioxide. Both reactions require water, namely steam, as reactant and both reactions demand the presence of a catalyst. In reality, reforming is a very complex combination of various reactions similar to the above mentioned.

22.4.1.1   Hydrogen production

The conversion of natural gas or biogas to high-purity hydrogen is performed in two major steps:

(1)   Steam reforming of CH_4 in a catalytic reactor (reformer).
(2)   Purification resp. upgrading of the reformate via adsorption.

Table 22.4 describes the process and Figure 22.10 shows a block scheme of the plant.

22.4.1.2   Steam reforming

The central installation in a steam-reforming plant is the reformer, a heated pipe reactor filled with a nickel catalyst. The educt, a mixture of raw gas with a high methane content and water steam, flows through this catalyst bed. In the presence of the catalyst, connected with high temperature and pressure, the educts are cracked and a mixture of hydrogen, carbon monoxide, carbon dioxide, methane and water is generated.

Steam reforming takes place at temperatures around 900°C. The reformer is heated by a special type of FLOX®-burner. This burner reaches extremely low outputs of nitrogen oxides, due to air preheating in the burning area and flameless oxidation of the burner gas. During start up the feed gas with a high methane content serves as burner gas, later the off-gas from the PSA for hydrogen purification can be used.

Table 22.4   Process description of CH_4 reforming and H_2 enrichment.

1. Pressure increase of methane-rich feed gas and demineralised water.
2. Blending of methane-rich feed gas and water followed by water evaporation in counter flow with reformate stream.
3. Conversion of the methane-steam mixture at approximately 900°C according to the following reactions:

$$CH_4 + H_2O \rightarrow CO + 3H_2$$
$$CO + H_2O \rightarrow CO_2 + H_2$$

4. Cooling of the reformate stream in counter flow with the incoming methane-steam mixture, followed by an external cooling step to ambient temperature.
5. Separation of water condensate from reformate stream.
6. Separation of H_2 from CO, CO_2, N_2, CH_4 und H_2O by means of adsorption into a pure hydrogen stream and a tailgas stream by a PSA plant.
7. Recycling of PSA tailgas as fuel gas for the reformer burner.

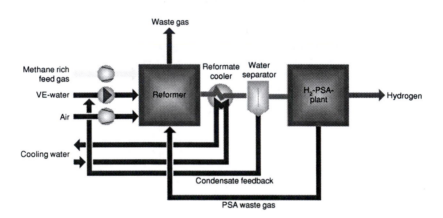

Figure 22.10   Process scheme of CH_4 reforming and H_2 enrichment.

Using this special type of reformer, no water steam is injected. The reformer is only charged with a mixture of methane-rich gas and, liquid, deionised water. The reactor is equipped with an inbuilt heat exchanger, heating up the educts and evaporating the water content. For the heating, the reformate product is used in counter current flow so that the product with a high hydrogen content reaches the reactor outlet quite cool.

The reformate leaves the reformer at temperatures of approximately 300°C. After a further cooling step down to ambient temperatures, a major part of the water steam content condensates and can be separated. The reformate can now be lead to a PSA plant, in order to remove the unwanted components and achieve a stream of high-purity hydrogen.

22.4.1.3   Hydrogen enrichment via adsorption

The removal of the undesired components from the synthesis gas is made by adsorption. Molecules with a high polarity, such as CO_2, CO, CH_4 or water, tend to adsorb on the surface of adsorbents, whereas molecules with lower polarities,

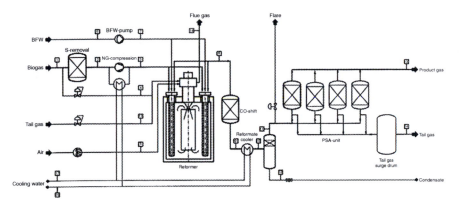

Figure 22.11   General flow sheet of CH$_4$ reforming.

such as hydrogen, have a smaller attraction towards adsorbents. This selective character of the adsorption capability can be used to produce the high-purity hydrogen from the reformate.

The PSA process operates on two pressure levels. Adsorption of the undesired components takes place at a higher pressure level. This increases the partial pressure of the undesired components and thus enhances the adsorption capability of the adsorbent. Regeneration of the adsorbers, also called desorption phase, takes place at a lower pressure level which decreases the adsorption capacity of the used adsorbent. Depending on the process, the upper pressure level is at approximately 8–16 bar g whereas the desorption takes place at a level close to atmospheric pressure.

To enable a continuous production of hydrogen, the PSA plant is of a four-bed type. Each adsorber is filled with layers of different adsorbents to optimise the adsorption capacity for each component of the synthesis gas. Similar to the methane-enrichment process, each adsorber is switched in a continuous cycle of adsorption, depressurisation, regeneration and repressurisation. During the regeneration phase the adsorber is purged with a small stream of the high-purity product hydrogen to ensure a better desorption.

During the depressurisation and the regeneration, an off-gas stream containing all undesired synthesis gas components with still a high hydrogen and carbon monoxide content is achieved. This off-gas is flammable and is used as fuel for the FLOX®-burner. The emissions from the plants can thus be minimised and the required amount of raw gas for heating of the reformer decreases.

22.4.2   Biogas-steam reforming

When using biogas as CH$_4$-feed gas for the steam-reforming process, higher impurity levels of sulfur components and oxygen have to be considered. Compared to natural gas as feed further conditioning stages as well as modifications of the catalyst and the reformer design are necessary, especially with regard to the higher levels of CO$_2$ and O$_2$ of the biogas. Figure 22.11 shows a simplified block diagram of the biogas-reforming process.

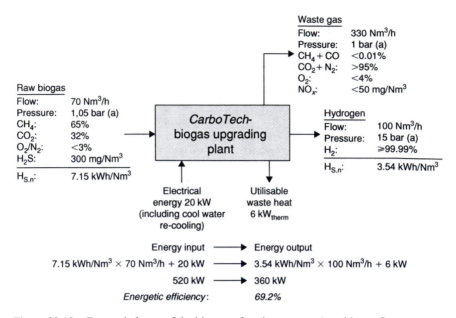

Waste gas
Flow:	330 Nm³/h
Pressure:	1 bar (a)
CH₄ + CO	<0.01%
CO₂ + N₂:	>95%
O₂:	<4%
NOₓ:	<50 mg/Nm³

Raw biogas
Flow:	70 Nm³/h
Pressure:	1,05 bar (a)
CH₄:	65%
CO₂:	32%
O₂/N₂:	<3%
H₂S:	300 mg/Nm³
Hₛ,ₙ:	7.15 kWh/Nm³

CarboTech-
biogas upgrading
plant

Hydrogen
Flow:	100 Nm³/h
Pressure:	15 bar (a)
H₂:	≥99.99%
Hₛ,ₙ:	3.54 kWh/Nm³

Electrical
energy 20 kW
(including cool water
re-cooling)

Utilisable
waste heat
6 kW$_{therm}$

Energy input ⟶ Energy output

7.15 kWh/Nm³ × 70 Nm³/h + 20 kW ⟶ 3.54 kWh/Nm³ × 100 Nm³/h + 6 kW

520 kW ⟶ 360 kW

Energetic efficiency: 69.2%

Figure 22.12   Energy balance of the biogas reforming process (raw biogas flow 70 Nm³/h).

Particularly of interest is here the correct choice and design of the adsorbers for removal of the sulfur components in biogas such as hydrogen sulfide or organic sulfur hydrides. In use are fixed bed adsorbers filled with special type activated carbons to remove these sulfur compounds.

When the raw biogas has been desulfurised in a biological desulfurisation reactor, the oxygen level in the gas can rise up to limits of volume percent. This requires the application of special types of catalysts in the reformer. Reforming on common nickel catalysts in the presence of higher levels of oxygen results in the formation of soot, which leads to progressive deactivation of the catalyst.

22.4.3   Energy and cost balance of the biogas reforming process

Comparing the energy balance of the biogas reforming (Figure 22.12) with the biogas upgrading (Figure 22.8) clearly indicates a much lower energy conversion efficiency, this is only 69.2% versus 94.2% in case of the biogas upgrading to natural gas.

Therefore, biogas reforming should preferably be used to produce hydrogen as process gas for industrial applications like heat treatment of metals or hydrogenation in the chemical industry or others.

A reliable statement about the economical and operating efficiency of the direct conversion of biogas to hydrogen in a steam reformer with enrichment of the reformate to high purities is, at the time being, not possible. Until today, this special process has been put in operation only once, integrated in a demonstration plant of the FAL Braunschweig in 2003 (Figure 22.13). The plant has been

Figure 22.13 Hydrogen generator, demo-plant at FAL Braunschweig, feed gas: biogas 2.5 Nm^3/h product gas/purity: >99.99% H$_2$.

developed, engineered and manufactured by RÜTGERS CarboTech Engineering GmbH and has been started up the end of 2003. Long-term experiences and measurements are presented in Chapter 25.

REFERENCES

Bayerisches Landesamt für Umweltschutz (1997) Energetische Nutzung von Biogas aus der Landwirtschaft, Ergebnisbericht. "Untersuchung der Biogaszusammensetzung bei Anlagen aus der Landwirtschaft", May 1997.

DVGW (2000) Deutsche Vereinigung des Gas- und Wasserfaches e.V. Arbeitsblatt G 260 "Gasbeschaffenheit", January 2000.

Weiland, P. (2003) Notwendigkeit der Biogasaufbereitung, Ansprüche einzelner Nutzungsrouten und Stand der Technik, Gülzower Fachgespräche, Band 21 Workshop "Aufbereitung von Biogas", FAL Braunschweig, June 2003, pp. 7–23.

Rieger, C., Ehrmann, T. und Weiland, P. (2003) Ergebnisse des Biogasmessprogramms zur Biogasqualität landwirtschaftlicher Kofermentationsanlagen, Gülzower Fachgespräche, Band 21 Workshop "Aufbereitung von Biogas", FAL Braunschweig, June 2003, pp. 24–31.

OTTI-KOLLEG (2002) 11. Symposium ENERGIE AUS BIOMASSE, Biogas, Pflanzenöl, Festbrennstoffe, November 2002, Dr.-Ing. U. Knopf, Dipl.-Ing. W. Schmidt. Verfahrenstechnisches Institut Saalfeld GmbH, "Biogasanlage mit direkter Kraft-Wärme-Kälte-Kopplung", pp. 191–195.

23

Treatment of hydrogen sulfide in biofuels

V.L. Barbosa and R.M. Stuetz

23.1 INTRODUCTION

The world biomass production on the land surface exceeds the energy consumption by a factor of 100 (Badin *et al.* 1999). Anaerobic conversion processes for the production of biofuels include biogas production at many sewage treatment plants and landfills; anaerobic processes applied to clean waste streams of agricultural processing as well as mesophilic and thermophilic digestion of organic waste. Although, the cost for using energy from biomass equals that of hydro and wind power, only a small part of the biomass available for energy production is currently utilised. Even though current efficiency conversion of biomass to useful energy is still low, it contributed 13% of all primary energy produced globally in 1990 (Klass 1999). In its 1997 White Paper "Energy for the Future – Renewable Sources of Energy" (European Commission 1997), the European Commission describes a strategy and action plan to enhance the use of renewable energy sources. In the EC White Paper, it is estimated that the amount of energy from biomass will treble between 1995 and 2010. It has also been estimated (Klass 1999) that in 2050 approximately 40% of direct fuel use and 17% of the power production will be provided from biomass. The US Environmental Protection Agency created the Landfill Methane Outreach Program (LMOP) in 1994 aimed at significantly reducing methane emissions from municipal solid waste, which is the largest human-generated source of methane emissions in the USA. It was estimated that the project would capture 60–90% of the methane emitted from landfills to produce electricity (EPA 2004).

In order to obtain pipeline quality gas, the biogas must pass through a cleaning process to remove any harmful components such as hydrogen sulfide (H_2S), carbon dioxide (CO_2) and nitrogen gas. Table 23.1 shows typical concentrations of H_2S, methane (CH_4) and CO_2 in processes producing or with the potential for producing biofuels. Carbon dioxide and nitrogen are inert gases, which reduce the heating value of the biogas. The most important characteristics of H_2S are its low odour threshold value (8.5–1000 ppbv), it is highly corrosive to plant equipment such as boilers, piping and engine parts, and it is highly toxic (Smet et al. 1998). Therefore, H_2S must be removed from waste gases not only owing to the corrosive effect and odour annoyance it causes, but also especially because of its serious health risks. Exposure to concentrations of around 15 ppmv may cause eye irritation, whilst those exceeding 320 ppmv carry a risk of death (WHO 1987). Its removal is also required due to pollution issues, since when H_2S is burned it gives out sulfur dioxide (SO_2), which is a known contributor to acid rain.

This chapter will describe the formation and composition of sulfurous waste gases emitted from processes producing waste gases, and will discuss the methods suitable for treatment of H_2S gas.

Table 23.1 Typical untreated biogas H_2S, methane and CO_2 concentrations for different processes.

Process	H_2S (ppmv)	CH_4 (%)	CO_2 (%)
Sewage treatment	10–40	55–65	Balance
Landfill	50–300	45–55	30–40
Biogas plant	10–2000	60–70	30–40

23.2 COMPOSITION AND FORMATION OF SULFUROUS WASTE GASES

The main volatile sulfur compounds (VSC) in waste gases comprise H_2S, dimethyl sulfide (Me_2S), dimethyl disulfide (Me_2S_2), methanethiol or methyl mercaptan (MeSH), carbon disulfide (CS_2) and carbonyl sulfide (COS). The properties of these VSCs are presented in Table 23.2.

The main source for the production of VSCs in biosolids is protein degradation, especially the degradation of the amino acid methionine (Higgins et al. 2004). Similarly, H_2S can be formed from the degradation of the sulfur containing amino acid cysteine. Once H_2S and MeSH are formed, they can be methylated to form Me_2S and MeSH and can be oxidised to form Me_2S_2. Therefore the main

Table 23.2   Properties of VSCs (Smet *et al.* 1998).

Compound	b.p. (°C)	OT (ppbv)	$H_{25°C}$	MAK (ppmv)
H_2S	−60.7	8.5–1000	0.41	10.0
MeSH	6.2	0.9–8.5	0.10	0.5
Me_2S	37.3	0.6–40.0	0.07	20.0
Me_2S_2	109.7	0.1–3.6	0.04	<20.0
CS_2	46.2	9.6	0.65	10.0
COS	−50.0	n.d.a.	1.94	n.d.a.

b.p.: boiling point; OT: odour threshold; $H_{25°C}$: Henry coefficient at 25°C;
MAK: maximum concentration value in workplace conditions; n.d.a.: no data
available.

starting point for production of VSCs is the degradation of protein to produce free
amino acids, and subsequent degradation of the amino acids to form the VSCs.
Higgins *et al.* (2004) found that bound methionine content to be a possible pre-
dictor of the VSC production potential for biosolids.

Waste gas emissions from anaerobic processes in wastewater treatment plants are
known to be generated by a number of sulfurous compounds (Vincent and Hobson
1998), the most significant being H_2S and MeSH. Under anaerobic conditions
sulfate-reducing bacteria produce H_2S from sulfate-containing wastewaters, and
also bacterial fermentation of organic matter (fats, polysaccharides and proteins)
occurs (Vincent 2001). Within a heated anaerobic digester, fermentation is the "acid
forming" stage of the process and volatile fatty acids (VFAs) are rapidly converted to methane. Hydrolysis of proteinaceous material, which contains the sulfur
containing amino acids: cysteine, cystine and methionine, and organic sulfur
compounds leads to the production of H_2S and organic sulfides and disulfides:

Organic matter→bacterial fermentation→VFAs→CH_4
Amino acids + bacteria→hydrolysis→H_2S + trace reduced S compounds

23.3   TREATMENT SYSTEMS FOR SULFUR WASTE GASES

Although many methods are available for the treatment of sulfurous compounds
in waste gases, not all are suitable for the selective removal of H_2S and other
sulfurous compounds. Available methods which are considered suitable for the
removal of H_2S without removing the useful biogas methane include physical and
chemical processes such as adsorption, chemical scrubbers, catalytic oxidation,
oxidation with ozone, membrane separation and biotechnological methods, such
as biofilters, biotrickling filters, bioscrubbers and membrane bioreactors (MBR).

23.3.1   Physicochemical methods
23.3.1.1   Adsorption
A typical adsorption treatment system consists of a vertical column, packed with
static beds of granular media, such as activated carbon, silica gel, zeolites, activated

aluminium and synthetic resins. The influent flows through the column and odorants are adsorbed onto the surface (Turk and Bandosz 2001). Activated carbon is the most commonly used adsorbent (Smet *et al.* 1998). It will adsorb organic molecules in preference to water vapour and can be tailored to provide optimum adsorption for the particular contaminant (Porteous 2000). The rate of adsorption for different constituents or compounds will depend on the nature of the constituents or compounds being adsorbed (non-polar versus polar) (Metcalf and Eddy 2003). It has also been found that the removal of H_2S depends on the concentration of the hydrocarbons in the waste gas. Typically, hydrocarbons are adsorbed preferentially before polar compounds such as H_2S are removed, as activated carbon is non-polar. Thus, the composition of the waste gases to be treated must be known if activated carbon is to be used effectively. When activated carbon is used for H_2S removal, chemical oxidation to elemental sulfur ($S°$) and sulfate (SO_4^{2-}) can also occur as well as adsorption, as activated carbon provides a catalytic surface for oxidation, which enhances the removal capacity (Smet *et al.* 1998). As the life of a carbon bed is limited, carbon must be regenerated or replaced regularly for continued H_2S removal (Metcalf and Eddy 2003). Adsorption could pose a problem for removing H_2S only and recovering methane for biofuel, as hydrocarbons are only slightly polar and therefore are adsorbed preferentially before polar compounds such as H_2S. However, when combined with an adsorber with affinity for H_2S (e.g. activated carbon) this method could be promising. Le Cloirec (1995) obtained a 99% removal efficiency for H_2S concentrations 7–70 ppmv in WWTP emissions, applying an activated carbon adsorber at a gas residence time of 6.5 s.

23.3.1.2   *Chemical scrubbers*

During chemical scrubbing, the pollutant is transferred from the waste gas stream to the scrubbing chemical through intense contacting (Smet *et al.* 1998). Two types of systems are commonly used, packed towers (Figure 23.1) and atomised mist systems. Chemical oxidation methods use oxygen with a catalyst, such as chelated iron, but to achieve better oxidation, chlorination, potassium permanganate treatment, hydrogen peroxide or ozonation are used (Janssen *et al.* 1999). Chemical scrubbing in packed bed towers is an established H_2S control technique. However, chemical scrubbing suffers from important drawbacks such as high operating costs, generation of halomethanes that are known air toxics (WERF 1996) and the requirement for hazardous chemicals, which poses serious health and safety concerns. Although the generation of halomethanes is a problem when using chemical scrubbers for the removal of VSC, this is what makes the technique a suitable method for cleaning waste gases while maintaining methane to be used as biofuels.

Chemical scrubbers specific for removing H_2S from biogases are commercially available. An example of such systems is the Apollo scrubber unit, which uses a scrubbing solution based on iron redox chemistry (Oceta 2004). The Apollo is reported to be up to 99% efficient at removing H_2S concentrations of up to 20,000 ppmv from biogas (Oceta 2004). Commercial chemical compounds solutions are also available for cleaning H_2S from sour gas using packed towers sprayed with the product. An example of such chemical solutions is the ISET process

Figure 23.1   Packed tower chemical scrubber (Northumbrian Water, UK).

(ISET 2004). The process involves dissolution of H_2S in an aqueous medium and ionisation to H^+ and S^{2-}. The sulfur ions can then be oxidised by polyvalent metal ions such as ferric (Fe^{3+}) and ferrous (Fe^{2+}) iron to elemental sulfur, which is precipitated. The gas scrubbing is a two-stage process in which the sour gas is passed through two packed columns, with the gas entering the column from the bottom and the cleaning solution from the top. The H_2S is absorbed by the cleaning solution and is converted to elemental sulfur. The sulfur is removed by filtering and the solution is regenerated in a packed regenerator column and re-circulated. However, the effectiveness of the ISET process is not well documented in the literature.

23.3.1.3   Catalytic oxidation

Catalytic oxidation is based on combustion of the waste gases. The efficiency of the technique for H_2S control is dependent on the level of combustion, as incomplete combustion could result in the generation of sulfurous compounds (Lens et al. 2001). Catalytic incineration is the complete chemical conversion of gaseous compounds with oxygen at a certain temperature (below or above 100°C) and pressure (1 or more atm) while the gases are in contact with a solid material (catalyst) to increase the oxidation rate. Gas phase catalytic incineration technology has many advantages over other gas phase treatment techniques. In the concentration range of 20–60 ppmv, it is sometimes cheaper than granular activated carbon adsorption. It requires less fuel and less expensive construction materials than high-temperature treatment processes such as incineration (Snape, 1977; Siebert et al. 1984). Most catalytic incineration operations still require elevated temperatures to ensure a complete reaction at a suitable fast rate. Therefore, either the catalytic bed or the feed stream is heated. Inlet gas streams are kept at 50–150°C higher than the ignition temperatures (Nakajima 1991). When the pre-heating uses an internal flame, the process is complex, involving both the products of combustion in the flame and the processes occurring at the catalyst. Lee *et al.* (1999) showed that platinum is a good catalyst for catalytic incineration of H_2S concentrations in the range of 15–40 ppm. Palladium was the best catalyst for methane oxidation, but was partially deactivated by H_2S (Meeyoo *et al.* 1998). This implies that Palladium should be avoided if methane is to be retained in the biogas mixture. Several sulfurous compounds can inhibit or poison the catalysts, which is a problem that can be partially overcome by operating the reactor at a higher temperature (>300°C). This method has a good potential for selectively removing sulfurous compounds, especially H_2S, when the right catalyst, such as platinum is used.

23.3.1.4   Oxidation with ozone

Sulfurous compounds can be oxidised by ozone (O_3) alone, without the need for a catalyst or UV light (Lens *et al.* 2001). Gas phase O_3 has the same preference for reaction with electron-rich parts of organic or inorganic compounds present. However, Kuo *et al.* (1997) showed that the reaction mechanism might differ by different order of reaction in both reactants in the gas phase compared to that in aqueous solution. Ozonation of offgas from wastewater pumping stations has worked successfully in the metropolitan water reclamation district of Chicago, USA (WEF and ASCE 1998). However, controlling the dosage of O_3 to ensure that overdosing does not occur is critical and expensive (WEF and ASCE 1998). Overdosing will result in the generation of an odorous O_3 offgas, although this can be overcome with an O_3 destruction system.

Ozonation becomes even more effective when combined with UV light. The UV light can dissociate the O_3 molecule, yielding reactive radicals that are capable of oxidising organic and inorganic compounds at higher rates than O_3 alone would do (Lens *et al.* 2001). UV light photolysis of sulfurous compounds is initiated by scission of the S—H bond (McClean *et al.* 1999):

$$H_2S + hv \rightarrow HS^\bullet + H^\bullet$$
$$CH_3SH + hv \rightarrow CH_2S^\bullet + H^\bullet$$

The radicals formed in this reaction can effectively be degraded further by reaction with air.

23.3.1.5   Membrane separation

The principle of membrane separation is that some components of the raw gas are transported through a thin membrane while others are retained. The permeability is a direct function of the chemical solubility of the target component in the membrane. Solid membranes can be constructed as hollow fibre modules, which give a large membrane surface per volume and thus very compact units. Typical operating pressures are in the range of 25–40 bar. The principle of membrane separation constitutes a conflict between high methane purity in the upgraded gas and high methane yield (Jonsson *et al.* 2003). The purity of the upgraded gas can be improved by increasing the size or number of the membrane modules, but a larger amount of the methane gas will also permeate through the membranes and will be lost.

23.3.2 Biotechnological methods

Biotechnological methods of waste gas treatment have gained support as efficient and cost-effective methods (Janssen *et al.* 1999), as most of the compounds associated with waste gases are biogenic in origin and, thus, are biodegradable. Biological techniques are increasingly being favoured for waste gas treatment and appear effective and promising (Frechen 1994). In Europe, this has occurred over the past decade, primarily due to their efficiency, cost-effectiveness and environmental acceptability (Burgess *et al.* 2001). The biotechnological processes in use include biofilters, biotrickling filters, bioscrubbers and MBRs. The use of biological treatment methods for waste gases cleaning for use as biofuel is only effective if H_2S and other sulfurous compounds can be selectively removed without the removal of the useful biogas methane. A way of making sure biofilters, biotrickling filters and bioscrubbers will selectively remove H_2S is by inoculation of the media with specialised bacterial species, which have affinity for H_2S. These are chemolithotrophic bacteria species, such as species from the *Thiobacillus* genus, which depend on the oxidation of H_2S to derive energy for growth, making them ideal candidates for the selective removal of H_2S. Examples of such systems have been reported in the literature and are given in Table 23.3.

23.3.2.1 Biofilters

Biofilters have been used extensively for H_2S control at wastewater treatment plants since the 1950s (Ergas *et al.* 1995). Biofiltration units are packed bed microbial systems made of a wide range of materials including peat, heather, bark, compost, soil or a combination of these on which the microorganisms are attached as a biofilm (Figure 23.2). Air containing gaseous pollutants is first humidified to facilitate microbial activity and then forced through the media, and the pollutants

Table 23.3 Removal capacities for H_2S in biotechnological systems with microbial inoculation.

Reactor type	Inoculation	Removal capacity ($kg\,H_2S\text{-}S\,m^3\,day^{-1}$)	Reference
Biotrickling filter (polypropylene)	*T. thioparus* TK-m	0.56 (95%)	Tanji *et al.* 1989
Biofilter (mearl)	*Thiobacilli* sp.	1.36–1.58 (99%)	Bonnin *et al.* 1994
Biofilter (gel beads)	*Pseudomonas putida* CHII	0.46 (96%)	Chung *et al.* 1996
Biotrickling filter (activated carbon)	*Thiobacillus* sp.	3.22 (100%)	Guey *et al.* 1995

Adapted from Smet *et al.* (1998).

are sorbed by the filter material and degraded by the biofilm microbes. Sulfurous compounds in the air stream provide an energy source for the biomass, while the moisture added is used to supply other nutrients (e.g. nitrogen, phosphorus and potassium) to the biofilm (Burgess *et al.* 2001). The specific performance of a biofilter is relatively low, owing to the low density of microbes present as a mixed culture, although removals of up to 90% for selective compounds were reported (WEF and ASCE 1995). Biofiltration for H_2S removal can be improved when combined with the specific bacterial inoculum, as discussed above. Oyarzún *et al.* (2003) showed 100% removal of H_2S concentrations up to 355 ppmv using a biofiltration system with peat as the solid support inoculated with *Thiobacillus thioparus* ATCC 23645. They reported that the inoculation system allowed an efficient colonisation of the peat with *T. thioparus* reaching 2.7×10^8 cells g^{-1} dry peat. The main disadvantages of biofiltration for waste gas treatment include requirement of regular media replacement and control of moisture content is required to compensate for the drying action of the gas stream.

23.3.2.2   Biotrickling filter

In a biotrickling filter (Figure 23.3), the waste gas is forced through a packed bed filled with chemically inert carrier material. This material is colonised by microorganisms in a similar way to trickling filters in wastewater treatment. The liquid medium is circulated over the packed bed and can be re-circulated continuously or discontinuously and in co- or counter-current to the gas stream. The pollutants are first taken up by the biofilm on the carrier material and then degraded by the microorganisms in the biofilm (Van Langenhove and De Heyder 2001).

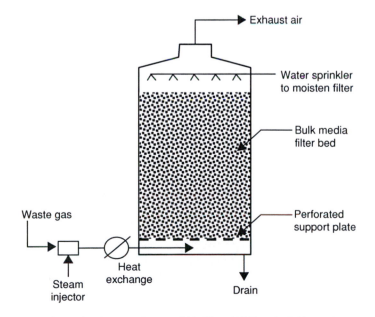

Figure 23.2   Schematic of a closed aerated biofilter (WEF and ASCE 1995).

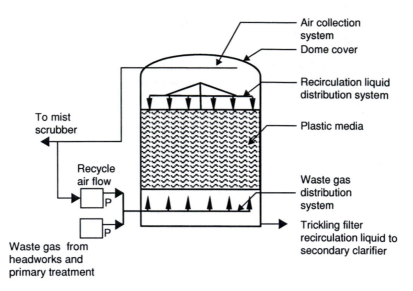

Figure 23.3   Schematic of a biotrickling filter for odour control (WEF/ASCE 1995).

Table 23.4   Microorganisms with H_2S removal capacity from waste gases.

Microorganism group	Species	References
Autotrophic	*Thiobacillus* spp.	Cho *et al.* 1991
		Bonnin *et al.* 1994
		Chung *et al.* 1996a
	Thioalkalivibrio versutus strain ALJ 15	Banciu *et al.* 2004
	Thioalkalivibrio spp.	Sorokin *et al.* 2001
	Thioalkalimicrobium spp.	Sorokin *et al.* 2001
Colourless sulfur bacteria	*Thiothrix* spp.	Lanting and Shah 1992
Phototrophs	*Chlorobium* spp.	Cork *et al.* 1983
	Chromatium spp.	Jensen and Webb 1995
	Ectothiorhodospira spp.	Then and Truper 1983
Methylotrophs	*Hyphomicrobium* spp.	Zhang *et al.* 1991
Cyanobacteria		Oren and Padan 1978
Fungi	*Sporormia concretivora*	Cho *et al.* 1994
Other heterotrophs	*Xanthomonas* spp.	Cho *et al.* 1992
	Pseudomonas putida CH11	Chung *et al.* 1996b

Owing to the presence of a re-circulating water phase, the applicability of biotrickling filters is also limited to pollutants with a low air-water partition coefficient (Herrygers *et al.* 2000). Guey *et al.* (1995) reported that the maximum H_2S elimination capacity for biofilters and biotrickling filters to be in the order of 3–3.5 kg H_2S -S m^3 day^{-1}. Herrygers *et al.* (2000) suggested a list of microorganisms that can be applied for H_2S removal from waste gases (Table 23.4).

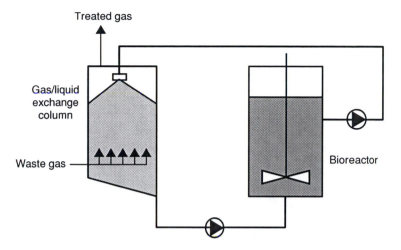

Figure 23.4   Schematic of a bioscrubber (Kennes and Thalasso 1998).

23.3.2.3   Bioscrubber

In a bioscrubber, the pollutant is absorbed in an aqueous phase in an absorption tower. The aqueous phase containing the dissolved compounds is then treated in a separate activated sludge unit (Smet *et al.* 1998) or upflow anaerobic sludge bed reactor (Sipma *et al.* 2003). The effluent of this unit is re-circulated over the absorption tower in a co- or counter-current way to the gas stream (Figure 23.4), which results in excellent cleaning of highly soluble pollutants. The pollutants are dissolved and degraded by the naturally occurring microorganisms in the bioreactor system (Kennes and Thalasso 1998). Nishimura and Yoda (1997) reported H_2S removal above 99% for an influent concentration of 300–2500 ppmv from anaerobic biogas using a bioscrubber at full scale for a 6-month period. Bioscrubbers have several advantages over media-based filtration. The process is more easily controlled because pH, temperature, nutrient balance and removal of metabolic products can be altered in the liquid of the reactor (Smet *et al.* 1998). One disadvantage of the bioscrubber is that the pollutant must be dissolved into the liquid phase during its short residence time in the absorption column, so it is best suited to contaminants with high solubility (Herrygers *et al.* 2000), which is a major consideration as many waste gases are volatile and exhibit poor solubility.

23.3.2.4   Membrane bioreactor

MBR contains a membrane, which separates the gas phase from the liquid phase (Herrygers *et al.* 2000). Sulfurous compounds and oxygen from the gaseous phase diffuse through the membrane to the biologically active liquid phase (Figure 23.5), where they may be degraded (Ergas and McGrath 1997). In an MBR, both the advantages of the biofilter and the biotrickling filter are combined. As there is no liquid phase between the pollutants and the biofilm, as in a biofilter, a high mass transfer rate to the biofilm is achieved. Control of the biofilm

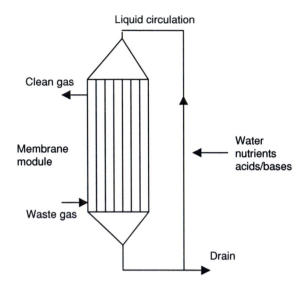

Figure 23.5   Schematic of a MBR for waste gas treatment, showing microporous membrane (Herrygers *et al.* 2000).

thickness is possible via nutrient addition to and removal of toxic products from the liquid phase.

Three types of membrane materials can be used for gas–liquid contact: hydrophobic microporous material (e.g. polypropylene, polycarbonate), dense material (e.g. silicone rubber) and composite membranes (which have a thin layer of dense material coated on top of a porous membrane). Microporous material has a higher permeability and has no selectivity towards different chemical compounds. A disadvantage of porous material is the possibility of blockage of the pores by microorganisms (Reij *et al.* 1998). Dense membranes (e.g. polydimethyl siloxane) are more advantageous if specific selectivity is required (Reij *et al.* 1998), as it has no pores the mass transfer of the compounds depends on their solubility and on the diffusion through the membrane. The mass transfer in these membranes is generally lower than in porous membranes. For a polypropylene porous membrane and polydimethyl siloxane dense membrane, this factor varies for different kinds of organic compounds, between 7 (propane) and 1600 (propanol) (Reij *et al.* 1998). Composite membranes have the advantage of a high mass transfer for the porous membrane and the thin layer of dense material located at the liquid site of the membranes avoids blockage of the pores by growth of microorganisms.

The advantage of MBRs is that they do not contain any moving parts, are easy to scale up, and the flows of gas and liquid can be varied independently, without the problems of flooding, loading or foaming commonly encountered in bubble columns (Reij *et al.* 1998). The disadvantages of MBRs are high capital and operational costs, and the unproven long-term operational stability (Reij *et al.* 1998). MBRs for H_2S removal to date have been limited mainly to laboratory and pilot tests.

23.4   CONCLUSIONS

The presence of sulfurous compounds in waste gas can impact upon the performance of the gases as a biofuel due to the corrosive nature of these compounds, particularly H_2S. Although most conventional gas treatment methods are suitable for removing H_2S from waste gases, not all methods are selective and might also remove the useful methane from the biogas. For this reason each method must be carefully examined before selection for H_2S scrubbing from biofuels. Examination of various methods reported in the literature showed that adsorption, chemical scrubbing and membrane separation are effective methods for the selective treatment of H_2S from sour gas. Biotechnological methods, such as the biofilter and biotrickling filter, which are becoming more favoured owing to their environmental acceptability, could also provide selective removal of H_2S from biofuels as biogas from wastewater processing, when inoculated with specific bacteria with affinity to H_2S.

REFERENCES

Badin, J., Kirschner, J. and McGuckin, R. (1999) *Biomass Energy – a Fuel Switching Strategy to Reduce Net Carbon Emissions and Enhance Asset Utilization*, ID No. 99/58, United Bioenergy Commercialization Association, Maryland, USA.

Banciu, H., Sorokin, D.Y., Kleerebezem, R., Muyzer, G., Galinski, E.A. and Kuenen, J.G. (2004) Growth kinetics of haloalkaliphilic, sulfur-oxidizing bacterium *Thioalkalivibrio versutus* strain ALJ 15 in continuous culture. *Extremophiles* **8**, 185–192.

Bonnin, C., Coriton, G. and Martin, G. (1994) Biodeodorization processes: from organic media filters to mineral beds. *VDI Berichte* **1104**, 217–230.

Boon, A.G. (1995) Septicity in sewers: causes, consequences and containment. *Water Sci. Technol.* **31**(7), 237–253.

Burgess, J.E., Parsons, S.A. and Stuetz, R.M. (2001) Developments in odour control and waste gas treatment biotechnology: a review. *Biotechnol. Adv.* **19**, 35–63.

Cho, K.-S., Zhang, L., Hirai, M. and Shoda, M. (1991). Removal characteristics of hydrogen sulfide and methanethiol by *Thiobacillus* sp. isolated from peat in biological deodorization. *J. Ferment. Bioeng.* **71**, 384–389.

Cho, K.-S., Hirai, M. and Shoda, M. (1992) Degradation of hydrogen sulfide by *Xanthomonas* spp. Strain DY44 isolated from peat. *Appl. Envrion. Microbiol.* **58**, 1183–1189.

Cho, K.-S., Orthani, T. and Mori, T. (1994) Newly found fungus helps to corrode concrete pipes. *Water Quarter Int.* **3**, 27.

Chung, Y.-C., Huang, C. and Tseng, C.-P. (1996a) Operation optimization of *Thiobacillus thioparus* CH11 biofilter for hydrogen sulfide removal. *J. Biotechnol.* **52**, 31–38.

Chung, Y.-C, Huang, C. and Tseng, C.-P. (1996b) Biodegradation of hydrogen sulfide by a laboratory-scale immobilized *Pseudomonas putida* CHII biofilter. *Biotechnol. Progress* **12**, 773–778.

Cork, D.J., Garunas, R. and Sajjad, A. (1983) *Chlorobium limicola* forma *thiosulfatophilum:* biocatalyst in the production of sulfur and organic carbon from a gas stream containing H_2S and CO_2. *Appl. Environ. Microbiol.* **45**, 913–918.

EPA (2004) *http://www.epa.gov/lmop/benefits.htm*

Ergas, S.J., Schroeder, E.D., Chang, D.P.Y. and Morton, R.L. (1995) Control of volatile organic compound emissions using a compost biofilter. *Water Environ. Res.* **67**, 817–821.

Ergas, S.J. and McGrath, M.S. (1997) Membrane bioreactor for control of volatile organic compound emissions. *J. Environ. Eng. ASCE* **123**, 593–598.

European Commission (1997) Energy for the Future: Renewable Sources of Energy – White Paper for a Community Strategy and Action Plan, COM (97) 599 final, Brussels.

Frechen, F.-B. (1994) Odour emissions of wastewater treatment plants – recent German experiences. *Water Sci. Technol.* **30**(4), 35–46.

Guey, C., Degorce-Dumas, J.R. and Le Cloirec, P. (1995) Hydrogen sulfide removal by biological activated carbon. *Odours VOCs J.* **1**, 136–137.

Herrygers, V., Langenhove, V.H. and Smet, E. (2000) Biological treatment of gases polluted by volatile sulfur compounds. In: *Environmental Technology to Treat Sulfur Pollution, Principles and Engineering* (eds. Lens, P. and Hulshoff, L.). IWA Publishing, London.

Higgins, M.J., Adams, G., Card, T., Chen, Y.-C., Erdal, Z. and Witherspoon, J. (2004) Relationship between biochemical constituents and production of odor causing compounds from anaerobically digested biosolids. In: *Proceedings of WEF/A&WMA Odors and Air Emissions Conference*, 18–21 April, Bellevue, WA, USA.

ISET (2004) Energy from waste: power generation in distillery effluent/sewage treatment plants using ISET process. *http://cgpl.iisc.ernet.in/iset.html*

Jensen, A.B. and Webb, C. (1995) Treatment of H₂S-containing gases: a review of microbiological alternatives. *Enzyme Microb. Technol.* **17**, 2–10.

Janssen A.J.H., Lettinga G., and de Keizer A. (1999) Removal of hydrogen sulfide from wastewater and waste gases by biological conversion to elemental sulfur – colloidal and interfacial aspects of biologically produced sulfur particles. *Colloid. Surf. A: Physiochem. Eng.Aspect.* **151**, 389–397.

Jonsson, O., Polman, E., Jensen, J.K., Eklund, R., Schyl, H. and Ivarsson, S. (2003) Sustainable gas enters the European gas distribution system. *International Gas Union 22nd World Gas Conference*, 1–5 June 2003, Tokyo, Japan.

Kennes, C. and Thalasso, F. (1998) Waste gas biotreatment technology. *J. Chem. Technol. Biotechnol.* **72**, 303–319.

Klass, D.L. (1999) *An Introduction to Biomass Energy Research – A Renewable Resource.* Biomass Energy Research Association, WA, USA.

Kuo, C.H., Zhong, L., Wang, J. and Zappi, M.E. (1997) Vapour and liquid phase ozonation of benzene. *Ozone Sci. Eng.* **19**, 109–117.

Lanting, J. and Shah, A.S. (1992) Biological removal of hydrogen sulfide from biogas. In: *46th Purdue Industrial Waste Conference Proceedings.* Lewis Publishers, Inc., Chelsea.

Laplanche, A., Bonnin, C. and Darmon, D. (1994) Comparative study of odours removal in a wastewater treatment plant by wet scrubbing and oxidation by chlorine or ozone. In: *Characterisation and Control of Odoriferous Pollutants in Process Industries*, (eds. S. Vigneron, Nermia, J. and Chaouki, J.), pp. 277–294. Elsevier Science B.V., Amsterdam.

Le Cloirec, P. (1995) Treatment of odors and VOC with gas-solid transfer: adsorption. *Odours VOC's J. 1*, 209–213.

Lee, J.H., Trimm, D. and Cant, N.W. (1999) The catalytic combustion of methane and hydrogen sulfide. *Catal. Today* **47**, 353–357.

Lens, P.N., De Poorter, M.-P., Cronenberg, C.C. and Verstraete, W.H. (1995) Sulfate reducing and methane producing bacteria in aerobic waste water treatment. *Water Res.* **29**, 871–880.

Lens, P.N.L., Boncz, M.A., Sipma, J., Bruning, H. and Rulkens, W.H. (2001) Catalytic oxidation of odorous compounds from waste treatment processes. In: *Odours in Wastewater Treatment: Measurement, Modelling and Control* (eds. R.M. Stuetz and F.-B. Frechen). IWA Publishing, London.

Makaly-Biey, E.M. and Verstraete, W. (1999) The use of a UV lamp for control of odour decomposition of kitchen and vegetable waste. *Environ. Technol.* **20**, 331–335.

McClean, J.C., Hamilton, D.J. and Clark, N. (1999) Odour abatement using enhanced UV photolysis: a new application for a proven technology. *Proceedings of the CIWEM and IAWQ Conference on Control and Prevention of Odours in the Water Industry*, London, UK, September 1999.

Meeyoo, V., Lee, J.H., Trimm, D.L. and Cant, N.W. (1998) Hydrogen sulfide emission control by combined adsorption and catalytic combustion. *Catal. Today* **44**, 67–72.

Metcalf and Eddy, (2003) *Wastewater Engineering Treatment and Reuse*, 3rd edn. (revised by G. Tchobanoglous, F.L. Burton and H.D. Stensel). McGraw-Hill Inc, Toronto.

Nakajima, F. (1991) Air pollution control with catalysis – past, present and future. *Catal. Today* **10**, 1.

Nishimura, S. and Yoda, M. (1997) Removal of hydrogen sulfide from an anaerobic biogas using a bio-scrubber. *Water Sci. Technol.* **36**(6–7), 349–356.

Oceta (2004) Environmental Technology Profiles Catalogue. *http://www.oceta.on.ca/profiles/apollo/scrubber.html*

Oren, A. and Padan, E. (1978) Induction of anaerobic, photoautotrophic growth in the cyanobacterium *Oscillatoria limmetica*. *J. Bacteriol.* **133**, 558–563.

Oyarzún, P., Arancibia, F., Canales, C and Aroca, G.E. (2003) Biofiltration of high concentration of hydrogen sulfide using *Thiobacillus thioparus*. *Process Biochem.* **39**, 165–170.

Porteous, A. (2000) *Dictionary of Environmental Science and Technology*, 3rd edn. John Wiley & Sons, Ltd., Chichester.

Reij, M.W., Keurentjes, J.T.F. and Hartmans, S. (1998) Membrane bioreactors for waste gas treatment. *J. Biotechnol.* **59**, 155–167.

Siebert, P.C., Meardon, K.R. and Serne, J.C. (1984) Emission control in polymer production. *Chem. Eng. Proc.* 86–176.

Sipma, J., Janssen, J.J.H., Hulshoff Pol, L.W. and Lettinga, G. (2003) Development of a novel process for the biological conversion of H_2S and methanethiol to elemental sulfur. *Biotechnol. Bioeng.* **82**, 1–11.

Smet, E., Lens, P. and Van Langenhove, H. (1998) Treatment of waste gases contaminated with odorous sulfur compounds. *Crit. Rev. Environ. Sci. Technol.* **28**, 89–117.

Snape, T.H. (1977) Catalytic oxidation of pollutants from ink drying ovens. *Plat. Metal Rev.* **21**, 90–91.

Sorokin, D.Y., Lysenko, A.M., Mityushina, L.L., Turova, T.P., Jones, B.E., Rainey, F.A., Robertson, L.A. and Kuenen, J.G. (2001) *Thioalkalimicrobium aerophilum gen. nov., sp. nov.* and *Thioalkalimicrobium sibiricum. sp. nov.,* and *Thioalkalivibrio versutus gen. nov., sp. nov., Thioalkalivibrio nitratis sp. nov.* and *Thioalkalivibrio denitrificans sp. nov.,* Novel obligately alkaliphilic and obligately chemolithoautotrophic sulfur-oxidizing bacteria from soda lakes. *Int. J. Syst. Evolut. Microbiol.* **51**, 565–580.

Tanji, Y., Kanagawa, T. and Mikami, E. (1989) Removal of dimethyl sulfide, methyl mercaptan and hydrogen sulfide by immobilized *Thiobacillus thioparus* TK-m. *J. Ferment. Bioeng.* **67**, 280–285.

Then, J. and Truper, H.G. (1983) Sulfide oxidation in *Ectothiorhodospira abdelmaleckii*. Evidence for the catalytic role of cytochrome c-551. *Arch. Microbiol.* **135**, 254–258.

Turk, A. and Bandosz, T.J. (2001) Adsorption systems for odour treatment. In: *Odours in Wastewater Treatment: Measurement, Modelling and Control* (eds. R.M. Stuetz, and F.-B. Frechen). IWA Publishing, London.

Van Langenhove, H. and De Heyder, B. (2001) Biological treatment of odours. In: *Odours in Wastewater Treatment: Measurement, Modelling and Control* (eds. R.M. Stuetz and F.-B. Frechen). IWA Publishing, London.

Vincent, A.J. (2001) Sources of odours in wastewater treatment. In: *Odours in Wastewater Treatment: Measurement, Modelling and Control* (eds. R.M. Stuetz, and F.-B. Frechen). IWA Publishing, London.

Vincent, A. and Hobson, J. (1998) Odour control. *CIWEN Monographs on Best Practice* No. 2. London, UK: Chartered Institution of Water and Environmental Management, pp. 31.

Water Environment Federation (WEF) and American Society of Civil Engineers (ASCE) (1995) Odour control in wastewater treatment plants. *Water Environment Federation (WEF) Manual of Practice No. 22, American Society of Civil Engineers (ASCE) Manuals and Reports on Engineering Practice No. 82.* WEF/ASCE, USA, 304 pp.

Water Environment Federation (WEF) and American Society of Civil Engineers (ASCE) (1998) *Design of Municipal Wastewater Treatment Plants,* 4th edn., Vol. 2. WEF/ASCE, VA, USA.

Water Environment Research Foundation (1996) *A Critical Review of Odour Control Equipment for Toxic Air Emissions Reduction.* WERF Project No. 91 – VOC2 (Water Environment Research, Alexandria, VA).

World Health Organisation (WHO) (1987) *Air Quality Guidelines for Europe.* WHO Regional Publications Series No. 23, Regional Office for Europe, Copenhagen.

24

Control of siloxanes

F. Accettola and M. Haberbauer

24.1 INTRODUCTION

The increase of the usage of biogas in power generation plants can be pursued only if the costs associated with using this resource are similar to the purchase of traditional fuels. In this framework, a cost-competitive biogas-cleaning unit plays an important role. The aim of gas cleanup is to reduce the effects of contaminants and to promote a higher degree of operational effectiveness. This is especially important when integrating biofuel usage and fuel cell technology.

Prior to its use in a fuel cell (as well as in other power generation systems) biofuels must be essentially free of harmful compounds like particulates, water (although a harmless product for the fuel cell itself, it damages for example the flow controls and it becomes corrosive in combination with ammonia (NH_3) and especially hydrogen sulfide (H_2S), NH_3, carbon dioxide (CO_2) (which dilutes the biofuel and which is only harmful for the Alkaline Fuel Cells), H_2S, halogenated hydrocarbons and volatile methyl siloxanes (VMSs). Gas cleanup technologies for removal of siloxanes will be discussed in more details in this chapter.

24.2   OVERVIEW OF SILOXANES

A major problem with the use of biogas for heat and power production is silicate deposits. The primary cause of silicate-based material deposition is the presence of organosilicon compounds in the biogas. These compounds (VMSs) originate from the hydrolysis of polydimethylsiloxane (PDMS), an organosilicon compound which is used in a wide range of household and industrial applications. In industrial applications, PDMS fluids functions as highly efficient process aids. They are used as softeners and wetting agents in textile manufacturing, for example, and as a component of many polishes and other surface treatment formulations (Allen *et al.* 1997). Other applications include electrical transformer fluids, mould releasing agents in the rubber industry, paint additives, water repellent apparel, shoe polish, medical anti-acids and anti-foaming agents. PDMS is characterized by low surface tension, resistance to chemical and thermal attack, water repellency and resistance to oxidation, and therefore often used in various adverse environments, such as those with corrosive chemicals, high humidity, high temperature, electrical fields, etc. (Watts *et al.* 1995). Silicones are also used in several

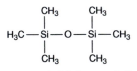

Hexamethyldisiloxane (L2)   Octamethyltrisiloxane (L3)

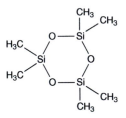

Hexamethylcyclotrisiloxane (D3)   Octamethylcyclotetrasiloxane (D4)

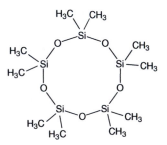

Decamethylcyclopentasiloxane (D5)

Figure 24.1   Structural formula of siloxanes found in sewage and landfill gas.

kinds of domestic products, such as detergents, shampoos, conditioner, deodorants, creams, gels, water repellents, and other products. In these products PDMS fluids functions as anti-foam, anti-flatulant, conditioning, carrier, aesthetics or water repellency.

After use these products are released into the environment through landfills and wastewater treatment plants. Insoluble in water, high-molecular-mass liquid siloxanes have a high adsorption coefficient. Therefore, they strongly bind with other organic molecules and they become part of the activated sludge, subsequently used for biogas production by anaerobic digestion (Chandra 1997).

The most important species detected so far in biogas (Figure 24.1) are: hexamethylcyclotrisiloxane (D3), decamethylcyclopentasiloxane (D5), octamethyl-cyclotetrasiloxane (D4), hexamethyldisiloxane (L2), octamethyltrisiloxane (L3).

The concentration values of siloxanes in biogas depends on the origin of the biogas. Table 24.1 gives the average concentrations of siloxanes measured in an experimental plant in Asten (Austria), where landfill gas as well as sewage gas are produced. Landfill gas is usually richer in siloxanes than sewage gas, with an average concentration of 24 mg/m³ against 7 mg/m³. Among VMSs, D4 is the compound with the highest concentration, in landfill gas it can be present with a concentration of 12 mg/m³, while in sewage gas an average concentration of D5 of 6 mg/m³ has been reported. Concentration of siloxanes in biogas can reach higher values (up to 50 mg/m³) than the ones here reported (Schweigkofler and Niessner 2001).

24.3   SILOXANE DAMAGES

Over the past several years, a lot of attention has been focused on siloxanes in biogas and the costly problems they create for operators of power generation equipment because of the formation of silicate-based deposits.

During combustion of biogas containing siloxanes, silicon is released and can combine with free oxygen or various other elements in the combustion gas. These deposits can ultimately build to a surface thickness of several millimetres and they are difficult to remove by chemical or mechanical means. They stack on pistons, cylinder heads and valves reducing compression and engine efficiency (Martin et al. 1996). The manufacturers of gas engines recently introduced a limit value for silicon of 1 mg/l measured in the oil of the gas engine in order to prevent premature engine failure due to silicon-induced damages (Prabucki et al. 2001).

Table 24.1   Siloxane concentrations in landfill gas and sewage gas (Asten, Austria).

Siloxanes	Landfill gas (mg/m³)	Sewage gas (mg/m³)
Hexamethyldisiloxane (L2)	6.07	0.02
Hexamethylcyclotrisiloxane (D3)	0.49	0.04
Octamethyltrisiloxane (L3)	0.32	0.02
Octamethylcyclotetrasiloxane (D4)	12.53	0.93
Decamethylcyclopentasiloxane (D5)	4.73	6.03
Total	24.15	7.04

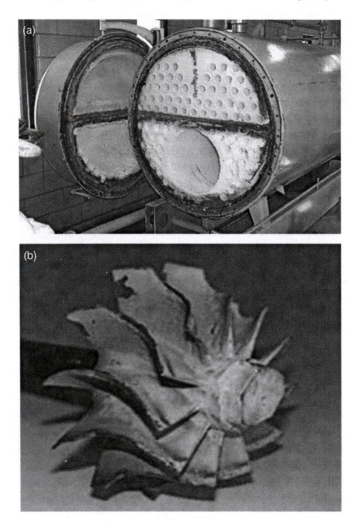

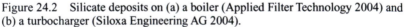

Figure 24.2   Silicate deposits on (a) a boiler (Applied Filter Technology 2004) and (b) a turbocharger (Siloxa Engineering AG 2004).

Silicate deposits result also in poor heat transfer in heat exchangers (Figure 24.2a). In turbines, silicates can cause severe abrasion of blades (Figure 24.2b). Additionally, the glassy residues are responsible for the inactivity of the surfaces of the catalytic system for the control of waste gas, thus reducing the efficiency of the catalyst for waste gas cleaning. Catalyst beds can fail in a few days and there is no way to recover them.

24.3.1   Siloxane damages in fuel cells

While the effect of siloxanes in gas engines and turbines is well known, there is still only limited data about the effect they can have on fuel cells. In a research project (AMONCO 2004), siloxane D5 has been injected into a single polymer

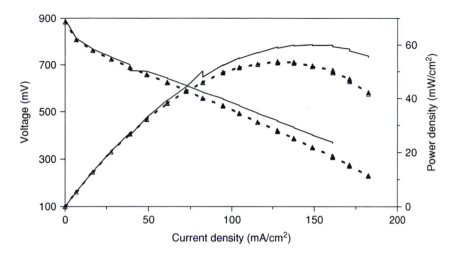

Figure 24.3   Polarization curves of single PEMFC before and after D5 injections (111 ppm) to the anodic O_2 flow with 5 min between (AMONCO 2004). — Before injection, ▲ after injection.

electrolyte fuel cell (PEMFC), operating at low temperature (80–120°C). Siloxane injection results in a lower voltage and power density, for a given value of the current density (Figure 24.3). This effect might be explained by supposing that silicate deposits stick on the catalyst blocking the active sites, as reported already for catalysts in the effluent gas treatment system.

24.4   PHYSICAL SILOXANE REMOVAL PROCESSES

Currently, a standard method for the elimination of siloxanes from biogas does not exist. Treatment systems reported by industrial sources are typically based on physical operations.

24.4.1   Physical removal processes

24.4.1.1   Full scale applications

The Thyssen Fuel cell plant in Cologne (Germany) has been operating the first Phosphoric Acid Fuel Cell in Europe producing heat and electricity from digester gas produced from the digestion of sewage sludge (Langnickel 2000). The cleaning device adopted in the plant consists of a two-stage basic cleaning unit (Siloxa Engineering AG 2004). The process gas is chilled at a temperature of −30°C. Condensation and ice formation occur due to the moisture contained in the device. The ice absorbs the siloxanes. The resulting gas is finally purified by activated carbon. The system can treat $110\,m^3/h$ gas, guaranteeing a purity level of $<1\,mg/m^3$ gas and no siloxanes were detected after the gas treatment in the Thyssen Fuel Cell plant in Cologne.

   Another reported method by industry sources is based on the physical adsorption employing polymorphous porous graphite sieves (by the company Applied

Filter Technology), which are very selective regarding siloxanes. This method has been successfully applied in several plants in USA, producing heat and power from biogas, which reported damage to engines, turbines and the catalytic oxidizer for carbon monoxide and hydrocarbon removal. The main advantage of the adsorption material is that it may be regenerated. The system can guarantee a purity level of $<0.3 \, mg/m^3$ gas.

In the USA, liquid absorption is used by a handful of landfill operators to treat biogas before using in combustion devices. One popular liquid is dimethylether of polyethylene glycol (SELEXOL®). It is designed to provide CO_2, H_2S, COS, mercaptans and BTEX removal from biogas, but it has also been found to be efficient in the removal of VMS (this solvent has been used at a number of USA landfills like Monterey Park, California; Calumet City, Illinois; Fresh Kills, New York). This solvent is both non-toxic and non-corrosive. Its main disadvantage is a relatively high cost of £4.40/l (Environmental Agency 2002). It can be regenerated using a series of flash depressurization and air stripping columns. One specific company which reportedly has succeeded in removing siloxane is Ecogas, Inc., a division of Getty Synthetic Fuels. Their system uses SELEXOL® in conjunction with a condenser and activated carbon. Reported removal efficiencies are high, and the processed gas in one case is used at a local university's cogeneration turbine which has catalytic oxidizers for carbon monoxide (CO) and hydrocarbon removal (Glus *et al.* 2001).

24.4.1.2   Novel developments

At the present time, there is no consensus regarding a siloxane treatment alternative for biogas. In addition, there is a lack of sufficient experience in dealing with siloxanes within the electrochemical processes occurring in Fuel Cells.

In the UK, a solvent liquid absorption system using a hydrocarbon oil, aimed primarily at scrubbing halogenated organics was tested (Stoddart *et al.* 1999). This system also achieved 60% siloxanes removal. The process was based on scrubbing landfill gas with a counter current of hydrocarbon oil. The trace components were adsorbed in a packed bed absorption column. The contaminated oil was heated up and pumped into a desorption tower where the absorbed contaminants were vaporized under vacuum. The gaseous contaminants were flared off and the condensate was collected from the separator into drums for further disposal. The regenerated and cooled oil was pumped back into the absorption tower. The efficiency of the removal is related to the boiling point of the trace compounds as well as the landfill gas to hydrocarbon oil ratio. The main disadvantage of this method is that it is not very efficient at removing some contaminants, especially those with low boiling points, for example freons (Stoddart *et al.* 1999).

Activated carbon is used in many studies for siloxanes removal (Hagmann *et al.* 2001; Prabucki *et al.* 2001; Environmental Agency 2002). Adsorption on silica gel and polymer beads (Schweigkofler and Niessner 2001) can also be used. In laboratory studies, several liquid and solid materials were evaluated for their siloxane elimination efficiency. The efficiency of hot concentrated sulfuric acid in eliminating gaseous siloxanes is already known (Huppmann *et al.* 1996). Practical application of this absorbent, however, seems problematic due to its corrosive potential. Therefore in the study of Schweigkofler and Niessner (2001), solutions of H_2SO_4 differing in concentration and working temperature were tested to determine the mildest conditions for quantitative siloxane removal. The siloxanes studied were eliminated almost quantitatively using half-concentrated H_2SO_4 (48 wt%) at a temperature of 60°C (Schweigkofler and Niessner 2001). With sulfuric acid at room temperature, only 56–70% of both L2 and D5 was removed. A further decrease to an acid concentration of 24 wt% resulted in a significant decline in siloxane elimination efficiency to 64–73%. Also concentrated nitric acid was found to be especially potent siloxane removing agent at elevated temperature (Schweigkofler and Niessner 2001). Although the efficiency of these materials is high, a practical application as absorbent media for biogas pretreatment should be performed very carefully because of their high potential risk for health and environment. Solid adsorbents tested include activated charcoal, carbopack B, Tenax TA, XAD II resins, molecular sieve 13X and silica gel. Apart from activated charcoal, silica gel showed especially high adsorption capacities of more than 100 mg/g siloxanes. Furthermore, excellent thermal regeneration of the loaded material is possible.

24.5   BIODEGRADATION OF SILOXANES

24.5.1   Biodegradation pathways

Initially siloxanes were treated as non-degradable in the environment and for many years, they were considered to be inert with respect to living organisms. Between the 1960s and the 1970s, the possibility of their chemical (environmental) and biodegradation was proven (Matsumura and Steinbüchel 2003). So far, there has been very little research concerning biodegradation of siloxanes, therefore only few data on the degradation and its products are available. One barrier in this research is given by the difficulty in measuring siloxanes, due to their low solubility and the high vaporization rate of VMSs.

Hydrolysis and re-arrangement of the silicon-oxygen bond in PDMS, catalyzed by contact with dry soil, has been observed (Lehmann et al. 1995). The products are mainly low-molecular-weight siloxanes (primarily D4 and D5 and linear VMSs) and silanols (mostly dimethylsilanediol (DMSD)). Upon this partial hydrolysis of PDMS to smaller molecules, biodegradation may proceed, giving inorganic water-soluble compounds as final product. Figure 24.4 shows a possible degradation pathway for D4.

24.5.2   Siloxanes biodegradation in soil

Micro-organisms responsible for siloxane degradation in soils were identified as *Arthrobacter* and *Fusarium Oxysporum* (Sabourin et al. 1996). Soil cores were collected every two weeks and analyzed for the decrease in total soil PDMS, and decrease in molecular weight of remaining PDMS (Lehmann et al. 1998a, 2000). PDMS concentrations decreased 50% in 4.5, 5.3, and 9.6 weeks for the low (215 μg/g), medium (430 μg/g), and high (860 μg/g) treatments, respectively. Degradation rates were 0.26 (low), 0.44 (medium), and 0.44 (high) g PDMS/m^2.day,

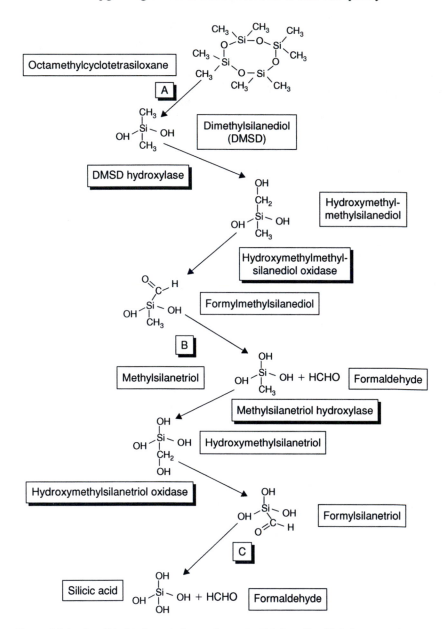

Figure 24.4　Possible biodegradation pathway for D4, http://umbbd.ahc.umn.edu/osi/osi_map.html

indicating that degradation capacity of the soil was exceeded at the high loaded treatment. DMSD, the main degradation product, was detected in most samples at <5% of the original PDMS. This is consistent with laboratory data showing biodegradation and volatilization of DMSD (Lehmann et al. 1998b). These studies suggest that PDMS will be degraded under field conditions.

Research conducted by Wasserbauer and Zadák (1990) demonstrated that bacteria from the genus *Pseudomonas* (*P. putida* and *P. fluorescens*) are capable of growing on oligomeric and polymeric dimethylsiloxanes, with the growth being a function of the increasing viscosity of siloxanes. Biodegradation occurs in aerobic conditions. Nevertheless, these results must be interpreted carefully because the increase of the degradation with the viscosity is a result unprecedented in literature and difficult to explain.

24.5.3 Bioreactor for siloxanes removal

Biodegradation of siloxane is an important field of study because, if further study will confirm the possibility to degrade VMSs and will as well identify the bacteria responsible and the optimum conditions, it might be possible to develop a biological system for biofuel upgrading, similar to the ones already successfully applied for the treatment of H_2S (Chapter 23) and halogenated hydrocarbons (Kleinheinz and Bagley 1998; Okkerse *et al.* 1999) for biogas. The main advantages of such a system are:

- low capital costs,
- low running costs,
- low maintenance,
- low noise,
- tolerance to fluctuations,
- no production of CO or other polluting substances.

All these features would therefore turn into a cost-competitive biofuel upgrading unit.

24.6 CONCLUSIONS

The level or nature of clean-up of a biofuel, and consequently its cost, depends on both the type of contaminants present and the type of technology adopted for power generation. One of the most important issues concerning biogas upgrading is related to the presence of volatile organosilicon compounds, which lead to the formation of silicate-based deposits, responsible for damages to equipment in power plants. So far, siloxanes are removed by means of expensive physical methods, but current research is addressed to find out less costly processes (e.g. redeemable materials) in order to encourage usage of biogas as energy source, alternative to traditional fuels.

REFERENCES

Allen, R.B., Kochs, P. and Chandra, G. (1997) Industrial organosilicon materials, their environmental entry and predicted fate. In: *Organosilicon Materials*, Vol. 3, Part H. Springer-Verlag, Berlin Heidelberg.

AMONCO (2004) *EU Project: Advanced Prediction, Monitoring and Controlling of Anaerobic Digestion Behaviour Towards Biogas Usage in Fuel Cells*. Contract No. ENK6-CT-2001-00518, 4th Progress Report from the institute CSIC.

Applied Filter Technology (2004) http://www.appliedfiltertechnology.com
Chandra, G. (1997) Organosilicon materials. In: *The Handbook of Environmental Chemistry*, Vol.3, Part H. Springer-Verlag, Berlin Heidelberg.
Environmental Agency (2002) *Guidance on Gas Treatment Technologies for Landfill Gas Engines*, December 2002, Bristol. www.environment-agency.gov.uk
Glus, P.H., Pirnie, M., Liang, K.Y., Ramon Li, P.E., Richard, P.E. and Pope, J. (2001) Recent advances in the removal of volatile methylsiloxanes from biogas at sewage treatment plants and landfills. *Paper presented at the Annual Air and Waste Management (AWMA) 2001 Conference*, Orlando, Florida.
Hagmann, M., Hesse, E., Hentshel, P. and Bauer, T. (2001) Purification of biogas – removal of volatile silicones. *Proceedings Sardinia 2001, Eighth International Waste Management and Landfill Symposium*, Vol. II, pp. 641–644.
Huppmann, R., Lohoff, H.W. and Schröder, H.F. (1996) Cyclic siloxanes in the biological waste water treatment process – Determination, quantification and possibilities of elimination. *Fresen. J. Anal. Chem.* **354**, 66–71.
Kleinheinz, G.T. and Bagley, S.T. (1998) Biofiltration for the removal and 'detoxification' of a complex mixture of volatile organic compounds. *J. Ind. Microbiol. Biotechnol.* **20**, 101–108.
Langnickel, U. (2000) First European fuel cell application using digester gas. *Fuel Cell Bull.*, 10–12.
Lehmann, R.G., Varaprath, S., Annelin, R.B. and Arndt, J.L. (1995) Degradation of silicone polymer in a variety of soils. *Environ. Toxicol. Chem.* **14**, 1299–1305.
Lehmann, R.G., Miller, J.R., Xu, S., Singh, U.B. and Reece, C.F. (1998a) Degradation of silicone polymer at different soil moistures. *Environ. Sci. Technol.* **32**, 1260–1264.
Lehmann, R.G., Miller, J.R. and Collins, H.P. (1998b) Microbial degradation of dimethylsilanediol in soil. *Water Air Soil Pollut.* **106**, 111–122.
Lehmann, R.G., Miller, J.R., and Kozerski, G.E. (2000) Degradation of silicone polymer in a field soil under natural conditions. *Chemosphere* **41**, 743–749.
Martin, P., Ellersdorfer, E. and Zeman, A. (1996) Auswirkungen flüchtiger Siloxane auf Verbrennungsmotoren. *Korrespondenz Abwasser* **43**, 1574.
Matsumura, S. and Steinbüchel, A. (2003) *Miscellaneous Biopolymers and Biodegradation of Synthetic Polymers: Biodegradation of Silicones (Organosiloxanes)*, Vol. 9, pp. 539–568. Wiley-VCH, Weinheim.
Okkerse, W.J.H., Ottengraf, S.P.P., Diks, R.M.M., Osinga-Kuipers, B. and Jacobs, P. (1999) Long term performance of biotrickling filters removing a mixture of volatile organic compounds from an artificial waste gas: dichloromethane and methylmethacrylate. *Bioprocess Eng.* **20**, 49–57.
Prabucki, M.J., Doczyck, W. and Asmus, D. (2001) Removal of organic silicon compounds from landfill and sewer gas. *Proceedings Sardinia 2001, Eighth International Waste Management and Landfill Symposium*, Vol. II, pp. 631–639.
Sabourin, C.L., Carpenter, J.C., Leib, T.K. and Spivack, J.L. (1996) Biodegradation of dimethylsilanediol in soils. *Appl. Environ. Microbiol.* **62**, 4352–4360.
Schweigkofler, M. and Niessner, R. (2001) Removal of siloxanes in biogases. *J. Hazard. Mater.* **B83**, 183–196.
Siloxa Engineering AG (2004) http://www.siloxa-ag.de
Stoddart, J., Zhu, M., Staines, J., Rothery, E. and Lewicki, R. (1999) Experience with halogenated hydrocarbons removal from landfill gas. *Proceedings Sardinia 99, Seventh International Waste Management and Landfill Symposium*, Vol. II, pp. 489–498.
Wasserbauer, R. and Zadák, Z. (1990) Growth of *Pseudomonas putida* and *P. fluorescens* on silicone oils. *Folia Microbiol.* **35**, 384–393.
Watts, R.J., Kong, S., Haling, S., Gearhart, L., Frye, C.L. and Vigon, B.W. (1995) Fate and effects of polydimethylsiloxanes on pilot and bench-top activated sludge reactors and anaerobic/aerobic digesters. *Water Res.* **28**, 2405–2411.

PART FIVE: Case studies

25

Energetical utilisation of biogas with PEM: fuel cell technologies

T. Ahrens and P. Weiland

25.1 INTRODUCTION

The utilisation of biogas with low-temperature fuel cells (FC) gives a new perspective to the usage of renewable energy out of agricultural residues, energy crops and renewable resources. Due to this, the corresponding process chain was realised and tested in a pilot plant at the Institute of Technology and Biosystems Engineering of the Federal Agricultural Research Centre (FAL) in Braunschweig (Germany). Major topics investigated were:

- Examination of different ensiled energy crops concerning anaerobic degradability and biogas quality (including trace gases).
- Hot testing and evaluating of different biological, chemical and physical technologies for upgrading of biogas into biogenous hydrogen.
- Hot testing and evaluating of utilisation of biogenous hydrogen in low-temperature FC.

25.2   ANAEROBIC DEGRADATION OF ENSILED ENERGY CROPS

Important aspects for implementation of biogas technology in connection with ensiled energy crops on farms are electrical and thermal efficiency of the process, economical aspects, technical aspects concerning the fermentation process, process requirements (gas quality/gas production) and substrate aspects.

In Germany, the energy production out of renewable resources is subsidised by the government. The "Erneuerbare-Energien-Gesetz" (EEG), which became operative in April 2000 and was amended in 2004, grants certain allowances depending on the kind of utilised renewable energy. The two main aims of this law are first stimulating renewable energy for multiplying the shares of renewable energy in the whole energy infrastructure in Germany and second doubling the contingent

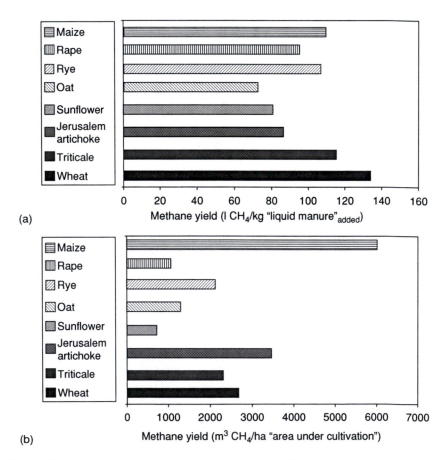

Figure 25.1   Methane yield of ensiled energy crops (a) referring to liquid matter and (b) referring to crop yield (Bley 2001; Tietjen 2002; Vogeler 2003).

of renewable energy referring to the whole energy consumption in Germany until year 2010. In Section 5 of the EEG an allowance per supplied kWh of electricity out of biomass is ensured to plant operators, depending on the installed electrical plant power. With a substrate constriction to energy crops and agricultural residues an additional allowance of 6€cent per supplied kWh of electricity is given. In addition to this, a further allowance is granted for the implementation of future heat and energy utilisation technologies. These are the most important impulses for biogas production and establishment of agricultural biogas plants on farms in Germany.

Due to these facts examinations with different ensiled energy crops were done. Figure 25.1a shows different results of anaerobic batch experiments. Under consideration of crop yields referring to area under cultivation, further varieties appear (Figure 25.1b). Figure 25.1b highlights the special position of ensiled maize as substrate for anaerobic digestion. But it is of high importance to consider that an excessive cultivation of maize in agricultural mono-cultures might lead into serious ecological and societal problems. Furthermore, a co-fermentation of a mixture out of different energy crops with or even without liquid manure leads to a more stable digestion process in comparison to digestion in mono-fermentation processes (e.g. with ensiled maize).

In conclusion, the digestion of energy crops in biogas processes leads to high-quality biogas mixtures, which allow a direct upgrading into biogenous hydrogen.

25.3   TRACE GASES IN BIOGAS MIXTURES

One main aspect in connecting low temperature FC to biogas plants is the strong demand in a high-quality process gas (clean hydrogen with a concentration of disturbing trace gas compounds below 10 ppm). Therefore, the spectrum of disturbing trace gas compounds in biogas mixtures, depending on different substrates for the fermentation process, was measured with a combination of gas chromatography with mass spectrometer technology (GC-MS). Depending on variation of substrates several trace gas compounds were detected (Table 25.1). The biogas plants listed in Table 25.1 work with mixtures of substrates, which include energy crops and manure as well as with different materials from waste material markets like fatty acids and glycerine for example.

The most important result of the trace gas examinations is a significant relationship between different substrates and certain trace gas compounds in the biogas and a lower concentration of trace gas compounds in biogas mixtures from renewable resources in comparison to biogas mixtures out of substrates from waste material markets.

The concentration rate of disturbing trace gas compounds in most biogas mixtures is significantly lower than 10 pp, except hydrogen sulfide (Table 25.2).

Most biogas mixtures from agricultural biogas plants are suitable for high-efficient energy converting processes based on catalysed reaction steps, like FC technology for example. It is of importance to consider that the trace gas spectrum might vary due to the situation of the anaerobic digestion process, the type of feedstock and that a biogas clean-up step is compulsory.

Table 25.1   Trace gases depending on substrate conditions (Brosien 2003; Vogeler 2004).

Trace gas compound	Manure/ maize	Biogas plant 1	Biogas plant 2	Biogas plant 3	Biogas plant 4
Carbonyl sulfide	X		X	X	
Propene	X	X	X	X	X
Carbon disulfide		X	X	X	
2-methyl-1-propene		X		X	X
Pentane	X	X	X	X	X
Methanthiol	X	X	X	X	X
2,3-epoxypropylester–acrylacid			X	X	
n.b.*	X		X		
2-methyl-1,3-butadiene		X	X	X	X
n.b.*			X	X	X
Ethanethiol		X	X	X	X
Heptane		X		X	X
2-methyl-furane	X	X	X	X	X
2-propanethiol		X	X	X	X
1-propanethiol			X	X	X
2-methyl-furane	X	X		X	X
1-bromo-2-methyl-cyclohexane	X				
2-methyl-hexane	X		X	X	X
2-butanethiol		X	X	X	X
n.b.*	X				
Toluene	X	X	X	X	X
n.b.*	X				
Ethylbenzol		X		X	X
2-butanone	X	X	X	X	X
2,3-butandion		X	X		X

* Not identified.

Table 25.2   Trace gases depending on gas upgrading step (Vogeler 2004).

Trace gas compound	Raw gas (ppm)	Desulfurised gas (ppm)
Hydrogen sulfide	1835	14.80
Dimethyl sulfide	0.57	0
Dimethyl disulfide	0.47	0
Carbonyl sulfide	0.38	0.41
Carbonyl disulfide	0	0
Methylmercaptane	0	0.61
Ethylmercaptane	0.09	0.16
Pentane	0	0.20
Propylene	0	0.15

25.4   PILOT PLANT

Figure 25.2 shows the whole process flow sheet of a polymer-electrolyte-membrane (PEM) FC pilot plant developed by the Farmatic GmbH (Germany), which is connected to an agricultural biogas plant located at the FAL (Braunschweig,

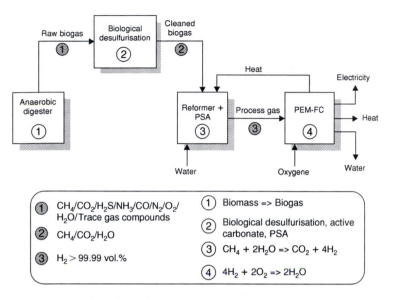

Figure 25.2   Process flow sheet of a biogas-PEMFC-pilot plant (Bergmann 2004). PSA: pressure swing adsorption.

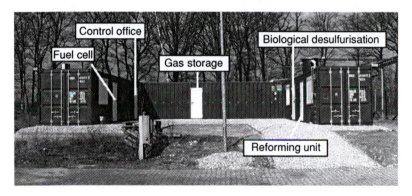

Figure 25.3   PEM-FC pilot plant (Farmatic Biotech Energy GmbH) at FAL.

Germany). Figure 25.3 gives a view to the PEM-FC pilot plant. In Figure 25.3 the different process steps are marked, Figures 25.4–25.6 present a zoom into the single process steps.

Process steps 1 and 2 are well established at a high number of agricultural biogas plants. Therefore, a representative comparison between a combined heat and power applications (CHP) unit and process steps 3 and 4 becomes possible. Furthermore, this scheme contains three variations for trend-setting utilisation of biogas in the future. The first is producing a clean gas mixture with a very low concentration of disturbing trace gas compounds (after step 2), the second is producing a hydrogen-rich process gas (after step 3) and finally the third, which is the electrical utilisation of biogenous hydrogen with FC technology (within step 4).

Figure 25.4   Biological desulfurisation unit (Farmatic Biotech Energy GmbH).

Figure 25.5   Hydrogen reforming system 2.5 (RÜTGERS CarboTech Engineering GmbH).

25.4.1   Desulfurisation unit

The desulfurisation unit (Figure 25.4) works at a temperature of 37°C under aerobic conditions with atmospheric oxygen, and reaches desulfurisation rates of 99% and higher (referring to H_2S). The average gas flow is $9\,m^3/h$ with an average H_2S content of 1600–2000 ppm. The electrical efficiency referring to the heating value of the cleaned biogas is 96% and with procedures to reduce the electricity consumption of the system an enhancement up to 98.5% is possible.

25.4.2   Reforming unit

Following the biological desulfurisation, the biogas enters the reforming unit (Figure 25.5).

Inside the hydrogen reforming system (HCG 2.5), a volume flow of $1\,m^3$ desulfurised biogas is transferred into a hydrogen-rich process gas with hot steam reforming at a temperature of 950°C and a pressure of eight bars. In addition this

Figure 25.6   PEM-FC test stacks (Farmatic Biotech Energy GmbH).

process gas is cleaned up with a pressure swing adsorption unit and becomes biogenous hydrogen with a pureness of at least 99.99%.

During the examinations several optimisations were done. First of all, a further catalyst bed for selective oxidation of methane was added into the reforming unit, due to leftover oxygen (approximately 1.5 vol.%) in the entering desulfurised biogas. This oxygen destroyed the hot-steam catalyst and the additional catalyst bed made sure that all oxygen is removed before the gas flow enters the hot-steam catalyst bed.

In a second step, the plant operation was extended with a catalyst bed regeneration circle in regular intervals. During these intervals the reforming unit operated with normal operation pressure and temperature but under a nitrogen atmosphere with addition of water. The result of this adapted operation scheme was the removal of accumulated carbon in the catalyst bed, which means a clear improvement to avoid blockages in the system due to this accumulated carbon.

The whole process chain reached electrical efficiency rates (referring to the heating value of the upgraded biogas) of only 20%. The reason for this is the pilot scale of the plant; through simple improvement steps efficiency rates of 75% and higher would become possible (Ahrens 2004; Bergmann 2004).

25.4.3   Fuel cell

Subsequent to the reforming unit the biogenous hydrogen enters the PEM-FC (Figure 25.6) and is converted to electricity, water and heat. Figure 25.6 shows test stacks with an electrical power of 25 W up to 100 W. These stacks were connected with different measurement equipment for sampling temperature, gas flow and humidity. The biogenous hydrogen was absolutely appropriate for the tested PEM-FC applications.

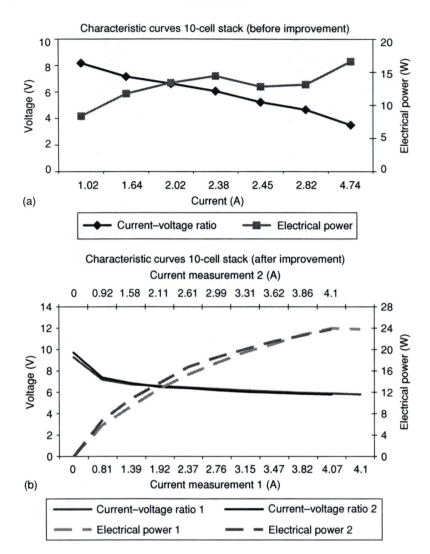

Figure 25.7   Stack performance before (a) and after (b) improvement (Ahrens 2004).

During the examinations several important proceedings for the operation of PEM-FC were reached and established, for example solutions for an appropriate humidity management inside the cells. Furthermore a sampling and calculation model was established for calculation of electrical efficiency and hydrogen exploitation rates of PEM-FC through a balance of incoming and outgoing humidity. The electrical efficiency of the tested FC systems reached 35% with hydrogen exploitation rates of 40% (Ahrens 2004; Bergmann 2004). The improvement of the stack performance in operation with biogenous hydrogen is shown in Figure 25.7a and b.

25.5 CONCLUSION

Ensiled energy crops are able to give a high specific biogas yield with a low amount of disturbing trace gas compounds in the produced biogas at the same time. Concerning FC technology further improvements are of high interest. One promising aspect for the future of PEM-FC technology is the development of new membrane materials with operation temperatures of 160°C and a CO tolerance of up to 2 vol.%. With the electrical utilisation of biogas with FC technology a new field of technology is entered. Within this, trace gas analytics of biogas mixtures, biogas upgrading, hydrogen reforming and testing of different FC types (high and low temperature) with new material solutions (e.g. for the membrane material) are matters of particular interest.

REFERENCES

Ahrens, T. (2004) Technologische Bewertung der Gasaufbereitung und Verstromung in einer Brennstoffzelle einschließlich der Beurteilung der Biogasqualität bei Einsatz unterschiedlicher nachwachsender Rohstoffe, Final Report of FNR Project 22009000; Federal Agricultural Research Centre of Germany, Institute of Technology and Biosystems Engineering, Braunschweig, Germany.

Bergmann, T. (2004) Verfahrenstechnische Beschreibung einer Versuchsanlage zur Herstellung von biogenem Wasserstoff aus Biogas zum Zweck der Verstromung in einer Polymerelektrolytmembran-Brennstoffzelle, Diploma Thesis; Faculty of Supply Engineering, Technical College of Braunschweig/Wolfenbüttel, Germany.

Bley, S. (2001) Untersuchungen zur anaeroben Vergärung von Silagen ausgesuchter nachwachsender Rohstoffe, Seminar Project; FAL Braunschweig, Germany.

Brosien, S. (2003) Analytische Bestimmung von Spurengaskomponenten in Biogasgemischen in Abhängigkeit von individuellen Betriebsparametern und der Substratsituation im Vergärungsprozess, Diploma Thesis; Institute for Waste Engineering and Environmental Monitoring, Technical College of Braunschweig/Wolfenbüttel, Germany.

Tietjen, C. (2002) Untersuchungen zur anaeroben Vergärung von Silagen ausgesuchter nachwachsender Rohstoffe, Seminar Project; FAL Braunschweig, Germany.

Vogeler, H. (2003) Optimierung der Betriebsführung von landwirtschaftlichen Biogasanlagen unter der Berücksichtigung des Einsatzes von nachwachsenden Rohstoffen, Diploma Thesis; Institute for Process Engineering, Technical University of Clausthal, Germany.

Vogeler, H. (2004) Spurengastechnische Bewertung einer Versuchsanlage zur biologisch-physikalischen Reinigung von Biogas und zur Produktion von biogenem Wasserstoff mittels Heißdampfreforming, Seminar Project; Faculty of Engineering, Process Engineering and Chemistry, Technical University of Clausthal, Germany.

26

Phosphoric acid fuel cells operating on biogas

K. Stahl

26.1 INTRODUCTION

26.1.1 Phosphoric acid fuel cell overview

The phosphoric acid fuel cell (PAFC) is the fuel cell type that has accumulated the greatest operation experience over a period of several decades. Since the early 1970s, more than 500 PAFC power plants have been installed worldwide. Their power has grown over the years and culminated in the largest fuel cell power plant ever built, a 11-MW unit for the Tokyo Electric Power Corporation (TEPCO) in Japan (Kordesch and Simader 1996).

Leading companies that have developed PAFC products are the United Technologies Corporation (UTC) in the USA and Toshiba (Joint-Venture with UTC), Fuji Electric and Mitsubishi Electric in Japan. This chapter focuses on the products and development lines of UTC in detail, because Japanese PAFC products were rarely used outside of Japan.

26.1.2 PAFC operation principle

The fundamentals of fuel cells are covered in detail in Chapters 16 and 17. Therefore, only a brief outline of special design features of the PAFC system will be provided here.

A single PAFC cell of today consists of two highly porous gas diffusion electrodes made of heat-treated graphite. Between the electrodes, a so-called matrix holds the liquid electrolyte, consisting of a silicon carbide fabric drenched with phosphoric acid (>90% H_3PO_4). The wet matrix separates the anode and cathode gas flows. Both electrodes are coated with carbon black and contain microscopical particles of the electrocatalyst based on platinum.

To run the fuel cell, hydrogen is supplied to the anode and air (oxygen) to the cathode. At elevated temperatures close to 200°C and in the presence of the electrocatalyst, hydrogen molecules split into H^+ ions and are dissolved by the phosphoric acid electrolyte. Free electrons lead to a voltage rise across the fuel cell, and when an electrical circuit is connected to it, electrons flow from the anode to the cathode, where they combine with oxygen and hydrogen ions and form water molecules.

The total fuel cell reaction can be expressed as:

$$H_2 + \tfrac{1}{2}O_2 \rightarrow H_2O \qquad \Delta H_0 = -285.8\,\text{kJ/mol}$$

The reversible potential of a hydrogen–air fuel cell is 1.23 V. Practical voltages between anode and cathode may rise close to 1 V in idle and 0.60–0.65 V at rated power.

The fuel cell generates electrical current as long as hydrogen and air are supplied to it, and it produces water in that process. At the operation temperature of PAFC systems, the water is in the vapor phase and is carried out of the cell stack with the cathode air. In state-of-the-art PAFC systems the water vapor is condensed and recycled internally to feed the steam reformer in the fuel-processing system that generates the hydrogen fuel. Hydrogen is formed using a two-stage steam reforming reaction of natural gas (methane) or methane-rich biogases. The steam reforming reactions can be described as:

$$CH_4 + H_2O \rightarrow 3H_2 + CO \qquad \Delta H_0 = +205.8\,\text{kJ/mol}$$

$$CO + H_2O \rightarrow H_2 + CO_2 \qquad \Delta H_0 = -42.3\,\text{kJ/mol}$$

The overall steam reforming reaction is endothermal, it requires a continuous heat source. Typically, the anode off-gas from the fuel cell that still contains more than 10% hydrogen is burned in the reformer vessel and provides the necessary energy for the reforming reaction.

26.2 BACKGROUND AND HISTORY OF PAFC

PAFC development at the UTC began in the mid-1960s, after the successful fuel cell program for the National Aeronautics and Space Administration (NASA) Apollo mission (International Fuel Cells, 2001). The Apollo spacecraft that was used for the first moon landing was powered by three alkaline fuel cells named PC3, made by Pratt & Whitney, a UTC company.

Figure 26.1   The PC10 fuel cell home experiment in Ohio (USA) in 1968
(courtesy of UTC Fuel Cells).

It was the military that showed interest in stationary and portable fuel cells
first. In the 1960s, several small fuel cell systems were developed in the load
range of 500 W up to 5 kW that were operated on "logistic fuels" such as diesel,
gasoline and propane. Such a system was later redesigned to run on natural gas
and it was tested by Columbia Gas to power metering equipment at a gas transmission station. In 1968, Columbia Gas ordered a 4-kW fuel cell that operated on
natural gas and provided grid-quality AC power for the home of one of their exec-
utives in Ohio, USA (Figure 26.1). In response to electricity companies who promoted the all-electric home, several gas companies founded Team to Advance
Research on Gas Energy Transformation (TARGET) and, together with UTC, the
development of stationary fuel cells began in earnest.

Due to the limitations of alkaline fuel cells in the presence of carbon dioxide,
the stationary fuel cell program focused on the phosphoric acid electrolyte. The
result was the PC11 fuel cell power plant, a PAFC system that generated 12.5 kW
of power (Figure 26.2). Between 1971 and 1973, 65 units were installed in the
USA and in Canada, some of them even drew the attention of Japanese gas
companies and were tested at Tokyo Gas and Osaka Gas.

In 1979, the PC18 was developed by TARGET, a 40-kW PAFC system that was
believed to better match market requirements (Figure 26.3). Many improvements
were made compared to the PC11: the fuel cell started automatically, it could
operate either connected to the grid or grid independent, and the waste heat from
the power plant could be used for domestic heating, achieving fuel utilizations of
above 80%. A total of 46 units were produced until 1986.

The field experience from the PC18 highlighted several weaknesses of the
system. Many unexpected problems arose from conventional parts, such as water
pumps, fans, heat exchangers and valves, resulting in poor power plant availability.
The cell life was limited to less than 10.000 h, too short for a stationary power plant.
The lifetime limitation was caused by slow evaporation of the phosphoric acid

Figure 26.2   The PC11 fuel cell, 12.5 kW, 1971 (courtesy of UTC Fuel Cells).

Figure 26.3   Two PC18 fuel cells, 40 kW each, installed at a health club in California (USA), 1984 (courtesy of UTC Fuel Cells).

electrolyte in the cells, which eventually dried out and ceased to generate voltage. The problem was solved in the end of the 1980s, when a new cell design was introduced that had a far higher amount of electrolyte in electrochemically inactive areas of the cell, from where it could wick into the active matrix to replenish evaporated electrolyte (Kordesch and Simader 1996).

The new cell design was first tested in a series of megawatt-scale power plants for the Tokyo Electric Power Co. in Goi (Japan). First, a 4.5-MW power plant was tested, followed by a 11-MW unit in 1992. The 11-MW system consisted of 18 cell stacks that were operated at elevated pressure of up to 8 bar (Figure 26.4). This largest fuel cell power plant ever built achieved an electrical efficiency of more than 44% and was operated until 1997 for more than 20.000 h. Plans existed to scale

Figure 26.4   The PC23 power plant, a 11-MW system, Tokyo Bay, Japan
(courtesy of UTC Fuel Cells).

up the system to 26 MW, but interest in electric utility companies was too low to expect substantial sales, and this development was suspended in the mid-1990s.

The attention again focused on distributed generation, continuing the achievements of the TARGET program. In 1987, UTC built a first in-house "breadboard" PAFC system with a power of 200 kW, and in 1988 four 200-kW prototype units were tested by major Japanese gas companies (Figure 26.5). These prototype units were the predecessors of the PC25 fuel cell that was introduced into the market in 1991.

Three major different types of the PC25 were developed over a time period of more than a decade. Fifty-six PC25A and 18 PC25B models were built, and since 1995 the PC25C (or "PureCell™ 200") is in production. About 200 PC25C units have been produced so far, and an ongoing improvement was witnessed every year.

The PC25 fuel cell is often referred to as the first commercially available fuel cell in the world. It came with a product warranty, customer support and service by the manufacturer, and fast response spare parts delivery. The latest PC25C power plants show major overhaul intervals exceeding 50.000 h, which is in line with competing technologies for distributed generation, such as reciprocating engines and microgas turbines. However, the PC25 systems remained costly, and market demand was lower than anticipated. It is expected that new fuel cell products by UTC will be introduced into the market soon.

26.3   PAFC OPERATION ON BIOGAS

26.3.1   Sources of biogas

Biogases are formed in the decay of organic matter in the absence of air. The resulting gas mixture mainly consists of methane and carbon dioxide, with some variation

Figure 26.5   A 200-kW "pre-prototype-plant" – predecessor of the PC25, 1988
(courtesy of UTC Fuel Cells).

depending on the composition of the organic matter that has been fed into the fermentation process. The term "biogas" is typically used for gas that is produced from
(animal) farm waste or composted biowaste. On wastewater treatment plants, the
resulting gas is referred to as *sewage gas* or *anaerobic digester gas* (ADG), and
on waste dumps it is called *landfill gas*. Biogas and ADG are quite similar in their
composition (biogas typically has far higher sulfur contents than ADG, though),
but landfill gas may differ significantly due to its often high nitrogen content.

26.3.2   PAFC operating on landfill gas

When the first commercial fuel cell product PC25A was introduced by UTC in the
early 1990s, soon the idea came up to use renewable fuel sources rather than fossil
fuels. It was probably Mr. Ronald J. Spiegel, an engineer at the US Environmental
Protection Agency (EPA), who has invented the concept to use landfill gas in fuel
cells (Figure 26.6; Spiegel *et al.* 1992).

Biogas (and landfill gas in particular) contains contaminants such as sulfur,
halides and other chemical elements bound in hundreds of organic molecule
species. Biogases are typically saturated with water and may contain particles, or
even bacteria. A gas pretreatment that removes the unwanted contaminants is
essential for a reasonable fuel cell life. Together with UTC, EPA has developed a
gas-processing unit (GPU) based on three functional groups: gas drying, gas
freezing and filtration on activated charcoal. A first demonstration plant was built
and tested between 1993 and 1995 at the Penrose Landfill, a waste dump in
Los Angeles, California (Masemore and Piccot 1998). The fuel cell of type PC25A
by UTC was operated for 707 h on landfill gas, with a maximum power of 137 kW.
At the Penrose Landfill site, the methane content was too low (44%), and it was
decided to continue the demonstration program on a different location.

At the Groton Landfill in Connecticut, a suitable site was found where landfill
gas with a methane content of close to 57% was available. The fuel cell power

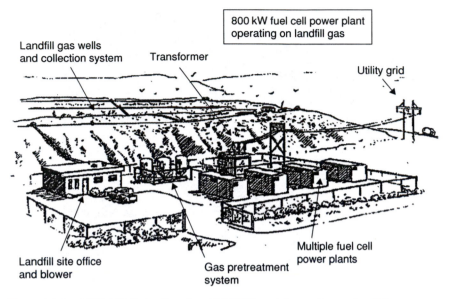

Figure 26.6   A 800-kW fuel cell system (4 PC25A units) operating on landfill gas
(EPA-Report "Demonstration of Fuel Cells to Recover Energy from Landfill Gas").

plant was relocated from California to Connecticut in June 1995, and was operated at its new site until July 1997 for a total of 3.313 h (Masemore and Piccot
1998). Due to the higher methane content in the landfill gas at Groton Landfill, a
power output of 165 kW was achieved (Figure 26.7).

Operating a PAFC on landfill gas had proven the feasibility of fuel cell power
plants running on biogas. However, landfill gas is the most challenging fuel
source in terms of contaminant removal, and gas processing is costly. In an EPA
study, a worldwide market potential of 11.000 MW of power from landfill gas was
determined (Spiegel 2001a), but today only a single fuel cell (PC25C) exists that
runs on landfill gas at the Braintree Landfill in Connecticut.

26.3.3   PAFC operating on ADG

After the landfill gas project, EPA turned its attention to ADG. Again, in cooperation with UTC, a GPU for ADG was designed in 1995. However, the Japanese
partner of UTC, Toshiba, had already begun to test a PC25B fuel cell power plant
on ADG in Yokohama City starting in 1994. The "AquaPower" project was based
on a pressure-swing adsorption system that enriched the methane content of the
digester gas to more than 90% (upgrading to natural gas quality).

As expected, the fuel cell power plant achieved rated power at methane levels
above 90%. Test runs at lower methane levels again showed the incapability of the
PC25B unit to sustain that power level. The "AquaPower" project's true achievement was the detection that the power limitations of the fuel cell were not caused
by the electrochemical reaction inside the cell stack, but rather by insufficient gas
supply. Further, the cross-stack pressure limits were exceeded when operating the

Figure 26.7   PAFC PC25A running on landfill gas, Groton Landfill, Connecticut, USA (*Source*: US EPA).

PC25B on ADG. A redesign of the fuel supply train and of the cathode air supply eliminated the need for methane enrichment, and the "AquaPower" fuel cell unit was successfully operated at rated power on ADG for more than 3.000 h.

In the next year, several PAFC units were installed at wastewater treatment plants that operated on ADG, in New York and Boston as well as in three Japanese breweries (Asahi, Sapporo and Kirin). Soon after, wastewater treatment plants in California and in Oregon purchased and installed fuel cells. Except for the Kirin Brewery that installed a fuel cell unit made by Mitsubishi Electric, all fuel cells were of the type PC25C made by UTC/Toshiba.

26.4   CASE STUDY: THE RODENKIRCHEN FUEL CELL

26.4.1   Site description and project initiators

The first (and only) large-scale fuel cell installation in Europe that utilizes a biogas is located on the Rodenkirchen wastewater treatment plant in Cologne, Germany (Figure 26.8). The Rodenkirchen WWTP treats wastewater from some 65,000 people, and is one of the smaller WWTPs of Cologne. It was the last site where a large portion of the ADG was still flared, and plans called for the installation of reciprocating engines to generate power.

The fuel cell project was initiated by the GEW RheinEnergie AG, the utility company for Cologne and by the wastewater department of the City of Cologne. The fuel cell power plant was installed and commissioned on a turn-key basis by the Technische Beratung Energie (TBE) GmbH early in the year 2000. Due to its unique nature, this fuel cell system operating on ADG was elected as a "world-wide project" of the world fair EXPO 2000 in Hannover, and it is also part of the Initiative for Energies of the Future by the federal state of North-Rhine Westphalia.

Figure 26.8   First large-scale fuel cell in Europe to utilize ADG gas WWTP
Rodenkirchen, Cologne (Germany), GEW RheinEnergie (2000).

26.4.2   Gas-processing unit

On the WWTP Rodenkirchen, a GPU was installed that was based on the concept
of gas freezing and gas filtration on activated charcoal. The low-temperature stage
cooled down the gas to below −25°C, lower than the dew point temperature of
several higher organic compounds, especially siloxanes, which had been found to
be the second major contaminant in the ADG stream, beside hydrogen sulfide.
Desulfurization occurs mainly in the activated charcoal filter beds, although some
20% are already eliminated in the pre-cooler stage, where some hydrogen sulfide
dissolves in the condensing water and is drained out of the ADG stream.

The GPU was tested for half a year before the fuel cell was connected to it. Gas
quality was monitored frequently (for hydrogen sulfide, a continuous monitoring
system is installed), and no laboratory analysis found anything harmful for the
fuel cell. However, although the GPU had proven its functional design in a continuous run in excess of 5.000 h, in the long term it became too maintenance-
intensive, and 2 years later it was redesigned to simplify the process.

26.4.3   Fuel cell operation

The fuel cell was successfully started in March 2000 and achieved rated power
from the beginning. No methane enrichment is required, as the fuel cell balance-
of-plant was modified to accept the higher gas flows of ADG compared to natural
gas. The fuel cell power output is set by the WWTP operators to match the ADG
production rate, which varies depending on external influences such as weather
conditions, ambient temperature and waste freight of the incoming water. The
average ADG flow is 65 m³/h, and it is converted by the fuel cell to generate
around 150 kW of power continuously.

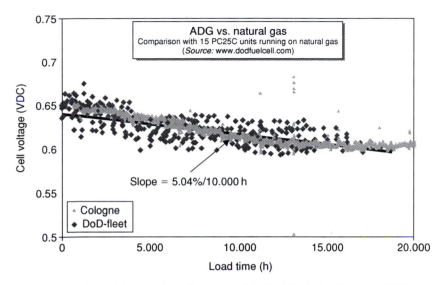

Figure 26.9   Effect of time on the cell voltage of the PC25C fuel cell at the WWTP
Rodenkirchen, compared to 15 PC25C units running on natural gas
(*Source*: US Department of Defense, data recordings at WWTP Rodenkirchen).

The electrical efficiency of the fuel cell remains constant over a wide load
range from 50% to 100%, and this makes it ideal for this WWTP site. When in
operation, all ADG available on site is converted into electricity, with no by-products
other than water.

The fuel cell is now in operation for 4 years and has proven its durability on
ADG. The average electrical efficiency is around 36%, which is 1–2% lower compared to operation of the same system on natural gas. The ADG contains close to
40% carbon dioxide, which is carried through the entire fuel-processing system and
accounts for additional heat requirement. Other than that, the carbon dioxide is an
inert to the fuel cell. It has a diluting effect on the hydrogen concentration at the fuel
cell anode, but the influence on the cell voltage is negligible, as seen in Figure 26.9.
Stack voltage degradation over time is similar to natural gas units, although it is
expected that the electrolyte loss is accelerated with the increased gas flows.

The fuel cell power plant covers about 50% of the power demand of the
WWTP Rodenkirchen. ADG is a renewable fuel, and the power generated by the
fuel cell receives a special rate according to the German Renewable Energy Law
(Erneuerbare-Energien-Gesetz, EEG). Compared to the average German power
plant, the utilization of ADG in the Rodenkirchen fuel cell saves around 600 tons
of carbon dioxide emissions per year.

Up to now, a total of 5 MW of fuel cell power has been installed and tested on
biogas around the world. The majority of these plants operate on ADG, similar to
the Rodenkirchen fuel cell, and most of them are PAFCs. This trend is expected to
continue – in 2003, in New York City eight PAFC PC25C units were installed and
commissioned on four municipal wastewater treatment plants. For the year 2004
another 1.5 MW of fuel cell power will be installed on various biogas applications,

although there is a shift of attention towards molten carbonate fuel cells (MCFCs). Unlike any other fuel cell type, MCFC performance improves when operated on CO_2-rich gases.

Early 2005, Germany's second fuel cell power plant that runs on ADG will be installed and commissioned in the municipal wastewater treatment plant of the City of Ahlen. This 250-kW MCFC will be supplied by the RWE Fuel Cells GmbH in Essen, and it is currently being manufactured by the MTU CFC Solutions GmbH in Ottobrunn (Munich).

26.5 CONCLUSION

PAFCs have demonstrated their capability in more than 500 systems installed around the world, including more than 250 PC25 units. PAFC systems have made stationary fuel cells a reality today, although their economics is still far from favorable. In an emerging fuel cell industry, many fuel cell products will soon compete, driving each other to improve the technology, achieve higher efficiency and above all lower the system cost.

The operation of fuel cells on biofuels such as biogas, ADG and landfill gas has been successfully demonstrated in more than 20 applications around the world. A gas-processing system to remove contaminants is required, and system designs may vary depending on the fuel source. However, economically feasible solutions for such gas-processing systems have been identified. Due to their low emissions, excellent part-load characteristics and high efficiencies, fuel cells and renewables are a perfect match and may become a significant market for fuel cells in the future.

REFERENCES

CADDET Japanese National Team (1998) Sludge gas powers fuel cells. *CADDET Renewable Energy Newsletter*, pp. 13–15.

CADDET Japanese National Team (1999) Fuel cell CHP using biogas from brewery effluent. *CADDET Renewable Energy Newsletter*, pp. 14–16.

EG&G Services Parsons, Inc. (2000) *Fuel Cell Handbook*, 5th edn. US Department of Energy.

Heiming, A., Hail, H.-J. and Wismann, G. (2001) Kraft-Wärme-Kopplung mit Brennstoffzellen – Erfahrungen aus einem 5-jährigen Versuchsbetrieb mit 200 kW-PAFC-Anlagen, Brennstoffzellen, 2. Auflage. C.F. Müller Verlag, Nürnberg.

International Fuel Cells, United Technologies Fuel Cell History (2001) International Fuel Cells Brochure.

Kordesch, K. and Simader, G. (1996) *Fuel Cells and Their Applications*. VCH Verlagsgesellschaft, Weinheim.

Masemore, S. and Piccot, S. (1998) *Electric Power Generation Using a Phosphoric Acid Fuel Cell on a Municipal Solid Waste Landfill Gas Stream*. Southern Research Institute, US Environmental Protection Agency.

Nakayama, Y. *et al.* (1996) FCPP application to utilize anaerobic digester gas. *Fuel Cell Seminar Program and Abstracts*.

RWE Fuel Cells (2004) *RWE Fuel Cells and the City of Ahlen Plan Europe's First MCFC Fuel Cell Project on the Basis of Sewage Gas*. RWE Fuels Cells GmbH Press and Public Relations, June 30.

Spiegel, R.J. (2001a) Fuel cell operation on landfill gas. *EPA Fuel Cell Workshop*, Cincinnati, Ohio, June 26–27.

Spiegel, R.J. (2001b) Fuel cell operation on anaerobic digester gas. *US DOE Natural Gas/Renewable Energy Hybrids Workshop*, August 7–8, National Energy Technology Laboratory, Morgantown, WV.

Spiegel, R.J. and Sandelli, G.J. (1992) Fuel cell power plant fueled by landfill gas. *Proceedings of the 1992 Greenhouse Gas Emissions and Mitigation Research Symposium.*

Trocciola, J.C. and Healy, H.C. (1995) *Demonstration of Fuel Cells to Recover Energy from an Anaerobic Digester Gas – Phase I. Conceptual Design, Preliminary Cost, and Evaluation Study.* US Environmental Protection Agency.

Trocciola, J.C. and Preston, J.L. (1998) *Demonstration of Fuel Cells to Recover Energy from Landfill Gas – Phase III. Demonstration Tests and Phase IV. Guidelines and Recommendations.* US Environmental Protection Agency.

Wismann, G., Ledjeff-Hey, K., Mahlendorf, F., Schieke, W. (1999) Post-Mortem-Analyse und Entsorgung einer phosphorsauren Brennstoffzellenanlage PC25A, GASWÄRME International, 48.

27

Applications of molten carbonate fuel cells with biofuels

J. Hoffmann

27.1 INTRODUCTION

27.1.1 Available fuels for the molten carbonate fuel cells

The molten carbonate fuel cell (MCFC) belongs to the most established and developed types of fuel cells. Up to now several MCFC-based facilities are under operation worldwide and several companies in Europe, USA, Japan and Korea are competing in their research and development activities. Recently one company, the MTU CFC Solutions GmbH (Munich, Germany), released a number of field test units and intends to commercialize its MCFC system in the near future, which is well known under the brand name HotModule (homepage: www.mtu-cfc.com). One big advantage of the HotModule is that it can be operated by a wide variety of fuels. The chemical enthalpy of nearly any organic hydrocarbon can be converted into electrical energy inside a HotModule system. Basically natural gas is the first option fuel due to its wide distribution and availability within the developed countries of the western hemisphere. Another reason for its utilization is the more or less constant composition of natural gas, where the CH_4 content is in a range of 80–98 vol.%.

The different fuels that can be consumed by a HotModule are listed in Table 27.1. Next to methane any other saturated hydrocarbon can be used. Applications of different fuels than methane are in the design status already. In areas, where natural gas is not available, predominately LPG (a mixture of propane and butane) is considered to operate a MCFC system and military applications of MCFCs set their focus on diesel and gasoline.

Other liquid fuels such as alcohols are also considered to operate a MCFC system. In 2004 the BEWAG, the municipal power utilities of Berlin, intends to operate a HotModule with methanol as fuel. Several scientific approaches are made to inquire the application of ethanol as fuel. Ethanol is little more critical since it tends to eliminate H_2O when heat treated under MCFC conditions. The remaining hydrocarbon, ethylene, is rather reactive and tends to form undesired graphitic material. It is obvious that there is still some room for development before ethanol can be used as fuel for the MCFC system.

The treatment of biomass, either by fermentation or by thermal gasification, delivers a methane rich fuel, which contains other components like carbon dioxide and in the case of the thermal gasification additionally nitrogen, hydrogen, steam and carbon monoxide. It is expected that these gas mixtures can be used to operate a HotModule fuel cell system without causing any problems. Especially the carbon monoxide, which is also present in the coal gas, is a suitable fuel for the operation of a HotModule. The fact that carbon monoxide can be used as fuel is another big advantage of the high-temperature fuel cells with respect to their low-temperature counterparts (e.g. phosphoric acid fuel cell (PAFC), proton exchange membrane

Table 27.1   List of different fuels for MCFC operation and their application[a].

Natural gas: several commercial units from FCE[b], MTU CFC Solutions, Ansaldo[c]
Gas from fermentation of biomass: several commercial units from FCE and laboratory-size units from MTU[d] and Ansaldo[d] since 2000
Gas from thermal gasification of biomass: design studies[e]
Landfill gas: laboratory-size units from MTU in 2004[d]
Mining gas: design studies[e]
Gas from gasification of coal: laboratory-size tests by MTU in 1994
Exhaust gases from chemical processes: design studies[e]
Methanol: operation of commercial unit from MTU in Berlin expected late in 2004; light house demonstration by FCE in 2001[b]
Alcohols (Ethanols and others): laboratory-size tests by MTU in 1996[e]
LPG: laboratory-size tests by MTU in 1995[e]
Gasoline and diesel fuel: design studies[e]

[a] For further details see specific publications, for example contributions to the Grove Fuel Cell conferences in 2003 and 2004
[b] homepage www.fce.com
[c] homepage www.afc.ansaldo.it
[d] European Fuel Cell and Hydrogen Projects 1999–2002, Project Synopses, Directorate-Generale for Research 2003; EUR 20718; pages 24–30, 104
[e] Unpublished results by MTU CFC Solutions GmbH.

fuel cell (PEM)), where carbon monoxide is deactivating the catalytic surfaces of the electrodes.

27.1.2   MCFC features

The high-exergy heat of 400–450°C enables a large variety of thermal applications. Production of process steam and process heat for industrial purposes, but also bottoming cycling are the present state. A most interesting application of the HotModule will be the combination with a high-temperature fed absorption cooler or a steam-injection chiller. The high-temperature feed will enable an increased coefficient of power compared with presently used thermal coolers. The revenues for cooling power are significantly higher than for heat, and the overlapping of heat and cooling power demands over the year enables a long annual operating time under full load, which will reduce the pay back period.

The combination of systems for biomass or waste material gasification with the fuel cell HotModule and adapted thermal fed coolers forms a basic building block for an integrated energy supply system fulfilling the requirements of industry, manufacturing, trade, municipal applications, utilities and possibly the private sector. Manufacturing plants, cold stores, office buildings, computer and telecommunication centers, super markets, sport facilities, residential areas and others will be the future sites of applications.

The decentralized utilization of decentralized available energy sources for production of consumable energy forms has many impacts on the economical and ecological situations by reducing transmission losses, reduced requirements to the infrastructure, reduced emissions of greenhouse gases and pollutants, reduced dependence on primary energy carrier imports with positive impacts to the balance of trade. Last but not the least decentralized consumable energy production and its decentralized utilization increases employment in rural economies and areas as well as it offers advantages for small and medium enterprises.

27.2   BIOFUEL UPGRADING

27.2.1   Contaminants

Most fuel gas mixtures may contain a number of different contaminants which are hazardous to nearly any fuel cell system. Sulfur compounds are most critical. They can damage the system in two different ways. On the one hand it is well known that sulfur is responsible for an accelerated high corrosion rate of the stainless steal inside the system. On the other hand the sulfur poisons the nickel based reforming catalysts. These catalysts will loose their reactivity rapidly and the hydrogen formation will slow down until it is terminated irreversibly. Similar to the sulfides but with a lower rate silicon-containing compounds, like the recently identified siloxanes, may decompose on the catalyst during the reforming reaction and irreversibly block the catalytic surface by a SiO_2 layer.

Even more critical are organic halides since they tend to form acids like hydrogen chloride (HCl) which can easily react with the strongly basic electrolyte of the MCFC (mostly an eutectic mixture of lithium and potassium carbonate in the molar ratio of 62 : 38). In this reaction mainly volatile potassium halides are formed, which may evaporate out of the cells and condense and precipitate on colder surfaces outside the cells. This becomes critical when the precipitation takes place within the pipes of a heat exchanger, since the pipes may get blocked by the condensed material.

Also oxygen, if present in the fuel gas, may damage the reforming catalysts irreversibly. The catalyst consists mainly of nickel which is oxidized rapidly at the elevated operation temperature of the reforming process (300–650°C). Oxygen is rather common in fermentation gases, since the access of air to the fermentation process cannot be excluded. For instance it is a popular method to add small amounts of air to the fuel in order to remove the sulfur from the gas phase.

Therefore the selective removal of any contaminants, like sulfur, halides, silicon and others is essential, since, if present, most of them are limiting the lifetime of the fuel cell system in general. All previously identified contaminants are listed in Table 27.2.

Table 27.2   Identified contaminants in gases used as fuels for MCFCs.

Type of contaminant	Formula	Origin
Oxygen	O_2	Peak shaving gas, air from cleaning process, landfill gas, mining gas, leaks
Sulfur-containing compounds		
Hydrogen sulfide	H_2S	Natural gas, fermentation gas, landfill gas, mining gas
Carbonyl sulfide	COS	Natural gas, fermentation gas
Mercaptanes	R—SH	Odorants
Thioester	R—CS—SR	Odorants
Thioethers	R—S—R	Odorants (most common THT[*])
Nitrogen-containing compounds		
Ammonia	NH_3	Fermentation gas, gasification gas
Amines	R—NH_2	Fermentation gas, gasification gas
Nitrogen	N_2	Leaks, gasification gas, landfill gas, mining gas
Nitroxides	NO_x	Gasification gas
Hydrocarbons		
Olefinic hydrocarbons, tar, aromatic compounds	$R_2C{=}CR_2$	Landfill gas, gasification gas, exhaust gas from chemical processes, cracking reactions
Halides	R—X (X = F, Cl, Br, I)	Landfill gas
Siloxanes	R_3—Si—O—Si—R_3	Fermentation gas, landfill gas
Moisture		Processes with high humidity condensates during compression, adsorption on activated carbon if humidity of gas is higher than 60% rel.
Heavy metal dust	Hg, other	Poisons catalysts

[*]THT: Tetrahydrothiophene.

27.2.2   Biofuel cleaning

27.2.2.1   Activated carbon

Several methods are employed to remove the contaminants from the biofuels. Most familiar is the application of activated carbon as filter material. Quite a number of vendors are offering a wide variety of different materials (manufacturers of activated carbons: CarboTec, Sued-Chemie, Sutcliff Carbon and others). They differ in their origin, some of the activated carbons are based on hard coal others on carburized coconuts or similar material. In order to increase their capacity towards certain contaminants the surface of the activated carbon can be modified by the addition of "catalysts", such as copper, potassium iodide, basic compounds like sodium carbonate or potassium hydroxide, and others.

Unfortunately, the activated carbon still remains rather unspecific and adsorbs most of the contaminants, which are present in the fuel. And it seems to be possible, that displacement reactions of certain compounds occur. For example the presence of higher hydrocarbons (propane, butane, etc.) is limiting the adsorption capacity of activated carbon for sulfur-containing contaminants. If hydrocarbons are present in reasonable amounts, they tend to wash off some of the sulfur compounds from the surface of the activated carbon.

Also the presence of vapor within the fuel gas, which is typical for the gases from fermentation processes, is critical. Normally the level of the relative humidity in these types of fuels is close to 100%. Unfortunately, the vapor is condensing on the surface of the activated carbon and is flooding the adsorber. In order to avoid this undesired condensation some vendors specify, that the relative humidity of the fuel gas should not exceed 60%.

27.2.2.2   Chemisorption of sulfur-containing compounds

In order to remove the sulfur compounds selectively, the fuel gas can undergo the so called hydrodesulfurization reaction (HDS). HDS is distributed by the known catalyst manufacturers and applied in the steam reforming of natural gas during the ammonia generation (Haber–Bosch process). For this, small amounts of hydrogen are added to the fuel gas. Then the gas is passing a catalyst bed, which transfers all sulfur compounds into hydrogen sulfide. In the next step the fuel is passing through an adsorber bed consisting of zinc oxide. There the hydrogen sulfide is trapped selectively in the following reaction (27.1):

$$ZnO + H_2S \Leftrightarrow ZnS + H_2O \qquad (27.1)$$

If vapor is present, this reaction may become reversible. Higher concentrations of steam prevent the complete formation of zinc sulfide. Therefore, the moisture has to be removed first before undergoing the HDS/ZnO treatment. Totally unknown yet is the behavior of the siloxanes during the HDS reaction. It is expected too that the surface of the catalyst may be blocked by the Si compounds, which are formed during the HDS reaction by the siloxanes.

Most recently catalyst manufacturers like BASF, Sued-Chemie and others are offering a new type of adsorbing material (some of them are still in state of development; BASF catalysts are available under the brand names R 3–12 and M 8–12). This

Figure 27.1   Laboratory-scale filters for hydrogen sulfide removal: (a) chemical filter and (b) see next page.

material, mostly a basing on a mixture of reactive metal oxides, can be used at room temperature and it is trapping selectively nearly any sulfur-containing compound from the feed gas. The new adsorbents should be inert towards the displacement reactions and keep the trapped sulfur compounds on the filter material, since it seems to be inert towards the condensation of moisture. According to the information of the manufacturers the capacity of these materials is somewhat higher than the capacity of the activated carbon. These materials are tested under field conditions at natural gas operation (MTU CFC result achieved in the EU EFFECTIVE-Project (ENK5-CT-1999-00007)).

In case, that hydrogen sulfide is the only sulfur compound within the fuel gas, it seems to be possible to remove the hydrogen sulfide by passing the fuel through a bed or an aqueous suspension of iron oxide (Figure 27.1a). In laboratory-scale experiments <0.2 ppm of hydrogen sulfide were detected after an iron oxide filter (Seaborne results achieved in the EU EFFECTIVE-Project). These filters can be operated in a regenerative mode, when the formed iron sulfide can undergo

Figure 27.1   (b) biological hydrogen sulfide removal.

the Klaus reaction, where iron sulfide is transferred into iron oxide and sulfur (27.3). Such a filter has been constructed and tested by Seaborne during the EU EFFEC-TIVE-Project (Figure 27.1a):

$$FeO + H_2S \Leftrightarrow FeS + H_2O \tag{27.2}$$

$$nFeS + n/2O_2 \Leftrightarrow nFeO + S_n \tag{27.3}$$

Recently a gas cleaning system has been set in operation in King County (Washington, DC, USA). There an MCFC-power plant, manufactured by Fuel Cell Energy Inc. (Danbury, Connecticut, USA), is operated with municipal digester gas as fuel. The gas cleaning consists of several process stages: an iron oxide bed for crude sulfur removal, and differently impregnated activated carbons for the removal of the remaining hydrogen sulfide, the halides and the siloxanes. The oxygen is removed by a combustion catalyst.

Of course it is known that other heavy metals and their oxides can also be applied to remove hydrogen sulfide. In contrast to this chemical removal there are trends to extract hydrogen sulfide by passing the fuel through a microbiological filter. Most recently tests were made in order to remove the hydrogen sulfide by bubbling the gaseous fuel through a biotrickling filter inoculated with *Thiobacillus*. These bacteria convert the hydrogen sulfide into elemental sulfur or sulfate. A prototype of such type of filter was constructed by Profactor and has been tested too during the EU EFFECTIVE-Project (Figure 27.1b).

27.2.2.3   Condensation

A different approach to remove at least the condensable contaminants is to reduce the temperature of the fuel down to 0 to $-30°C$. At these temperatures most of the contaminants tend to condensate and can be collected in a condensation trap. This is rather effective, since most of the higher hydrocarbons, the siloxanes and the moisture are removed from the fuel gas by this method. The disadvantage of the low condensation temperature is, that the cooling process requires extra energy which reduces the overall performance of the fuel cell system. Such a filter, which is applied often at gas engines, has been designed by Siloxa for a fuel cell project in Cologne (Germany).

27.2.2.4   Fuel cleaning step

There are still numbers of unmentioned methods to remove the contaminants from the fuel gases. It is beyond the scope of the chapter to discuss all these methods in detail. Most of the gas cleaning processes remove the contaminants down to a level of 1–10 ppm (e.g. for sulfur). It is important to point out that this concentration level is still far too high by one order of magnitude. The reforming process, which will be discussed in the next section, requires purities in a range of 20–100 ppb. Actually this extremely low concentration can be achieved only by passing the pre-filtered fuel gas through a "fine"-filter consisting of activated carbon. In a HotModule system, two 400-l containers filled with activated carbon are used for this final cleaning step. The frequency, in which the filter containers have to replaced, is indicated by the break through of one or more initially trapped contaminants. The break through depends as mentioned above on the concentration and possible displacement reactions.

27.3   STEAM REFORMING

27.3.1   Steam-to-carbon ratio

In the next process step, the purified fuel is heated up and humidified in order to undergo the steam-reforming reaction. For pure methane the required amount of steam is given in the famous steam-to-carbon ratio (S/C, mole steam vs. mole methane). For an MCFC system with internal reforming the experimentally determined value for the lower limit is close to 2.0. In case the steam concentration gets beyond a critical value, the formation of graphitic carbon as product of the equilibrium reactions cannot be avoided. This graphitic carbon blocks the surface of the reforming catalysts. As a consequence, the catalyst is deactivated slowly and the rate of the reforming reaction decreases until it is terminating at all.

In case other carbon-containing compounds are present in the fuel gas, the S/C ratio has to be adjusted with respect to the nature of these compounds. For mixtures consisting of methane and carbon dioxide, typical for biomass fermentation, this S/C ratio should correspond better to the steam-to-methane ratio. In case of the presence of higher hydrocarbons the calculation of S/C ratio has to consider the total number of carbon atoms present in the molecule, for example for methane and propane:

$$CH_4 + H_2O \Leftrightarrow CO + 3H_2 \qquad (27.4)$$

$$\underline{CO + H_2O \Leftrightarrow CO_2 + H_2} \qquad (27.5)$$

$$CH_4 + 2H_2O \Leftrightarrow CO_2 + 4H_2 \qquad (27.6)$$

$$C_3H_8 + 6H_2O \Leftrightarrow 3CO_2 + 10H_2 \qquad (27.7)$$

27.3.2   Reforming stages

Inside a HotModule system this reforming reaction takes place at several stages. First of all the humidified gas is heated up to a temperature close to 500°C. Then it enters a reactor which is called the Pre-Reformer. This is an adiabatic operating reformer. This reactor contains commercially available nickel catalyst and its main purpose is to convert the higher hydrocarbons into hydrogen, carbon dioxide and methane. Additionally a smaller portion of the methane (approximately 20%) is undergoing the reforming process too. Due to the adiabatic nature of this process the gas mixture will cool down to a temperature where the kinetics of the reforming reaction is too slow to continue. Typically the exhaust-gas leaves the reactor with a temperature of 300–350°C. In the next step the pre-reformed mixture is heated up close to the process temperature of 650°C and undergoes several internal reforming steps inside the stack. Here, the remaining methane is transferred completely into hydrogen, which is consumed by the electrochemical fuel cell reaction (MTU CFC Solutions data).

27.4   THE FUEL CELL SYSTEM

27.4.1   MTU CFC Solutions's HotModule system

The HotModule is a complex fuel cell system, which has been developed during the last years by MTU CFC Solutions GmbH, which is located south of Munich in Ottobrunn (Germany). The HotModule system consists of five main components (Figure 27.2):

- the vessel which contains the fuel cell stack;
- the fuel processing unit (gas fine cleaning, humidification and pre-reforming);
- an interface, where the customer can connect his heat converting process (steam generator, cooling machine, etc.);
- the DC/AC-converter, which is connected to the electrical grid;
- the process control system with remote access for operation support.

The stack and most of the hot components are located inside an insulated vessel, whose inventory is kept on the temperature level by the released heat of the ongoing fuel cell reaction. This design allows reducing the efforts for the thermal insulation to a minimum. Figure 27.3 shows the vessel equipped with an internal heat exchanger. The heat exchanger is charged with the released process heat and heats up the incoming pre-reformed fuel gas mixture.

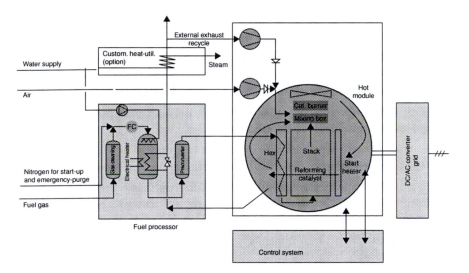

Figure 27.2   Process flow diagram of the HotModule system.

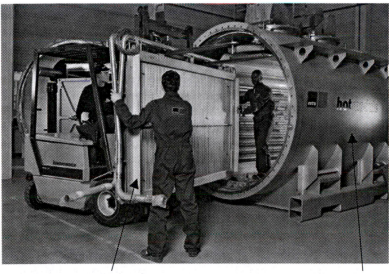

Heat exchanger   Vessel

Figure 27.3   Vessel and heat exchanger of a HotModule.

Figure 27.4   (a) Turning the stack from its assembly position to the operation position and (b) MCFC stack after complete installation.

27.4.2   Assembly of the MCFC stack

The fuel cell stack, which is introduced into a HotModule, is assembled by piling up one cell on top of the prior cell. When the desired number of cells is reached the stack will be compressed by endplates. Then the stack is laid down on its longer side. This turning process can be seen in Figure 27.4a. Here, the stack is leaving its upright position and is laid down slowly.

Once the stack is in this position it is equipped with several sensors, like thermocouples and potential sensors (Figure 27.4b). Then the stack is pulled into the vessel, where the installation is completed.

Later the vessel is closed and connected to the above-mentioned components of the system. Big gas pipes are linked to the fuel processor, which provides on the one hand the pre-reformed fuel for the electrochemical process inside the vessel. On the other hand, the fuel processor is consuming part of the process heat from the exhaust within heat exchangers. This heat transfer is necessary to provide the required amount of steam for the steam-reforming reaction and to heat up the humidified fuel to the initial temperature of the pre-reforming process. Both, the vessel of the HotModule and the container of the fuel processor (on the right), are presented in Figure 27.5. After completing the installation the fuel cell stack undergoes an initial start-up and formation procedure, which is performed normally before the HotModule is delivered to the customer.

During the conditioning procedure the temperature of the system is increased slowly from room temperature to the designed operation temperature. This happens in several steps and the gas composition is adjusted to the requirements of the internal reactions, which are appearing on the materials. After reaching the final operation temperature the performance of the stack is tested. If the stack passes the performance tests it is cooled down and delivered to the customer by truck.

After delivering the HotModule to the customer's location and completion of the installation it takes approximately 3–5 days until the system reaches its operation conditions (Table 27.3). Finally, the system is set on load and delivers

Figure 27.5   Vessel of the HotModule (left) and fuel processor (right).

electrical energy to the grid. Up to now quite a number of HotModules are in operation worldwide (Table 27.4), some of them are presented in Figure 27.6.

27.5   MCFC OPERATING ON BIOGAS

27.5.1   Location of test units

Most of the operating HotModules are fueled by natural gas (Table 27.4). In order to achieve a better experience about the operation MCFC systems under biogas conditions, MTU performed several experimental tests in the laboratory scale as well in full-HotModule size (Figure 27.7). The laboratory-scale experiments were performed in two different research programs. One of them was granted by the EU, in which small laboratory stacks are operated at different locations with different fuels. The other project was granted by the Fachagentur für Nachwachsende Rohstoffe, in which one stack was operated on a farm close to Munich. The key points of these projects are listed in Table 27.5.

27.5.2   Biogas cleaning

The aim of both projects was to establish the process chain from biogas production, biogas cleaning and operating a MCFC stack with the biogas. The results are rather promising. It was expected, that the gas cleaning units should reduce the concentration of hydrogen sulfide from its original value (approximately 100–3000 ppm) down to the level, which is known from the natural gas operation (<20 ppm). It seems that all tested gas cleaning units (Table 27.5) fulfilled this demand. In the national project it was attempted to reduce the lower limiting value down to <0.1 ppm.

This is about the level which should be achieved after MTU's gas fine cleaning. Here too the available results are rather promising and it seems that there is an option

Figure 27.6   HotModule field test units operating at (a) the Meteorit in Essen (Germany), an RWE facility and (b) IZAR, a shipyard in Cartagena (Spain).

to avoid the gas fine cleaning in future. This means, that the system can be simplified and it consists of only one gas cleaning step between the fermentation process and the fuel cell system. Then the activated carbon filter for the gas fine cleaning, which is rather sensitive towards condensation of the humidity, can be excluded from the system (unpublished results achieved by Schmack Biogas AG).

27.5.3   Effect of biogas contaminants on MCFC operation

In order to study worst case conditions, the effects of several contaminants on the fuel cell system have been examined experimentally. Sulfur-containing compounds, like hydrogen sulfide (H_2S), carbonyl sulfide (COS), tetrahydrothiophene (THT),

Table 27.3   Technical data of the HotModule system under natural gas conditions.

Efficiency	
Maximum total efficiency at Tmin ≅50°C	Approximately 90%
Electrical stack efficiency	Approximately 55%
Electrical plant efficiency	approximately 47%, target 50%
Stack	
Maximum electrical power, DC	280 kW
System	
Electrical power delivered to the grid	240 kW
Thermal power at Tmin = 80°C	180 kW
Depleted air temperature	400–450°C
Emissions	
SO_2	<0.01 ppm – not detectable
NO_x	<2 ppm – not detectable
CO	<2 ppm
Classified according to national codes and standards (BlmSchG)	

Table 27.4   List of MCFC systems under operation.

Operator, location	Application
Ruhrgas AG, Dorsten	Operation of MTU's first system demonstrator
University of Bielefeld	First field trial unit
Rhön Klinikum AG, Bad Neustadt, Bad Berka	Emergency power supply and combined heat and power (CHP) for hospitals
DaimlerChrysler, Tuscaloosa	Power supply for the Mercedes-Benz M-class plant
LADWP, Los Angeles	Power supply for the Los Angeles Department of Water and Power – supported by FCE
Marubeni, Japan	Utilization of biogas for CHP supply at the Kirin Brewery near Tokyo, supported by FCE
US Coastguard, Massachusetts	CHP for Cape Cod maritime air rescue station supported by FCE
IZAR, Cartagena	Power supply for the shipbuilding industry
DeTeImmobilien, Munich	Direct current application and air conditioning for Telekom headquarters
Otto-von Guericke University, Magdeburg	Power and process steam for hospital supply
Michelin, Karlsruhe	Power and process steam for tire production
Meteorit, RWE, Essen	Demonstration in fuel cell pavilion at RWE's "Meteorit Energy Park"
Pfalzwerke, Grünstadt	CHP system for hospital in Grünstadt
BEWAG, Berlin	CHP system fueled with natural gas or methanol started in October 2004

one siloxane, hexamethyl disiloxane (HMDS), and ammonia (NH_3) were used in test runs. The effect of the contaminants on the reforming catalyst is demonstrated in Figure 27.8. The reforming catalyst looses its activity as soon as it gets in contact with the sulfur-containing compounds. The poisoning by the siloxane is far less critical but observable (Chapter 24).

Table 27.5   Laboratory-scale operation of MCFC stacks fueled with biogas.

Gas suppliers	Seaborne (Owschlag: D) agricultural biogas Linz AG (Linz: A) gas from wastewater treatment Urbaser (Madrid: E) landfill gas Univ. Nitra (Nitra: SK) agricultural biogas Farmer in Munich area
Gas cleaning systems	Seaborne (Owschlag: D, Figure 27.1a), Profactor (Steyr: A, Figure 27.1b) Different systems designed by Schmack Biogas AG (Burglengenfeld: D)
Test beds and fuel cells Scientific support	MTU CFC Solutions GmbH (Munich: D, Figure 27.7) Studia (Schlierbach: A) Ciemat (Madrid: E) Rent-A-Scientist (Regensburg: D)
Industrial partner	EON AG (Munich: D)

Number of test runs in total seven laboratory stacks at five different locations.

Figure 27.7   (a) Laboratory-scale stack and (b) test unit used during the biogas projects.

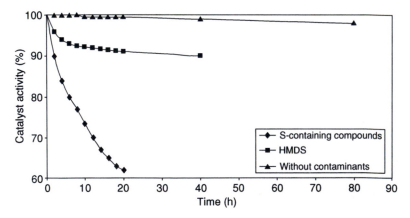

Figure 27.8   Effect of sulfur containing contaminants (20 ppm) on the activity of the reforming catalyst.

The reactivity of ammonia has been determined in laboratory-scale tests as well as in full-size HotModule tests. Similar results were obtained in both test runs, the added ammonia was decomposing more or less completely while passing through the fuel cell system (unpublished result achieved by MTU CFC Solutions GmbH). During the full-size test (HotModule) ammonia was mixed to the natural gas to reach an inlet concentration of about 5000 ppm. The NH_3 concentration was monitored at different locations inside the system. It turned out that ammonia was decomposing stepwise in each reforming step. Finally the concentration came down to <20 ppm (<1% of the initial value). It is assumed that ammonia decomposes in a retro-Haber–Bosch reaction to nitrogen and hydrogen:

$$2NH_3 \Leftrightarrow N_2 + 3H_2 \qquad (27.8)$$

The formed hydrogen can be consumed as fuel in the electrochemical fuel cell reaction. This may also give an explanation to the observation that the decomposition of ammonia was sensitive towards the applied fuel utilization. At lower fuel utilizations, the equilibrium of the decomposition reaction is shifted slightly to the left side of the upper equation, since hydrogen is present in excess.

27.5.4   Effect of CO_2 and CH_4 on MCFC operation

In a different test run the effect of carbon dioxide and nitrogen on the performance of an operating HotModule was examined. Both gases were added to the natural gas in order to simulate two different types of biogas. On the one hand the addition of carbon dioxide should result in a mixture, which is typical for the fermentation process. On the other hand the addition of nitrogen should simulate a gas which is representing the thermal gasification of biomass. An effect of the carbon dioxide addition was expected, since the performance increased slightly. In contrast to the other types of fuel cells carbon dioxide is participating actively in the electrochemical MCFC reaction (see above). Therefore, it should have the beneficial effect on the overall performance of the HotModule system, which can be seen in Figure 27.9. The amount of carbon dioxide was increased stepwise from more or less pure methane (natural gas) up to a ratio of $CH_4 : CO_2 = 2 : 3$. The average cell potential rose immediately by nearly 7 mV after the first addition. The effect of further CO_2 additions was smaller than the one which was observed during the first step.

This experiment was repeated but nitrogen was mixed to the natural gas instead of carbon dioxide. During the auto-thermal gasification of biomass the organic material is oxidized partially by the oxygen of the added air, which is removed completely. The remaining nitrogen then is diluting the formed process gas, which consists mainly of methane, hydrogen, carbon monoxide, carbon dioxide and steam. In order to learn more about the influence of the presence of nitrogen it was mixed stepwise to the natural gas. Surprisingly, the potential of the stack remained unchanged even when the ratio of $CH_4 : N_2$ increased up to a value of $1 : 2$. During the experimental run, the average stack temperature did not change too (Figure 27.10).

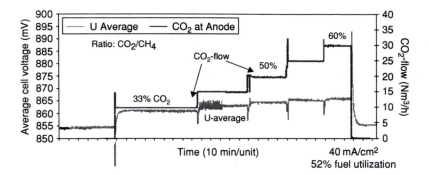

Figure 27.9   Effect of CO_2 addition on the performance of a HotModule-simulated biogas operation feeding the stack with CH_4 and CO_2.

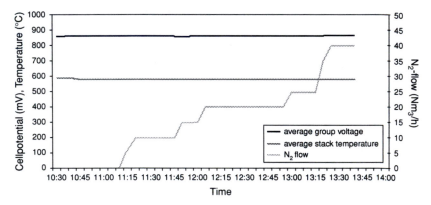

Figure 27.10   Effect of nitrogen addition to natural gas on the performance of a HotModule system.

27.6   CONCLUSIONS

The fuel cell HotModule of MTU CFC Solution is a highly integrated and reliable carbonate fuel cell system for stationary applications (240 kW, AC). Due to the electrochemical operational principle under contribution of carbon dioxide, the carbonate fuel cell is not only highly suitable for the utilization of all gaseous hydrocarbons, in particular natural gases and synthesis gases, but it is the preferred transformer to electricity of low-heating value gases with high amounts of inert components. Even such lean gases will be transformed with an electrical efficiency in the range of 50%. Hydrocarbons, carbon monoxide and mixtures of them with hydrogen are welcome fuels in all composition ratios. Addition of components, for example CO_2, N_2 and others, do not impact the efficiency up to a range of approximately 50%. These characteristics recommend the fuel cell HotModule of MTU CFC Solutions for utilization of secondary fuels, which are fuels from decentralized and regenerative sources, in particular from biomass and waste material, for example gases from anaerobic fermentation (biogas, sewage gas, landfill gas, coal mine gas, etc.) and gasified products from wood, paper, waste material, etc. This saves high value primary energy carriers and recycles greenhouse gases.

28

Application of molten carbonate fuel cells for the exploitation of landfill gas

R. Iannelli and A. Moreno

28.1 CASE STUDY PRESENTATION

The scope of the present case study was double: firstly, it intends to verify the feasibility, specially in terms of costs, of a project of energy recovery from landfill gas using molten carbonate fuel cells (MCFCs). Secondly, it presents a rational evaluation criterion to choose, from several project alternatives, the best in economic and environmental terms.

Among the various fuel cell technologies, the most appropriate for stationary applications in the power range 200 kW to 1 MW are the phosphoric acid (PAFC), the MCFC and the solid oxide (SOFC) fuel cells. The first one (PAFC) is the most mature of the three and the third (SOFC) the most promising in terms of efficiency. But the MCFC technology represents the most interesting for the near future: it

Table 28.1   Main polluting agents in landfill gas.

Classes of pollutants	Pollutants	Potential effects
Particulates	Coal powder, ashes	Pore obstruction
Sulfur compounds	H_2S, COS, CS_2, C_4H_4S	Voltage losses Electrolyte reaction (by mean of SO_2)
Halogen compounds	HCl, HF, HBr, $SnCl_2$	Corrosion Electrolyte reaction
Nitrogen compounds	NH_3, HCN, N_2	Electrolyte reaction (by means of NO_x)

offers greater efficiency than PAFC and notable simplifications in the use of fuels as natural gas, biogas or coal gas. Moreover, in contrast to SOFC, the MCFC is now at the end of the experimentation phase and near the first commercialisation. For these reasons, this chapter focuses on the MCFC technology to compare a possible solution of biogas exploitation from a real landfill site recently closed in Calabria (southern Italy) with the present prevalent technology of the gas engine.

The method proposed, developed by multicriteria analysis, allows to obtain easy comprehensive results and could be a valid support for landfill managers to develop energy recovery projects.

28.1.1   Features of MCFC fed by landfill gas

In the MCFC, object of the present case study, the electrolyte is a combination of alkaline carbonates in a ceramic matrix of γ-LiAlO$_2$ (Ronchetti and Iacobazzi 2002). The operation temperatures range from 600°C to 700°C, which allows the feeding of the cell with traditional fuels (natural gas or biogas) avoiding the use of noble metals as catalysers: nickel–chrome is used at the anode and lithied NiO at the cathode. Landfill gas is one of the most promising fuels for MCFC, thanks to its high CH_4 concentration, but the presence of contaminants can generate undesired effects (Table 28.1; see Chapter 18).

The presence of H_2S can be harmful even at very low concentration (<10 ppm), because it reacts with the electrolyte diminishing the cell efficiency (Figure 28.1). Nevertheless, otherwise then in gas engines, the effect is temporary, and the introduction of a clean gas restores the normal efficiency.

Despite these problems, the MCFC in general offers many advantages:

- high efficiency and independence on the electric load;
- reduced emission of CO_2, CO and H_2S;
- low noise and reduced running costs because of the absence of moving parts.

These properties were considered in the definition of the necessary pre-treatments, that influence the capital and running costs used to compare the MCFC exploitation with the traditional landfill gas utilisation in gas engines.

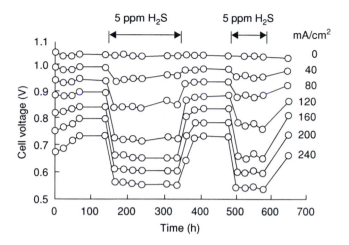

Figure 28.1   Influence of 5 ppm of H_2S on the MCFC potential (EG&G Services Parsons Inc. 2000).

28.2   CASE STUDY: "TUFOLO/LAMPARANELLO" LANDFILL SITE

28.2.1   Landfill site characteristics

Since the data already available allow planning exactly the exercise of an MCFC plant fed by landfill gas, this work is aimed to carry out an accurate technical–economical comparison with the prevalent technology, represented by the gas engine. This will be done with reference to a municipal solid waste (MSW) land-fill under final closing and put-in-safety intervention: the "Tufolo/Lamparanello" landfill, located south of the town of Crotone (Calabria, Italy).

The site is situated in a calanque area 2 km far from Crotone, and begun its activity in 1970. It provided for the disposal of MSW from Crotone till its closing in 2001. The volume of the embankment is about 1,000,000 m³, with a compaction ratio of 0.7–0.8 Mg/m³ (Figure 28.2). For the extension of the area and the volume of the embankment, the landfill was included in the list of national interest sites to undergo soil remediation by Italian Law 426/1998 on the remediation of contaminated soils.

In the last 15 years, about 430,000 Mg of wastes were dumped, with a maximum annual value of 35,000 Mg in 2000. Considering that in the area, until the closing, there were no other forms of solid waste disposal (separate collection, incineration), a notable organic fraction, accessible to the anaerobic degradation, is expected. Table 28.2 reports the MSW production of Crotone, assumed to represent the progressive embankment of the landfill. Table 28.3 reports the main waste typology taken from previous studies (CNR 1980) and from the Regional Waste Plan.

28.2.2   Estimation of biogas production

The biogas production was estimated by applying three literature models and a modified version of one of them: the three literature models are identified as

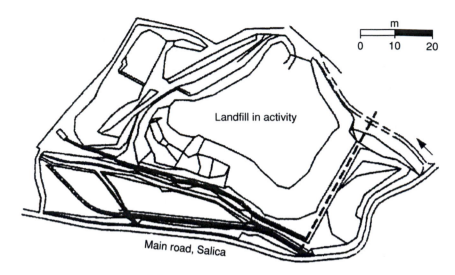

Figure 28.2   Site mapping during the exercise period.

Table 28.2   Solid waste produced in Crotone.

	Unit	1970–1980	1981–1990	1991–2000
Resident population		55,000	58,000	60,000
Specific MSW production	g/(PE/day)	730	920	1350
Annual production	Mg/year	14,655	19,491	29,200

PE: population equivalent.

Table 28.3   Classification of MSW from Crotone.

Component	1970–1980	1981–1990	1991–2000
Organic refuse (%)	45.1	40.2	38.0
Green refuse (%)	3.7	4.0	5.0
Paper (%)	18.6	22.2	25.0
Plastic (%)	6.2	6.5	10.0
Metal (%)	2.3	8.0	4.0
Wood (%)	5.0	3.0	3.0
Clothes (%)	2.0	2.0	2.0
Glass (%)	7.0	7.0	7.0
Undersieve and inert (%)	10.1	7.0	6.0

Tchobanoglous (Tchobanoglous *et al.* 1993), Basse di Stura (Conte 1999) and Cossu–Andreottola (Cossu and Andreottola 1988). The last model, derived from the Cossu–Andreottola model, but after redefining decay constants to better suit the waste composition, will be called "modified Cossu–Andreottola" (see further). In all the models, the evaluation was performed dividing the site in 31 elementary cells, related to each year of activity and estimating the gas production by an accurate hydrological balance of every cell. For the period of activity prior to the enforcement

Table 28.4   Peculiar features of the four estimation models.

	Tchobanoglous	Basse di Stura	Cossu–Andreottola	Modified Cossu–Andreottola
Readily biodegradable fraction	Organics, paper, green	Organic and green refuse	Organic refuse	Organic refuse
Middly biodegradable fraction	NC	NC	Green refuse	Green refuse
Slowly biodegradable fraction	Wood, clothes and leather	Paper, wood, clothes and leather	Paper, wood, clothes and leather	Paper, wood, clothes and leather
Decomposition law	Linear with time	Exponential decrease with time	Exponential decrease with time	Exponential decrease with time
Decay constants	NC	$k_1 = 0.693/$ year $k_3 = 0.046/$ year	$k_1 = 0.693/$ year $k_2 = 0.139/$ year $k_3 = 0.046/$ year	$k_1 = 0.185/\text{year}$ $k_2 = 0.100/\text{year}$ $k_3 = 0.030/\text{year}$
Temperature	NC	NC	30°C	30°C

NC: not considered.

of the Italian Law DPR 7/7/84, the total producible biogas volume per kg of embanked waste was corrected by the methane correction factor (MCF) proposed by the IPCC (Colombari *et al.* 1998). This factor reduces the theoretical production as a function of the embankment height and the landfill classification (controlled or not controlled). The peculiar features of the four models are shown in Table 28.4.

The three literature models differ in the supposed waste composition, the stoichiometry of decay and the kinetic constants adopted to simulate the decay process. The Cossu–Andreottola is the most complete of the three, because it considers the biodegradable waste divided into three fractions with different decay rates, and seems to be the most suitable to the specific situation. But the original version of this model supposes high-humidity level at the interior of the embankment, while, in southern Italy, the climatic conditions allow significantly lower values. This parameter, according to Drees (2001), is the most affecting one. To take it correctly into account, the decay constants were modified using the values measured in a landfill near to the one considered in this study (Conte 2002). Thus the modified Cossu–Andreottola model was obtained. Preliminary evaluations recently carried out in the Tufolo/Lamparanello site confirm the adopted values.

In fact, applying the continuity equation to each cell and taking into account the morphology of the landfill, the cultivation procedures, the drain surfaces, the leachate recirculation and the mean annual rainfall (580 mm/year) and temperature (~ 17°C), a theoretical humidity level of $<40\%$ was obtained. This is far from the maximum field capacity of 70% related to the compaction degree of the site (~ 0.75 Mg/m^3).

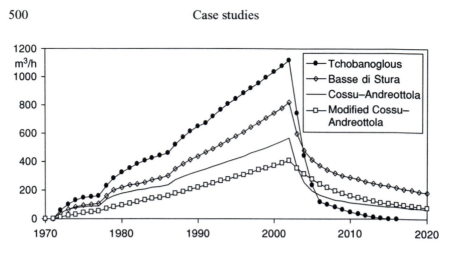

Figure 28.3   Landfill gas production curves of the four models.

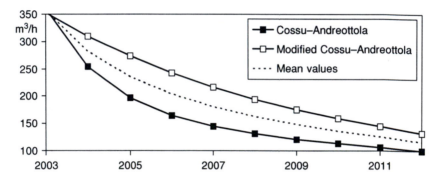

Figure 28.4   Confidence interval of landfill gas production and mean values in the after-care phase (adopted models).

This value significantly decreases the decay compared to the original Cossu–Andreottola model that considers humidity close to the maximum field capacity.

Figure 28.3 shows landfill gas production curves according to the four models. It shows that in the post-closure phase, the two Cossu–Andreottola models, compared to the others, present intermediate values and the modified version, thanks to the lower decay rate, presents higher gas production rates and a longer after-care phase than the original model. Consequently, it was decided to consider these two models as the boundaries of a confidence interval. The final evaluation was done with the mean values of the two models (Figure 28.4), with a procedure similar to those proposed by Belfiore *et al.* (2001) and Muntoni and Polettini (2002).

28.3   BIOGAS EXTRACTION AND COLLECTION

The European Directive on Landfills in force today (99/31/CE), invites to minimise the environmental impact due to biogas dispersion, but does not impose any specific limit. Consequently, till today, the design of the biogas extraction system is driven mainly by economic considerations.

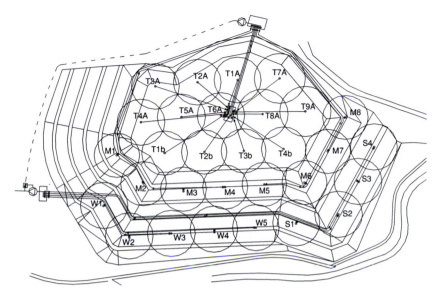

Figure 28.5   Map of the designed network for the extraction and collection of landfill gas (80% extraction efficiency option).

A possible further incentive to increase the extraction efficiency comes from the application of the obligations assumed with the subscription of the Kyoto Protocol in 1997: any demonstrated reduction of greenhouse gas production can be valued in the form of the so-called "carbon certificates". A better extraction efficiency involves an increase of the amount of methane oxidised to carbon dioxide, with a relevant advantage in terms of reduced greenhouse effect.

To augment the extraction coefficient (ratio between the quantity of biogas really extracted and the total produced quantity), it is possible to increase the density of the extraction wells and to enhance the quality and technology of the capping with several consequences in terms of costs and environmental impact.

For these reasons, several possible biogas extraction options were studied, all commonly based on the present best technique of extraction, the drilling of vertical wells performed after the dump filling. The various choices were diversified in type of capping and number of wells, with the purpose of defining four options differenced by the increasing predictable extraction efficiency (50%, 60%, 70% and 80%). For each of the increasing extraction efficiency options, the optimal sizing of the following aspects were designed:

- the capping of the closed landfill (David 1993);
- the number, sizing and positions of the extraction wells (Conte 1999);
- the structure and sizing of the gas collection and transport network;
- the aspiration station with the related security and biogas quality control systems.

Figure 28.5 shows the network designed for the most complete solution: the 80% extraction efficiency option. Finally, an accurate evaluation of all the related capital and running costs was carried out (Galea 2002; Iannelli et al. 2003).

Table 28.5   Requested fuel quality for MCFC and gas engine fed with landfill gas.

Component	Limits for gas engines	Limits for MCFC
Fraction of CH_4	>48%	Not limiting
Net heat input	>4,125 kcal/m^3	Not limiting
Total sulfur (H_2S included)	<550 ppm	<0.5 ppm
Total chlorine (HCl, HHC)	<60 ppm	<10 ppm
Relative humidity	<8 g/m^3	Not limiting
Silicium compounds	0	0

Table 28.6   Gas engine and MCFC main features in the power range 200–500 kW.

Generator	Electric efficiency	NO_x emission	CO emission	Specific cost
Gas engine	32–35%	<450 Mg/m^3	<500 Mg/m^3	1050 €/kW
MCFC	48–52%	<5.45 Mg/m^3	<1.8 Mg/m^3	1500–1800 €/kW

28.4   BIOGAS PRE-TREATMENT PLANT

The requested biogas pre-treatment plant will be significantly different for the two alternative energy exploitation fittings, owing to the peculiarities of the two systems (Table 28.5).

For the gas engine, a moisture removal system was provided through cooling and post-heating, able to remove particulates and water soluble gases (mainly H_2S) up to adequate levels for the operational conditions and the prescribed gaseous emissions (Cossu and Reiter 1996).

The MCFC needs a strict H_2S removal (S < 0.5 ppm) obtainable by iron oxide pellets adsorption after condensate separation (Rettenberger 1996). A removal of halogenated hydrocarbons (HHC) by activated carbons was also provided (Henning *et al.* 1996).

28.5   COMPARISON BETWEEN MCFC AND GAS ENGINE

28.5.1   Multicriteria analysis

For the final comparison, three commercial gas engines of increasing nominal power (230, 330 and 460 kW) and two MCFC modules (250 and 500 kW) were chosen in the set of unit-sizes included in the first commercialisation programme. Table 28.6 shows the main features of these options and Table 28.7 gives the calculated running costs of the required pre-treatment plants.

To compare the various options, the multicriteria analysis (MCA) method was adopted (Lazzari and Mazzetto 1992). This technique allows the evaluation of several project alternatives with respect to non-homogeneous judgement criteria.

This method requires the definition of an $n \times m$ evaluation matrix (where n is the number of project alternatives and m is the number of judgement criteria) and a vector of *m* weights that quantify the relative importance that the decision-maker attributes to those criteria in his classification.

The procedure was carried out by the following phases:

(a) *Definition of the project alternatives*: the first line of the evaluation matrix is compiled with the five considered project alternatives (230-, 330- and 460-kW gas engines; 250- and 500-kW MCFC modules).

(b) *Definition of the judgement criteria and attribution of the relative weights*: two economic (initial investment (II) and net present value (NPV)) and two environmental criteria (NO_x emissions and biogas extraction efficiency) were adopted.

For the definition of the relative weights higher values for the two economic parameters (35% for II and 40% for NPV) and lower values for the two environmental parameters (5% for NO_x emission and 20% for extraction efficiency) were adopted to be conservative.

These weighing factors were selected based on: (1) all the considered solutions respect all the environment protection laws and regulations in force today in Italy; and (2) the environmental parameters are of advantage for the MCFC solutions while the economic parameters are of advantage for the gas turbine. The preference was to compare the solution with the maximum objectivity.

For the last parameter (extraction efficiency) a relatively high value was adopted because the economic advantages obtainable by the related reduction of greenhouse gas emission in terms of carbon certificates were included in it.

(c) *Evaluation of costs and benefits* (Table 28.8): at this phase the cumulate NO_x emissions during the whole duration of the plant and the two economic parameters were evaluated (ENEA 1998). The running costs of the gas engines were determined considering the maintenance intervals observed for a number of working unities fed with landfill gas: two interventions after 20,000 and 40,000 h of operation. For the MCFC the running costs were determined taking into account the goals of the first 250 kW demonstrative projects that consider the duration of the MCFC of about 40,000 h. Only the substitution costs

Table 28.7   Costs and consumptions of treatments for gas engine and MCFC.

Type of generator	Refrigerating machine (€/kW)	Refrigerating machine (kW-h/m³)	Post-heating (kW-h/m³)	Secondary treatments (€/kW)	Secondary treatments (kW-h/m³)
Gas engine 230 kW	13.8	0.0046	0.004	NR	NR
Gas engine 330 kW	12.0	0.0046	0.004	NR	NR
Gas engine 460 kW	9.4	0.0046	0.004	NR	NR
MCFC 250 kW	11.5	0.0046	0.004	500	0.002
MCFC 500 kW	8.6	0.0046	0.004	500	0.002

NR: not required.

Table 28.8   Main factors affecting the costs–benefits analysis.

Investment costs	• Extraction plant adequate to improve efficiency over 50%
	• Generator and electric equipment
	• Refrigerating machine and secondary treatments
Running costs	• Personnel
	• Ordinary servicing and lubricating oil
	• Depreciation of the generator
Benefits of energy production	• 0.124 €/kW-h during first 8 years, then 0.062 €/kW-h

Table 28.9   Example of sub-analysis matrix (230-kW gas engine).

Gas engine 230 kW	Extraction up to 50%	Extraction up to 60%	Extraction up to 70%	Extraction up to 80%	Weight (%)
Initial investment (€)	244,370	266,529	332,857	373,055	35
NPV (€)	82,867	176,958	196,376	230,417	40
Total NO$_x$ emission (kg)	1576	1860	3117	3599	5
Biogas extraction efficiency (%)	50	60	70	80	20

for the exploitation options that reached a sufficient production for a duration superior to the life of the cell were taken into account.

(d) *Compilation and standardisation of the sub-analysis matrix*: the extraction efficiency criterion was defined by a sub-analysis matrix calculated for every option with positive NPV to choose the best of the four considered possibilities (50%, 60%, 70% and 80%). As an example, see the sub-analysis matrix developed for the 230-kW gas engine option in Table 28.9. To homogenise and standardise the scores of the various benefit and onerous criteria the following transformations were adopted:

$$\hat{e}_{ij} = \frac{x_{ij} - \min_j(x_{ij})}{\max_j(x_{ij}) - \min_j(x_{ij})}$$

$$e_{ij} = \hat{e}_{ij} \text{(benefit criterion)} \quad \text{or} \quad e_{ij} = 1 - \hat{e}_{ij} \text{(onerous criterion)}$$

where x_{ij} is the gross score (*i*-th criterion; *j*-th alternative), $\hat{e}_{ij}$ is the homogenised score and e_{ij} is the standardised score.

(e) *Assembling of the final matrix and comparison of the main alternatives*: after application of the weights vector, the final matrix, as shown in Table 28.10, was obtained, that contains a further economic criterion (return on investment) aimed to appreciate the solutions with a faster capital amortisation.

28.5.2   Evaluation using multicriteria analysis

The generalised concordance analysis method (Pratelli 1998) requires the calculation of a concordance index, CI, comparing the benefit criteria, and a discordance

Table 28.10   Final evaluation matrix of the main alternatives.

	Gas engine 230 kW	Gas engine 330 kW	Gas engine 460 kW	MCFC 250 kW	MCFC 500 kW	Weight (%)
Initial investment (€)	266,529	479,185	615,685	599,719	1,282,658	30
NPV (€)	138,258	340,445	123,762	227,557	126,515	30
Return on investment (year)	4	4	5	6	8	10
Total NO_x emission (kg)	1872	2676	2721	45	64	15
Biogas extraction efficiency (%)	60	80	80	60	80	15

index, DI, comparing the onerous criteria, for every couple of alternatives, and traces the final classification based on the total index, TI = CI − DI. The best solution is the one with the higher TI index. The results of the comparison are presented in Table 28.10.

The final classification presents the best score for the 330-kW gas engine, followed by the 250-kW MCFC and the 230-kW gas engine. These results entail some final considerations:

- The scores of the different solutions are driven much more by the nominal power than by the technology: the differences between unities of the same type and different power are much higher than those of unities of the same power and different type. The best classification of the 330-kW gas engine derives only by the unavailability of an MCFC unity of the same power.
- The fuel cells maintain their efficiency at the decrease of the utilisation rate much more than the gas engines. This preserves the investments from possible overestimations of the production curves and prolongs the duration of the plants beyond the reaching of production levels corresponding to the minimum exploitation limits of the gas engines.
- Even with a non-mature high-cost technology, the MCFC seem to be comparable or even better than traditional engines of the same power. The foreseeable decrease of cost with the enlargement and development of the production will further improve this advantage of MCFC.

To demonstrate this, the TI indexes were calculated for several values of the specific cost of MCFC units, from the estimated first commercialisation (1800 €/kW) to the mature technology cost (1200 €/kW expected in 2010). The results (Figure 28.6) are presented in three graphs drawn for MCFC efficiencies of 48%, 50% and 52% to consider the uncertainty of the expected on-field efficiency. Figure 28.6 shows better scores for the MCFC than for the traditional units of the same power, even for the minimum expected efficiency of 48%, with the 250-kW MCFC that results even better than the 330-kW gas engine for specific costs lower than 1250 €/kW.

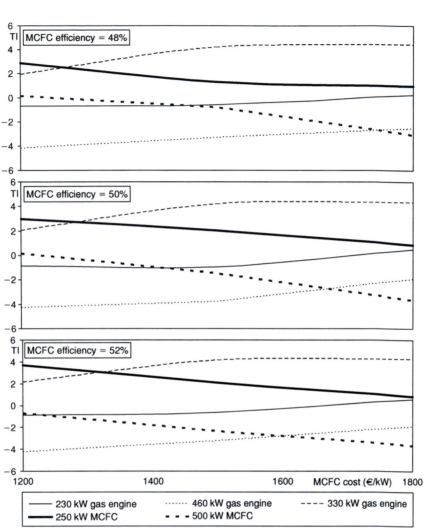

Figure 28.6   TI index as a function of the unitary cost and the efficiency of the MCFC.

28.6   CONCLUSIONS

This chapter presented a possible evaluation criterion for the choice of energy exploitation systems for low gas-productive landfills. This field of application will become more relevant with the complete enforcement of Italian Laws Decree Legislation No. 22/97 and No. 36/03 that will determine a significant reduction of the organic fraction of the dumped waste: in such conditions, the gas exploitation will be possible only with high-efficiency and -flexibility systems. The peculiar features of the MCFC – near-zero emissions and high efficiency largely independent on the load and the net heat input of the gas – make this technology particularly suitable for the future scenarios of landfill management. Moreover, these small-scale high-efficiency gas exploitation possibilities will allow the planning of smaller landfill sites, with several advantages, especially in areas with high population density, where adequate sites are less and less available.

ACKNOWLEDGEMENTS

The authors wish to thank first of all Dr. Mario Galea that contributed to the elaborations of this study during his Graduate Thesis. Then G. Aloisio and M. Praticò of ASPS for the documentation on the site under study; I. Conte and G. Petraglia of AMIAT for the data on forecast models and cost parameters, S. Amadori and V. Lami for the useful advice, S. Dotti of CGT, G. Gavioli of JENBACHER and Mr. Picotti of IVECO-AIFO for the technical–economical data on gas engines.

REFERENCES

Belfiore, F., Magnano, E. and Clister, W.E. (2001) Control of landfill gas excursions based on a selected gas production model. In: *Proceedings of 8th International Waste Management and Landfill Symposium*, Sardinia.
CNR (CONSIGLIO NAZIONALE DELLE RICERCHE) (1980) *Indagine sui rifiuti solidi urbani in Italia – libro bianco sulle ricerche condotte per la determinazione delle quantità e qualità dei RSU*, internal publication, Rome.
Colombari, F., De Lauretis, R., De Stefanis, P. and Gaudioso, D. (1998) *Le emissioni di metano dalle discariche di rifiuti in Italia: stima e scenari futuri*, ENEA Research Centre, Casaccia, RM.
Conte, I. (1999) *Progetto di revisione ed ottimizzazione del sistema di estrazione forzata del biogas con potenziamento della centrale di recupero energetico*, AMIAT, Torino.
Conte, I. (2002) *Relazione di Consulenza Tecnica relativa alla produzione di biogas della discarica di RSU sita in località Colombra, Crotone.* Azienda Speciale Pubblici Servizi – Crotone.
Cossu, R. and Andreottola, G. (1998) Modello matematico di produzione del biogas in uno scarico controllato, *RS Rifiuti Solidi* **2**(6).
Cossu, R. and Reiter, M. (1996) Condensate from landfill gas: production, quality and removal. In: *Landfilling of Waste: Biogas* (eds. Christensen, Cossu and Stegmann), pp. 484–494. E & FN Spon, London.
David E. Daniel (1993) Geomembrane liners: design models and examples. In: *Geotechnical Practice for Waste Disposal*, pp. 172–176. Chapman & Hall, London.
Drees, K.T. (2001) Accelerated discharge of pollutants from reactor landfills: results and consequences. In: *Proceedings of 8th International Waste Management and Landfill Symposium*, Sardinia.
EG&G Services Parsons Inc. (2000) Effect of impurities. In: *Fuel Cell Handbook*, 5th edn. U.S. Department of Energy. pp. 153–157.
ENEA (1998) Modalità di realizzazione di un piano di finanziamento. In: *Gli strumenti di finanziamento di progetti energetico-ambientali. Conf. naz. Energia e Ambiente.* pp. 234–256.
Galea, M. (2002) *Confronto con l'analisi multicriteria tra celle a combustibile e tecnologie convenzionali nell'utilizzo del gas da discarica.* Graduate Thesis, rapp. Iannelli, Cavazza, Moreno, University of Pisa.
Henning, K.D., Schafer, M. and Giessler, K. (1996) Landfill gas upgrading: removal of halogenated hydrocarbons and other trace organics. In: *Landfilling of Waste: Biogas* Christensen (eds. Cossu, Stegmann), pp. 569–583. E & FN Spo.n, London.

Iannelli, R., Galea, M. and Moreno, A. (2003) Application of molten carbonate fuel cells in the exploitation of land-fill gas. A case study in southern Italy. In: *Proceedings of the 9th International Waste Management and Landfill Symposium*, Sardinia, S. Margherita di Pula, Cagliari. pp. 157–159.

Lazzari, M. and Mazzetto, F. (1992) Modello decisionale ad approccio multicriteriale per la valutazione di tecnologie energetiche in aziende agricole. In: *Atti Convegno Informatica in Agricoltura*, pp. 117–130. Accademia dei Georgofili, Firenze.

Muntoni, A. and Polettini, A. (2002) Modelli di produzione del biogas – Limiti di applicazione e sensitività. In: *Gestione del biogas da discarica, recupero e monitoraggio*. Univ. la Sapienza, Roma.

Pratelli, A. (1998) Cenni di analisi multicritera. In: *Ingegneria dei sistemi di trasporto* (ed. Pitagora), pp. 124–134.

Rettenberger, G. (1996) Lanfill gas upgrading: removal of hydrogen sulphide. In: *Landifilling of Waste: Biogas* (eds. Christensen, Cossu and Stegmann), pp. 559–568. E & FN Spon, London.

Ronchetti, M. and Iacobazzi, A. (2002) *Celle a combustibile – stato di sviluppo e prospettive della tecnologia*, ENEA, Roma.

Tchobanoglous, G., Theisen, H. and Vigil, S. (1993) *Integrated Solid Waste Management. Engineering, Principles and Management Issues*, McGraw-Hill, New York.

PART SIX: Glossary

1 BIOMASS FERMENTATION AND BIOFUELS

Aceticlastic   Microorganism able to split acetate into methane and inorganic carbon.

Acetogen   Microorganism that converts volatile fatty acids into acetic acid and other low molecular weight compounds.

Acetylation   Linking of acetyl/acetic acid to another chemical compound.

Acidogenesis   Microbial process in which organic matter is transformed into organic acids. The organic acids appear as anions. The corresponding cation is the hydrogen ion, which is neutralized if the buffer capacity is sufficiently high. If not, the pH of the medium decreases. Acidogenesis is therefore often mixed up with acidification.

Activated sludge   Sludge consisting mainly of active microorganisms.

Active biomass   Mass of microorganisms, which are metabolically active.

Anaerobic digestion   Process in which organic matter is microbially transformed into methane or other reduced organic compounds (e.g. ethanol or lactic acid) in the absence of oxygen.

Anaerobic microorganisms   Microorganisms that live and reproduce in the absence of oxygen. Obligate (or strict) anaerobes are unable to live at even low oxygen concentrations. Facultative anaerobes are able to live at low or even normal oxygen concentration. All anaerobes are single celled organisms such as bacteria, archaea and some fungi and protozoa.

Archaea   *Archaebacteria*, microorganism(s) derived from an ancestor common to prokaryotes and eukaryotes, though forming a different kingdom. They are typically found in extreme environments (e.g. in hot springs, salt lakes, etc.). Archaea are the only methane-producing organisms (methanogens).

Batch process   Process where the feeding material is added once at the beginning and the end products are removed at the end of the process.

Biodegradable   Degradable by living organisms.

Biodiesel   Diesel oil or compound with similar properties produced from biomass.

Bioethanol   Ethanol produced by fermentation of mainly low value biomass for fuel purpose.

Biofuel   A gas or liquid fuel made from biomass.

Biogas   Mixture of methane and carbon dioxide produced from organic matter by microorganisms in an anaerobic digester. Usually restricted to gases where the major component is methane. Biogas typically consists of 50–70% methane and 25–40% carbon dioxide and minor amounts of various other gases, such as nitrogen, hydrogen sulfide, ammonia, hydrogen and water vapor.

Biogas plant   A biotechnological facility producing biogas from organic matter. The biogas plant typically consists of a storage system for biomass, one or several anaerobic digesters, a storage system for the effluent and a storage/treatment/utilization system for the produced biogas.

Biomethanation   Technique leading to the formation of biogas from organic matter, using microorganisms.

Biorefinery   A refinery in which the raw material is biomass, and where the majority of the processes are biological and/or biochemical.

Biosolids   Waste material from animal or plant sources. Mainly consisting of carbon and hydrogen.

BOD   Biological oxygen demand. Quantity of oxygen needed to oxidize one unit material by microbial processes.

Catalyst   A material that can increase the rate of a chemical reaction between other chemical species without being consumed in the process.

Cellulase   An enzyme family able to hydrolyze cellulose (e.g. endoglucanase).

Cellulose   Polymeric structure of the carbohydrate glucose.

C/N ratio   Mass ratio of carbon to nitrogen in a material.

COD   Chemical oxygen demand. Quantity of oxygen needed to oxidize one unit material by chemical processes.

Co-digestion   Anaerobic digestion of mixed materials of different origins.

Corn stover   Semi-dry waste product from the production of corn.

Crystallinity   Molecular organization of cellulose complicating the degradation by enzymes.

CSTR   Continuously stirred tank reactor. Digester in which organic matter is continuously added and removed, and homogeneously mixed by a stirring (mixing) device.

Dark hydrogen fermentation   Fermentation which can be operated under dark or light conditions (i.e. there is no need for light energy).

Denitrification   Microbial process by where nitrate and nitrite are microbially removed from an aqueous liquid as gaseous nitrogen.

De-polymerize   The breaking down of a polymeric structure (e.g. cellulose).

Detoxification   Process for the removal of one or more inhibitory compounds.

Effluent   Waste end product, usually liquid, of an industrial or microbial process. Any liquid discharged from a source into the environment.

Energy carriers   Energy rich compounds that can be utilized to store and transport energy (e.g. biogas, hydrogen and ethanol).

Energy conversion efficiency   Combustion enthalpy of products as percentage of combustion enthalpy of substrates.

Energy crop   Agricultural crop, which is grown with the main purpose of subsequent conversion into energy.

Enzymatic loading or dose   The amount of enzymes supplemented to a given process.

Eubacteria   Microorganisms (about 10 μm) lacking a nucleus and other membrane-enclosed organelles, usually having their DNA in a single circular molecule.

Eukaryotes   A cell or an organism, including plants, animals and fungi, which have a membrane-enclosed nucleus and usually other organelles.

Fed batch   Process where the feeding material is added progressively over a given time but where the end products are removed once at the end of the process.

Feedback inhibition   Inhibition of a specific enzyme caused by a product that is produced by the same enzyme.

Fermentation   A series of anaerobic biological transformation processes where organic matter is broken down into more reduced and more oxidized small compounds to maintain the redox balance. As a technological process, fermentation is considered an aerobic or anaerobic process in which microorganisms are used to produce a useful end product.

Fixed film   Film of microorganisms fixed on an inert material.

Flexible fuel vehicle (FFV)   Vehicle that is able to drive on gasoline containing in the range of 0–85% ethanol.

Floc   Aggregate of microorganisms usually a few millimeters in diameter.

Fluidized bed   Particles of inert material inside a digester, which is kept fluidized by a continuous liquid up-flow.

Fossil fuels   Formed from the decomposition of ancient animal and plant remains. A major concern is that they emit carbon dioxide into the atmosphere when burnt, a major contributor to the enhanced greenhouse effect.

Granule   Solid particle up to a few millimeters in diameter consisting of microorganisms embedded in a matrix produced by the microorganisms.

Hemicellulase   An enzymatic family able to hydrolyze hemicellulose (e.g. xylanase).

Hemicellulose   Polymeric substrate containing different carbohydrates, primarily pentose (e.g. xylose).

Hexose   Carbohydrate sugar containing six carbon atoms.

Holocellulose   The combined structure of cellulose and hemicellulose.

Homoacetogenic   Microbial process where organic matter is transformed solely into acetate.

Hydrocarbons   Family of organic compounds, composed only of carbon and hydrogen. They may be emitted into the air by natural sources (e.g. trees) and as a result of fossil and biomass combustion, fuel volatilization or solvent use.

Hydrogen (H_2)   Flammable non-toxic colorless gas. Hydrogen is considered as the fuel of the future because combustion produces only water.

Hydrogenotrophic   Property of a microorganism to utilize hydrogen for growth.

Hydrolysate   Liquid produced from pre-treatment and hydrolysis of biomass material.

Hydrolysis   Catalytic breakdown of large molecules into smaller molecules with water as reagent.

Hygienization   Process by which pathogenic microorganisms are killed.

Influent   Material, usually a liquid, which enters a process, for example, anaerobic digestion.

Inoculum   Small mass of microorganisms used to start a process where the microorganisms are either not present or present in very low numbers.

Interspecies hydrogen transfer   Process by which hydrogen is produced in one microorganism and then transferred and used by an adjacent microorganism of another species. Often a syntrophic relationship.

Laccase   Enzyme able to degrade phenolic compounds.

Landfill gas   Product of the natural decomposition process occurring in a landfill. Consists of 50–60% methane, 40–50% carbon dioxide and less than 1% hydrogen, oxygen, nitrogen and other trace gases.

Lignin   Complex highly branched organic molecule of high molecular weight. Heterogeneous compound based on a phenol-propene backbone.

Lignocellulose   The combined structure of cellulose, hemicellulose and lignin found in almost all plant derived biomass materials.

Mass balance   The relation between influent masses and effluent masses of a process.

Maximum achievable yield   Yield, which is in reality obtained taking into account utilization of substrates for the production and maintenance of biomass.

Maximum theoretical efficiency   Efficiency of conversion according to the reaction equation only.

Mean retention time   Average period, a unit volume of liquid remains in a vessel operated under continuous conditions.

Mesophilic   Biological processes with an optimum temperature between 25°C and 40°C. Mesophilic methanogenesis occurs optimally at 35°C.

Metabolism   Transformation of compounds by organisms.

Metabolite   Product of a transformation carried out by an organism.

Methane (CH_4)   Colorless, non-poisonous, flammable gas created by anaerobic decomposition of organic compounds.

Methanogenesis   Process in which organic matter is transformed to methane by microorganisms in the absence of air. Methanogenesis is mediated by archaea called methanogens.

Micronutrient   Nutrient utilized in small amounts to support growth (e.g. metals like nickel and cobalt or vitamins).

Oxygenate   Oxygen containing compound added to gasoline to obtain a cleaner combustion.

Pentose   Carbohydrate sugar containing five carbon atoms.

Phenolics   Compounds formed during the breakdown of lignin.

Plasmid   A small, circular piece of DNA found outside the chromosome in bacteria. Plasmids are the principal tools for inserting new genetic information into microorganisms or plants.

Plug flow system   Vessel in which a continuous process proceeds without mixing, typically in a horizontal tube-like fermenter.

Pre-treatment   Physical, chemical or biological process applied to biomass material. Can also be a combination of two or more processes.

Primary crops   Crops normally produced for food purposes.

Productivity   The amount of a product produced in a given volume and a given time period (typically stated as g-product/l-reactor volume/time).

Prokaryote   Small microorganism (about 10 μm) having no membrane around its genetic material (i.e. no nucleus).

Psychrophilic   Biological process occurring at temperatures between 0°C and 20°C.

Putrefaction Process by which organic matter is transformed by microorganisms into malodorous nitrogen- and/or sulfur-containing end products.

Redox potential A quantitative expression of the oxidizing or the reducing power of a substance or a solution.

Renewable resources Resources for industry and energy, which do not lead to exhaustion of fossil reserves and do not contribute to increase of atmospheric CO_2 concentration.

Retention time Average time, a unit volume of liquid remains in a vessel operated under continuous conditions.

Saccharification Hydrolysis of sugar polymers into sugar monomers.

Simultaneous saccharification and fermentation (SSF) Process occurring in a fermenter where added enzymes hydrolyze polymeric sugars (starch, cellulose) to monomers, which are simultaneously fermented by microorganisms.

Sludge bed Mass of microorganisms trapped and retained in a digester by gravitation.

Slurry A liquid mixed with a large proportion of solid material.

Specific growth rate Parameter giving the rate of increase in mass of microorganisms or of division of cells per unit mass of microorganisms.

Stationary bed Inert immobile material inside a digester, which supports growth of attached microorganisms.

Steam explosion A process in which biomass is treated at high temperature and pressure terminated by a quick pressure release (explosion).

Suspended solids (SS) Solid particles in suspension in a liquid.

Symbiosis Relationship in which microorganisms that belong to different species live together for their mutual benefit.

Syntrophic Characterizes an obligate symbiosis in which two microorganisms, belonging to two different species share one substrate. Typically the substrate is partially metabolized by one organism while the other organism consumes one or more of the residues making the process thermodynamically possible.

Theoretical yield The yield that can be obtained if a given substrate is completely converted into a given product.

Thermophilic Microbial process taking place at temperatures between 55°C and 65°C.

UASB Upflow Anaerobic Sludge Bed/Blanket: An up-flow digester in which the active biomass (microorganisms) is retained as dense granules having a sedimentation velocity higher than the liquid up-flow velocity.

Ultra filtration Filtration through a grid with very tiny holes.

Upgrading Process by which a gas is cleaned of useless and/or inhibitory compounds and impurities.

Vitamin Chemical compound essential in minor amounts for the life of some organisms.

Volatile fatty acid Acid compound made of a chain of two to eight carbon atoms.

Volatile solids (VS)   Matter, which disappears from a material by heating from 105°C to 550°C or 600°C. Consists typically of organic compounds, carbonates, nitrogen and sulfur compounds, which are oxidized to volatile oxides.

Wet oxidation   Process in which biomass material is treated at high temperature and pressure with the addition of an oxidizing agent (e.g. oxygen).

Working volume   Portion of a digester available for a microbial process.

Xenobiotic   Characterizes a compound, usually man-made, which is not found in unpolluted nature.

2   BIOFUEL CLEAN-UP/UPGRADING

Biogas reforming   Conversion of biogas to hydrogen by a reforming process.

Biogas upgrading   Process to clean and upgrade biogas to natural gas quality.

Calorific value   The amount of energy obtained by burning one unit of a compound.

Carbon dioxide (CO_2)   Colorless, odorless gas that occurs naturally in the earth's atmosphere. Significant quantities are also emitted into the air by fossil fuel combustion and deforestation. It is a greenhouse gas of major concern in the study of global warming.

Carbon monoxide (CO)   Colorless, odorless gas resulting from the incomplete combustion of hydrocarbon fuels.

Co-generation   Gas-driven turbines producing heat during electricity generation. The heat is exploited in steam generators for generation of more electricity.

Clean up   Removal of a contaminant from gaseous or liquid feed streams by mechanical, biological or chemical processes.

Emission   The release of a substance (usually a gas when referring to the subject of climate change) into the atmosphere.

Gas purifier   Device that removes H_2S from biogas.

Green gas, bio-natural gas   Gas with natural gas quality achieved by cleaning and upgrading biogas from fermentation of biomass.

Halocarbon   Halocarbon is a compound consisting only of carbon, one or more halogens and sometimes hydrogen.

Halogens   The halogens are the elements: fluorine (F), chlorine (Cl), bromine (Br), iodine (I) and astatine (At). They are non-metals and compounds of these elements are called halogenides or halides.

Hydrogen sulfide (H_2S)   A flammable, highly poisonous gas having a pungent unpleasant odor.

Mercaptans   Strong-smelling compounds of carbon, hydrogen and sulfur.

Nitrogen oxides (NO_x)   General term pertaining to compounds of nitric oxide, nitrogen dioxide and other oxides of nitrogen. Nitrogen oxides are typically created during combustion or combustion processes.

ppm   Parts per million.

Pressure swing   Process in which a gas undergoes a high pressure followed by a low pressure.

Pressure swing adsorption (PSA) Process to clean or separate gases.

Reformer Device for reforming.

Reforming Catalytic conversion of light hydrocarbons (biomass, fossil energy carriers, e.g. natural gas) producing a hydrogen rich gas in a steam process or by CO_2 (dry reforming).

Saturated gas Gas containing maximum water vapor at a given pressure and temperature. If more water is added to the gas stream, or the pressure increases or temperature drops, condensation will occur.

Scrubbing Process by which a gas is put into contact with a liquid, which removes components of the gas as soluble matter, usually in a counter-flow manner.

Sewage gas Mixture of methane and carbon dioxide produced from municipal wastewater or sewage sludge by anaerobic digestion.

Siloxanes Siloxanes are saturated silicon–oxygen hydrides with unbranched or branched chains of alternating silicon and oxygen atoms (—Si—O—Si—).

Stripping Process by which a compound is removed from a liquid in a gaseous form.

Sulfur dioxide (SO_2) Strong smelling, colorless gas formed by the combustion of fossil fuels.

Volatile organic compounds (VOCs) Group of organic compounds characterized by their tendency to evaporate easily. They are emitted through evaporation or combustion. The term VOC is usually used in regard to stationary emission sources.

3 FUEL CELLS

AFC Alkaline fuel cell, characterized by alkaline electrolyte. The electrolyte is composed of potassium hydroxide. The energy carrier is OH^-. Operation temperature can vary from room temperature to approximately 250°C.

Anode One of the two porous electrodes constituting the fuel cell, where oxidation occurs. The anode is usually fed with a hydrogen rich gas or a hydrocarbon to be internally reformed. For direct methanol fuel cells and direct hydrocarbon fuel cells, the fuel is directly oxidized, without any reforming process.

Balance of plant (BoP) It is composed of all the devices composing the system, excluding the stack itself. A BoP includes the fuel processing section, the power conditioning, as well as all the auxiliaries.

Bipolar plate Conductive plate separating the anode from the cathode of the adjacent cell in a fuel cell stack containing several individual cells. It ensures electrical contact and usually incorporates flow channels for gas feeds and may also contain conduits for heat transfer.

Cathode One of the two porous electrodes constituting the fuel cell, where reduction occurs. The cathode is usually fed with air or pure oxygen.

Combined heat and power system (CHP) Usually refers to a system that can simultaneously produce heat and electricity.

Co-generation The simultaneous on-site production of electric energy and process steam or heat from the same power source.

Current collector   Conductive material in a fuel cell that collects electrons (anode side) or disburses electrons (cathode side).

Current density and power density   The amount of current and power, respectively, that is obtained from a unit surface area of fuel cell.

Direct methanol fuel cell (DMFC)   A particular type of PEM fuel cell capable of oxidizing methanol directly, without the need of a reforming process.

Electrolyte   A medium (acidic or alkaline) in liquid or solid form through which ions are conducted to create an electrochemical reaction. The electrolyte must present very high ionic conductivity, while the electronic conductivity must be as close as possible to zero (high electronic resistance). The name of the fuel cell type is usually referred to the type of electrolyte and the useful operating temperature range at which it runs.

Equilibrium cell potential   The maximum theoretical (thermodynamic) voltage that can be achieved by the reaction of a fuel and oxidant in a fuel cell.

Finite element method (FEM)   It is a particular mathematical method used to numerically integrate partial differential equations.

Fuel cell   Electrochemical device in which hydrogen (or a hydrocarbon-based fuel) and oxygen combine isothermally, without following the Carnot's theorem (in contrast to combustion or explosion) to directly produce electrical energy and heat.

Fuel cell power density   The power in kilowatts (or equivalent) that can be achieved from a unit volume or unit mass of fuel cell.

Hybrid fuel cell system   The definition is commonly referred to a system, composed of a high temperature fuel cell and a thermal energy system (typically a gas turbine or a steam turbine).

Interconnect   Used in SOFC systems, see Bipolar plate.

MCFC   Molten carbonate fuel cell. A particular type of fuel cell, characterized by a liquid molten carbonate, forming the electrolyte. The operating temperature is in the range of 600–700°C. The energy carrier is CO_3^{2-} that flows from the cathode to the anode, thus water is produced at the anode side of the cell. A peculiarity of MCFC is that CO_2 is needed at the cathode side, thus this is usually recycled from the anode outlet.

Membrane electrode assembly (MEA)   A unit composed of membrane (polymer electrolyte), anode and cathode.

Open circuit voltage (OCV)   It is the voltage established between the anode and the cathode of a single cell, or of a stack, when no electric load is connected.

PAFC   Phosphoric acid fuel cell. A particular type of fuel cell, characterized by liquid phosphoric acid composing the electrolyte. The operating temperature is about 160–220°C. The energy carrier is H^+ (proton) that flows from the anode to the cathode, thus water is produced at the cathode side of the cell.

PEM   Proton exchange membrane or polymer electrolyte membrane fuel cell, characterized by a solid polymer membrane. The energy carrier is H^+ that flows from the anode to the cathode, thus water is produced at the cathode side of the cell. This is usually based on Nafion, or a variation, thereof. The operating temperature is typically 60–100°C. A characteristic of PEM fuel cells is that they are not CO tolerant (because of platinum catalysts at both anode and cathode), thus a shift reactor is needed to convert CO to CO_2.

Regenerative fuel cell   A particular type of fuel cell system, where a fuel cell section produces electricity from hydrogen and oxygen and an electrolyzer, powered by a renewable energy system (usually a solar or a wind system), produces hydrogen through water electrolysis. This peculiarity allows the system to be independent from any kind of fuel infrastructure.

Single cell   A unit composed of electrolyte, anode, cathode and current collector.

SOFC   Solid oxide fuel cell. A particular type of fuel cell, characterized by a ceramic electrolyte. The operating temperature is in the range of 800–1000°C. Currently, there is a common effort to reduce the operating temperature to 500–800°C. These kinds of fuel cells are usually referred to as an intermediate temperature SOFC. The energy carrier is O^{2-} that flows from the cathode to the anode, thus water is produced at the anode side of the cell.

Stack   A bank of fuel cells or electrochemical cells connected in series.

Index

Printed in the United Kingdom
by Lightning Source UK Ltd.
130017UK00001B/16/A